DATE DUE

APR 1 2 2011			

THE OHIO HOPEWELL EPISODE

Paradigm Lost and Paradigm Gained

THE OHIO HOPEWELL EPISODE

Paradigm Lost and Paradigm Gained

A. MARTIN BYERS

The University of Akron Press   Akron, Ohio

All inquiries and permissions requests should be addressed to the publisher, The University of Akron Press, Akron, OH 44325-1703

Manufactured in the United States of America

First Edition 2004

07 06 05 04   5 4 3 2 1

LIBRARY OF CONGRESS CATALOGING-IN-PUBLICATION DATA

Byers, A. Martin, 1937–
   The Ohio Hopewell episode : paradigm lost and paradigm gained /
A. Martin Byers.—1st ed.
     p. cm. — (Series on Ohio history and culture)
   Includes bibliographical references and index.
   ISBN 1-931968-00-4 (cloth : alk. paper)
    1. Hopewell culture—Ohio River Valley.   2. Hopewell architecture—
Ohio River Valley.   3. Earthworks (Archaeology)—Ohio River
Valley.   4. Mounds—Ohio River Valley.   5. Ohio River Valley—
Antiquities.   I. Title.   II. Series.
E99.H69B94 2004
977.1'01—dc22

                                        2004007246

The paper used in this publication meets the minimum requirements of American National Standard for Information Sciences—Permanence of Paper for Printed Library Materials, ANSI Z39.48-1984.
∞

This book is dedicated to the two people who are closest to me, my wife Joy and my daughter Jennifer. Without their constant support of my work, this book could not have been possible.

CONTENTS

FIGURES

TABLES

xiv   TABLES

ACKNOWLEDGMENTS

I wish to acknowledge the important role that my archaeology and anthropology students of Vanier College, Montreal, played in the formation of this book. They allowed me to test my ideas and their responses often forced me to revise and modify both the ideas and my expression of them. I have a long-term debt of thanks I wish to make to the Department of Anthropology of the University at Albany (SUNY), especially to my dissertation committee, Gary Gossen, Robert Jarvenpa, Dean Snow, and Gary Wright, who gave me the support and freedom to explore what was then an almost unknown territory in North American archaeology, the role of symbols. It was out of this initial exploration that I pursued further studies in meaning, symbols, pragmatics, intentionality, and social constitution, thereby forming the basis for applying the symbolic pragmatic approach to the patterning of the archaeological record. I especially wish to express my gratitude to the Anthropology Department of McGill University for granting a research affiliate position to me. This has made it possible to undertake the research that was required for this book and to pursue the research for my next book.

The works of many prehistoric archaeologists who have particularly focused on the Eastern Woodlands have figured importantly in this book, and I wish to extend my thanks to them. There is no question that without their work in the field, laboratory, and library this book would not have been possible. Among these, however, there are a number whose works have played a key role in the shaping of this book. To these I wish to extend my deep thanks: James Brown, Jane Buikstra, Douglas Charles, Robert Connolly, William Dancey, Warren DeBoer, N'omi Greber, Bradley Lepper, Robert Mainfort Jr., Paul Pacheco, Mark Seeman, Robert Riordan, William Romain, and Bruce Smith. Those of us who have been particularly drawn to the possible astronomical alignments embedded in the Ohio Hopewell embankment earthworks owe a debt to the pioneering work of Ray Hively and Robert Horn at Newark and High Bank, and I wish to use this opportunity to thank them.

Finally, I wish to emphasize the great importance that Robert Hall's work on historic and prehistoric symbolic expression in Native American peoples has had in the shaping of the core symbolic pragmatics of mortuary practices that this book has explored. I was able to appreciate the implications that his work had for mortuary studies only when I wedded it with the deontic ecological perspective on which this book hinges. I see that my drawing out some of the implications of his insights is one concrete way by which I can express the great value that his work has had for me and for many others drawn to the study of Native American culture and society.

PROLOGUE

WHAT COUNTS AS THE OHIO HOPEWELL?

A view of the Newark earthworks as they were mapped by the surveyor Charles Whittlesey just prior to the middle of the nineteenth century is shown in fig. P.1. Newark is a small city in Licking County, Ohio, in the upper Licking drainage and about 50 km east of Columbus, the state capital. It is on the northern margin of the Central Ohio Valley, a region that is outstanding for many major prehistoric embankment earthworks similar to those of Newark. The Newark site is one of the most complex of these earthwork sites, and even though the growth of the city during the nineteenth century did serious damage to it, two major components have survived relatively intact: the Newark Circle-Octagon, shown in the left sector of fig. P.1, and the Fairground Circle, the large eastern circle shown in the lower right sector. The magnitude of these and similar earthworks can be appreciated by noting that the Circle-Octagon alone incorporates most of a nine-hole golf course (also see fig. 4.2 and fig. 4.3). The diameter of the monumental Circle component is almost exactly 320 m, and the Octagon is based on a square that has sides of 320 m.[1] The embankments average about 2 m in height and their bases are about 8 m to 9 m wide. The southwest sector of the embankment of the Circle component is partly covered by a large mound built at a tangent to its perimeter. Usually referred to as the Observatory Mound, it is about 50 m long and 5 m high.

The Fairground Circle in the southeastern sector of the site is

I

equally impressive, having a diameter close to 400 m and an embankment varying from 2 m to 5 m high. There is a broad ditch running inside the base of the Fairground Circle and having a depth that only roughly corresponds to the height of the contiguous embankment. The name of the circle derives from its use as the county fairground.[2] There are several other major embankment components that tie the total site together, as well as some that reach outward, as will be discussed in much greater detail later.

Between ca. 100 B.C. and A.D. 400, the Central Ohio Valley was characterized by the systematic construction and use of many prehistoric embankment earthwork complexes as impressive as those of Newark. These demarcate a major prehistoric episode of this region generally referred to as the Ohio Hopewell of the Middle Woodland period (fig. P.2). Many of these were mapped in the mid-nineteenth century by Ephraim G. Squier and Edwin H. Davis, two amateur scholars. Their compilation of these plans, along with those that they commissioned or "borrowed" from other publications, was published in 1848 as volume 1 of the Smithsonian Contributions to Science and titled *Ancient Monuments of the Mississippi Valley*.[3]

Some of the best known of these works are Liberty Works (fig. P.3) and Seip (fig. P.4), and the eponymous Hopewell earthwork site (fig. P.5). These three, along with a number of other major works, are loosely clustered around Chillicothe, the seat of Ross County, in the middle sector of the lower Scioto River, where it is joined by the eastern-flowing Paint Creek (fig. P.2). There are also very impressive non-"geometrical" earthworks, such as Fort Hill (fig. P.6) in Highland County, Ohio, southwest of Chillicothe, and Fort Ancient in the Little Miami Valley in Warren County (fig. P.7), southwestern Ohio, among others.

The embankment earthworks often incorporate equally impressive mounds, many but not all of these covering floors on which were constructed mortuary features and facilities, including crematory basins, prepared burial platforms, and crypts. Many of the latter contained the remains of human deceased often associated with elaborate artifacts (fig. P.8). Together these make up an assemblage that is in many ways unique in North American prehistory. The purpose of this book is to use this complex of embankment earthworks, mounds, floors, and the mortuary features and facilities, along with the remains of the human deceased and the associated assemblage, as the primary evidence for initiating a comprehensive reconstruction of current archaeological understandings of the Middle Woodland Ohio social systems. The theoretical framework will be constructed around a symbolic perspective developed as an integrated set of ecological, territorial, social organizational, and mortuary models.

THE CURRENT ARCHAEOLOGICAL STATUS

North American archaeologists generally consider the Ohio Hopewell to be the "classic" expression of the Middle Woodland period of the Eastern Woodlands. The earthworks are located on both the floodplain terraces and the higher bluffs and tabletop ridges of the main rivers of the Central Ohio Valley, particularly but not limited to the drainages of the Muskingum River in the eastern sector of the central valley, the Scioto River in the central sector, and the Great and Little Miami Rivers in the western sector. Ohio Hopewell ceremonial construction can be divided into three major categories: embankment earthworks (although stone was also sometimes used), mound earthworks covering prepared floors, and timber constructions, many of the latter having been ritually destroyed and covered by the mound earthworks. Not all Ohio Hopewell ceremonial sites include all three categories of construction. However, the most impressive Ohio Hopewell sites usually have major samples of all three categories.

The major mounds were usually constructed in layers of selected earths, clays, stones, and gravels. The timber structures whose residue these mounds covered were built only after major prepared floors were constructed of layers of gravels and different earths, "topped off" usually by an extremely durable mixed sand and clay surface. The mortuary features typically built on and sometimes into these floors included clay basins displaying marks of heavy burning, hearths, burnt deposits, log cribs, and so on. The burials and the associated artifacts often found in these log cribs were usually either extended burials or cremations, with some variation in between. The associated artifacts came in many categories: shell beads, pearl beads, animal canine pendants, projectile points of exotic or selected cherts, stone and copper ear spools, copper "breast" plates and headdresses, copper axes and adzes, stone pipes, copper, silver, and even iron "pan flutes," and so on. Much of this varied array of raw materials was procured or derived from distant regions: obsidian from the Rocky Mountains, copper from the Lake Superior region, silver from northern Ontario, mica and galena from the Appalachians, local freshwater pearls, exotic flints from various regions, including the well-known Flint Ridge quarries southeast of Newark, ocean shells from the southeastern Florida and Gulf coasts, and so on.[4]

Both exotic resources and stylistic parallels in artifacts and mortuary treatment are used by archaeologists to link the Ohio Hopewell sites to mound sites in many parts of the Eastern Woodlands: southern Ontario, Indiana, Illinois, Missouri, New York, Tennessee, Florida, Louisiana, and so on. In quantity and variety of artifacts and raw materials, the Ohio Hopewell is without peer, but it is the monumental embankment earthworks, such as Newark and those clustered around Chillicothe and in southwestern Ohio, that make it so outstanding. There were embankment earthworks in other regions of the Eastern Woodlands caught up in the Hopewellian episode, of course. However, embankment earthworks equivalent in scale to those of the Ohio Hopewell are found in only a few of these regions, for instance, the Pinson Mounds site in Tennessee.[5] It is somewhat surprising, therefore, that the embankment earthworks received rather scant systematic attention from twentieth-century archaeologists until late in the century.[6] In contrast, general and scholarly interest was first drawn in the nineteenth century to the Central Ohio Valley largely because of the earthworks. Squier and Davis, though not the first to publish on these earthworks, are widely regarded as responsible for promoting the study and speculation about them.[7] Despite a number of empirical inaccuracies, their work is valuable as a permanent record of many of the major Ohio Hopewellian embankment constructions that have been nearly or effectively destroyed by the encroachment of modern agriculture and urban expansion.

Following this initial interest in the embankment earthworks, archaeologists shifted their focus to the great mounds. It was through analyzing the contents of the mounds that they were able to discount the "mound builder myths" that had proliferated during the nineteenth century, these being various claims that the earthworks were the production of migrant populations from one or another nonindigenous civilization, usually specifically presumed to be derived from Old World roots. These myths held in common the claim that the nonindigenous "mound building civilization" had been destroyed by later indigenous invaders, deemed to be the immediate ancestors of the resident Native American peoples Europeans encountered on entering the Ohio Valley. By branding the indigenous occupants as the descendants of peoples who had conquered and displaced prior occupants of "advanced civilizations," these myths served as contrived justification for the removal of the Native Americans and the occupation of these regions by invading Europeans. Once these embankment earthworks and their associated mounds were demonstrated archaeologically to be the works not of a "lost civilization" with which the American public had closely identified—the Lost Tribes of Israel, adventuring Vikings or Phoenicians—but of the ancestors of displaced Native American peoples, the general American public, perhaps not surprisingly, lost interest in following the study of these monuments.

It is still puzzling, however, that the emerging archaeological profession was not more attentive to the embankment earthworks as a source of data for constructing and verifying cultural models. No doubt their very size was part of the problem. Archaeological investigation was costly and had to maximize its gains. Also, many of these earthworks were either in expanding urban areas or on private estates and farms that were not easily accessible to archaeologists. As a result, many were well on the way to being destroyed by the time Squier and

Davis did their work. A few such sites were deliberately preserved by local action, such as the great Newark Circle-Octagon and Fairground Circle. Some were preserved because they were on high tabletop ridges, largely inaccessible to most modernizing incursions. Fort Hill (fig. P.6), Fort Ancient (fig. P.7), and Miami Fort (fig. P.9) are examples of these. It is also notable that when nineteenth-century excavation of the embankments was done, they often received the "death knell" comment that "nothing of interest was found." In comparison, the rather rough-and-ready excavation of the great mounds revealed the mortuary and associated artifactual data described above and, to early archaeologists, these served as the real measure of their archaeological value.[8] Therefore, other than the type of measuring and mapping of the embankments already mentioned, systematic excavation was rather low in priority, even during the first half of the 1900s.

It was only in the 1960s that the embankment earthworks again came to figure in archaeological analysis.[9] However, this analysis did not focus on the embankment constructions as a source of cultural information about their builders, but rather on the objective conditions that must have existed for these monuments to have been built: prehistoric demographics, settlement patterns, and subsistence practices. These concerns will figure prominently in this book, but first to be addressed is the symbolic nature that made building and using them necessary in the eyes of their builders. They will then be used as indicators of the social, ecological, and ritual conditions that made building them possible.

What the above implies is the absence of an adequate archaeological theory of prehistoric monumental architecture that can treat earthworks as an intrinsic manifestation of the culture of those who built and used them. An adequate theory would not be only about labor costs and resource uses, although these conditions certainly must be figured in. It would also analyze and interpret the embankment earthworks in terms relevant to those who built and used them since, without this field of relevance, they would not have built them in any case. This means a cultural theory that affords the archaeologist entry into the social world in which their being built and used made sense. Such a theory would use terms that, if appropriately translated, would be relevant to the prehistoric subjects who built them.

This is a major claim and, quite properly, it should be immediately challenged. The typical challenge might be to ask: "Since the builders are dead, how are we ever to know what terms they would use if asked to account for their earthwork construction and related activities?" Legitimate skepticism of this sort will be addressed both by logical argument and by pointing out the relevant aspects of the earthworks and the mounds that can be used as evidence to support claims about the intentions and reasons of the builders for building them. These aspects, however, can be identified as relevant only because of the general theory of material culture that will be presented in this book. If this theory adequately characterizes the general way humans use material culture for symbolically constructing their behaviors as social acts, then it logically follows that, until demonstrated otherwise, the earthwork data that will be cited serve as an expressive signature of the collective beliefs, duties, and intentions of the builders, and, therefore, as a verifiable cultural entry into their social world.

In short, the central argument will be that it is this cultural set of beliefs, duties, and intentions, and the related set of social relations that realized them, that is at the root of the earthworks. Therefore, as the first level entry into an explanation of the earthworks, a teleological stance will be required. Nonintentional factors are not to be eschewed, however. The earthworks may have been among the primary conditions by which collective duties of the builders were fulfilled, but building them would also have generated conditions that were not among those intended or expected, although these consequences were extremely important for the continuity of the practices, particularly by generating a sense of social integration through experiencing a sense of communitas. Nonetheless, these unintended consequences, both of social integration and of social conflict and disintegration, would not have occurred but for the exercise of the collective intentions, duties, and beliefs that brought about the construction of these monuments.

It is for this reason that we cannot discount human purposes in pre-history. In causal terms, they are necessary but not sufficient.

OUTLINE OF THE BOOK

The goal of this book, to reconstruct the Ohio Hopewell episode and the social system that was responsible for it, will be accomplished by using the embankment earthworks as the initial empirical base and then extending the emic insights derived from this analytical study to the empirical data marking the subsistence, settlement, and ceremonial practices, with special focus on the mortuary aspect of the latter. Part 1 is a systematic regional analysis of the embankment earthworks in terms of their formal variation and patterning. From this analysis a model will be postulated of the Central Ohio Valley embankment earthwork traditions of the Early and Middle Woodland periods. This will form the basis on which an initial chronological framework will be constructed. Following this, and in preparation for the symbolic and social interpretation of the meaning and uses of the embankment earthworks, with their associated mounds and ceremonial contents, a theoretical characterization of the meaningful nature of material culture will be presented. The general approach, termed symbolic pragmatics, will be used to develop the Warranting Model, which will serve as a heuristic device for applying the symbolic pragmatic perspective to the archaeological record. Using this general theoretical framework, the World Renewal Model of the Ohio Hopewell embankment earthworks will be elucidated and grounded on the embankment earthworks selected for their particular relevance. The core finding of this model is what is termed here the Sacred Earth Principle. This principle will serve as the cognitive-normative portal into the world as it was experienced by the builders.

Part 2 will build on the implications of the Sacred Earth Principle for understanding Ohio Hopewell by elucidating an ecological framework that integrates practical subsistence and settlement with ceremony and ritual. In short, the technical and the normative will be integrated through a focus on the normative-ethical implications for ecological practices arising from the Sacred Earth Principle. The general approach will be termed deontic ecology, deontics being understood as the normative, ethical, and moral dimension of social life. Out of this theoretical framework a symbolic integrative account of the ritual and practical ecological practices manifested by the total range of archaeological data—subsistence, settlement, and ceremonial—will be presented. It will rest on three linked claims postulated as being at the core of the Middle Woodland cultural framework, and by extension, of the prehistoric Eastern Woodlands. The first claim is that, in virtue of their cultural beliefs, humans took themselves to be occupying a world that was immanently sacred. The second claim follows from the first in that, for these populations, the pursuit of biological survival was experienced as entailing an ongoing obligation to minimize the impact that these necessary subsistence and settlement practices had on the immanently sacred natural order of the land and to actively rectify this impact through what might be appropriately called renewal ritual. The third related claim is that, as a logical cornerstone of the Sacred Earth Principle and the deontic ecological framework it grounded, the resources of the land were available to all groups according to their recognized needs, subject only to their fulfilling the traditional modes of appropriation by which to minimize the polluting consequences of their subsistence and settlement interventions while maximizing the sanctifying consequences.

These three claims postulate three core principles as the basis of the integrative framework of this book, which will be termed the Inclusive Territorial/Custodial Domain Paradigm of the prehistoric Eastern Woodlands. Central to this paradigm is the principle of inclusive territories, as expressed in the third claim above, namely, since the land was immanently sacred, no one could possibly own the land or its resources. Rather these were available to all as required. A group's territory, therefore, was the range it habitually occupied and practically exploited.[10] Local groups shared among themselves the responsibility to act as custodians of the land, this custodianship being the obligation to use the land in accordance to views articulated under the Sacred

Earth Principle. This Inclusive Territorial/Custodial Domain Paradigm will be contrasted to the alternative general account that is broadly accepted in current North American archaeology, which will be termed here the Exclusive Territorial/Proprietorial Domain Paradigm. At the core of this latter paradigm is the premise that prehistoric exploitation of resources was based on exclusive territories that were constituted as boundaried, proprietorial domains. Thus regions were divided into spatial modules, and their collective owners as proprietorial corporations would "naturally" defend their claims to exclusive control of the land and its resources against incursions by "outsiders."

These alternative paradigms are elucidated and critiqued by taking a deep historical perspective both to anchor the basic premises of the Inclusive Territorial/Custodial Domain Paradigm and to articulate what was the most probable cultural historical background of Hopewell, this being a set of Eastern Woodland traditions that were emerging by the Middle Archaic (ca. 6000 B.C.), if not earlier. Out of this paradigm is presented a global theoretical framework termed the Proscriptive/Prescriptive Ecological Strategy Model. Although this model is designed to be applied to the archaeological record of the Ohio Early and Middle Woodland periods, it is shaped by critically reviewing current accounts of the Archaic and Woodland prehistory under the Exclusive Territorial/Proprietorial Domain Paradigm. The latter half of part 2 and part 3 are devoted to developing this global framework by elucidating and empirically grounding a series of auxiliary and ancillary models necessary to reconstruct the social organizational dimension as manifested in the Ohio Hopewell mortuary record. Part 4 serves as the culmination of this complex analysis by taking an ideological factional approach to interpreting the variation of the Ohio Hopewell archaeological record. This provides an account of the dynamics of the Ohio Hopewell episode and its rather abrupt termination, demarcated by what has been called the Middle Woodland-Late Woodland transition in this region.

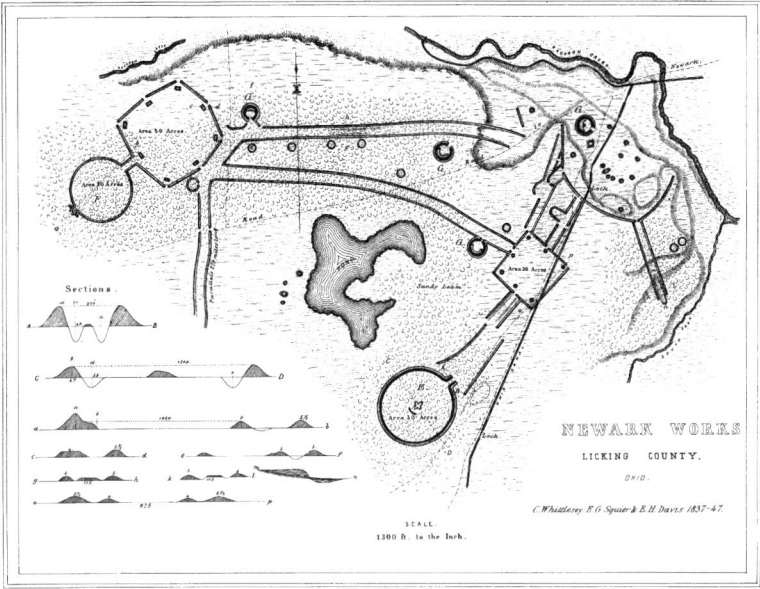

FIG. P.I. Newark Works (Squier and Davis [1848] 1973, plate XXV)

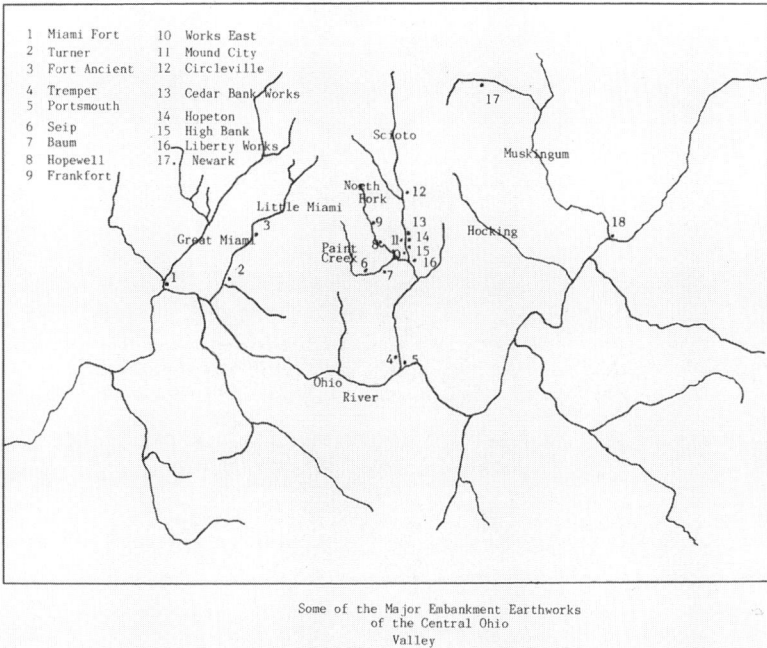

1	Miami Fort	10	Works East
2	Turner	11	Mound City
3	Fort Ancient	12	Circleville
4	Tremper	13	Cedar Bank Works
5	Portsmouth	14	Hopeton
6	Seip	15	High Bank
7	Baum	16	Liberty Works
8	Hopewell	17	Newark
9	Frankfort		

Some of the Major Embankment Earthworks
of the Central Ohio
Valley

FIG. P.2. Some of the Major Ohio Embankment Earthwork Sites

X X .

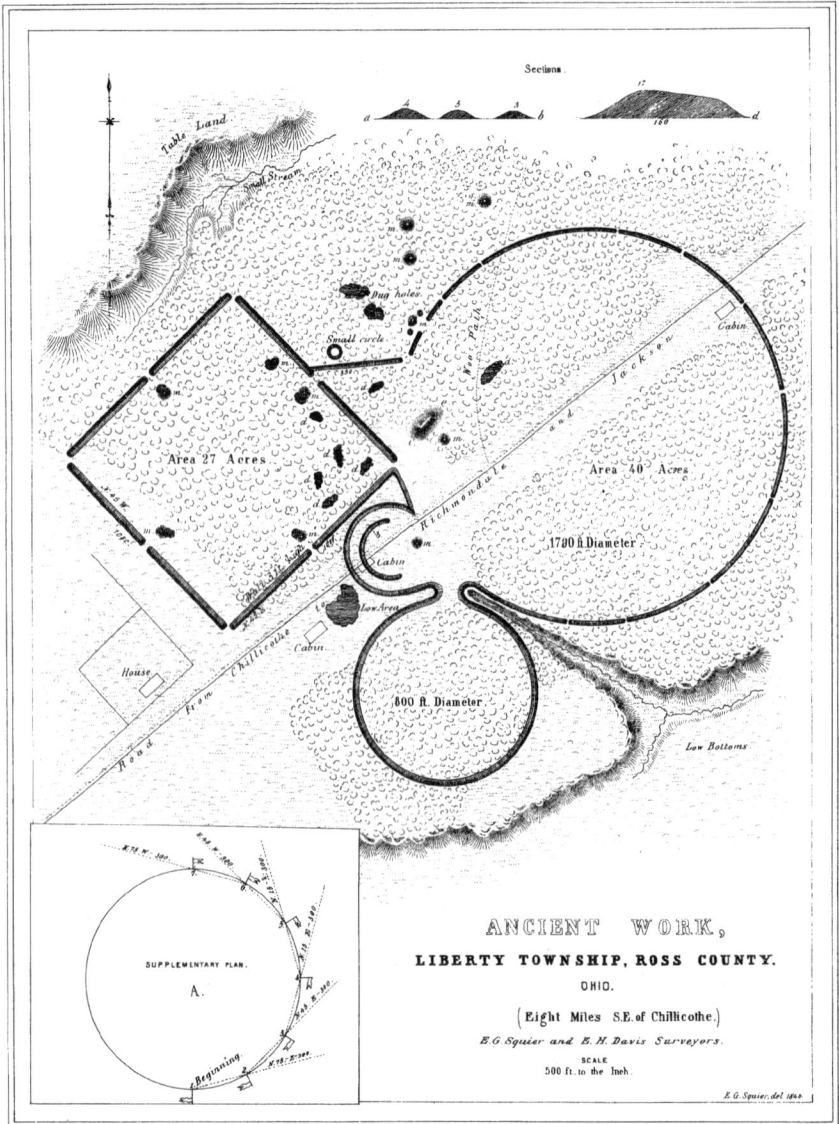

FIG. P.3. Liberty Works (Squier and Davis [1848] 1973, plate XX)

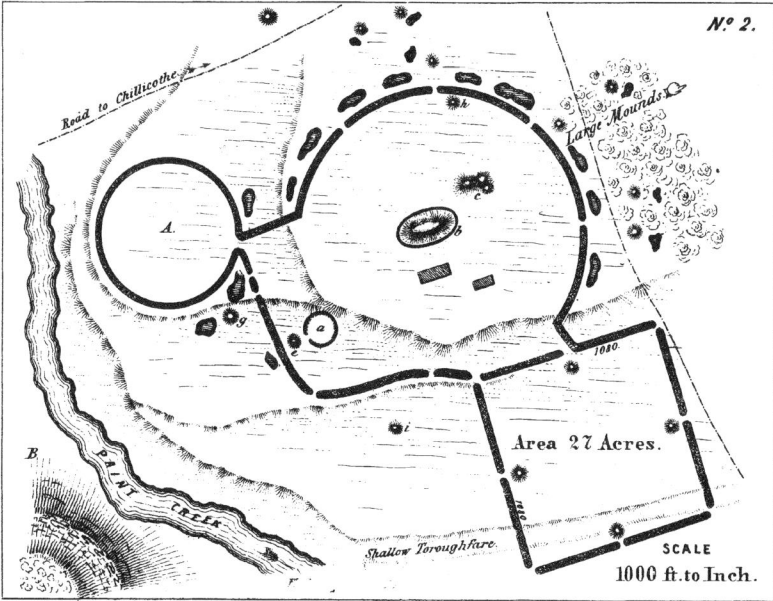

FIG. P.4. Seip (Squier and Davis [1848] 1973, plate XXI, no.2)

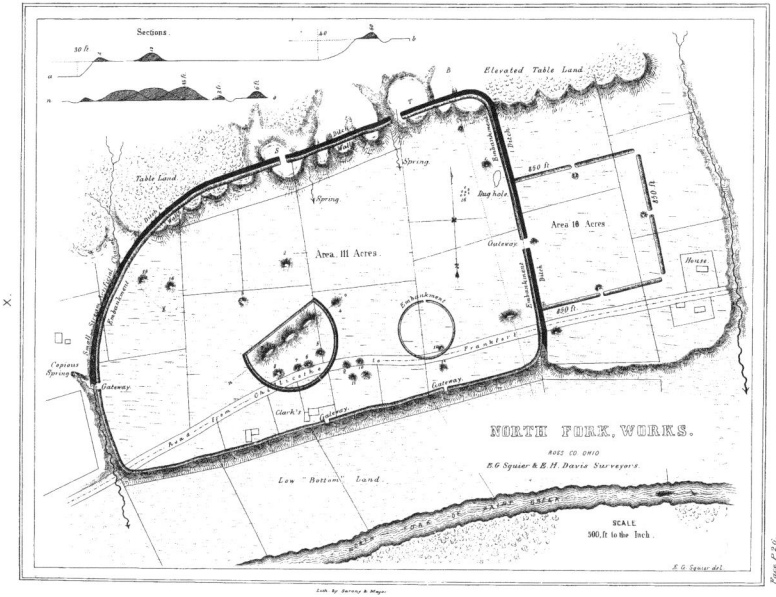

FIG. P.5. Hopewell (Squier and Davis [1848] 1973, plate X)

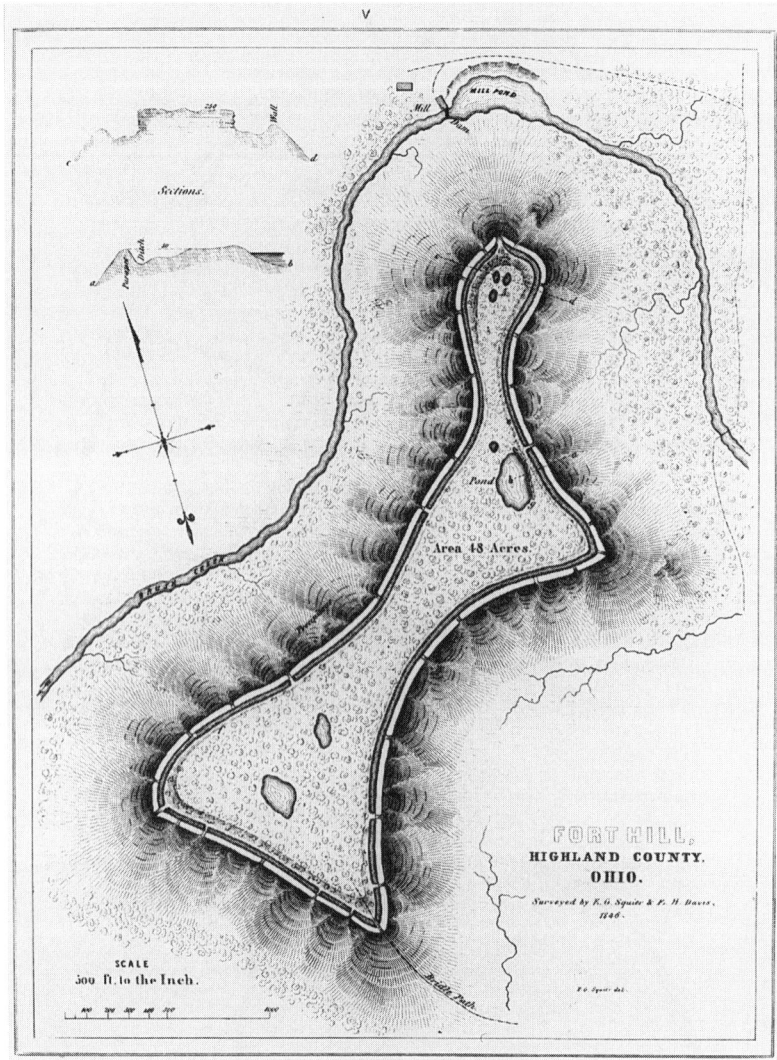

FIG. P.6. Fort Hill (Squier and Davis [1848] 1973, plate V)

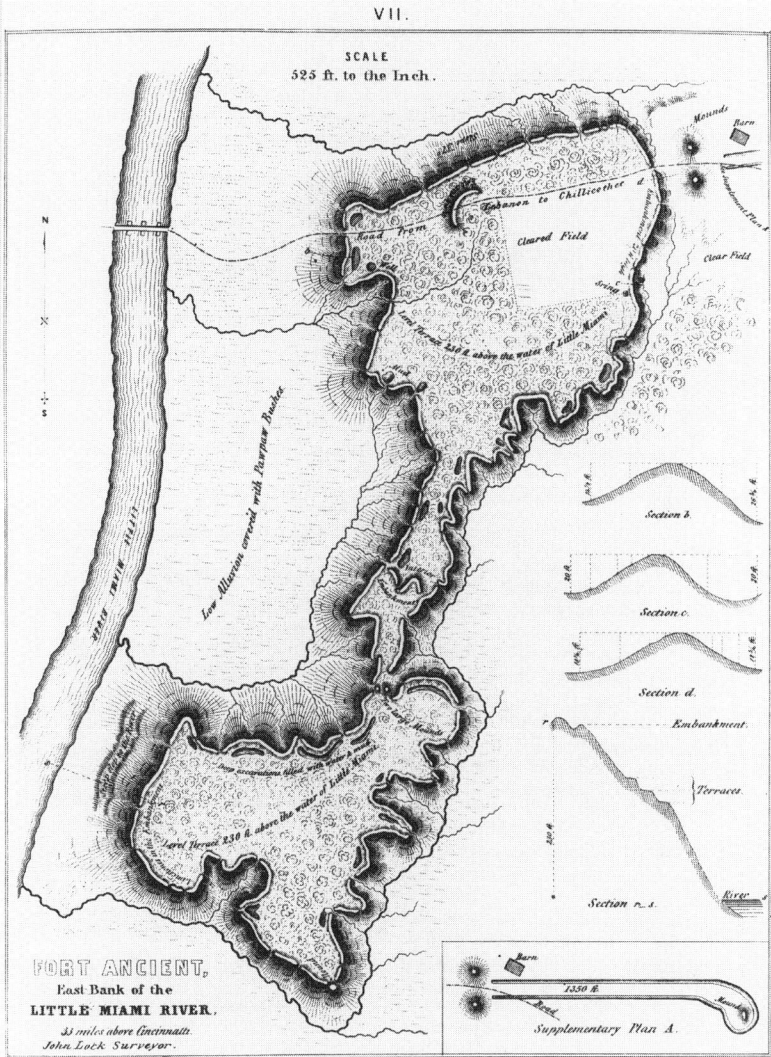

FIG. P.7. Fort Ancient (Squier and Davis [1848] 1973, plate VII)

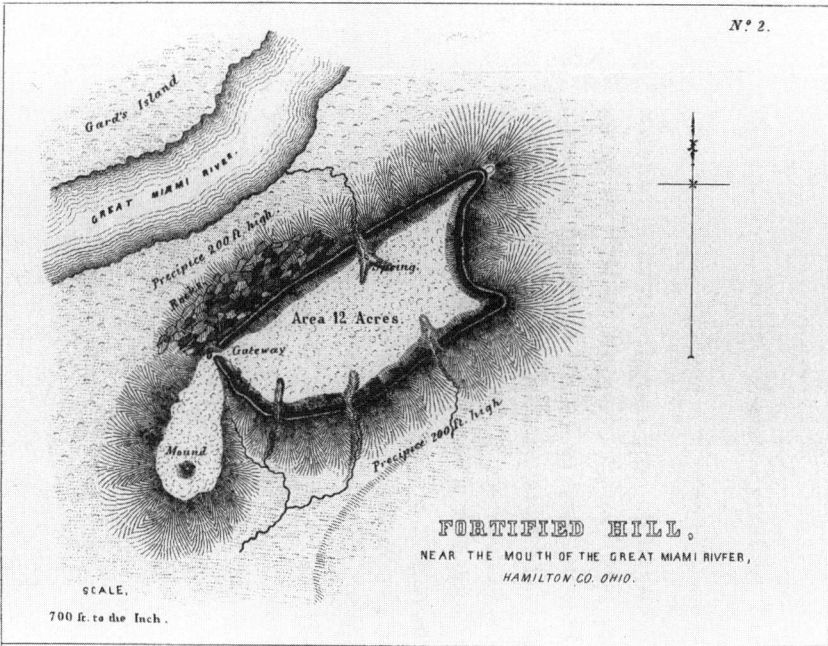

FIG. P.9. Miami Fort (Squier and Davis [1848] 1973, plate IX, no. 2)

FIG. P.8. (OPPOSITE PAGE) Illustrations of "Classic" Ohio Hopewellian Assemblage (Griffin 1967, 182, figure 4) [Copyright American Association for the Advancement of Science]
(*a*) Turner Earthworks, Hamilton County, Ohio (diameter of elevated circle, 140 meters); (*b*) designs in thin copper plate from Hopewell site, Ross County, Ohio (about X⅓); (*c*) bone engraving on proximal half of human femur, Hopewell site, (length of bone, about 13 centimeters); (*d*) engraved stone tablet from Cincinnati (length, about 13 centimeters); (*e*) painted pottery male effigy from Knight site, Calhoun County, Illinois (height, 11 centimeters); (*f*) *Busycon* cup, Hopewell site (about 30 centimeters); (*g*) mica effigy of eagle (?) claw, Hopewell site (28 centimeters); (*h*) different types of copper earspools from Seip Mound, Ross County, Ohio (each spool about 5 centimeters); (*i*) obsidian ceremonial spear from Hopewell site (23 centimeters); (*j*) Havana Zoned Dentate vessel from Havana Mound 6, Mason County, Illinois (height, about 15 centimeters); (*k*) embossed flying eagle (?) design on copper plate, Mount City, Ross County, Ohio (length, 30 centimeters); (*l*) sheet copper over cane panpipe, Hopewell site (length, 11½ centimeters); (*m*) Hopewell style vessel from Crooks site, La Salle Parish, Louisiana (height, about 15 centimeters); (*n*) Hopewell style vessel from Crooks site; (*o*) sculptured and engraved hawk effigy pipe from Tremper Site, Scioto County, Ohio, and plain platform pipe from Hopewell site (both about 10 centimeters long); (*p*) engraved roseate spoonbilled-duck design, Hopewell site (length, about 10 centimeters)

Part One

THE WORLD EMBODIED

Chapter 1

THE OHIO HOPEWELL EMBANKMENT EARTHWORK SYSTEMATICS

The embankment earthworks encountered by Europeans entering the Central Ohio Valley in the late 1700s were magnificent monuments made of locally procured earth (and some stone). For immediate purposes, two major formal types will be addressed, what are often called the "geometricals" and the "fortifications." The former, typified by Newark (fig. P.1), Liberty Works (fig. P.3), and Seip (fig. P.4), can also be subdivided into symmetrical and subsymmetrical types. The "fortifications," exemplified by Fort Hill (fig. P.6), Fort Ancient (fig. P.7), and Miami Fort (fig. P.9), were built as great wall-like embankments on isolated tabletop plateaus and ridge tops. The term "fortification" may be a serious misnomer, since there is now growing evidence that the embankments served as major ceremonial rather than as defensive features. The alternative "fortification" view is still represented in the literature.[1] The distribution of these two types correlates roughly with two different regions of the Central Ohio Valley, so they seem to be manifesting some real social distinctions. But there is some intermixing. Several representatives of the ridge-top earthworks, such as Fort Hill near Chillicothe, are in the zone where the "geometricals" dominate, and a number of "geometricals" are in the zone where "fortifications" dominate. Because of this intermixing, and further reasons discussed later, both sets will be treated as ceremonial locales.

EMBANKMENT ATTRIBUTES

An overview of the known earthworks reveals a great deal of variation and complexity of form that cannot be adequately characterized by the two terms "fortification" and "geometrical." Moreover, such functional-descriptive terms prefigure particular prehistoric reconstructions, thereby implicating cultural backgrounds that presuppose particular types of social systems. The approach in this book attempts to minimize this prefiguring by avoiding "loaded" functional-descriptive terms and, instead, classifies the earthworks in terms of the variation in the primary attributes, these being the attributes making up horizontal or *plan* form and vertical or *profile* form.

Because the plan form, or the "bird's-eye-view" aspect, is the primary mode of representing them in the literature, the earthworks tend to be thought of as two-dimensional. However, they are three-dimensional, having horizontal and vertical extension.[2] In principle, horizontal and vertical forms can vary independently. Therefore, by analyzing these two forms as separate and, in principle, independent attributes, correlations of the variations in these two attributes can be made to establish whether significant and mutually exclusive patterns exist. If they do, then the possibility of a real or emic typology of embankment earthworks can be established. If these types can be further noted as sorting regionally and temporally, this becomes reasonable grounds to postulate sets of construction protocols that varied historically and geographically across the Central Ohio Valley. This analytical procedure was performed, and it resulted in empirical evidence useful for postulating embankment earthwork traditions. These were also used to ground a chronological model based on the combination of three analyses: (1) a form of "horizontal stratigraphy" based on the analysis of contrasting vertical profile and horizontal plan forms combined in several separate complex earthworks; (2) a formal seriation analysis of the major attributes earthworks; and (3) an analysis using the available radiocarbon dates.

The Horizontal Plan Attribute

The horizontal plan form has three basic variants, here termed the G-Form, T-Form, and C-Form. Traditionally, only the first two have been recognized, those called "geometricals" and "fortifications," respectively. The identification of the third type, the C-Form, resulted from the formal analysis referred to above and summarized below.

1. The G-Form

Whether symmetrical or subsymmetrical, the G-Form (i.e., Geometrical Form) embankment earthwork can be subdivided into two varieties, the *GC-Form* (Circle) and *GR-Form* (Rectilinear). G-Form sites can be further classified into *Simple* and *Complex*. A *Simple G-Form* embankment earthwork displays only one of the two G-Form attributes, circulinear or rectilinear. A *Complex G-Form* has two or more Simple G-Form components, and typically at least one example each of the GC-Form and the GR-Form is either physically linked or spatially associated. The Fairground Circle at Newark and the Newark Circle-Octagon are good examples of the Simple and the Complex G-Form, respectively (fig. P.1).

2. The T-Form

Many of the earthworks referred to as "fortifications," "forts," "stockades," and so on, were constructed on isolated tabletop ridges with the embankment closely following the ridge edge. For this reason, they can be called *T-Forms* (Topographical Forms). It appears that often parts of the embankment were deliberately placed so as to "exclude" certain elements of the ridge, forming what Connolly has called "spurs."[3] The earth for the embankment was procured from "inside," by contiguous ditching, by surface scraping, or by borrow pitting; the borrow pit was often transformed into a contiguous pond-like depression ballasted with flat stones. Connolly also argues that considerable labor was expended in removing surfaces and filling in eroded areas on which to build the embankments. Fort Ancient (fig. P.7) represents this type.

3. The C-Form

The third category of horizontal plan form, the C-Form, was not previously recognized. Rather it was assimilated into either the "fortification" or the "geometrical" type, either because a particular C-Form embankment earthwork shared the subsymmetry of some of the G-Form embankment earthwork sites or because it shared with the T-Form the appearance and positioning of being some form of "fort" or "fortification," often (not always) being built on a rather high terrace. In fact, a number of archaeologists have argued that this type of embankment locale lacks an embankment along the terrace edge because the "precipitous" drop to the valley floor acted as a natural barrier against an attacking force.[4] In any case, typically the C-Form was built on an upper terrace overlooking the bottom lands. The embankment was always built in a large "C," either with a rather wide or sometimes with a more constrained "open" side, and each end of the C terminated at or near the terrace edge. This resulted in the "open" part of the "C" being the terrace edge overlooking the valley. Although carefully constructed, the overall horizontal form of the embankment would often be subsymmetrical. The main embankment of the Hopewell site (fig. P.5) illustrates the C-Form type, as does Colerain (fig. 1.1).

This constellation of attributes—being built on an upper terrace overlooking the lower terraces of the floodplain; subsymmetry; having irregularly placed "gateways," much like the T-Form; and so on—gave it an anomalous position: somewhat like a "fortification," but also somewhat like a subsymmetrical "geometrical." However, the correlation of horizontal form and vertical profile gives empirical grounds for treating this large embankment earthwork as a third horizontal form category having ceremonial and not military relevance.

The Vertical Profile Attributes

A neutral formal terminology is also required in order to delineate the vertical profile attribute and its variations. Typically, the latter are characterized by the presence/absence of ditches and, if ditches are present, then the relative positioning of these ditches, whether "inside" or "outside" the embankment. The "inside/outside" distinction, however, implicates the notion of an "enclosure." This is not a neutral term since, of course, it implies that the earthwork was built to close off a space, separating this from the "outside," and this further implies exclusion, a select in-group separated from an amorphous out-group, and so on. The nature of the social system that was responsible for this "enclosure," then, is prefigured in the description, since the system would be based on principles that promote mutual exclusion of components. To avoid this prefiguring, terms were innovated by using an observant-relative approach, as described below.

1. The S-Profile

Imagine yourself as an observer standing on the west side of a 50 m diameter G-Form earthwork having a continuous and attached "inner" ditch. The embankment of the circle does not connect, leaving a space in the eastern sector of about 5 m—an "open" sector. Since you are standing on top of the western sector of the embankment and facing north, then the eastern "opening" is on your right-hand side. This would mean that the contiguous ditch is also on your right-hand side. Now imagine cross-sectioning the embankment/ditch where you are standing, starting from the far side of the ditch, through the embankment to its far side. This cross-section could be graphically depicted as a sine-like wave and it will be called the vertical S-Profile form attribute. This is a compound attribute, the positive half of the sine wave depicting the embankment wall and the negative half, the ditch. Since the ditch is on your right side, this can be called the *SR-Profile* variant. If it were on your left side, that is, "outside" the embankment earthwork, it would display the *SL-Profile* variant.

Objectively, there are several embankment/ditch S-Profile variations that are possible: (1) a ditch on both sides (SLR-Profile); (2) a ditch
only on one side and not the other of the embankment (SL-Profile *or* SR-
Profile); (3) ditches alternating from one side of the embankment to the
other (SL/SR/SL-Profile). In fact, the vast majority of actual S-Profile
earthworks fall into either the SR-Profile or the SL-Profile variant.

2. The K-Profile

There are many important earthworks that have no apparent
contiguous ditches or only dispersed borrow pits. In effect, the cross-
section vertical profile would form a cone with a rounded top. Since
the letter "C" has been used to name the C-Form, the phonic route is
taken by using the letter "K." The K-Profile embankment attribute is
the rounded-conical profile that is formed by a cross-section through
an embankment wall that has no visible contiguous ditch. But the K-
Profile is still treated here as a compound attribute since the procurement of the fill necessarily made local topographical surface changes.
There are two possibilities: borrow pitting and surface scraping. An
embankment earthwork with no apparent borrow pits "displays" the
surface-scraped attribute. This can be termed the KS-Profile. Those
displaying borrow pitting can be characterized as displaying the KP-
Profile attribute. The latter can be further distinguished as proximal
and distal pits, referred to as KPp-Profiles and KPd-Profiles, respectively.
Although these latter variations appear to be real variants, it is not
necessary to distinguish routinely between them. Table 1.1 summarizes
the critical profile attributes. These are the S-Profile in two important
versions, the SR-Profile and the SL-Profile, and the K-Profile in two
important versions, the KS-Profile and the KP-Profile, the latter having
two subvariants, the KPp-Profile and the KPd-Profile.

DISCUSSION

Assuming no particular physical constraints, the null-hypothesis
would claim that the correlation of these horizontal plan and vertical

profile attributes should be random. However, a correlation of these attributes established significant associations, giving empirical grounds for rejecting the hypothesis and concluding that real emic categories of embankment earthwork types existed.[5] Typically, (1) Simple G-Forms are correlated with SR-Profiles or K-Profiles. (2) C-Forms are correlated almost exclusively with SL-Profiles. (3) T-Forms and Complex G-Forms are correlated with K-Profiles (Table 1.2), generating four formal types of embankment earthworks: (1) the Simple G-Form/SR-Profile embankment earthwork type (there are exceptions, e.g., the Simple G-Form/K-Profile); (2) the C-Form/SL-Profile embankment earthwork type; (3) the T-Form/K-Profile embankment earthwork type; and (4) the Complex G-Form/K-Profile embankment earthwork type. One of the few excavated embankment sites displaying the Simple G-Form/SR-Profile combination is Mt. Horeb (fig. 1.2).

The Complex G-Form/K-Profile embankment earthworks figure prominently in this book and are well illustrated by Liberty Works (fig. P.3), Seip (fig. P.4), and High Bank (fig. 1.3), these three being within a few kilometers of each other.

These are not hermetically sealed categories, since there are a number of sites that do not fully conform. Either they display morphological attributes that are unique variants of the four already outlined, or, more commonly, they combine morphological components of two or more of these four.

High Bank (fig. 1.3) illustrates this well. While the major component consists of a circle and an octagon, (a Complex G-Form/K-Profile type), the southern G-Form/SR-Profile elements are also considered to be part of this total site. The Hopewell site (fig. P.5) is another example which displays a "cross-fertilization" in that it combines the C-Form/SL-Profile type with components of the Complex G-Form/K-Profile type. It turns out that this mixing is important for chronological analysis based on "horizontal stratigraphy" and seriation dating, and these will be used to ground models of intra-site development and inter-site interaction.

The distribution of these four types within the Central Ohio Valley defines embankment traditions in either two or three configurations, depending on the particular attributes that are emphasized. Based on the G-Form, C-Form, and T-Form horizontal attribute types, three traditions could be argued for. There are two problems with this. First, it subsumes the Simple/Complex variation of the G-Form. This variation is critically important since it has both regional and chronological implications. Second, it does not accord well with the distribution of these different types. Three traditions imply three different regions. While the Simple G-Form is almost ubiquitous in the Central Ohio Valley, a dichotomized east-west distribution of the Complex G-Form, T-Form, and C-Form is fairly clear. The Complex G-Form types are found concentrated in south-central, central, and eastern Ohio with a focus on the central and lower Scioto River, particularly in the region surrounding Chillicothe, Ross County. The T-Form and C-Form types are concentrated in southwest Ohio, particularly in the Great and Little Miami drainages, as well as spilling over into parts of southeast Indiana and west Kentucky.[6] Of course, as already noted, there are important examples of Complex G-Forms in southwest Ohio and T-Forms and C-Forms in the Chillicothe area, as well as near Newark. However, this relatively limited intermixing does not detract from the overall tendency toward a dichotomized distribution.

The Simple G-Form/SR-Profile embankment earthwork type, or what is commonly referred to as the Adena Sacred Circle, is generally recognized as demarcating the Early Woodland period of the Central Ohio Valley, at least the later Early Woodland period. For this reason it is treated here as defining the *Mt. Horeb Tradition*. This term was chosen following Warren DeBoer's (1997) use of it. The Mt. Horeb Tradition would include both circular and rectilinear G-Form/SR-Profile earthworks (Simple GC-Forms and Simple GR-Forms, although the latter are a distinct minority). It is now recognized, however, that the Mt. Horeb Tradition not only largely defines the later Early Wood-land period but also extends into the Middle Woodland in limited parts of this region, terminating around A.D. 250.[7]

The other three types, the Complex G-Form/K-Profile, the T-Form/K-Profile, and the C-Form/SL-Profile, will be treated as defining the Ohio Hopewell Tradition, and the generally east-west dichotomous distribution of these as described above will be treated as delineating two co-traditions. Because the Complex G-Form/K-Profile type is par-ticularly richly represented in the Chillicothe region, it defines what will be termed here the Chillicothe Tradition. The C-Form/SL-Profile and T-Form/K-Profile types of southwest Ohio and, to a lesser degree, southeast Indiana and Kentucky define what will be termed here the Miami Fort Tradition. This latter term was chosen since "Fort An-cient," the near neighbor of Miami Fort, has already been used to name the Late Prehistoric period of the Central Ohio Valley region.

The Mt. Horeb Tradition is represented by Sacred Circles throughout the Central Ohio Valley. In contrast, the earthworks of the Ohio Hopewell tradition are more limited in distribution. This suggests that the Mt. Horeb Tradition emerged as one of this region's original expressions of embankment earthwork construction. However, since the SR-Profile and SL-Profile attributes appear to be complementary opposites, these have been interpreted as overlapping in time, with the SL-Profile attribute possibly emerging later than the SR-Profile. Since the SL-Profile attribute has been identified with the Miami Fort Tra-dition, it would be one indicator demarcating the emergence of Ohio Hopewell. However, there is the possibility that SL-Profile embank-ments may actually predate the Sacred Circle type (Simple G-Form/SR-Profile). The Peter Village site (fig. 1.4) displays the SL-Profile at-tribute, junctured with a subsymmetrical G-Form (oval). It has recently been excavated and analyzed by Berle Clay who interprets it as possibly early Early Woodland, that is, pre-Adena.[8] This suggests that, although the full C-Form/SL-Profile embankment earthwork, per se, may have been among the earliest of the southwest Ohio Hopewell embankment earthworks to emerge, the SL-Profile embankment attribute itself may be at least contemporary with or even earlier than the SR-Profile attribute characteristic of the Mt. Horeb Tradition (Adena).

Unless the discussion requires reference to the profile attribute, only the horizontal plan form terms will be used. Therefore, instead of C-Form/SL-Profile, the term C-Form will be used and the normally associated SL-Profile attribute will be assumed. The C-Form, in fact, is found as an important embankment earthwork of mid-southern Hopewellian sites, for example, the Marksville site in Louisiana, the Spanish Fort and the Little Spanish Fort sites in the Yazoo Basin, and possibly the Old Stone Fort of Eastern Tennessee although Bacon argues that it is more akin to the "fortification" type or the T-Form, in the terminology used here.[9] Therefore, the possibility is left open that, the C-Form, rather than originating in the Central Ohio Valley, may have emerged in the mid-southern region of the Eastern Woodlands and moved north, showing up in Kentucky and southwestern Ohio.

Not only are there examples of the C-Form type that are distributed well outside the Miamis, the main embankment component of the Hopewell site (fig. P.5) displays all the key attributes of this class, even though this eponymous site is at the "heart" of the region dominated by the Complex G-Form earthworks defining the Chillicothe Tradition. Importantly, it also displays the key elements of this latter tradition, that is, the GC-Form (circle) and the GR-Form (rectilinear), and both of these display the K-Profile attribute. Another C-Form earthwork of this region is Cedar Bank (fig. 1.5) on the east side of the Scioto River just north of Chillicothe. As stated earlier, rather than seeing this spill-over of earthworks as a "spoiler," it can be profitably treated as a historical record of interaction constituting a dynamic process of alliance building and breaking.

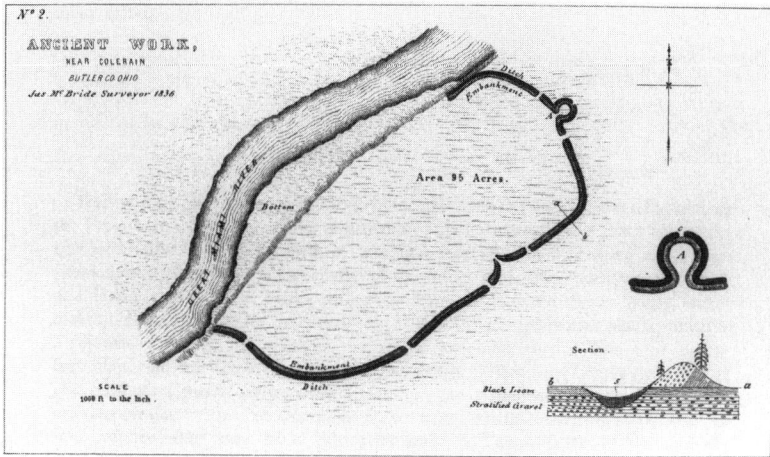

FIG. 1.1. Colerain (Squier and Davis [1848] 1973, plate XIII, no. 2)

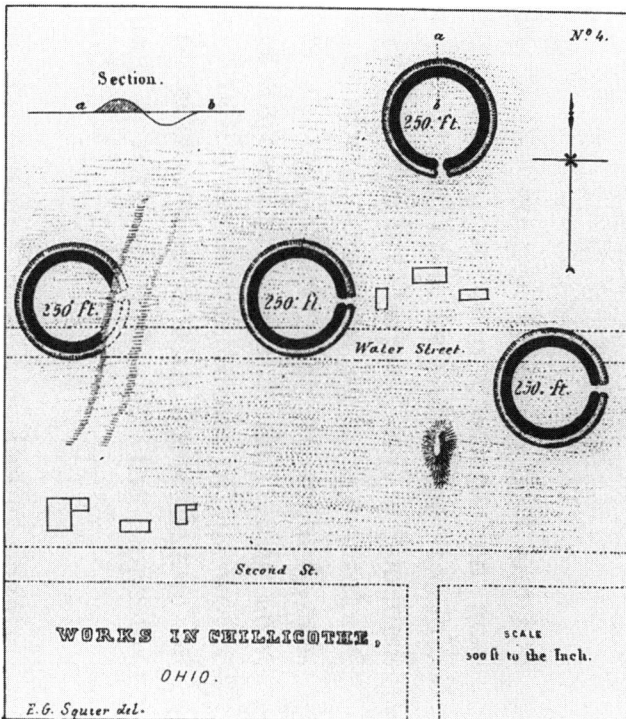

FIG. 1.2. Sacred Circle of Mt. Horeb Tradition (Squier and Davis [1848] 1973, plate XXXII, no. 4)

XVI.

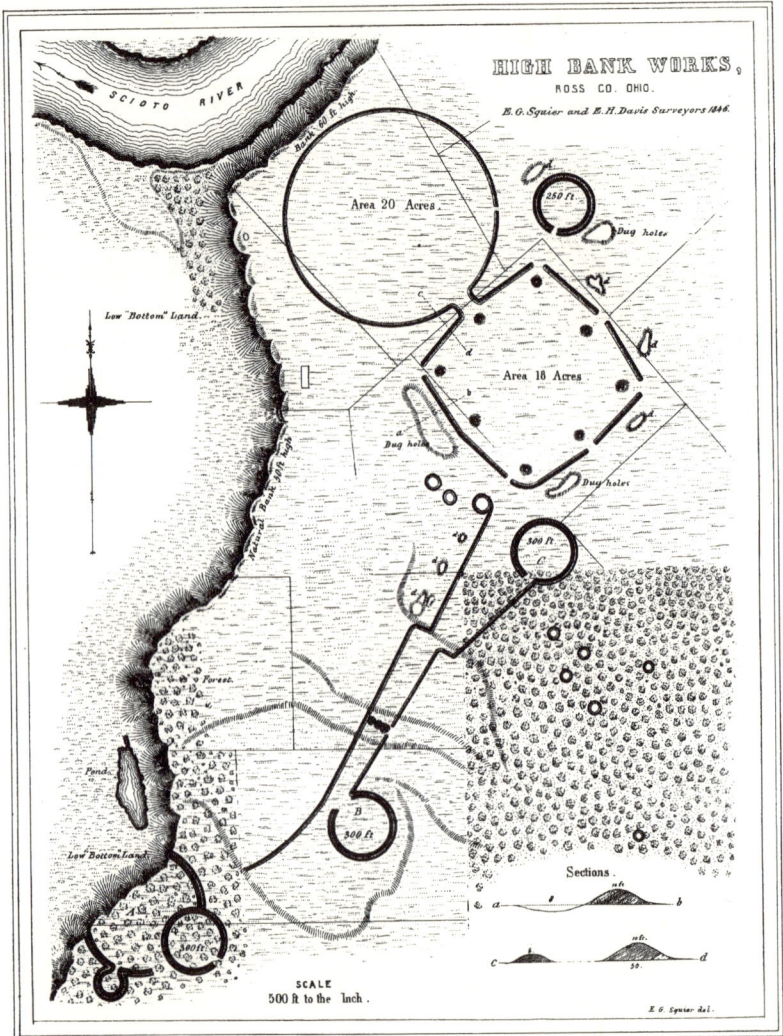

FIG. 1.3. High Bank (Squier and Davis [1848] 1973, plate XVI)

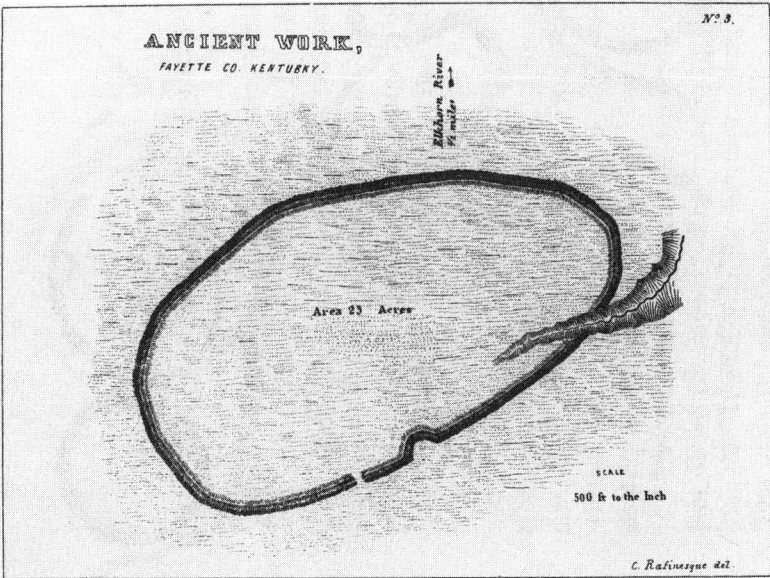

FIG. I.4. Peter Village (Squier and Davis [1848] 1973, plate XIV, no. 3)

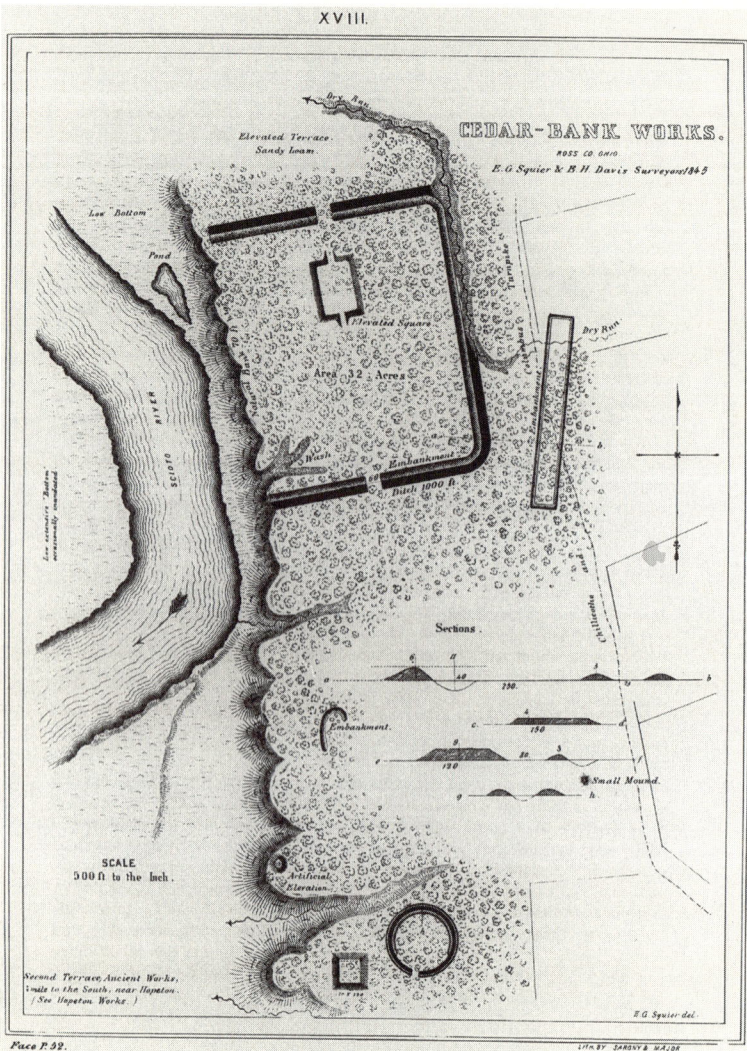

FIG. 1.5. Cedar Bank (Squier and Davis [1848] 1973, plate XVIII)

Table 1.1 Vertical Profile Attributes

Ditch-Contiguous		Ditch-Absent	
S-Profile		K-Profile	
Left	Right	Surface	Borrow Pit
SL-Profile	SR-Profile	KS-Profile	KP-Profile
			Proximal Pit Distal Pit
			KPp-Profile KPd-Profile

Table 1.2 Ohio Hopewell Embankment Earthwork Types
Based on Profile and Plan Correlations

	G-Form	C-Form	T-Form
SR-Profile	Simple G-Form/ SR-Profile	Very Rare	Rare
SL-Profile	N/A	C-Form/SL-Profile	N/A
K-Profile	Simple and Complex G-Form/K-Profile	Very Rare	T-Form/K-Profile (includes KS, KP and SR Profiles)

Chapter 2

THE C-R CONFIGURATION (CIRCLE-RECTILINEAR)

The Complex G-Form of the Chillicothe Tradition appears to be underwritten by a basic dual patterning of conjoined circular and rectilinear earthworks. This will be termed here the C-R Configuration (Circle-Rectilinear Configuration). There appear to be two basic modes by which the C and the R components are conjoined. These conjoining modes are represented by the hyphen (-) in the C-R acronym. While the rectilinear components vary between symmetrical and subsymmetrical squares, and other rectilinears, such as octagons, and one rhomboid, the circle components are almost always symmetrical. The variation in the form of the rectilinears and the number and sizes of the circles appear to be correlated with the two different conjoining modes, thereby constituting the Chillicothe Tradition as being defined by two C-R Configuration variants. These two C-R variants are well illustrated by the High Bank Works (fig. 1.2) and the Liberty Works (fig. P.3). The Circle-Octagon of the High Bank site can be used as the type site for what is termed here the High Bank C-R Configuration variant. Liberty Works is a slightly deviant example of what will be termed the Paint Creek C-R Configuration variant, a name chosen because there are five major embankment sites in the central Paint Creek-Scioto region that display the basic attributes of this C-R Configuration variant: Liberty Works, Seip (fig. P.4), Works East (fig. 2.1),

Baum (fig. 2.2), and Frankfort (fig. 2.3). Arguably, the Hopewell site (fig. P.5) also displays the basic components of the Paint Creek C-R Configuration. Adding this site to the above five means that there are at least six Paint Creek C-R Configuration embankment earthwork sites in this immediate region, all within Ross County, Ohio.

THE HIGH BANK C-R CONFIGURATION

The High Bank site (fig. 1.3) is one of at least three clearly recognizable versions of this Complex G-Form type. Another is the Newark Circle-Octagon (fig. P.1), and a third is the Circleville site (fig. 2.4), on the Scioto River, about halfway between Chillicothe and Columbus, Ohio. Among several other sites that can be considered as falling within this category are Dunlaps Works (fig. 2.5), north of Chillicothe; the concentric quartered circles of Portsmouth, termed the Indian Fort (fig. 2.6); and the circle-rectilinear at Marietta (fig. 2.7). The northern component of Fort Ancient (fig. P.7) may have critical properties embedded in it that would allow classifying that component as the equivalent of a High Bank C-R Configuration.[1] If these are treated as subvariants, then there are at least seven known sites that display all or some of the elements of the High Bank C-R Configuration: High Bank itself, Newark, Circleville, Dunlaps Works, and elements of Portsmouth, Marietta, and Fort Ancient.

The C (Circle) and R (Octagon) of High Bank are conjoined by a set of parallel K-Profile embankments that will be termed here the aggregation neck or, more simply, the neck. If a line were drawn parallel through the neck and extended in both directions to the far perimeter of the Circle and the far side of the Octagon, it would bisect the total C-R Configuration into almost perfect mirror-image halves. This mirror-image symmetry is an important criterion of the High Bank C-R Configuration type. Also it will be noted that the Octagon component is constructed by straight embankments that, with one exception, do not connect, in this way forming what appear to be "gates." Again, the use of this or related terms, such as "entryway," or "gateway," is misleading since they imply that the embankment earthworks functioned as "enclosures" for the builders. Previously, in the attempt to avoid prejudging their sense, they were referred to as "hyphenations."[2] The latter strategy may not be much more illuminating. Therefore, the term "gate" will be used here, with the caveat that using this and related terms could be and probably are misleading. Typically each "gate" is associated with an earthen "gate" mound inset from the perimeter of the Octagon. The High Bank Octagon, in fact, has eight "gate" mounds but only seven "gates," further suggesting that these do not express the sense that we attach to this term. Finally, according to Hively and Horn, the C and R components of the High Bank C-R Configuration are geometrically related.[3]

While Octagons are found at High Bank and Newark, the equivalent rectilinears of Portsmouth, Marietta, and Circleville would be more appropriately classed as Squares. Because the rectilinear at Dunlaps Works forms a rhomboid or trapezoid, it displays balanced rather than mirror-image symmetry and it has a second aggregation element that incorporates a distal mound into the complex. It also lacks "gates" and "gate" mounds. However, in other respects it conforms to the High Bank C-R Configuration type. Abstracting from such variation, the key defining attributes of High Bank C-R Configuration status would seem to be the following: It usually has one unbroken C component and one regular R component having "gates" and inset "gate" mounds and these two components are proportionally related to each other through a common unit measure. The C and R components must be conjoined, normally by a short neck, although direct joining is possible. The mode of the High Bank C-R Configuration juncture is for the conjoined side of the R component to be centered and at a tangent to the perimeter of the C component such that a single axis parallel through the neck creates near mirror-image symmetry (Dunlaps Works being one apparent exception). There are other properties that a High Bank C-R Configuration type usually possesses, but those listed appear to be the primary constituent ones. Others will be discussed later. The clearest expressions of the High Bank C-R Configuration type are at High Bank and Newark. The former has been chosen here as the eponymous site for this variant of the C-R Configuration because it is in Ross County, the region in Ohio that contains the richest concentration of Complex G-Form sites.

THE PAINT CREEK C-R CONFIGURATION

The second variant of the Complex G-Form of the Chillicothe Tradition is what was earlier referred to as the Paint Creek C-R Configuration. Frankfort (fig. 2.3) will be used here as the exemplary representation. Among other terms for this form are "keyhole" form, tripartite (or quadripartite) form, "snowman" form, and "Casper the Ghost" form. Indeed, the tripartite term might appear the most appropriate since this form has three distinct components: small circle, large circle, large square. However, this tripartite pattern is a surface appearance that obscures a deeper structural duality that is equivalent to the High Bank C-R version. It is simply more complexly expressed.

Ideally, the three components of the Paint Creek C-R Configuration, as exemplified by Frankfort, individually display mirror-image symmetry, while the overall configuration does not. That is, unlike the tangent/centered juncture mode of the C and R components of the High Bank variant, the C and R components of the Paint Creek variant relate to each other in a distinctly skewed and off-centered manner. Therefore, I argue that the tripartite patterning is, in fact, the result of the formal skewed/off-centered juncture mode used to conjoin the C and R components. If this is the case, the large square, generically referred to here as a Paint Creek Square, can be treated as the complementary equivalent of the R component of the High Bank C-R Configuration. The Paint Creek Square also typically has eight "gates," four corner and four medial "gates," but only four "gate" mounds inset at the medial "gates." However, as impressive as it is, I claim that the large circle at Frankfort *is not* the equivalent of the large circle at High Bank. Rather, the *small circle* attached to the perimeter of the large circle at Frankfort is the complementary equivalent of the large circle of High Bank. This means that just as the neck at High Bank conjoins the C and R components of this site, so the *large circle* at Frankfort serves to link the equivalent C and R components of the Paint Creek C-R Configuration, and these components are (1) the small perimeter circle and (2) the large Paint Creek Square, respectively. In addition, William Romain has demonstrated that these three components are both proportionally related and based on the same unit measure as mentioned above for High Bank.[4] In this case, the side of the Paint Creek Square is about 320 m, the diameter of the large conjoining circle is equal to the diagonal of the Paint Creek Square, and the diameter of the small perimeter circle is equal to the side of a square having its diagonal equal to about 320 m.[5]

In sum, while there is a basic, underlying unity, the actual expression of the C-R Configuration distinctly varies between the Paint Creek and High Bank C-R Configurations, largely, but not only, in terms of the mode of juncture. In the High Bank C-R case, it is tangential/centered and normally effected by the aggregation neck. In the Paint Creek C-R case, it is skewed/off-centered and normally effected by the "inner" circle, which will be termed here the infix component. This term has been borrowed from Old English grammar. Its Anglo-Saxon roots have left their residue in English in the formation of the plural 'geese' and 'mice' from the singular 'goose' and 'mouse'. This morphemic change is effected by changing the internal vowel of the word and this type of inflection is termed the infix. The large circular component of the Paint Creek C-R Configuration is imagined here to be "like" an infix. In effect, the tangent/centered juncture of the High Bank C-R Configuration as mediated by the aggregation neck is transformed by the infix into the skewed/off-centered juncture of the Paint Creek C-R Configuration. The infix juncture and the aggregation neck juncture, therefore, constitute the Paint Creek and High Bank C-R Configurations as a complementary oppositional duality.

DISCUSSION

The Liberty Works (fig. P.3) and Seip sites (fig. P.4) are also classic examples of the Paint Creek C-R Configuration. The former site is about 2 km south of the High Bank site and on the same (east) bank of the Scioto River, just opposite the Paint Creek juncture. Seip is found about 6 km west on the Paint Creek. Despite their "classic" status, each has an anomaly that, in a sense, prevents them from being used for exemplary purposes. Initially N'omi Greber treated both Liberty Works and Seip as displaying the tripartite pattern. However, she later noted that each of these two classic expressions has a "fourth" component, the "lobed" circle at Liberty Works and the "amorphous" embankment at Seip, and reinterpreted each as displaying a quadripartite pattern.[6] In each case, the anomaly is found between the C component (the small perimeter circle component) and the R component (the Paint Creek Square). These "fourth" components will be treated here not as intrinsic structural elements of the Paint Creek C-R Configuration but as modifications that had to be made during the original construction programs. That they may have been allowable deviations from the norm reinforces the view that the primary purpose of the infix was to serve as a conjoining component. The extra or lobed component at Works East might, however, be different since it may have resulted from using Liberty Works as the authoritative model or template. This implies close cultural and historical relations between the two, points that will be addressed later.

It was earlier noted that the Hopewell site (fig. P.5) is a variant expression of the Paint Creek C-R Configuration. Is this a valid claim? The core component of the Hopewell earthwork is the C-Form embankment. Attached to the east side of this large embankment, however, is a version of the Paint Creek Square, and "inside" the C-Form there is a small circle. Both of these G-Forms display the K-Profile attribute. Furthermore, an attribute of the C-Form is for it to be open along the terrace edge while an attribute of the infix is to have a complete embankment with asymmetrically spaced "gates." Initially the

Hopewell site had the two ends of the open "C" terminating at the terrace edge overlooking the floodplain. It is clear that this terrace edge was subsequently closed, not by building an earth embankment but, instead, by building a *stone* embankment. This combination is very unusual. It suggests that, in fact, the Hopewell site was initially built in the standard C-Form / SL-Profile manner with the terrace edge open. Subsequently, this edge was "closed" by the construction of the stone wall. This later construction transformed the C-Form earthwork into the *equivalent* of an infix element, including asymmetrically spaced gates. Then by adding the square to the east side of the C-Form and building the circle "inside" the C-Form, now an "infix," the total earthwork was transformed into a unique expression of the Paint Creek C-R Configuration. It is notable that the small circle and square are oriented toward each other in a skewed manner / off-centered, almost exactly replicated by the C and R components of Frankfort.

There may be several other variant examples of the Paint Creek C-R Configuration. For example, Stubbs Works (fig. 2.8), Milford (fig. 2.9), and Camden (fig. 2.10) are all in southwest Ohio. A very unique variant, also in southwest Ohio, might be the Turner site (fig. 2.11), which will be discussed in considerable detail in Part 3. Finally, Newark (fig. P.1) may also have a variant of the Paint Creek C-R Configuration embedded in it, a point that will be discussed shortly. If Stubbs Works, Milford, Camden, and part of Newark (and possibly Turner, although this will be left out for the moment) are added to the above six (Liberty Works, Seip, Works East, Baum, Frankfort, and Hopewell), this makes at least ten examples of sites displaying recognizable properties of the Paint Creek C-R Configuration (the Seal site south of Chillicothe, even though lacking an infix component, may be another example). Add these Paint Creek C-R Configuration sites to the seven High Bank C-R Configuration sites (High Bank, Newark, Circleville, Dunlaps, Marietta, Portsmouth, and possibly, elements of Fort Ancient), and this gives a total of seventeen known C-R Configuration sites (nineteen if the Seal site and the Turner site are included). These do not exhaust the total known number but are sufficient to demonstrate a real dual patterning.

The seventeen are all equally large. The components that make up the five most prominent Paint Creek C-R Configuration sites, Liberty Works, Baum, Works East, Seip, and Frankfort, are almost all the same size. Typically the embankments of the High Bank C-R Configuration sites are the higher of the two, standing 2 m to 3 m before modern reduction processes diminished or, in most cases, effectively leveled them. The embankments of the Paint Creek C-R Configuration sites are typically lower, only about 1 m to 1.5 m high. Nevertheless, these earthworks share the same overall areal magnitude and clearly entailed significant collective labor.

CHRONOLOGICAL RELATIONS
Hopewell Site

Although many of the embankment earthworks in the core region of the Chillicothe Tradition display patterning that fits comfortably within this dual C-R Configuration, not all are so easily accommodated. The embankment earthworks of Mound City (fig. 2.12), Tremper (fig. 2.13), and Hopeton (fig. 2.14) are not neatly classifiable as either Paint Creek or High Bank types. However, based on associated mortuary materials, Mound City and Tremper are recognized as major Ohio Hopewell locales. These sites are particularly important since the contents of the associated mounds, along with the contents of the mounds at the Hopewell and Seip sites, were used to define the Ohio Hopewell artifactual and mortuary traits.[7]

How can these embankment earthworks be fitted into the C-R Configuration—or can they be? Hopeton and Mound City can be partly assimilated to the C-R Configuration by taking a historical approach. A form of "horizontal stratigraphy" can be carried out. The primary working assumption is that embankment earthworks that have the same horizontal plans and vertical profiles fall within the same time horizon. Accordingly, K-Profile embankment earthworks are treated as effectively contemporary with each other and are equally noncontemporary with S-Profile earthworks. S-Profile embankments that contrast on the SR-Profile and SL-Profile attributes would also form a largely contemporary set so that, in chronological terms, both the G-Form/SR-Profile type of the Mt. Horeb Tradition and the C-Form/SL-Profile type would at least partially overlap in time while being noncontemporary with K-Profile types. Therefore, embankment earthwork sites that combine two or more discretely different horizontal plan/vertical profile patterns may allow for establishing the relative chronological order of these different sites.

Critically, only embankment earthwork sites that display contrasting plan and profile attributes linked by horizontal construction extension can be used for this analytical method. The operating assumption is that the attributes of the add-on components clearly demarcate a later construction event. To determine which construction event came first and which was the add-on event, alternative construction sequences can be postulated and critiqued on the assumption that each set of similar components was part of a coherent, unified construction program. Therefore, the sequence that generates the fewest anomalies is accepted and those that generate the most anomalies are rejected.

The Hopewell site fills the primary requirement of displaying contrasting plan and profile attributes linked by horizontal construction extension. If this site were to be "dismantled" according to plan and profile components by conceptually removing the small circle, the square, that is, the K-Profile elements, and the stone wall running along the terrace edge, what is revealed is its "pure" C-Form/SL-Profile embankment pattern, an anomaly-free construction being the material outcome of a single, coherent construction program. This would support the conclusion that, since the G-Form/K-Profile plan/profile attributes were *added on* to the C-Form/SL-Profile component, the latter was built before the former. To confirm this claim, imagine the reverse sequence by conceptually removing the C-Form/SL-Profile Hopewell site components. This would leave an incomplete set of G-Form/K-

Profile components: the spatially isolated Circle, a three-sided or incomplete, Paint Creek Square east of it, and, of course, the single and isolated stone embankment with its western end terminating at a small erosion ditch.

Particularly interesting here would be the incomplete state of the Paint Creek Square. It is notable that the actual fourth side is the eastern "wall" of the C-Form/SL-Profile embankment, except that in this postulated sequence, this embankment would not yet have been built. Therefore, in contrast to the first sequence scenario, which revealed a unitary C-Form/SL-Profile embankment earthwork as the first construction event, in this second sequence scenario, the first construction event forms a pattern that is not replicated elsewhere among the C-R Configuration sites: three isolated K-Profile components, a three-sided "square," an anomalous "gate" mound where there is no "gate," and an anomalous stone wall.

It might be argued, however, that the "square" was simply not completed before the postulated change in construction plan that resulted in the construction of the C-Form/SL-Profile embankment. If no change had occurred, then it follows that the fourth wall would have been built to complete the Paint Creek Square and a large infix constructed to link together this R component and the small C component, thereby completing the basic morphology of the Paint Creek C-R Configuration. However, according to this scenario, the original plan would have been abandoned part way to completion and replaced with the large C-Form/SL-Profile embankment, thereby creating a hybrid earthwork as we see it today.

Is this a plausible scenario? No. It requires claiming that just as the builders were at the point of completing a *typical* Paint Creek C-R Configuration, requiring merely the addition of the fourth side to the Paint Creek Square to complete it and building the infix component, they chose to make a radical digression by abandoning it and building a C-Form/SL-Profile embankment. Because this would have required integrating it with the incomplete C-R Configuration, *and* closing the terrace with the stone wall, the result would be an embankment locale that satisfied neither type and had never been in a state that could have satisfied the builders.

A much simpler and more coherent move is to reject this second scenario and accept the first. That is, the C-Form/SL-Profile component was constructed as a rather standard earthwork of the Miami Fort Tradition type and used for an indefinite period. Then, an interesting event occurred. The K-Profile elements were added. If this scenario is accepted, two logical conclusions follow. First, the chronological priority of the C-Form/SL-Profile over the Complex G-Form/K-Profile is confirmed, and, second, since an infix of the sort associated with the total Paint Creek C-R Configuration is missing, it must be that the preexisting C-Form/SL-Profile embankment earthwork was co-opted for this purpose. This would explain building the stone embankment as a method of "closing off" the terrace edge. Thus, the builders perceived no contradiction in using a C-Form/SL-Profile element in the same way as the infix was used, as long as it was modified so as to be "closed." From this a third conclusion can be drawn: *whatever the meaningful nature of the infix of the Paint Creek C-R Configuration was, it corresponded to the meaningful nature of the C-Form*, or, at the least, the meanings of these two were not incompatible. Therefore, in the understanding of those who were responsible for this site, the transformation of the Hopewell site was not a transformation of its purpose and meaning but of the expressive aspect by which this purpose was realized.[8]

The Mound City/Hopeton Complex

The chronological priority of the C-Form/SL-Profile over Complex G-Form/K-Profile can be more firmly established by analyzing other examples of sites combining contrasting embankment plan/profile variants. This is the case with the Hopeton site (fig. 2.14). Although different from the combination of attributes of the Hopewell site, Hopeton displays a number of the properties of both the Simple and the Complex G-Form patterns. It has both a large K-Profile Circle and a large K-Profile Rectilinear component, satisfying the C-R contrast. However, these are not conjoined by a neck, nor is the Rectilinear component symmetrical, and, noticeably, it has "gates" but no "gate" mounds. Despite these differences, it clearly does display a C-R pattern with strong tendencies to mirror-image symmetry. It also differs in a third manner from the C-R Configuration, particularly the High Bank version. Hopeton has two Simple G-Form/SR-Profile components (Adena Sacred Circles) associated with it. One is partly incorporated into the Rectilinear component, while the other is spatially related by having its entry opposite a "gate." Furthermore, the conjoining mode of the large Circle and the large subsymmetrical Rectilinear also suggests that the former was already built when the subsymmetrical Rectilinear was constructed. This is clear since the southeastern sector of the Circle effectively forms the northwest wall of the rectilinear.

Thus, in sequential terms, it is reasonable to conclude that the Sacred Circles, correlating with Middle Adena, ca. 150 B.C.–A.D. 1, were already in place when the large K-Profile Hopeton Circle was constructed, confirming the relative temporal ordering of the S-Profile and K-Profile attributes that has already been noted at the Hopewell site. Subsequently the Rectilinear was constructed, incorporating the southeast sector of the Circle as its northwest wall, and the positioning of its northeast wall integrating with the two Adena Sacred Circles. (The Adena Sacred Circle will now be referred to simply as the sacred circle). A very important point here is that the sacred circle components were left intact and incorporated "into" the Rectilinear itself. This suggests that whatever the sacred circle meant to the builders, it did not contradict the meaning and purpose of this new, albeit subsymmetrical, C-R Configuration construction. Indeed, as at Hopewell, the purposes of the sacred circle and the Rectilinear may have been not only compatible but simply variable expressions serving the same purposes.

Mound City (fig. 2.12) is on the bank of the Scioto River immediately opposite but southwest of Hopeton. It should be noted that there is actually a set of three embankment earthworks opposite Hopeton. These will be referred to here as the Mound City Cluster.

The subsymmetrical K-Profile embankment earthwork that surrounds the twenty-three mounds of this site is here termed Component A of the Mound City Cluster. The mounds make up what will be called Mound City Proper. To the northwest of Component A is a sacred circle, referred to here as Component B of the Mound City Cluster. The large, subsymmetrical SL-Profile embankment with a mound at its epicenter, often referred to as the Shriver Works, is the southwestern component, and will be referred to as Component C of the Mound City Cluster. In the terms already argued, Component C (SL-Profile) and Component B (SR-Profile) would have preexisted Component A (KPp-Profile). This does not mean that some or all of the mounds of Mound City Proper could not also have preexisted the K-Profile embankment earthwork built around them since, in terms of this horizontal stratigraphy, some of these may have been built before and some after Component A was constructed. The timing of the embankment earthwork construction and the mounds is discussed in fuller detail in chapter 22.

Both Component A of the Mound City Cluster and the Hopeton embankments (excluding the two sacred circles) are K-Profile types, suggesting these belong to the same time horizon. Hopeton has a long aggregation element at the point where the Hopeton Circle and Rectilinear are conjoined. The southwest end of this aggregation element leads to the bank overlooking the Scioto River. If the aggregation element were extended straight across the Scioto River and at this angle, it would terminate at the base of a walkway that extends from the riverbank to the eastern "gate" of the Component A embankment of the Mound City Cluster.

> In the case of the east gap (of Component A of the Mound City Cluster), its position in the Squier and Davis plat places it directly east of Mounds 2 and 14 and in line with the graded cut in the west bank of the Scioto and *the end of the long 'graded way' leading southwest from the Hopeton works 3/4ths of a mile to the east*. The gaps themselves have not been scrutinized by archaeological techniques and they can only be interpreted as entrances into the earthworks. (Brown and Baby 1966, 10, emphasis added)[9]

This construction orientation suggests that the aggregation element was intended to integrate the Hopeton earthwork site and the Mound City Cluster, or at the minimum, Component A and Mound City Proper, into a single complex.

Mound City Proper, it is postulated, started as an open site of timber structures with the large C component to its southwest and Component B to the north. Both Component A of the Mound City Cluster and the Hopeton Circle were then built, more or less at the same time, since each displays the same K-Profile attribute, although priority might go to the Hopeton Circle, for reasons given below. The addition of the Rectilinear to the large Hopeton Circle *may have constituted the earliest expression of the C-R Configuration in its High Bank version*, well before the construction of the latter earthwork itself. Subsequently, Component A would have been constructed and this was a part of an overall plan that was finalized with the construction of the aggregation element, designed to "integrate" Component A and Hopeton, along with the Mound City Proper structures. Of course, the aggregation element and its orientation may have achieved other goals. However, it seems reasonable to interpret the purpose of the aggregation element as integrating these earthworks, separated by the Scioto River, into a single complex, which can be called the Mound City/Hopeton Complex.

What is notable is that the northeast end of the aggregation element is at the juncture of the Hopeton Circle and Rectilinear. If the Component A embankment and the mounds of Mound City Proper that it surrounds were to be physically moved and positioned at the northeast end of the aggregation element, it would be positioned in a virtually equivalent manner to the infix component of the Paint Creek C-R Configuration, as described earlier, thereby constituting Component A as the equivalent of the infix of the Paint Creek C-R Configuration. In support of this claim it is notable that whenever mortuary mounds are found associated with the Complex G-Form earthworks, they are typically situated in the infix component or its equivalent (e.g., Seip, Liberty Works, Frankfort). This reinforces the possibility that, in the variant forms of Paint Creek C-R Configuration, such as the Hopewell site and the Component A of the Mound City Cluster, the embankment components containing mortuary and related mounds are the equivalent of the infix component.

While it is likely that Hopeton was initially constructed to constitute what may have been the first, albeit subsymmetrical, material expression of the High Bank C-R Configuration, a further stage was developed. This entailed building the southern aggregation element at Hopeton and Component A around Mound City Proper. The result embodied both the High Bank and Paint Creek C-R Configuration motifs into the Mound City/Hopeton Complex, making it possibly the first expression of a dual C-R Configuration.

Two other C-R Configuration sites on the Scioto River and in the Chillicothe region are Dunlaps and Circleville. Various temporal scenarios could be given, to none of which do I wish to be committed. Dunlaps, which is only several kilometers north of the Mound City Cluster, is interesting in that it displays what might be called geometrical angles and "perfect" circularity characteristic of the classic Complex G-Form, but without the mirror-image symmetry. It is almost as if the builders wished to avoid the latter by emphasizing a form of balanced symmetry. This subtle blend of properties might suggest a fairly late construction, after the high point of the "classic" C-R Configuration sites, or else an initial experimenting with geometrical angles and circles preliminary to High Bank itself. The other site is Circleville (fig. 2.4). It should be noted that this site is about 50 km north of Chillicothe and, therefore, probably implicates a construction program that was autonomous of what was going on at Chillicothe. Circleville presents interesting problems in that it deviates in three ways from the "standard" High Bank C-R Configuration form. First, instead of the Circle component being a single K-Profile embankment, it is a concentric pair of embankments. Second, its neck is reduced to a nominal form since the Rectilinear and Circle appear to be almost directly conjoined. Third, while the Rectilinear has the full complement of eight "gates" and "gate" mounds it is a square, deviating, for example, from High Bank and Newark, which display octagons. These deviations were probably deliberate on the part of the Circleville builders and suggest that, in a sense, they were carving out their own version of the High Bank C-R Configuration, possibly introducing variant properties because of new forms of ritual.

Table 2.1 summarizes the chronological scheme that the seriation and "horizontal stratigraphic" analyses allow. Its confirmation will ultimately depend on further archaeological research, particularly through the development of a suite of radiocarbon and other forms of dating, some of which will be reviewed and presented in later chapters. The next major step in this analysis is to invest this typology and the regional and chronological models it grounds with cultural meaning, social significance, and historical content, effectively constituting a substantive interpretation of the earthworks as organized by this typology. Chapter 3 lays out the theoretical approach to material culture in its symbolic aspect upon which the subsequent development of this book will rely.

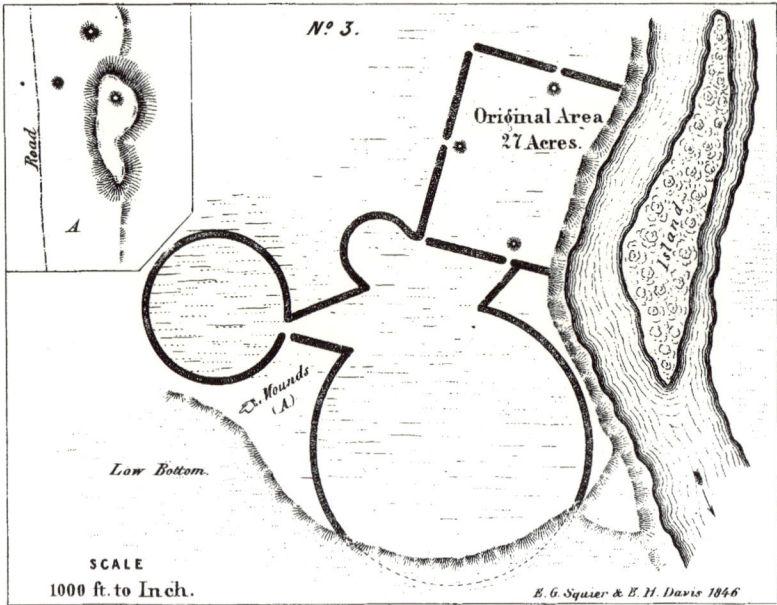

FIG. 2.1. Works East (Squier and Davis [1848] 1973, plate XXI, no. 3)

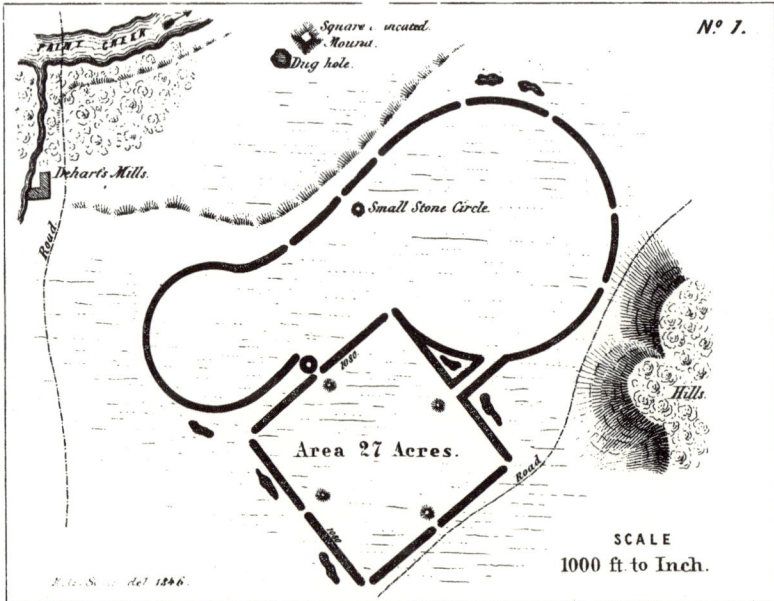

FIG. 2.2. Baum (Squier and Davis [1848] 1973, plate XXI, no. 1)

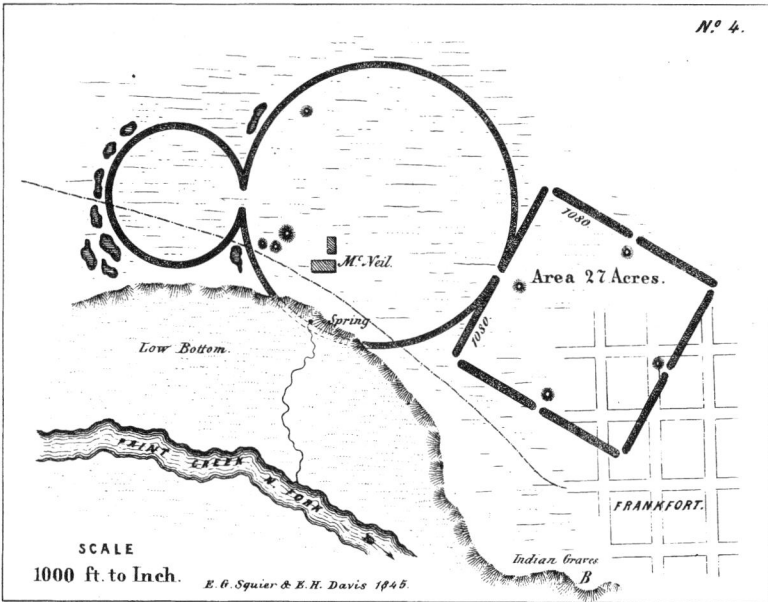

FIG. 2.3. Frankfort (Squier and Davis [1848] 1973, plate XXI, no. 4)

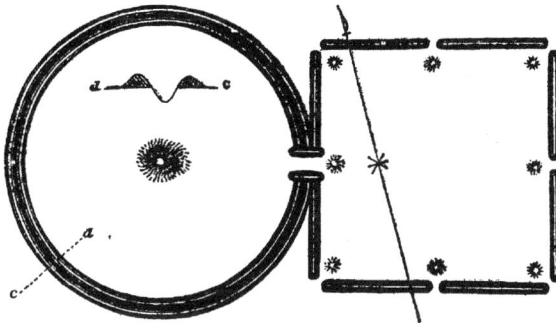

FIG. 2.4. Circleville (Squier and Davis [1848] 1973, 60, figure 10)

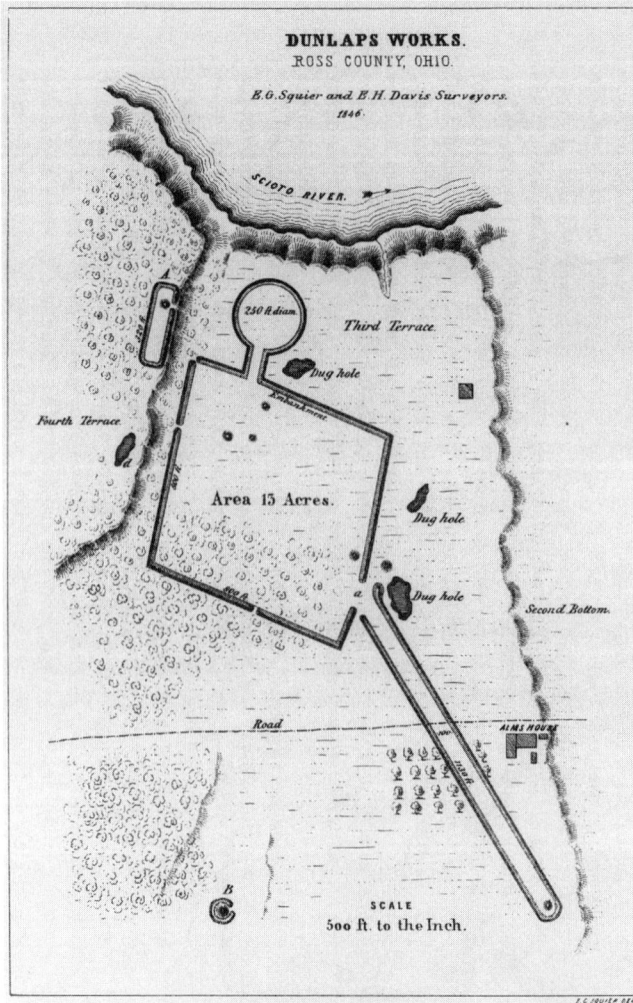

DUNLAPS WORKS.
ROSS COUNTY, OHIO.

E.G.Squier and E.H.Davis Surveyors.
1846.

SCIOTO RIVER.

250 ft diam.

Third Terrace.

Dug hole

Fourth Terrace.

Area 15 Acres.

Dug hole

Dug hole

Second Bottom.

Road

ALMS HOUSE

B

SCALE
5oo ft. to the Inch.

FIG. 2.5. Dunlaps Works (Squier and Davis [1848] 1973, plate XXIII)

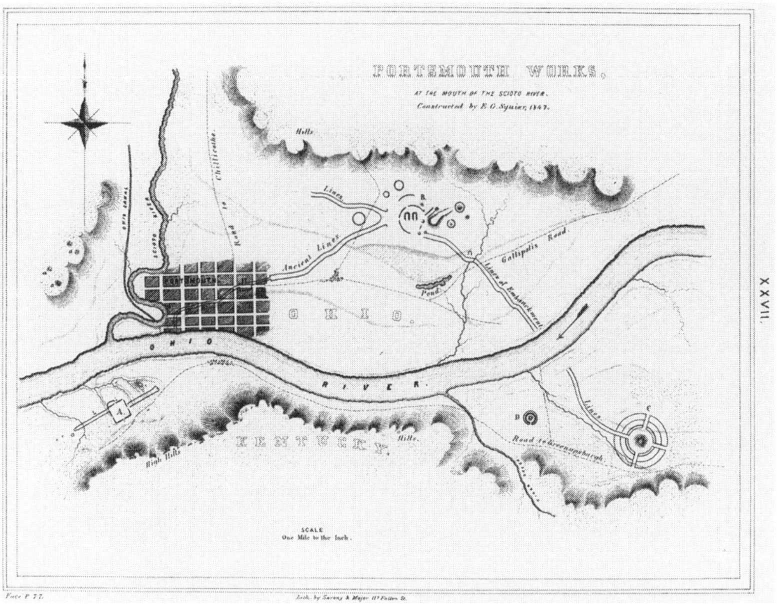

FIG. 2.6. Portsmouth (Squier and Davis [1848] 1973, plate XXVII)

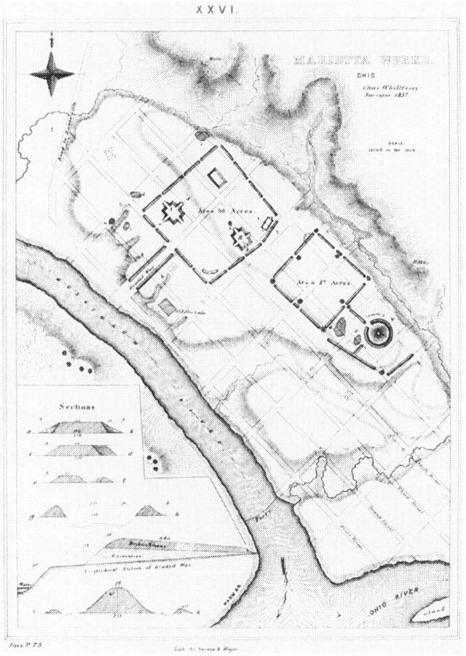

FIG. 2.7. Marietta (Squier and Davis [1848] 1973, plate XXVI)

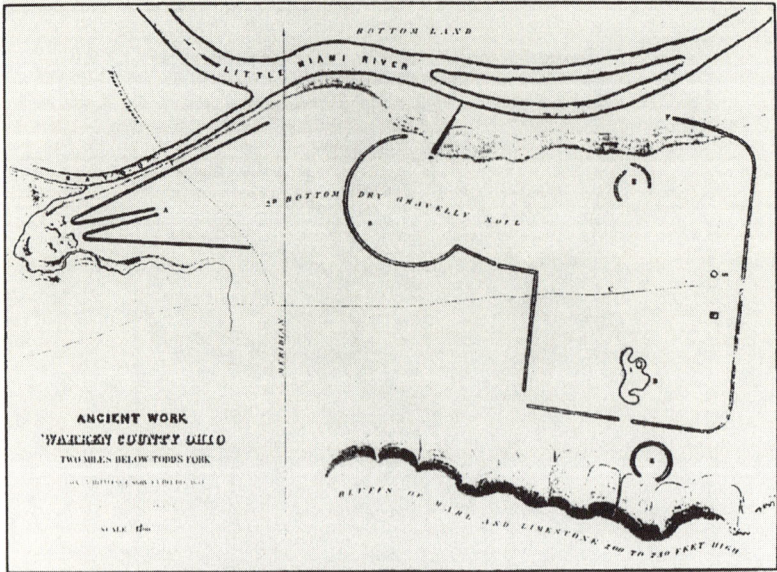

FIG. 2.8. Stubbs Works (Genheimer 1997, 297, figure 11.2)

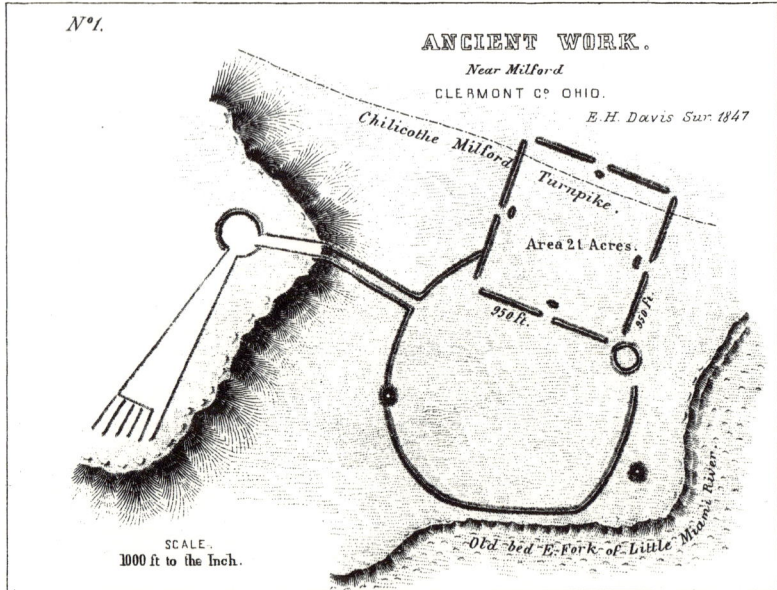

FIG. 2.9. Milford (Squier and Davis [1848] 1973, plate XXXVI, no. 1)

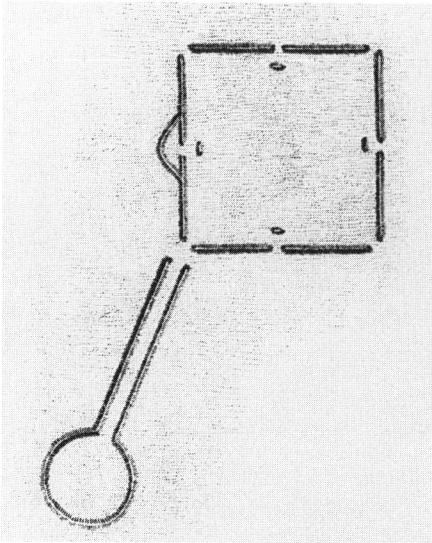

FIG. 2.10. Camden (Squier and Davis [1848] 1973, plate XXXIV, no. 2A)

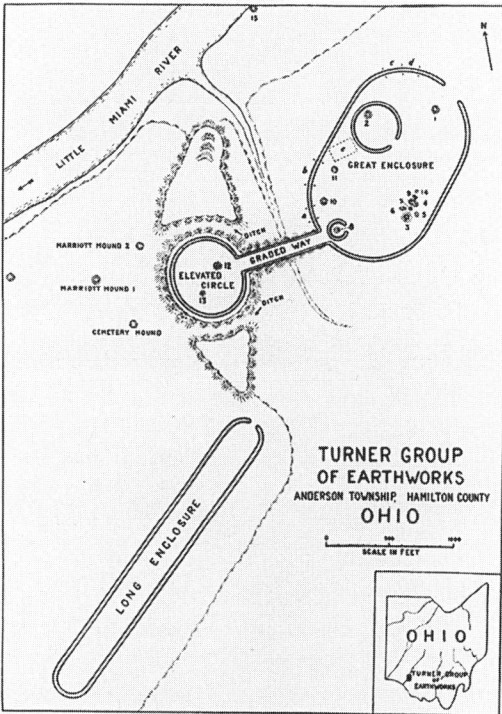

FIG. 2.11. Turner (Willoughby and Hooton 1922, plate I)

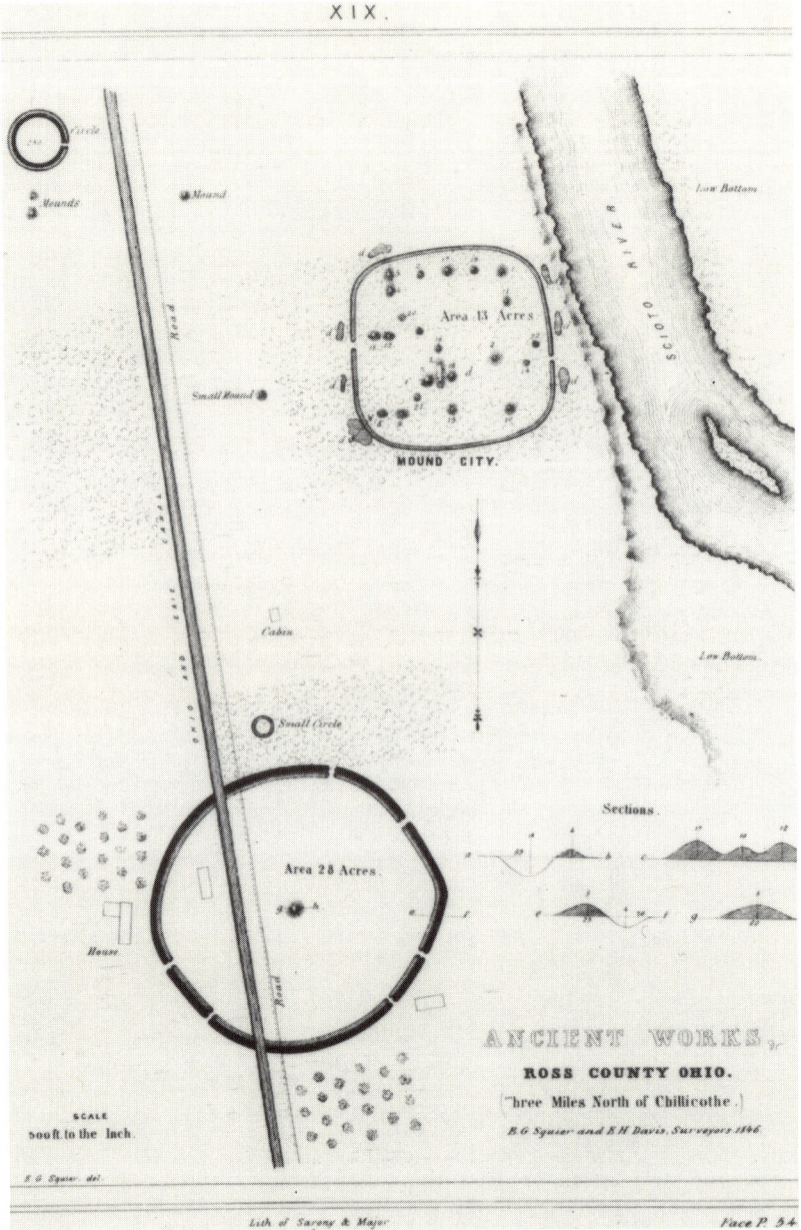

FIG. 2.12. Mound City (Squier and Davis [1848] 1973, plate XIX)

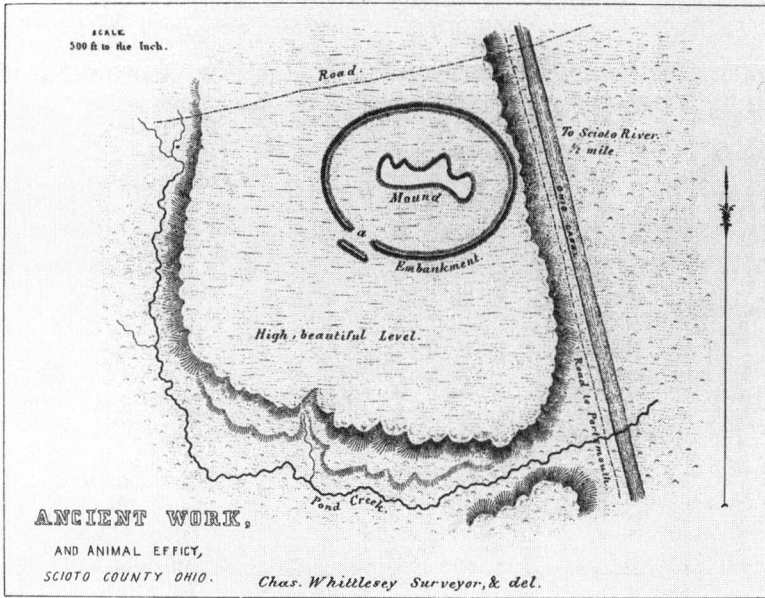

FIG. 2.13. Tremper (Squier and Davis [1848] 1973, plate XXIX)

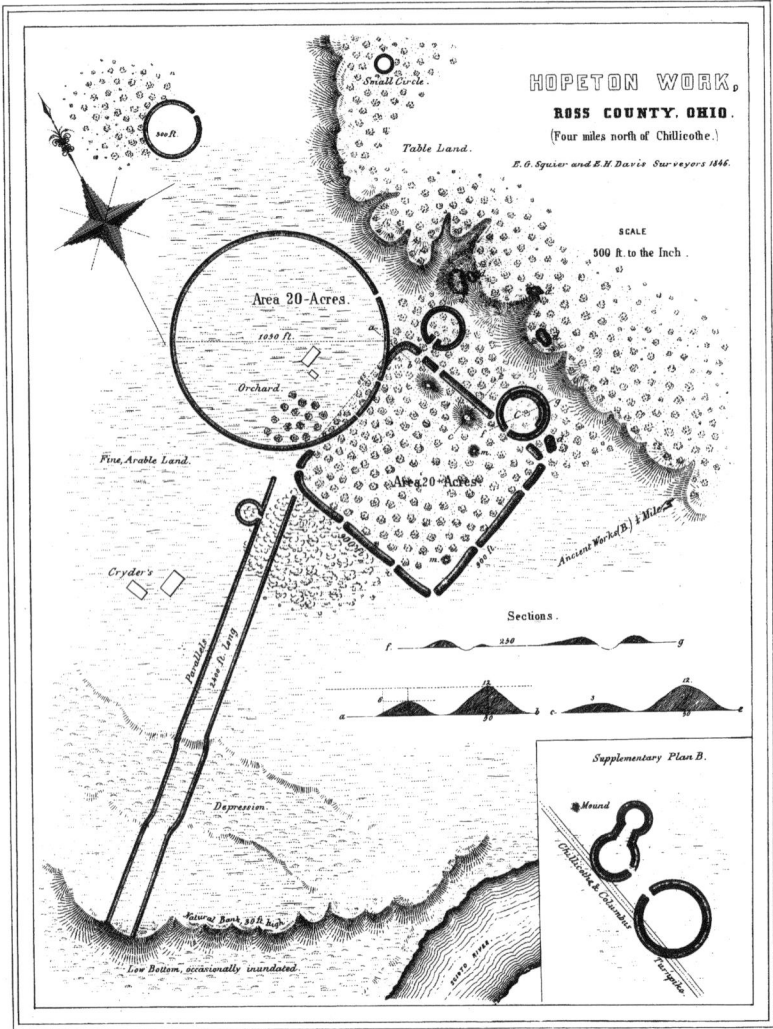

FIG. 2.14. Hopeton (Squier and Davis [1848] 1973, plate XVII)

Table 2.1 Initial Chronological Ordering of Ohio Hopewell in the Chillicothe
Region

Phases	Hopewell Site	Central Scioto Region
Late Ohio Hopewell	C-R Configuration added to C-Form/SL-Profile	High Bank, Seip, Liberty Works, Works East
Middle Ohio Hopewell	C-Form/SL-Profile	Mound City/Hopeton Complex Tremper
Early Ohio Hopewell	C-Form/SL-Profile	Mound City: Component B and Component C
		Hopeton: Sacred Circles Cedar Bank Works

Chapter 3

MEANING, SYMBOLIC PRAGMATICS, AND OHIO HOPEWELL STYLISTICS

Traditionally, in interpreting material artifacts and features, prehistoric archaeologists give priority to the type of physical actions that they probably mediated. The operating assumption is that the properties of material cultural items, whether artifacts, facilities, or features, manifest the action purposes of those who used them. That is, the physical properties "fit" the users' intentions to achieve certain goals by means of these particular material instruments. In a fundamental way, this is the approach taken here. It is an interpretive approach, directly addressing the possible purposes the users intended these material items to serve, and these purposes link the items to the meanings they held for the producers/users. Therefore, this approach also presumes to characterize material cultural categories in terms that broadly replicate the way those who were responsible would have experienced using them. This is to say that we are getting at the sense these things had for their users, and this sense is part of why they would produce them in the manner that they did.

Applying this action approach to interpreting the earthworks raises the question: "Why would it make sense, to those who were responsible, to build and use such large-scale constructions?" Sketching out an answer to this question requires using the patterning of the earthworks as empirical evidence to ground claims about the core of the collective beliefs and values that moved the builders, as well as the rules and protocols that gave direction to their undertaking these tasks. Developing claims about the collective consciousness—in particular, about the collective intentionality—manifested in these earthworks, and then empirically grounding these claims, creates a portal by which to enter into the builders' social and cultural world. Once in, all the factors that made the construction of these earthworks possible can be explored. The method to use is to develop models that allow using the empirical data to map the interplay between cultural values, social imperatives, labor availability, and material resources.

At this point, however, the skeptic can be expected to raise an objection. "Hold on. It's one thing to take an action approach to interpreting such practical items as stone tools, hearths, and storage pits and quite another to address the action nature of monumental earthworks. Stone tools implicate practical purposes and practical beliefs about how to survive in the world. Earthworks do not. Their meaning, surely, is symbolic and, unlike practical meaning that can be inferred from the empirical facts marking practical activities, symbolic meaning is arbitrary. Therefore, we can never infer from the empirical facts what the objects or 'referents' of these symbols were."

To answer the skeptic's legitimate concern, the first necessary step is to present a general theory about the meaningful nature of material culture that can, in fact, encompass both the practical and symbolic dimensions of materially mediated actions. To do this, however, it is necessary to go beyond the notion that symbols are exhaustively characterized in referential terms. This broader suggestion for understanding how humans use symbols can then be used as a reasonable, a priori guide in constructing models that can be tested by applying or "fitting" them to the empirical data, whether these data implicate practical or symbolic activities. This wedding of the practical and the symbolic will be termed symbolic pragmatics.[1]

In archaeology, as in many of the social sciences, there is a basic assumption that to symbolize is to refer by arbitrary signs. Thus, the

symbol is understood as a conventional or arbitrary sign and the meaning of the sign is the object that the sign is used to designate. Usually
this object is termed the referent of the sign. This notion of words as
referential symbols is then commonly used to characterize the symbolic
nature of material culture. I will call this symbolic referentialism. Symbolic referentialism encourages treating the symbolic aspect of material
cultural items as a concrete medium for referencing or designating the
identity of the user or users. This symbolic aspect is termed material
cultural style or stylistics, often thought of as the "decorative" aspect
of things, whether utilitarian in nature, such as projectile points or pots,
or non-utilitarian, such as "ritual" blades or headdresses, and so on.
Besides reducing the use of styles to being a sort of second-class type
of "speech," symbolic referentialism encourages dividing the uses of
material culture into two radically different and separate types of activity, stylistic or symbolic referential uses and practical or concrete
actions.

In strong contrast to this symbolic/func-
tional dualism, radically separating our materially mediated behaviors
into two separate action spheres: stylistic or referential actions, like
speech actions, being ephemeral descriptions of the world of humans,
and the practical or utilitarian actions, those that make concrete or "real"
changes. The concrete, practical nature of the utilitarian aspect of items,
such as tools, allows the archaeologist to move more or less directly to
defining the actions they mediated by simply "seeing" what kind of physical changes their properties made possible. Stylistics, however, are symbolic, conventional in their meaning, and, therefore, difficult to interpret
since their referents are culture-specific and rule-based.

In strong contrast to this symbolic/functional dualism whereby
symbolic and practical activities are separated into mutually exclusive
spheres, symbolic pragmatics characterizes the symbolic aspect of material culture, mediated by style, as central to all material action, even
the most practical.[2] It is central because it constitutes the *action nature*
of the material behaviors that it is used to perform. In these terms, all
the regular activities performed in a social context and through

material cultural mediation are constituted by symbolic means, whether these are concrete, practical actions or ephemeral, expressive actions. To help understand this, think of the activities we label "hunting" and "poaching." Objectively speaking, there is no difference between them. However, if we eliminate one or the other term, say, "poaching," this immediately eliminates the possibility of speaking of poaching and poachers. Yet, poaching is an activity in many societies. The perpetrators are termed poachers, and they are treated harshly in most societies, even though, objectively, they do the same type of killing as those deemed to be hunters. Indeed, the harsh treatment is often promoted by those recognized in the society as *hunters*. Given that both activities are grounded on the same range of behavioral interventions, the difference that these two terms mark, then, is not objective but *social* and meaningful in nature, and the only way this social difference can be marked is by symbolic means, making the symbolic, conventional moment as basic to practical social activities as it is to ritual social activities.

To be very clear about this, it is not that one activity, hunting, is symbolically marked and the other, poaching, is not. The material moments of both activities are constituted symbolically. Of course, poaching may be constituted by the absence of the appropriate hunting symbols, but this absence is itself a symbolic expression. In effect, displaying the hunter's symbols as manifested in the stylistics of the hunting gear constitutes one *as a hunter* in that social world. This is not an identity that the stylistics "refer to." It is an identity that is manifested, expressed, and socially constituted by means of the stylistics. Part of what makes it this identity is the recognition by relevant others that the legitimate user of these stylistics occupies the social position of hunter, and that this position has a whole range of rights, duties, and obligations that subjects must fulfill and knowledge that they must have to be proper hunters. This can only work, of course, if it is understood that the expressive meaning of the style is the social and intentional properties that this same style expressively manifests. The person who pursues and kills game without displaying the appropriate hunting stylistics, then, may well be assumed to be misappropriating the animals that are being killed.

Critically important here is the notion that the same objective behaviors can be different social actions. Humans as social agents can generate different social actions while behaving in similar ways, e.g., butchering and sacrificing. This will be termed the behavior/action duality. The insufficiency of behaviors to define the type of actions they are simply by their physical outcomes entails the necessity of symbolic expression. Style, as the symbolic aspect of material culture, functions to fulfill this pragmatic or action constitutive requirement or imperative. This constitutive pragmatic role of style is why I term this the symbolic pragmatic approach. Style is a symbolic expressive (not referential) medium. It mediates a transformative, perceptual moment by which the behavioral interventions count as the types of social actions intended. It works expressively because it manifests the users' intentions and social positions. Intentions must be made manifest by symbolic expression, and positions can exist only when subjects are perceived by relevant others to be occupying them. Thus, particular intentions (i.e., hunting rather than poaching intentions and harvesting rather than pilfering intentions) and particular positions (hunter rather than poacher; harvester rather than pilferer) must be manifested in the behavioral moment itself.[3]

Clearly, this places style squarely at the core of social interaction, and, for this reason, it is claimed that its primary function is as a *communicative medium*. This does not deny critical importance to the instrumental properties of material culture. The producers/users design tools so as to enhance their own limited physical capacities in intended ways. However, style is equally important since the tools can then conventionally communicate or make publicly manifest the socially appropriate intentions and the social standing of their users. This is critical in eliciting collective understanding, which constitutes the behavioral interventions as the types of social activities the responsible doers intended. Of course, using the correct styles is not a guaranteed "fail-safe" method, since mistakes and errors of understanding are always possible. Still, there is little doubt that if agents use material cultural forms that are not publicly recognized as appropriate, or if they appear to be "borderline" to relevant others, this will guarantee that questions will be raised and puzzlement will grow about the real intentions and actual social positions of their users. This doubt and skepticism will promote withholding full recognition of the doers' positions and their behavioral intentions and this means that the behaviors will not count as the types of social activities or practices that their doers intended. For these reasons a viable community requires that its material cultural assemblage adequately mediates the standard repertoire of material behaviors required for survival and, simultaneously, that it bears the appropriate styles to constitute these behaviors as the social activities intended.[4]

THE PRAGMATICS OF EVERYDAY SPEECH

Despite the rejection of material cultural style as a conventional form of referencing, symbolic pragmatics does recognize a very important relation between style and speech. In fact, the pragmatic dimension is a critical aspect of speech also and it has been much more thoroughly theorized for verbal behavior than for material behavior.[5] Traditional or nonpragmatic approaches to speech and communication, of course, typically treat speech as essentially about referencing by means of words. Certainly, this referential use of speech cannot be denied. In focusing on the information conveyance aspect, however, what has been largely ignored is that speech is the primary way humans interact through "pure" symbols.[6] By means of grammatically ordered words, we make promises to each other, swear oaths of allegiance, declare war and peace, give orders, make suggestions, apologize, offer condolences, give reports, theorize, explain, bless, curse, lie and deceive, and so on.

The study of speech as real action plays a central role in linguistic pragmatics. Speech acts are always social actions and they can have limited or broad impact. The scope of these actions and the powers they manifest are symbolic and real. This action power is a property of the social system, a social power, and, in being exercised through the medium of speech, it simultaneously reproduces the social positions that endow those who occupy them with the symbolic powers to make their verbalizations count as the potent social actions they are.

While this social power is both expressed and constituted in and through the medium of vocal symbols, it is important to stress that much of our speech activity depends on the simultaneous mediation by the appropriate material things, e.g., Bibles for taking oaths, passports for crossing borders, and visas for residency in foreign countries, as well as money and credit cards to order purchases, and so on. In short, symbolic pragmatics dissolves the symbolic/practical, expressive/functional dichotomy in both speech and material activity.

THE WARRANTING MODEL OF MATERIAL CULTURAL STYLE

As a bridge to applying the symbolic pragmatics of style to interpreting material culture, treating the court warrant as a symbolic pragmatic device can be very useful. The warrant is a conventional document by which the authority of the court that issues it is expressed and made manifest in the person who legitimately bears it. This expressive form also makes manifest the society on behalf of which the courts are mandated to act. As a symbolic pragmatic device, the warrant constitutes its legitimate bearer as an officer of the court, a bailiff or a sheriff. The range of behaviors she/he routinely performs through its mediation and by way of carrying out her/his official duties is constituted as the standard repertoire of legal acts: arrest, property seizure, evidence gathering, and so on. The transformative power of the warrant, it is important to stress, does not rely on individual attitudes or feelings toward it but on the collective understandings of its purpose. This is what underwrites institutional structures and these are expressively manifested and reproduced in the instance of the warrant's legitimate usage. Thus, even should the bailiff happen to hold private views that do not correspond to the duties that the warrant conventionally empowers her/him to carry out, it is the public and collective symbolic pragmatic meaning of the warrant that is relevant to the action constitutive moment. Independently of her/his personal opinions, the bailiff's proper use of the warrant transforms her/his physical interventions, and the physical interventions of those whom she/he may be directing, into legal acts, and this usage simultaneously participates in reconstituting the social system that this usage realizes.

Transferring this notion of the warrant so that it applies to material culture in general is the basis of the Warranting Model of material cultural style. To justify this transfer, Douglas's discussion of the nature of "primitive money" in nonliterate societies is illuminating. As she argues, artifacts referred to in the anthropological literature as "primitive money" are often used in a way that is equivalent to the way we use licenses and coupons.

> Some primitive currencies are perhaps more like systems of licensing than like rationing. . . . Both are instruments of social policy, but whereas rationing is egalitarian in intent, licensing is not. The object of licensing is protective, and to promote responsible administration. The object of rationing is to ensure equal distribution of scarce necessities. One of the objects of licensing is to ensure responsible use of possibly dangerous powers. . . . Licensing pins responsiblity, so we have marriage licences and pet licences. (Douglas 1982, 67–68)

If Douglas's insight is generalized to treating material cultural items bearing styles as having different types of warranting powers, both licensing and rationing, it now becomes rather straightforward to relate the symbolic pragmatic perspective to the complex set of earthwork locales and associated artifacts with their distinctive styles that characterize the Ohio Hopewell. Bear canines; pearl, shell, and copper bead necklaces and decorated blankets; plates, celts, and ear spools of copper; elaborate copper-based headdresses; obsidian bifaces; even chert bladelets, and so on, can be postulated as different formal and informal warranting devices constituting their users/wearers as occupying different social positions endowing them with a differentiated range of action powers. These powers enabled them to transform the behaviors these things characteristically mediated to count as the repertoire of activities (often ritual) by which they could discharge social duties (often sacred). The habitual locales of usage, the earthworks and their mortuary facilities, would be primary monumental warrants that expressed and manifested the source of powers from which these same features and facilities, as well as the associated artifacts, derived their capacities so that the conventional morphology of both artifacts and earthworks constituted them as action warrants, that is, as symbolic pragmatic devices.

When the symbolic pragmatic capacity of a material item is taken by its producers/users to emerge from the item's participating in the essential powers of what it represents, this will be termed an iconic warrant. The symbolic pragmatic powers of artifacts, facilities, and features that are perceived as iconic in nature can be deliberately produced by "fitting" their forms, raw materials, decorative elements, color, texture, and other tangible properties, to constitute expressive relations with what they are perceived to represent. This expressive fitting of the object to what it represents and fitting what it represents to the object will be termed symbolic congruency. Examples of earthwork construction as symbolic congruency practices will be presented in the following chapter, the thesis being that earthwork construction was the realization of strategies of iconic construction that constituted this collective building behavior as congruency rituals by which the cosmos was revitalized in the ritual act of the earthworks being built. Through the medium of these monumental warrants, the building behavior was constituted as the type of renewal ritual intended.

THE BELIEF-ACTION RELATION

This symbolic pragmatic treatment of material culture and the socio-intentionalist approach that goes along with it counter two serious criticisms to which cognitive-normative perspectives are often subjected, and this particularly includes the symbolic, of course. The first is the claim that explanations of the archaeological record in these terms are not adequate because they cannot account for the variation in the archaeological record. In this view, norms are deemed to suppress variation.[7] Hence, variation must be accounted for in objective terms as humans pursue survival in constantly varying conditions. The second criticism, often considered deadly, is epistemological in nature. Critics recognize that humans act for two broadly general reasons: to satisfy their survival needs and to achieve goals defined by their cultural beliefs and non-material values, the latter being the focus of cognitive-normative accounts. No matter how varied, the former activities constitute a pursuit that is based on universal drives of survival: hunger, thirst, need to rest and maintain livable conditions, and so on. Therefore, the aspect of the archaeological record that manifests this pursuit of survival can be validly used to map past conduct and verify claims about these concrete activities quite independently of those who actually carried out these survival practices.

The pursuit of goals defined by abstract beliefs and values, however, is necessarily arbitrary and historically contingent and, therefore, the symbolic or stylistic aspect of the archaeological record that manifests it is conventional and unique. The only way to verify any cognitive-normative accounts, then, is to be directly in touch with those whose pursuits were the outcome of these abstract beliefs and values. This presents an imponderable obstacle to prehistoric archaeologists since, of course, all these people are dead and they left no written record from which their abstract beliefs and values might be inferred. Therefore, the epistemological critic concludes, the residual archaeological data are inadequate to verify such cognitive-normative accounts. The rest of this chapter will address the first criticism and this will prepare the way to address the second criticism in the following chapter.

The first criticism is based on the largely unstated assumption that the human mentality is largely a monolithic, unitary consciousness by which beliefs are directly linked to actions and, indeed, determine actions. A religious consciousness might be characterized as a belief about a particular god and this belief is largely assumed to include the attitudes, wants, values, and intentions, as well as the range of actions, that would make up the appropriate worship of that god. Therefore, intrinsic to learning this religious belief would be learning the range of intentions, attitudes, specific acts, and so on, that go along with it. In learning the belief well, the believers cannot help but act in accordance with it. This unitary, monolithic view of consciousness and its relation to action can be called belief→action determinism.

Indeed, if this monolithic view were an adequate characterization of the belief→action relation, then it is a significant criticism and it makes good sense to avoid such accounts. However, consciousness is not monolithic and the belief→action relation is, in fact, contingent. This is because consciousness is structured into a limited range of *types* of mental states and these are interrelated but autonomous. In such a structured consciousness, beliefs and intentions are autonomous mental states. All human actions, in the normal sense of the word, are intentional. They are also, of course, presupposed by relevant knowledge or beliefs. One cannot hunt without knowing a great deal about what one is hunting. But one can have a great deal of background knowledge that any hunting requires and still not hunt. Furthermore, a person who has all the knowledge that hunting requires need not, on these grounds, form any intention whatsoever to go hunting. Therefore, actions are not prefigured by beliefs, nor are intentions simply a fixed, predetermined aspect of these beliefs. Instead, actions are the outcome of the exercise of intentions that presuppose the relevant autonomous wants and beliefs and these are historically constructed and modified.

To illuminate this view of consciousness as structured autonomous mental states, it is necessary to narrow in on that aspect most relevant to this book, this being intentionality. In philosophical tradition, intentionality is a mental property that is directed at, toward, or about objects, states of affairs, events, and processes in the world. Thus, a conscious subject cannot be said simply to "intend." Rather, to intend always entails intending to do something. One cannot be said simply to believe. Rather, one's believing is always that such-and-such is the case. To have a want entails wanting something or wanting to do something. Intentionality is the behavioral-cognitive link between the conscious agent and the world. Searle speaks of four basic forms of intentionality: beliefs (knowledge, speculations, theories, dogma, guesses, implications, and so on), wants (or desires, wishes, values), perceptions (visual, taste, aural, and so on), and intentions (prior intentions and intentions in action).[8] In these terms, there is a real difference between a want to do X and an intention to do X. The want may be a motive for doing X. But we have many wants that we never intend to act upon. In contrast, to intend to do X is the conscious state one has in planning and/or actually doing X.

However, beliefs, wants, and intentions, although autonomous, are also mutually related and any particular action presupposes all three mental states. To perform an action entails exercising a particular intention or complex of intentions. This complex presupposes particular and relevant wants and both the wants and intentions entail beliefs about the world that make it possible to have these intentions and wants. This action→intention→want→belief relation is necessary rather than contingent. However, the reverse belief→want→intention→action relation is contingent and not necessary. Particular beliefs can be held without the believers having any wants or intentions to act in ways that having these beliefs makes possible. There is even a form of intention that we can hold that does not entail immediate action. These are long-term or prior intentions, intentions we formulate that guide us into channels that will allow us to act in the specifiable future, whether this relates to subsistence hunting or religious ritual. Thus, prior intentions involve the active formation of strategies that involve formulating scheduling of events and activities to be carried out in order to achieve the goals as specified by the particular strategies.

THE COSMOLOGY/IDEOLOGY DUALITY

Because the belief→action relation is mediated by wants and intentions, and both wants and intentions are relatively autonomous intentional states, neighboring peoples can share the same set of beliefs while having practices that radically differ. Using a religious example for illustration, Societies A and B can share the same cosmological beliefs X, Y, and Z about the world-cosmos. Because of the contingent nature of the belief→action relation, the shared belief structures do not determine the contents of the religious strategies of either society. These strategies can vary such that behaviors in A that count as sacrificial ritual to the cosmos can be radically different from the behaviors in B that count as sacrifice. To mark this belief/intention contingency and the distinction it entails, it can be said that A and B share the same *cosmology* while not, or only partially, sharing the same *ideology*. That is, the cosmology/ideology duality is simply the mutual autonomy of belief and intention "writ large." Cosmology can be defined as the set of collective beliefs, implicit and explicit, that has the world and its nature as its object. Ideology will be treated here as the collective commitments, intentions, and strategies that a cosmology makes possible. The content of these religious strategies is made up of formal and informal know-how rules and protocols, including the stylistics of material culture, and the standard rationales which are evoked in performing the activities that fulfill these commitments.

To illustrate, assume that A and B above are two different Ohio Hopewell embankment earthwork groups. For Group A, in order for a behavioral stream to count as a fully felicitous ritual X, Y is required as a sacrificial offering, along with an elaborate material cultural assemblage, a, b, c, as the symbolic warranting ritual assemblage. Critical to this assemblage is having embankment earthworks that conform to the protocols of the C-R Configuration. In Group B, the same basic symbolic warranting ritual assemblage is required, although some variation between the kits is likely, but its earthworks must conform to the protocols of the C-Form/SL-Profile Configuration. Both the earthwork similarities and differences are the result of historical contingency. While being aware of each other, each group followed and developed a different ideological trajectory in this regard. This mutual awareness of each others' earthwork activities would have been part of the historical development of their autonomous ideological strategies. For this reason, it would be likely that each would perceive the ritual of the other as only partly felicitous, or even quite infelicitous, even though the rest of the associated symbolic warranting assemblages that they used would be nearly identical. Importantly, it is not that they would not understand each other. To the contrary. Since they would have shared the same religious cosmology, and much of the same ideology, they may have understood each others' differences only *too well*.

Socio-religious hostility is fueled by such ideological differences and overlaps between peoples who, however, broadly share the same cosmology. A similar, although usually lesser, ideological bifurcation could and probably would exist *within* each group. That is, two or more ideological schemes could exist simultaneously in the same embankment earthwork group, even though the group held to a single coherent religious cosmology. Each ideological scheme would have a variant model of the ideal religious sacrificial kits, from ritual tools to ritual constructions. This ideological bifurcation within a single organization would constitute different factions, each claiming that its ideological forms were "true" and those of other factions were "false," or, in some way, faulty.

By treating the Ohio Hopewell in these terms, the great variation in the embankment earthwork forms becomes the outcome of the historical variations in ideology in the context of the continuity of the belief structures that these varied earthworks presuppose. If the cosmology remains largely constant while ideology varies, and if the ideological contents of action intentions are conditioned by the social system, then the primary causal factor to account for this variation would be social. However, this does not mean that every variation in the material record marks a radical social change. Just as action forms can vary while beliefs remain constant, they can also vary while the social structure remains largely constant. That is, social structure→action determinism is to be avoided as much as belief→action determinism.

In sum, the characterization of human consciousness as a complex, structured set of mental states, processes, and capacities not only refutes the first criticism against cognitive-normative approaches, namely, that norms suppress variation, it initiates the refutation of the second or epistemological criticism, namely, that the abstract nature of beliefs and values that cognitive-normative accounts invoke makes it impossible to verify them by the archaeological record. It initiates the refutation by claiming that the contingent nature of the belief→action relation means that the action→belief relation is necessary. Therefore, if the pragmatic or action nature of the behaviors that generated the archaeological record is correctly identified, we can reconstruct the abstract beliefs that made these actions possible. This partial refutation clarifies precisely what the final problem is, this being how symbolic pragmatic claims about the nature of the activities performed can be adequately grounded in the empirical data. This problem is directly addressed in the next chapter by articulating a method that might be appropriately termed "arguing from empirical anomaly." This "empirical anomaly" method will be applied to the Newark earthworks, serving to empirically ground a claim termed the World Renewal Model of Ohio Hopewell earthworks. This model will serve as the portal into this social world.

Chapter 4

THE NEWARK WORLD
RENEWAL RITUAL CENTER

A major component of the Newark site is the Newark Circle-Octagon as depicted in the upper left of fig. P.1. This feature displays the central set of attributes of the High Bank C-R Configuration. Except for the larger size of the Octagon, it is almost a mirror image of the High Bank Circle-Octagon. If the Newark Circle-Octagon were to be placed over the High Bank site without changing the current axial orientations of the two sites, Newark's southwest-northeast axis would be at an angle of 90 degrees to High Bank's northwest-southeast axis.[1] Despite the similarities in form and layout that link these two spatially distant earthworks (separated by about 100 linear miles), there are a number of differences. One of the most important differences is that, while the High Bank and Newark Circles are effectively the same size, the Newark Circle has a large mound constructed at a tangent across the southwest sector of its perimeter. This large semi-platform mound is about 50 m long by 5 m high. It is commonly referred to as the Observatory Mound, and the associated circular embankment earthwork is often referred to as the Observatory Circle. No feature equivalent to this Observatory Mound exists at High Bank. There are two additional features unique to the Newark Circle-Octagon: these being two low, parallel embankments that jut southwest from the base of this mound. The mound and these two low embankments will be

treated as a single complex that will be termed here the Feature A/
Observatory Mound Complex (fig. 4.1).

The central aim in this chapter is to present and demonstrate the
World Renewal Model of the Ohio Hopewell embankment earthworks.
To do this it will focus on the claim that the Feature A/Observatory
Mound Complex is the consequence of an avoidance action that
emerged as a result of a change in the construction program of the
Circle-Octagon. In accordance with the view of intentionality as structured presented in the last chapter, the material residue of an avoidance
action entails the exercise of a proscription that might be expressed as
"Under condition X, do not do Y." From this finding, it will be argued
that the particular proscriptive rule being implemented could only exist
given a particular core belief principle that guided the work of those
who were responsible for this construction.

THE WORLD RENEWAL MODEL OF OHIO HOPEWELL EMBANKMENT EARTHWORKS

The World Renewal Model of the Ohio Hopewell embankment
earthworks is summarily presented below as four major postulates, two
characterizing the core of Ohio Hopewell cosmology (Cosmology 1
and Cosmology 2), and two characterizing the core ideology underwriting the construction and use of the earthworks (Ideology 1 and
Ideology 2).

> Cosmology Postulate 1: The builders took the cosmos to be immanently
> sacred. This will be expressed as the Sacred Earth Principle.

> Cosmology Postulate 2: The immanently sacred cosmos was structured
> vertically into at least three strata, the Heavens, the Middle World, and
> the Underworld, and horizontally into at least four quarters. The build-
> ers also took it to be ordered into at least two cyclic phases, the Lunar
> and the Solar.

> Ideology Postulate 1: The way the earthworks were built ensured that
> their formal properties would make it "fit" or be congruent with the
> cosmological categories, thereby constituting them as monumental

iconic warrants of world renewal ritual, that is, their construction was itself what can be termed a congruency ritual;

Ideology Postulate 2: They served as the ritual context by which to transform all subsequent collective behaviors performed in these locales into the types of world renewal and related rituals that the users intended to perform.

There are three converging empirical arguments to confirm this model: (1) the C-R Configuration Analysis; (2) the Construction Sequence Scenario Analysis; and (3) the Astronomical Alignments Analysis. These will be presented in the enumerated order.

1. The C-R Configuration Analysis

The basic components of the C-R Configuration were postulated and analyzed in chapter 2. This structural analysis is particularly relevant to the World Renewal Model because it establishes the range of normal variation that can be expected among these earthworks, quite independently of any particular symbolic pragmatic meaning that this configuration may have had. By establishing the range of variation, the C-R Configuration analysis establishes independent grounds for revealing any anomalies that may exist in particular sites. To say this is to claim that the analysis is itself independent of the symbolic pragmatic theoretical framework and the World Renewal Model.

The Newark Circle-Octagon displays all the key attributes of the High Bank C-R Configuration, namely, a large K-Profile Circle, a Rectilinear, and an aggregation neck that physically connects the first two in the tangent/centered mode. The Octagon has eight "gates" and "gate" mounds. However, with respect to the C-R Configuration norms, it also has a number of anomalies, the major one of which is the Feature A/Observatory Mound Complex itself. This can be judged to be an anomaly since none of the other known C-R Configuration sites, whether High Bank or Paint Creek type, displays this feature. According to the C-R Configuration analysis, in the initial stages of building the Newark Circle-Octagon, this could have been made to conform by simply removing the Feature A/Observatory Mound Complex, and filling in any gap exposed in the Circle. If this were to be done now, first by removing the Observatory Mound, it is postulated that it would expose Feature A as joined to the Circle. If this is the case, then Feature A would be an aggregation neck, very similar to the aggregation neck that conjoins the Newark Circle and Octagon, except that the latter is on the opposite side of the Newark Circle and is larger and wider.

This possibility does not eliminate the anomalous nature of Feature A. Instead, it suggests that the Newark Circle-Octagon has at least *three* anomalies (it has more): the Observatory Mound, Feature A as an aggregation neck, and a *second* neck, diametrically opposite from Feature A, termed here Feature B. Of course, it could be argued that Feature A could be an aggregation neck only if it were, in fact, connected to the base of the Circle under the Observatory Mound. At this point, this has not been demonstrated. However, this postulate can be used as a future test of the World Renewal Model through excavation or core boring. Based on the methodology of hypothetically reconstructing the Hopewell site and the Mound City/Hopeton Complex, and given the C-R Configuration analysis, it would seem very unusual, although not impossible, to find that Feature A, displaying the attributes of a typical aggregation component, is not connected to the Observatory Circle in the same manner as Feature B. Feature A, then, is postulated as an aggregation neck typical of the normal High Bank C-R Configuration. Empirical support for this claim arises from the fact that if a straight line were to be drawn through the center of Feature A, across the Circle and through the center of Feature B, and continued to the far northeast vertex of the Octagon, it would form the axis of symmetry characteristic of the High Bank C-R Configuration.

Establishing the range of normal variation among the earthworks of this type, the C-R Configuration analysis has also formed the background for identifying and analyzing these anomalies. The question this raises, of course, is why are there two aggregation necks when one was standard? Exploring the answer to this question through a series of postulated construction scenarios reveals the fundamental rules of the construction strategy, thereby not only demonstrating that the core of this strategy was governed by symbolic pragmatic rules and protocols but also revealing the cosmological core that this strategy presupposes, namely, the Sacred Earth Principle (Cosmology Postulate 1).

2. The Construction Scenario Analysis

Using an analytical procedure similar to the one applied earlier to analyzing the Hopewell, Mound City, and Hopeton sites, a series of hypothetical construction scenarios is postulated below, labeled as Construction Scenarios 1, 2, and 3. These form a logically exhaustive set of alternative accounts of how the Feature A/Observatory Mound Complex came about. By presenting and critically arguing each scenario, rationales can be constructed that eliminate two while retaining one and its possible variations as the most probable. From this analysis the Sacred Earth Principle can be directly deduced.

Construction Scenario 1

This scenario postulates that the building of Feature A initiated the construction of the Newark Circle-Octagon by building the two parallel embankments in a southwest to northeast direction about 30 m. This would have produced Feature A. Then each embankment was rotated away from the other at 90 degrees and continued in two large, complementary arcs. The plan would have been to join these two arcs more or less opposite Feature A, thereby completing the Circle. If this had been completed, the Octagon would then have been built onto the southwest end of Feature A, completing an embankment earthwork displaying the standard High Bank C-R Configuration. Instead, this construction scenario postulates that a change of plan occurred sometime after Feature A was built and prior to connecting the two arcs so as to complete the Circle. Because of this change of plan, rather than joining the two embankments in the northeast sector of the Circle, the builders left a gap and built two parallel embankments making up

Feature B, the second aggregation neck. They then built the Octagon to the northeast end of Feature B. Since the Feature A aggregation neck was no longer required, it would have been covered with a mounding of earth.[2] With the completion of this first phase, the embankments would probably have been only about 1 to 1.5 m high, the same as Feature A where it "emerges" from the base of the Observatory Mound, allowing for slumpage. Subsequent construction phases would have raised the embankments to the height and breadth that we see now, minus the reductions caused by erosion, slumpage, and historical modification.[3]

Construction Scenario 2

This scenario postulates that the first step of construction was building the Circle as a single, unbroken embankment and leaving a gap in the southwest sector to accommodate the building of the Feature A aggregation neck. Feature A was then added. Before adding the Octagon to the southwest or distal end of Feature A, as would be normal for completing a High Bank C-R Configuration type, the change of plan occurred. The builders abandoned Feature A by covering it in the manner described in Scenario 1, resulting in the first stage of the Observatory Mound. Since the Circle was completed, the builders would then have dismantled the embankment sector diametrically opposite Feature A to open a gap in order to add Feature B. With this done, as in Scenario 1, they would have added the Octagon to the northeast end of the Feature B aggregation neck.

Construction Scenario 3

Scenario 3 postulates that the Octagon was the first step in construction. The builders then added the Feature B aggregation neck. The Circle was constructed, probably starting at Feature B, and just before completing it at the spot opposite Feature B, a change of plan occurred that required building Feature A. Since it has been postulated that this was an aggregation neck, then this scenario implicates the probability that a second rectilinear was to be added to the southwestern end. Before doing this, another change of plan occurred, this being to "scrub" the second rectilinear. Therefore, Feature A was covered with a mounding of earth, completing the first phase of the Newark Circle-Octagon. Subsequent additions would have produced the earthwork largely as we see it now.

CRITICAL ANALYSES

These three construction sequence scenarios seem to exhaust the logical possibilities accounting for the Feature A/Observatory Mound Complex, allowing for some minor variations in each case. Of the three, Scenario 1 is the most probable. To demonstrate this, first Scenario 2 will be critiqued in its own terms and then Scenario 3. The critique of Scenario 1 then becomes largely a matter of showing how it resolves all the inconsistencies that Scenarios 2 and 3 raise and cannot resolve, thereby allowing the conclusion that Scenario 1 is the most probable construction sequence.

First, it is significant to note that all three scenarios agree that Feature A was *covered* by earth. Unlike Scenario 1, Scenario 2 entails construction practices that are logically inconsistent with this claim. The inconsistency that Scenario 2 reveals is that two diametrically opposite construction practices occurred. While Feature A was "closed" by having earth *added*, the embankment of the Circle was *dismantled* in the northeast sector. This would have entailed removing the earth in order to add on Feature B. If the embankment of the Circle could be dismantled to add Feature B, then why would the builders not have used the same tactic by dismantling Feature A and using the earth to fill the gap in the southwestern sector of the Circle? A dismantling strategy would have preserved symmetry, an important value that has been revealed by the C-R Configuration analyses. Furthermore, dismantling would probably have been less costly in labor than covering the gap with a mound that was much larger in area and volume than the earth that would have been removed. These inconsistencies in Construction Sequence Scenario 2 are grounds for rejecting it.

Scenario 3 effectively summarizes Squier and Davis's original analysis of the way the Newark Circle-Octagon was built. They describe it in similar steps, starting with the Octagon and terminating with the claim that builders changed their minds by adding "a crown work," the Observatory Mound, after building Feature A. "It would almost seem that the builders had originally determined to carry out parallel lines from this point; but after proceeding one hundred feet, had suddenly changed their minds and finished the enclosure, by throwing up an immense mound across the uncompleted parts."[4]

There are two serious inconsistencies with this scenario. The first inconsistency is that, while Scenarios 1 and 2 only require postulating one change of plan, Scenario 3 requires two. The first would have entailed adding a second aggregation neck, Feature A; the second would have entailed abandoning it. The second more serious inconsistency is that, in terms of the C-R Configuration analysis, if the sequence of construction postulated by Scenario 3 was correct, octagon→neck→circle, there would have been no need to add Feature A. The builders would have been close to completing the earthwork in conformity with the normal High Bank Configuration protocols when they would have drastically deviated from these by adding a second aggregation neck, Feature A. An aggregation neck conjoins Circle and Rectilinear components and since under Scenario 3 these latter would have just been completed, then adding Feature A as a second aggregation neck would implicate the plan to build a second rectilinear. In terms of the C-R Configuration analysis, there is no precedent for this. All this give grounds to reject Scenario 3.

By postulating Feature A as the initial step in construction, Scenario 1 avoids the inconsistencies of the other two. Unlike Scenario 3, it does not require two changes of plan and it does not require postulating an embankment complex that does not fit under the C-R Configuration variations. Also, Scenario 1 does not require postulating contradictory construction practices, as in Scenario 2. In short, Scenario 1 is the most consistent of the three, given the C-R Configuration analyses. From this it can be concluded that Construction Scenario 1—

along with any of its possible variants—is the most plausible construction sequence.

DISCUSSION

The most reasonable immediate conclusion of the Construction Scenario analyses is that the Feature A/Observatory Mound Complex was the result of *not* dismantling Feature A in circumstances that would seem to have called for just such a removal, particularly since there appear to be no physical impediments preventing dismantling. From this a second logical conclusion follows, namely, that dismantling Feature A was deliberately avoided. An alternative way of putting this would be to say that dismantling was simply inconceivable, not because of the labor involved but because dismantling would be tantamount to contradicting the point of the construction process itself. If this is the case, then, in the circumstances of a change of plan, covering Feature A must have been the only viable option—and the loss in symmetry and the greater construction costs were simply necessary sacrifices.

This leads to a third logical conclusion. Some real impediment that was and is still not directly observable must have been perceived by the builders to have precluded dismantling of Feature A, or, alternatively stated, to have forced their closing of Feature A by mounding instead of removing earth. If this is the case, then the fourth logical conclusion is that the builders avoided dismantling Feature A because the impediment was immanent in Feature A itself. This would have to have been some intangible but potent embankment property. Such an intangible, potent property might be adequately characterized as *sacredness*. Thus, the Feature A/Observatory Mound Complex was an anomaly that came about as a result of the sacred property that the builders experienced as being embodied in the embankments themselves.

To generalize, embankment earthworks displaying the range of physical properties characteristic of the C-R Configuration would also have been experienced as possessing the same intangible but real sacred property. In terms of the contingent nature of the cosmology→ ideology relation, earthwork construction was the realization of an essential *ideological strategy*. Therefore, those embankment earthworks of the general region that do not display the C-R Configuration patterning would have been expressions of alternative ideologies presupposing the same cosmology and, therefore, manifesting the same notion of sacredness.

On first sight, positing sacredness as a property of the embankment earthworks is not a particularly startling conclusion. What is important, however, is that this analysis has demonstrated that the Feature A/Observatory Mound Complex is the material manifestation of an avoidance practice. This conclusion constitutes an empirically confirmed entry into the cultural world of the builders. Furthermore, it goes beyond simply stating that these earthworks were sacred. It actually characterizes the nature of the sacredness and how it came about. While this characterization is not something that could have been predicted independently of the analyses, it has been established independently of confirmation by those who were responsible, the prehistoric builders. This follows from the characterization of intentionality that was presented in the last chapter. As stated then, contingent on correctly identifying the emic action nature of a behavioral intervention or its material effects, it necessarily follows that the requisite emic intentions were exercised, that relevant emic wants were fulfilled, and, of course, that both the background of emic beliefs and the social system that made having these wants and intentions possible must have existed.

In summary, the Feature A/Observatory Mound Complex entails the performance of an avoidance practice motivated by the perception of sacredness immanent in the earthwork. This necessarily entailed the discharging of sacred wants and duties by avoiding polluting the construction itself; and both entailed the existence of the relevant cosmology and, more precisely, the existence of a core principle of that cosmology, namely, the Sacred Earth Principle (Cosmology Postulate

1), that made the avoidance intention possible and necessary. It is not simply that the earthworks would have been experienced as sacred but that, given the avoidance practice, sacredness would have been experienced as being present from the moment that construction started. This is a critical conclusion, and it follows logically from Scenario 1, which claims: (1) Feature A was the first step of construction; and (2) the Circle was not completed when the change of plan created the need to abandon Feature A. (What this change of plan may have been will be discussed later.) Therefore, it must have been that the act of embankment construction was itself the realization of an ideological strategy of symbolic congruency, thereby making the embankment construction the performance of a sacred ritual (Ideology Postulate 1). Thus, as expressive monumental warrants, to dismantle them at *any* step in the construction process, including the very first, would have been tantamount to "dismantling" that which they represented, the cosmos.

The Sacred Earth Principle would characterize the world in the experience of the builders as immanently sacred. Human intervention into the natural order, then, would be transitive in the sense that, for the builders, this intervention would simultaneously intervene directly into the sacred order. This contrasts with a transcendent view of the cosmos, the type of cosmology that is common to the cultural background of Euro-American archaeologists. A transcendental cosmology envisions the natural order as intransitive with respect to the sacred order. The natural order is simply a "referential" reflection or facsimile of the "real" world, the sacred world. An asymmetry would exist between the transcendental sacred and the mundane natural orders in that while the sacred→natural relation would be transitive, the natural→sacred relation would normally be intransitive. In short, in a transcendentally sacred cosmos, the sacred powers would control the natural order but intervention into the natural order by humans would normally have no effect on the sacred order. In an immanently sacred cosmos, however, the relation between the sacred and natural orders was mutually transitive, a reciprocal, sacred order⇄natural order relation. Therefore, humans physically intervening in the natural order necessarily intervened directly in the sacred order.

CONFIRMATION

Before moving to the third line of evidence, the Astronomical Alignments analysis, it is important to specify the range of empirical evidence that can be marshaled to support the Sacred Earth Principle. Since this emic principle has been generalized as applying to all the Ohio Hopewell earthworks, much would appear to hinge on a single feature of a single embankment earthwork, although both the feature and the earthwork are outstanding examples of the general category. If the Sacred Earth Principle is valid, however, it should explain other noted construction anomalies. As illustrated at Baum (fig. 2.2) and Frankfort (fig. 2.3), whenever possible the "ideal" form of the infix of a Paint Creek C-R Configuration site should be a symmetrical circular element. As noted in chapter 2, this symmetry is "spoiled" at Liberty Works (fig. P.3) by an anomalous "lobed" circle located on the northern periphery of the infix and between the Circle and Paint Creek Square components. Inside this lobed circle there is a semicircular embankment, another apparent anomaly. This latter embankment can justifiably be claimed to be an *incomplete* Simple G-Form/K-Profile embankment earthwork since at Seip (fig. P.4) there is an embankment asymmetry in the infix component in the equivalent position, that is, between the C (small Circle) and R (Paint Creek Square) components, where there is also a small, but *complete*, Simple G-Form/K-Profile embankment earthwork.

Using Baum and Frankfort as the norm, both Seip and Liberty Works would probably have been built without the above anomalies if at all possible. A postulated scenario for Liberty Works claims the embankment found inside the lobed circle anomaly was an incomplete Simple G-Form circle in the process of being built when a change in construction plan occurred, causing its construction to be abandoned. This change of plan was probably what led to building the version of the Paint Creek C-R Configuration that we now call the Liberty Works.

This change would have presented a problem of what to do with the incomplete circular embankment. If the Liberty Works infix had been built as a symmetrical circular component, it would have been necessary to (1) dismantle the unfinished embankment, or (2) shift the whole construction further north or south on the terrace so as to exclude this latter feature, or (3) build around it. Apparently, the first was not a viable option; otherwise, this incomplete circle would not exist. Postulating a proscription implicated by the Sacred Earth Principle can explain this. Even though this small embankment was an incomplete earthwork, it was already an expressive icon and, therefore, it was part of the sacredness that it represented. Option 2 was also avoided, possibly because isolating an incomplete embankment earthwork may have been no less damaging than dismantling it. Furthermore, the infix contains the very important Edwin Harness mortuary structure and this probably was already in existence and being used when the decision was made to build the Liberty Works C-R Configuration. Therefore, if Liberty Works had been built to the north or south, it would have been excluded.

These are not mutually exclusive possibilities. Therefore, either or both may have forced the third option, which was to build the embankment earthwork precisely where it is found while avoiding dismantling the small incomplete circle. The solution, then, would be to build around the latter, simultaneously incorporating it into the ambit of the infix by means of the lobed circle, even though this sacrificed symmetry. It would seem appropriate to speak of this solution as avoiding dismantling by *circumvention*. It is reasonable to conclude that circumvention was a form of construction avoidance equivalent to the avoidance that resulted in the Feature A/Observatory Mound Complex, and the same Sacred Earth Principle is being manifested in both cases. The Seip case reinforces this conclusion. The equivalent asymmetry of the infix can be understood as the result of avoiding dismantling and/or excluding the associated small G-Form embankment circle. Furthermore, while at Liberty Works the Edwin Harness mortuary structure may also have figured in the strategy, at Seip there are two similar structures, also found in the infix. If either or both preexisted the construction of the Seip C-R embankment earthwork, then the same constraints would have applied at Seip as at Liberty Works.

The Hopewell site, as described earlier, also manifests several anomalies. In this case, the Circle and Rectilinear components are related spatially to each other in an appropriate manner, given the standard skewed/off-centered Paint Creek C-R Configuration juncture, but the location of the Circle is anomalous. It is "inside" the C-Form. As argued earlier, if the latter is the equivalent of the infix, then the Circle ought to be located "outside" and attached to the C-Form at an appropriate gate, much as the Paint Creek Square is. But the western gate is the only one that would preserve the relative spatial orientation of the Circle and Rectilinear, and there is a small creek at this point that may have precluded building the Circle there. Other gates are either on the raised, upper terrace or overlooking the bottom land, both locations that were probably unacceptable. Therefore, placing the Circle inside was the only acceptable alternative.

Another anomaly is the Hopewell Paint Creek Square itself. Of course, it is much smaller than the standard Paint Creek Square. The smaller size does not relate to the Sacred Earth Principle, per se, but there is a formal anomaly in this earthwork that does. This is the west wall of the Square, which actually consists of the original SL-Profile embankment making up the eastern wall of the Hopewell C-Form. As an SL-Profile embankment, the western wall of the Square has an "internal" ditch. All known Rectilinears of the C-R Configuration have K-Profile embankments. Indeed, the other three walls of the Hopewell Paint Creek Square are standard K-Profile types. Therefore, filling the ditch on that part of the C-Form embankment that makes up the west wall of the Paint Creek Square would have been a simple way of producing a formally correct Rectilinear component. Furthermore, it would have been a matter of adding rather than subtracting earth. However, the fact that this did not occur suggests another avoidance, despite the formal sacrifice entailed in having an "improper" Paint Creek Square. If this is the case, then it would reinforce the earlier claim, namely, that the S-Profile embankment, whether SL-Profile or SR-Profile, was a compound feature. Therefore, the ditch was as important as the embankment itself. It follows that deliberately filling in this ditch, even when it would have allowed for the completion of a formally correct Paint Creek Square, may have been no less taboo than dismantling the associated embankment.

Thus, by adding these anomalies of three major C-R Configuration sites to the Feature A/Observatory Mound Complex anomaly, this considerably broadens the empirical grounds of the claim for the existence of the Sacred Earth Principle. Others could be added, including the multiple truncated embankments that make up what appears to be "extraneous" elements of the Newark earthwork site, and, interestingly, the mode of building "gates" in the T-Form earthworks, which also appears to avoid dismantling preexisting embankments, and so on.[5] However, the above examples would seem to be sufficient.

3. Astronomical Alignments Analysis

According to the World Renewal Model, the embankment earthworks are postulated to be expressive iconic warrants of world renewal ritual. Since icons participate in the nature of what they represent, it would be expected that the earthworks would display anthropogenically produced properties that their builders would take as manifesting the powers of the cosmos. Since construction congruency is the primary mode by which monumental features are "fitted" to the phenomena they represent, the mode of spatial orientation of the earthworks would be a prime candidate. In this regard, the third line of evidence in support of the World Renewal Model focuses on claims that the embankment earthworks embodied astronomical alignments. These alignments would have worked to "fit" the earthworks to the relevant aspects of the cosmos that they were believed to represent and in which they were believed to participate.

Ray Hively and Robert Horn's analyses of the Newark and High Bank Circle-Octagons empirically established that the Octagon embankments were intentionally laid out to embody horizon azimuth alignments of the 18.6-year lunar cycle. They also established that the Wright Square, the large Paint Creek Square in the eastern sector of the Newark site, is aligned with the solstices and the equinoxes of the solar cycle. Confirmation of celestial alignments embedded in the layout of the rectilinears of the C-R Configuration earthworks has been made by the careful empirical work of a number of scholars.[6] Hively and Horn noted that the dimension of the Observatory Circle, effectively 320 m (i.e., 321.3 m), seemed to constitute a basic unit of measurement. They referred to it as the Observatory Circle Diameter unit, or the OCD unit. They postulated that the spatial organization of the Newark site was based on this measurement since it underwrote the proportions and placement of the different components of the total site.

There has been some debate over the significance of this or some other length as a particular measure.[7] It is not this debate that is relevant for immediate purposes, but the way Hively and Horn established that the OCD measure may have been used for this purpose. A schematic diagram of the Octagon (fig. 4.4) shows that the vertices A-C-E-G form a square with sides of 1 OCD. They suggest a simple geometric method by which this square could be transformed into the Octagon. For example, by using vertex A and the diagonal AE as the radius, an arc can be drawn through opposite vertex E. The same method can be repeated at vertices C, E, and G. This would create four pairs of arcs cutting at points B, D, F, and H. By joining all points, an octagon having the four pairs of opposite sides parallel would be generated, displaying full mirror-image symmetry.

However, as they critically note, the Octagon incompletely conforms to this layout.[8] There are several deviations from those "predicted." They claim that these were not the result of errors or miscalculations. If they are correct, then these deviations would be anomalies much in the sense expressed earlier. As such, they would result from sacrificing symmetry to achieve more fundamental goals. The major deviation is that only two of the four pairs of opposite sides are parallel. The positioning of Vertex F 6 m short of where it would have to be placed in order to ensure octagonal symmetry is the primary cause, they claim. Because of this "failure," opposite sides AB/EF and BC/FG are not parallel. That the other seven vertices are positioned as predicted by their method immediately raises the question for them as to why the builders would not have ensured mirror-image symmetry when they could so easily have done so.

They explored the possible astronomical rising and setting turning points that might be embedded in the Octagon as horizon alignments and noted that the Octagon effectively incorporates all eight of the turning points of the 18.6-year lunar cycle. In fact, without "mispositioning" Vertex F, two out of these eight alignments would not have been marked. To confirm that deviance from "ideal" geometry was deliberately motivated by lunar alignment concerns, they point out that but for several other deviations in the lengths of the embankments and the placement of some of the "gate" mounds, several other alignments marking important lunar turning points would have been absent. In short, they conclude that these deviations from the ideal were deliberate and that this patterned irregularity was most reasonably accounted for as the builders' ensuring that all eight turning points of the lunar cycle were embodied in the Octagon primarily through alignments embedded in the placement of its key features. They go on to examine other possible anomalies, focusing in this case on the High Bank Octagon, and again demonstrate certain anomalies and asymmetries that could be explained in the same terms. They conclude that the Octagon was deliberately constructed as part of a strategy to ensure that its formal properties would be aligned with the eight minimal and maximal rising and setting turning points of the moon defining its 18.6-year cycle. In effect, this supports Ideology Postulate 1.

Since the World Renewal Model postulates that the Newark Circle-Octagon was an iconic warrant of the cosmos, if the object of these alignments was the moon and its cyclic pathways, then this configuration was an expressive icon not of the total cosmos but only of the moon. Is there further evidence that the earthworks were, in fact, iconic warrants of the cosmos and not simply of the moon and its cycles bracketed by these alignments? Evidence that would support the holistic cosmic perspective would be data patterning that was superfluous to lunar alignment needs, as such. This evidence would not diminish the importance of the lunar alignments. Rather, it would suggest that the object that this monumental expressive symbol was representing was the cosmos in its totality, but under the lunar aspect.

In fact, if the Newark Circle-Octagon were simply a Lunar earthwork, it displays a major anomaly not noted before, this being the Newark Circle itself. In terms of lunar alignment needs, this major component is quite superfluous. Subtract the Circle and the lunar alignments would still be embedded in the Octagon, as described above. It is clear, however, that the Circle was an intrinsic part of this embankment earthwork. Not only were the Circle and Octagon contrasting geometrical expressions of the same OCD unit, they were directly connected by the tangent/centered juncture. Therefore, despite its being anomalous in terms of lunar alignments, the Circle must be a critical symbolic pragmatic constituent component of their material interventions by which the builders could achieve their construction goals. The complementary nature of the Circle/Octagon contrast based on the common OCD unit suggests that the earthworks manifest a complementary opposition, a duality representing the Heavens/Underworld duality of the cosmos, but under the lunar aspect. To speak of the High Bank C-R Configuration as the Lunar Motif, then, is a bit misleading unless it is understood that this is an elliptical way of saying that it represented and participated in the dual Heaven and Underworld powers of the cosmos under the lunar aspect.

When seen in this light, the different formal components of the High Bank C-R Configuration, here spoken of as the Lunar Motif (in the above sense), can be identified with different and complementary cosmic components, Cosmology Postulate 2 of the World Renewal Model. This postulate claims that, in the collective understanding of the builders, the cosmos was structured vertically into at least three sacred strata, the Heavens, the Middle World, and the Underworld, and was structured horizontally into at least four sacred quarters. This tripartite vertical sectoring and quadripartite horizontal sectoring of the cosmos, of course, is known to be pervasive in Native American cosmology.[9]

This can be made more concrete now by specifying that the Circle and the Octagon embodied and participated in the essential sacred powers of the Heavens and the Underworld, respectively, under the lunar aspect. Of course, while it is reasonable to claim that the C-R Configuration manifests a complementary duality, beyond this, there is no special reason that the Rectilinear should be associated with the Underworld and the Circle with the Heavens. It could as easily be the other way around. In fact, Romain does put it this way, arguing that the R component represents the Heavens and the C component, the Underworld.[10] Therefore, additional empirical grounds must be given. These will be detailed in the following chapter.

Assuming for the moment that the R component represents the Underworld, it follows under Ideology Postulate 2 that, as the complementary opposite of the Octagon, the Circle represents and participates in the sacred powers of the Heavens, under the lunar aspect. Hence, in the collective understanding of the builders, the Octagon embankments, "gates," and "gate" mounds, which were critical elements of these alignments, participated in the equivalent sectors on the horizon where the events of the moon rising from below and setting back into the Underworld were perceived as occurring at the maximum and minimum turning points in the lunar cycle. In terms of iconic logic, these features would represent and participate in the essential nature of the lunar portals. Furthermore, as mentioned earlier, alignments with the solar equinoctial and solstice turning points have been demonstrated to be embedded in the Paint Creek Square.[11] If the lunar alignments are accepted as evidence that the High Bank C-R Configuration constitutes the Lunar Motif (the cosmos under the lunar aspect), then it is logical to postulate that the Paint Creek C-R Configuration constitutes the Solar Motif (the total cosmos under the solar aspect), and therefore, the embankments, "gates," and "gate" mounds of the Paint Creek Square that marked these alignments would have been experienced as participating in the nature of those sectors on the

horizon where the events of the sun rising from below and setting
back into the Underworld were perceived as occurring at the turning
points of the annual solar cycle. Hence, the Paint Creek Square would
represent and participate in the Underworld under the solar aspect. It
is also reasonable to conclude that the small Circle on the perimeter
of the infix would be the complementary opposite of the Paint Creek
Square and would represent and participate in the sacred powers of
the Heavens under the solar aspect.

As pointed out earlier, the variation between the High Bank and
Paint Creek C-R Configurations seems to be tied to the nature of the
conjoining elements. The aggregation neck correlates with a tangent/
centered C-R juncture while the infix correlates with the skewed/off-
centered C-R juncture. If the C and the R represent the complementary
Heavens-Underworld duality of the cosmos, from this it follows that
the contrasting conjoining components, aggregation neck and infix, represent the Middle World under the lunar and solar aspects, respectively.
If this is the case, then the infix would manifest the Middle World
under the solar aspect, thereby being the interface of the Heavens and
the Underworld. The aggregation neck, then, expresses the Middle
World under the lunar aspect.

THE CHANGE OF PLAN

Having made an initial grounding of the central claims of the
World Renewal Model, it is now appropriate to explore the possible
circumstances that promoted the change of plan postulated in Construction Scenario 1. According to this scenario, the initial plan would
have seen the completion of the Circle and Feature A and then the
construction of the Octagon to the southwest end of Feature A. The
change of plan required building the Octagon in the sector northeast
of the Circle. Since it would appear that the lunar alignments were
neither improved nor hindered by this tactical change, the question
then is, Why was the change made? The answer would reveal an
important concern since, as argued earlier, the Sacred Earth Principle
proscribed simple removal of previous constructions.

It is postulated that at the time the Newark Circle was initiated, the master plan did not include the construction of a Paint Creek C-R Configuration, as such. As pointed out above, the construction of the High Bank site at Chillicothe probably predated the construction of Seip or Liberty Works, these being the "classic" Paint Creek C-R types. This would suggest that the Newark Circle-Octagon was initiated probably shortly *after* High Bank was built and *before* either Seip or Liberty Works was built.[12] It was probably after Feature A was built and before the Circle was completed that the construction of the first Paint Creek C-R Configuration site in the Chillicothe region was started or at least was first noted at Newark. Since they were already emulating the High Bank C-R Configuration of the Chillicothe region, it is plausible to suggest that Newark decided to expand its own program by including a Paint Creek C-R Configuration.

Newark has several standard-sized sacred circle embankment earthworks, as can be seen in fig. P.1, fig. 4.2, and fig. 4.3. But there is also the very large Fairground Circle. Recently, minimally invasive excavations of this earthwork have provided radiocarbon dates that place its initial construction in the final century B.C., 2,110 ± 80 B.P.[13] The sample was acquired from intact soil immediately below the embankment and the date certainly fits the chronological scheme as presented earlier. On the basis of the SR-Profile attribute, it would be roughly contemporary with Component B (sacred circle) and Component C (Shriver Works, SL-Profile) of the Mound City Cluster, as well as Cedar Bank.

This suggests that all the K-Profile embankments are post-SR-Profile constructions. These would include the Newark Circle-Octagon; the Wright Square; the large oval or C-Form/K-Profile embankment incorporating the mortuary mound cluster referred to as the Cherry Valley mounds; the three sets of parallel aggregation embankments that converge on the east and south sides of the Octagon, one from the Cherry Valley terrace, one from the large Wright Square in the eastern sector, and one from the south (this will be discussed in more detail in the next chapter); and the multiple truncated embankment units that make this huge site so complex, or, one might say, complicated.

It is claimed here that the Newark Circle-Octagon was the first

K-Profile construction at this site. To add a second K-Profile component, namely, a Paint Creek C-R earthwork, at Newark, would be difficult, given the preexisting set of sacred circles and the proscription of dismantling earthworks. Some of the truncated K-Profile embankment units might very well indicate initial but abandoned attempts to build this second large K-Profile earthwork, indicating considerable equivocation arising over just how to solve this Paint Creek C-R Configuration problem. It would not simply be a matter of building around these previous constructions. The Paint Creek C-R Configuration had quite specific requirements. It needed an infix that came between the C and R components, and, of course, as the Solar Motif, the rectilinear clearly had to differ from the Octagon of the High Bank type. As stated earlier, when terminal mortuary features are associated with the C-R Configuration, they normally are found within the infix component. It is likely that a number of the mounds of the Cherry Valley cluster containing mortuary residue were already in existence, even though the contents of some of these appear to be similar to the contents of the mortuary features at Seip, Liberty Works, Mound City, and Hopewell. Unfortunately, these were destroyed by canal construction and expansion of Newark during the nineteenth century, and only cursory reports were made of them. Lepper has carefully researched these and concludes that a number of the Cherry Valley mounds contained Hopewell-styled mortuary remains, including partial cremation, and major Hopewellian caches and artifactual deposits.[14]

All this suggests that, in adding a Paint Creek C-R earthwork to Newark, the builders would need to ensure its infix encircled the Cherry Valley mortuary mounds and facilities. This further suggests that, depending on which map is examined (fig. P.1, fig. 4.2, fig. 4.3), the C-Form or Oval K-Profile embankment earthwork that partly or fully encircles these mounds was built as an infix component. Certainly the formal parallels between this component and the C-Form at Hopewell are very similar (e.g., two-leveled construction incorporating the higher terrace into the lower terrace with its mortuary mounds). This may have been why the Wright Square is found south and west of these mortuary mounds. It displays all the properties of the Paint Creek Square, except that it has a total of eight "gate" mounds, both corner and medial, instead of the usual four medial mounds found in the Paint Creek Square of the central Scioto region. Thus, the Wright Square is positioned southwest of this postulated infix, roughly similar to the way that the Paint Creek Squares at Seip and Liberty Works, for example, were positioned relative to their respective infix components.

To complete a Paint Creek C-R Configuration, then, required the Circle component, which should have been a small Circle attached to the perimeter of the infix and conjoined in a skewed/off-centered manner to the Wright Square. However, achieving this may have been impeded by the set of small preexisting sacred circle embankment earthworks, which could not have been removed, according to the proscription against dismantling embankments. The Fairground Circle would have been available, except that either its position was inappropriate, or alternatively, as with the smaller sacred circles, the symbolic pragmatic meaning of the Fairground Circle may not have been compatible with the meaning of a Paint Creek C-R Configuration. In any case, the solution arrived at may have been to recruit the Newark Circle, just then being constructed, to serve two purposes, as the C component of the Newark Circle-Octagon, and as the C component of a Paint Creek C-R Configuration.

Although this seems to be a bit of a stretch, it actually is supported by the earlier assessment made of the role of the C component in the C-R Configuration. If embedding lunar and solar horizon rising and setting points is an important form of expressive iconic congruency, and if the C component plays little or no role in these alignments, then, under the significant constraints outlined above, recruiting a preexisting Circle to do double lunar and solar "Heavenly" duty seems plausible. Its symbolic pragmatic function would simply be to represent the Heavens in order to complete the Heavens/Underworld duality. Based on the C-R Configuration analysis, however, this double duty could not have been performed if the Octagon was southwest of the Newark Circle, postulated here as being under the original plan. This position would mean that the Circle would require two juncture components, Feature A conjoining it to the Octagon on its southwest end, and a second aggregation component extending from the northeast sector of the Circle to the Cherry Valley infix. The possibility of a Circle's having two aggregation necks has already been explored and rejected in the earlier Construction Scenario analyses. Apparently, for the builders the less radical solution was to position the Octagon northeast of the Circle, as outlined in Scenario 1. This would ensure a single direct conjoining of the Circle with its proximal Octagon. Then the Octagon, the Wright Square, and the Cherry Valley infix could be related by separate aggregation necks.

Extended parallel aggregation components, such as at Hopeton, have already been postulated as serving to "collapse" or "dissolve" the space separating two related components. Thus the two east-west aggregation components converging on the Octagon, one each from the Cherry Valley infix and the Wright Square (the postulated Paint Creek Square), may have been built to serve in this way. Conceptually the layout of these aggregation components probably expressed the intention to have the space between the Newark Circle and the Cherry Valley infix, on the one hand, and the Newark Circle and the Wright Square, on the other, "collapsed" while the orientation of these components relative to each other was to be held constant.

Imagine the aggregation elements progressively shortening from east to west, thereby "dragging" both the Cherry Valley infix and the Wright Square westward.[15] The Cherry Valley complex, as the infix, accompanied by a slight imagined skewing, would be conceptually positioned to "cover" the Octagon. This would situate the infix as the Middle World "over" the Octagon as the Underworld. The Wright Square would be "dragged" to the point where the western end of the aggregation element conjoins with the Octagon. Using the Newark Observatory Circle as the C component, this would result in a circle-infix-square pattern that would be equivalent to Seip, Liberty Works, and the other classic Paint Creek C-R Configuration sites. That is, this imaginary collapsing would generate all the attributes of the Paint

Creek C-R Configuration, except that the Circle would be out of proportion. But the latter would be accounted for as a sacrifice to the achievement of major iconic congruency.

Thus, the multiple attributes that make up the uniqueness of Newark can be accommodated to a *dual C-R Configuration*, making this site truly unique in that it may display more coherently than any other set of embankment earthworks the complex that was earlier referred to as the Sacred Dual C-R Motif. This is constituted by an actual or conceptual integration of a High Bank C-R Configuration site and a Paint Creek C-R Configuration site, or else the key attributes of these two configurations. Given the complementary oppositional dualities that have already been outlined, suggesting that the animating powers of the cosmos are manifested in the Paint Creek C-R Configuration under the solar aspect and in the High Bank C-R Configuration under the lunar aspect, it is reasonable to conclude that a full suite of world renewal rituals would require both types of monumental iconic warrants. Newark may be understood as just that, a monumental iconic warrant constructed in order to embody the totality of the cosmos in accordance with its Solar/Lunar sacred temporal structuring and its vertical/horizontal Heavens/Underworld sacred spatial structuring.

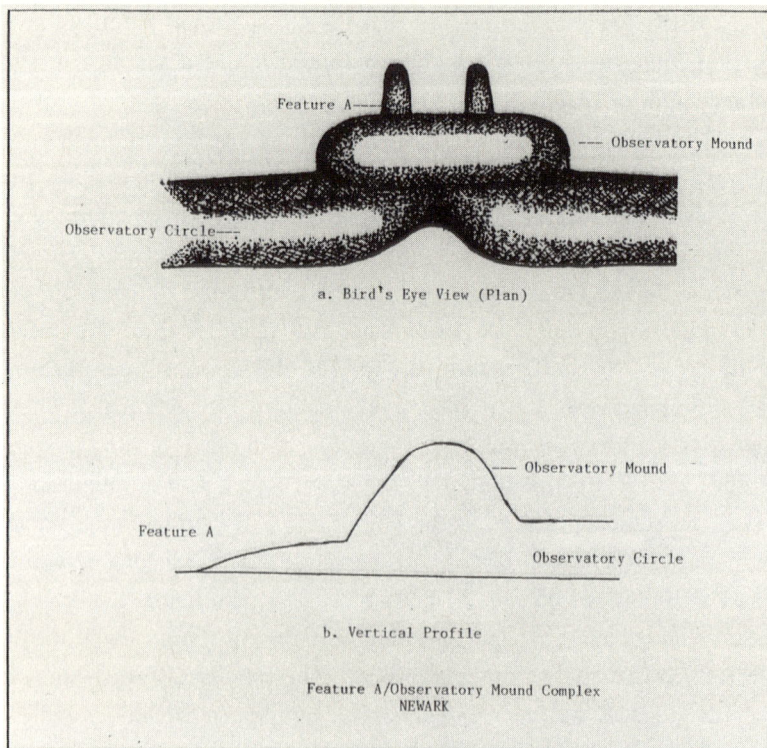

a. Bird's Eye View (Plan)

b. Vertical Profile

Feature A/Observatory Mound Complex
NEWARK

FIG. 4.1. The Feature A/Observatory Mound Complex

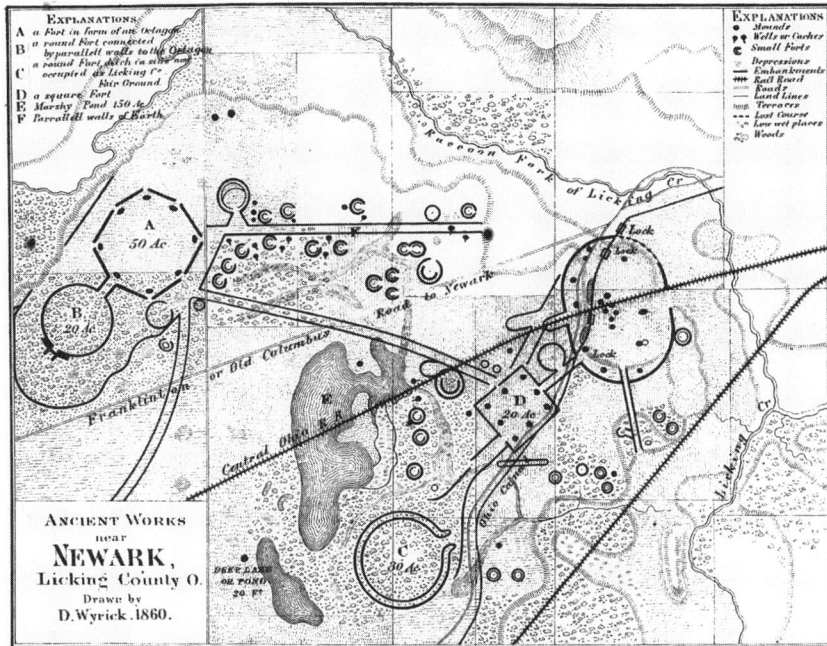

FIG. 4.2. The Wyrick Map of Newark (courtesy, Ohio Historical Society)

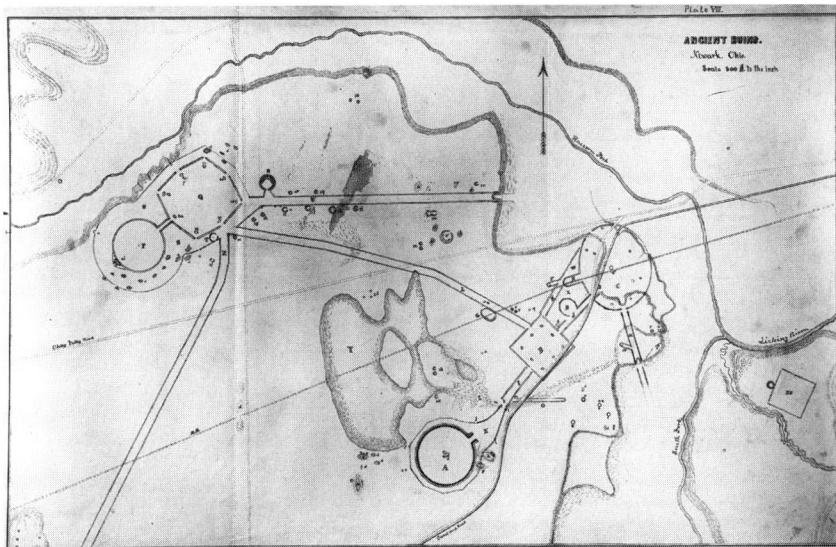

FIG. 4.3. The 1862 Salisbury Map of the Newark Earthworks (courtesy, American Antiquarian Society)

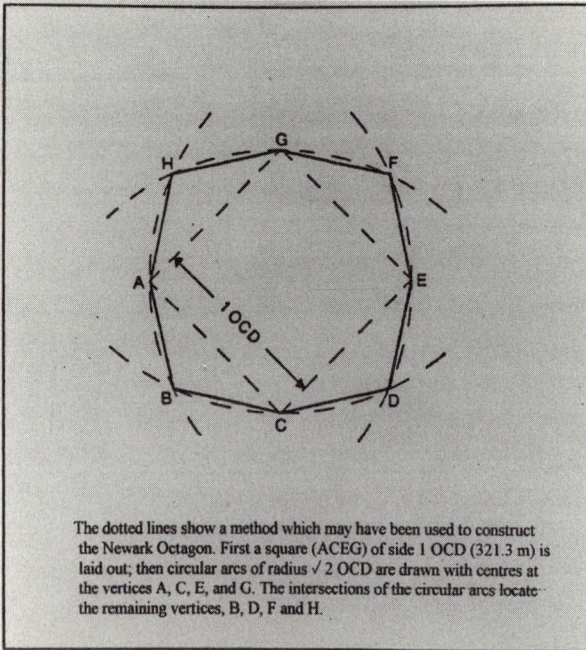

The dotted lines show a method which may have been used to construct the Newark Octagon. First a square (ACEG) of side 1 OCD (321.3 m) is laid out; then circular arcs of radius √2 OCD are drawn with centres at the vertices A, C, E, and G. The intersections of the circular arcs locate the remaining vertices, B, D, F and H.

FIG. 4.4. Schematic Diagram of the Transformation of the Newark Square to the Newark Octagon (Hively and Horn 1982)

Chapter 5

CRITIQUE OF THE WORLD
RENEWAL MODEL

The World Renewal Model was presented in the previous chapter to initiate a symbolic pragmatic account of the Newark embankment earthworks and, by extension, of the embankment earthwork traditions of the Early and Middle Woodland periods of the Central Ohio Valley. However, the findings of this model account for the earthworks only in terms of the immediate purposes they served. Of course, this is important since it reveals the core cosmology implicated by the ideology that it expresses. Furthermore, symbolic pragmatics argue that the intentionality manifested by the earthworks entails a social and material environment that made it possible to have and exercise that intentionality. Therefore, the findings of the World Renewal Model can be used to launch an exploration that might quite adequately characterize both this social world and its relations with its natural environment in terms that would account for the earthworks as critical constitutive media of these relations. These findings are ideal for this purpose since the Sacred Earth Principle implicates cosmology, ideology, society, and the environment in a systematic manner. This means, however, that new models must be presented that directly address the social organizational and ecological dimensions that correspond to the premises of the World Renewal Model. Of course, this also means expanding the theory into the social and ecological spheres and expanding the relevant empirical bases to support this elaboration.

Because of the importance of the claims of the World Renewal Model, they must be assessed and tested. The method that will be used here will be both critical and developmental. It has already been applied in a limited manner in the presentation and critical assessment of alternative construction scenarios and in reconstructing the Hopewell and Mound City/Hopeton earthworks. The same critical method will be applied to the social and ecological models that are presented to expand our understanding of the world of the earthwork builders. Following the presentation of a model and its application to the relevant empirical data, the conclusions are critically assessed through comparison with alternative accounts and their theoretical frameworks. In short, the process of model construction is also a process of model modification and confirmation through interpretive feedback in a process of producing, testing, and rejecting or modifying alternative models.

This method is sometimes referred to as the hermeneutic circle, or the circling of meaningful interpretation and explanation. However, since the process is, in principle, knowledge building and enhancing, it might be better to use the analogy of the spiral, calling it the hermeneutic spiral method.[1] As a mode of knowledge construction, the hermeneutic spiral promotes the critical and reasoned challenging of knowledge claims. The efficacy of a challenge entails participating in the spiral process so that a challenge results in enhancing rather than denying knowledge. This enhancement can be achieved in different ways. It can be a matter of showing that some of the analytical steps of a knowledge claim do not accord with the logic of the model or its theoretical background. That is, the interpreter is at fault and the correction improves the initial interpretation and enhances the model. Or, at the other extreme, it can be a matter of presenting a radically different theoretical model that contrasts with the original model on several and possibly all major points, establishing the two models as alternative paradigms that address the same empirical data.

For a model to be considered a paradigm, it requires a broad and deep scope, and this means that its universe of study requires multiple supporting models to address different aspects of the complex reality being explored. It is appropriate to treat the paradigm as the global model of the complex reality and the supporting models that address the component aspects as auxiliary models; for example, a mortuary model and a social systems model might form key auxiliary models of a global model of a prehistoric episode. The auxiliary models relate to each other in a mutually reinforcing manner. Aspects of the components addressed by an auxiliary model can in turn be addressed by what might be appropriately termed ancillary models. For example, an ancillary of the mortuary model might be a rank order social model. In effect the ancillary model is nested within the auxiliary model, and the auxiliary model is nested within the global model or paradigm. It is not really what the model addresses that determines its status but the way the model relates within the paradigm. As a paradigm is constructed and empirically grounded, this can change.

Alternative paradigms stand in a largely mutually exclusive posture toward each other with regard to the same empirical data. They particularly contrast in terms of basic principles and, for this reason, the auxiliary and ancillary models of each paradigm will also often contrast. However, if they were to contrast at every level this would suggest that these two paradigms are addressing quite different phenomena. Hence, while they will contrast, they also share considerable common ground. The contrasting auxiliary mortuary models of each paradigm, for example, must agree that the data constitute the residue of mortuary practices. Where they might radically disagree is over the type of mortuary practices being manifested (e.g., ritual cannibalism, human sacrifice, funerary burial, and so on). Also, the same paradigm can have alternative auxiliary and/or ancillary models, making it necessary to qualify the level at which models contrast. Resolving differences among auxiliary and ancillary models will both preserve and modify the paradigm that they cooperatively make up. However, resolving the differences of radically contrasting paradigms often requires long scientific, spiraling debate and the only resolution, at times, may appear to be the rejection of one in favor of the other. A much more satisfying resolution would be a synthesis arising from modifying the contrasting theoretical frameworks, thereby resulting in a fuller under-standing of the data. A new synthetic paradigm constitutes a major breakthrough, but one that would not have been possible without this critical process.

In keeping with the hermeneutic spiral method, two challenges of the World Renewal Model will be discussed and debated. The first has been put forward recently by Romain and, rather than being a critical challenge, it is more in the nature of a fine-tuning of ancillary alternatives.[2] He presents an alternative interpretation of the C-R Con-figuration to the one presented in the previous chapter with regard to the cosmic aspects its components represent. Addressing the differences necessitates reconsidering the claims of the World Renewal Model, not so much the cosmological postulates, but specific elements of the ide-ological postulates, and either defending and retaining them or chang-ing them. In either case, the result should be positive in terms of knowledge growth and enhancement. Therefore, Romain's findings can be used as a positive illustration of the benefits of the hermeneutic spiral.

The second challenge tests the model in the hypothetico-deductive manner. Lepper has re-presented empirical data that had been "forgotten" and concludes that the "initial inference, and the in-tricate web of speculation [Byers] spins from it [i.e., the World Renewal Model], are without foundation."[3] This is a particularly important chal-lenge to the World Renewal Model, not because the empirical data that he presents undermine the "foundation" of the World Renewal Model, but because they can be used inferentially to enhance the model.

There is a third challenge in the form of a global model, men-tioned earlier, termed here the Exclusive Territorial/Proprietorial Do-main Paradigm. In its terms, the cosmology and ideology postulated by the World Renewal Model cannot be the case since the social system model that can be most reasonably inferred from the Sacred Earth

Principle contradicts the social systems models that are central auxiliaries of the Exclusive Territorial/Proprietorial Domain Paradigm. This challenge is directly addressed in part 2. Put simply, addressing the Exclusive Territorial/Proprietorial Domain Paradigm challenge will take up the rest of this book.

SACRED GEOMETRY

Romain has focused on using the embankment earthwork data to reconstruct the geometrical knowledge that the builders of the Chillicothe Tradition must have possessed in order for the C-R Configuration to exist in its many known particular expressions. To this end, he has carried out a geometrical analysis of the same idealized set of components that are fundamental to the C-R Configuration, although he does not formally recognize these earthworks as falling into two contrasting but complementary categories.[4] He demonstrates that the proportions of the different components, i.e., small circles (C), large circles (infix), and rectilinears (R), are geometrically related and, in parallel with Hively and Horn's findings, claims these are based on a standard measuring unit. Broadening the base that Hively and Horn's work largely pioneered, Romain has examined the range of different ways that the lunar and solar alignments were embedded in the features of the rectilinears. Based on three lines of argument, (1) analogy with Native American cosmological beliefs and ritual practices, (2) the geometrically proportioned linear measurements of the C-R components, and (3) the astronomical alignments, he concludes that the earthworks were the context and media of world renewal ceremonies.

> By reconsecrating these monuments, or by adding layers to a central burial mound, or by performing sacred dances within such enclosures, the Ohio Hopewell reaffirmed their place in the universe, and at the same time, helped renew their cosmos, their universe, and their resources. Such world renewal ceremonies are still found today among many North American peoples. (Romain 1996, 208)

In particular, his claims that the builders performed "sacred dances within such enclosures," thereby reaffirming "their place in the

universe" and renewing "their cosmos, their universe, and their resources," are perfectly compatible with the World Renewal Model. Indeed, after recognizing that the earthworks could have been used for
various social activities attracting large numbers, such as being centers
for the exchange of valued goods among distantly related people, he
proffers the point that

> the geometrical enclosures were primarily used for ceremonies relating
> to passage from this world to the next, death and rebirth, world renewal
> and creation. Ultimately, though, *it was the geometrical symbolism . . . as*
> *expressed in the earthworks, that gave these earthen creations their power.*
> (Romain 208, emphases added)

His careful analyses have provided empirical support for the role of
geometry as a form of expressive symbolic congruency that complements the construction of the astronomical alignments, the manifestation of binary opposition in the C-R Configuration, and the formation
of the Sacred Dual C-R Motif as discussed earlier. Therefore, his claims
and the World Renewal Model are mutually reinforcing. This is not
simply a replicative reinforcement. His conclusion that "it was the geometrical symbolism . . . as expressed in the earthworks, that gave
these earthen creations their power" can be taken as among the symbolic pragmatics of material cultural style. His implicit reliance on the
symbolic pragmatic theoretical framework is shown in his speaking of
"adding layers to a central burial mound," and "performing sacred
dances within such enclosures" as, in fact, social activities of reconse-
cration by which the builders/users "helped renew their cosmos, their
universe, and their resources" or that "reaffirmed their place in the
universe." In effect, this paragraph expresses the symbolic pragmatic
perspective as applied to material behavior in the sense that these behaviors could only be the renewal ceremonies intended by being performed in the context and through the mediation of these earthworks.

However, this pragmatic framework must be made explicit. Otherwise the symbolic interpretation is open to the two primary skeptical
criticisms noted earlier to which normative accounts have been sub-

jected, namely, the deemed inability to account for variation and the epistemological criticism that normative claims can never be empirically demonstrated. Romain relies primarily on analogy with historically known Native American peoples. As he argues, these carried out complex world renewal rituals. The skeptic could, however, point out the limits of this analogy. That is, historically, most Native American peoples performed their rituals without the use of monumental embankment earthworks of the sort produced by Ohio Hopewell. From this it could be concluded that religious belief and practice cannot account for the Ohio Hopewell embankment earthworks. The World Renewal Model can counter this type of criticism because it is explicitly placed within the socio-intentionalist theory and the symbolic pragmatic approach and this specifically acknowledges that ideological strategies can vary while cosmology can be constant. The ideological strategy of a people is manifested in their material culture. Since it also recognizes the contingent nature of the cosmology-ideology relation, so that ideology can vary while cosmology can remain constant, the fact that historic Native American peoples did not build Hopewellian-type embankment earthworks cannot be used to undermine the analogy.

There is one clear divergence between Romain's interpretations of the earthworks and the interpretation as given by the World Renewal Model, and they mark alternative accounts of the same empirical data. Although these two readings agree that the infix represents the earth or Middle World, the meanings of the C and R components are reversed. In the World Renewal Model, the R represents the Underworld. In his view, it represents the Heavens—the Upper World. First, he notes that many historic Native American cosmologies characterize squares as sky symbols. Second, of course, he notes that the squares embed the lunar alignments that designate the horizon turning points where the moon rises into and sets from the sky. Third, he notes the "statistical association between the general orientation of the square enclosures and the prevailing wind" thereby reinforcing his claim that the squares are associated with the wind-sky phenomena.[5] Finally, he caps this interpretation by further drawing on ethnographic analogy, pointing out (1) that many historic Native Americans perceive the horizon as a circle so that they conceive of the earth or land, rather than the sky, as a flat circular island; and (2) that southeastern Native American peoples used "the square . . . to represent the four world quarters, four winds, and four directions."[6]

The contrasts between these two accounts highlight the limitations of analogy. While Romain is correct to point out that Native American peoples have been historically known to manifest their concept of the sacred Underworld by using a circle and a square for the sacred Heavens, because of the contingency of the cosmology→ideology relation, the very same pragmatics could have been expressed prehistorically by reversing this, while the cosmology could remain constant. Furthermore, as stated earlier, the C-R contrast alone does not resolve the problem since it serves primarily to express a complementary opposition. Therefore, there does not seem to be any a priori reason for associating the Underworld with the Circle and the Heavens with the Octagon or Square, or vice versa. Instead, these alternative claims must be adjudicated empirically. There are empirical data that can be used to this effect. Presenting them and demonstrating their relevance can be most effectively done, however, by turning to Lepper's critique of the World Renewal Model since, in fact, this hinges on the same data. Therefore, countering Lepper's critique of the World Renewal Model will include resolving these alternative claims about the cosmological representations expressed by the C and R components.

EMIC CATEGORIES AND ICONIC CONGRUENCY

In the original presentation of the World Renewal Model, I expressed the notion that the earthworks would be built so as to manifest iconic congruency.

> If the C-R motif is, indeed, an iconic model of the 'umwelt' (cosmos) which has the structural properties of binary opposition, it follows then

that the two units ought to be constructed in contrasting ways with contrasting resources. These contrasts can be based on different types of resources, different types of procurement or different types of vertical profiles, or all of these. (Byers 1987, 307)

Now Fowke noted two deep borrow pit excavations immediately north of the Octagon and, with regard to the Observatory Circle, he noted "slight depressions along the walls as well as at some little distance away, both within and without the enclosure whence the earth for the embankments was taken."[7] These observations suggested to me that (1) the earth for building the Observatory Circle probably was procured by scraping the surface stratum soils in the proximity of that construction, and (2) the earth fill for the Octagon probably was derived from the two, deep proximal borrow pits immediately north and northeast of the Octagon, thereby grounding my claim that the two earth fills were procured by contrasting methods, one from surface scraping and the other from deep pitting. I used this surface scraping/deep borrow pitting contrast correlated with the Circle/Octagon contrast to postulate that surface stratum earth and deep stratum earth were real emic categories. In terms of the Sacred Earth Principle, since the surface stratum/deep stratum natural stratification simultaneously embodied the sacred order, I postulated that it would have been appropriate for the builders to use these two different earth strata to replicate the Heavens/Underworld stratification in the Circle-Octagon pattern, using the surface stratum earth for the Circle (Heavens) and the deep stratum earth for the Octagon (Underworld).

Lepper has recently brought two nineteenth-century maps of Newark to the attention of archaeologists, one made by Wyrick (fig. 4.2), and one by the Salisbury brothers (fig. 4.3), along with their accompanying commentary. These two maps had, unfortunately, dropped from Ohio Hopewell archaeological memory. In his view, these maps serve as evidence that contradicts the above claims, and that justify rejection of the total model. The Salisbury map is particularly germane in that it confirms that there were borrow pits associated with the

Octagon, as noted above. However, it further shows that "extensive borrow pits also were associated with the circle" (230). These can be seen "outside" the southeastern half of the Circle in the zone made by the embankment of the Newark Circle and a low embankment that this map indicates was built around the west and south sectors of the C component of the Circle-Octagon. This low embankment is also a significant and new datum and will be commented on shortly. From this distribution of "borrow pits" parallel to but distal from the south-eastern sector of the circumference of the Circle, he concludes that my "initial inference, and the intricate web of speculation he spins from it, are without foundation" (230).

Lepper's bringing forward such important, previously overlooked data is certainly valuable. However, are these data grounds for rejecting the World Renewal Model, as he claims, or, more importantly, can they be used to correct and extend, or even affirm, the model? It will be argued here that these map data can be construed as evidence for both confirming and elaborating the World Renewal Model. Given what I actually stated in the above quotation, "the two units ought to be constructed in contrasting ways with contrasting resources," it is clear that I was arguing that one of the possible ways that a complementary opposition could be expressed in construction is by contrasting modes of procuring fill and that this fill could be reasonably treated as being emically perceived to contrast. If such contrasts can be logically grounded, they might also be grounds for characterizing the nature of the complementary opposition that they are expressing. Therefore, to undermine the claim that the Circle/Octagon contrast is oppositional in this sense, it is first necessary to demonstrate that there is no empirical basis for claiming a categorical distinction in earth fill procurement methods between the Circle and the Octagon. If this cannot be done and, instead, there is evidence of variable procurement, this evidence could then be examined to see if it supports or undermines the second claim, namely, that the procurement variation manifests an emic distinction of the above nature. These structural congruencies are schematically presented in table 5.1.

In these terms, then, the null hypothesis would be that, in the absence of any objective constraints, and given that the Circle and the Octagon required the same type of "brute" or objective labor, the residual patternings of the earth procurement processes should be equivalent for both features. This still would not be definitive since it is quite possible that culturally constituted categories of earth were based on some criterion that would not leave observable clues (e.g., a behavioral criterion differentiating earth fills in terms of whether they were procured by women or by men). Possibilities of this sort illustrate the problem with using the hypothetico-deductive approach by itself. We cannot develop a theory that could predict all the possible arbitrary ways people could use to constitute their behavior and the products of this behavior as the type of activities they intended. The best that might be done is to argue retro-inferentially that, given the symbolic pragmatic view, a structuring of the world in complementary oppositions is likely to be manifested in major construction contrasts.

Fortunately, however, quite different patternings in the Octagon and Circle "borrow pittings" are clearly discernible in the Salisbury map. These can be used to assess whether the null hypothesis can be rejected. In the case of those Lepper associates with the Circle, it is clear that they are the result of procuring earth that entailed more than surface-scraping, but less than deep pitting: they are moderate depressions, and they are symmetrically distributed along the "outer" zone and somewhat distally from the Circle, about halfway between the Circle and the outer embankment shown on the map. This outer embankment, also indicated on the Wyrick map (fig. 4.2), "embraces" the total Newark complex, including the Fairground Circle. As Lepper points out, the excavations that he, Dee Ann Wymer, and William Pickard carried out revealed a residue of this low embankment.[8] Thus, while the original claim was to account for the procurement of earth fill for only the Circle and Octagon, these maps indicate that there was another extensive, although low, embankment that required earth fill. In this case, since the Salisbury map indicates that these shallow and linearly dispersed "borrow pits," or what might be better termed "borrow depressions," are about midway between the Circle and the outer embankment, it is possible that they served both purposes.

In quite distinct contrast with the form and distribution of these linearly dispersed "borrow depressions," the residue of earth procurement associated with the Octagon consists of a set of relatively tightly clustered and distinctly deep depressions, which can be appropriately characterized as *proximal borrow pits*. They are spatially concentrated outside the northern sector of the Octagon. According to Fowke, there were only two, while according to the Salisbury map there were at least four or five. Hence, the two spatially distinct sets of pits and depressions vary substantially, sufficiently to be treated as the results of meaningfully different forms of earth procurement. The Octagon procurement pattern can be treated as a focused, deep proximal borrow pit mode, and that of the Circle and outer embankment as a dispersed, shallow, distal borrow depression mode.

Does this variation justify rejecting the null hypothesis? Since the latter recognizes that both the Circle and Octagon would demand the same objective labor force to perform the same type of earth moving, and since there were no discernible objective constraints that made it necessary to vary the resource procurement mode, the same pattern ought to be replicated. This was clearly not the case. The null hypothesis can be rejected. Does this leave a knowledge vacuum? The new data certainly do not contradict the original claim. Indeed, they appear to enhance it, particularly now that the procurement needs for the outer embankment must be added. First, it is reasonable to conclude that these two patterns mark complementary strategies for procuring soils that would minimize the degree of intervention into the sacred natural order while achieving the quantities and types of soils required to embody the appropriate sacredness in the Circle and the outer embankment (surface stratum soil) and the Octagon (deep stratum soil). By serving as evidence in support of dispersed and focused modes, the data suggest that procurement was more nuanced than originally suggested. In this case, the variation in patterning can be reasonably explained in terms of a proscriptive procurement strategy, one that minimized the extent and degree of disordering of nature while fulfilling the needs of construction, and this strategy would be consistent with the sense of the world as being immanently sacred. The dispersed mode would be proscriptive in minimizing the depth of disturbance by spreading this surface scraping over an area sufficient to procure the appropriate type and quantities of soil, and, in similar terms, the focused borrow pit mode would minimize the areal surface disturbance while procuring sufficient deep stratum fill to serve the symbolic pragmatic needs of the Octagon construction. The focused nature of this pattern is particularly revealing. If these were the only deep borrow pits, and all the earth for the Octagon—or at least a significant amount of it—was derived from them, then the labor involved in procuring earth fill would necessarily expand as the distance between the embankment construction and the borrow pits would progressively increase with the building of the Octagon. Minimizing this labor would have been assured by simply distributing borrow pits at logistically strategic locales around the Octagon, as we can see was the case at High Bank (fig. 1.2).

Rather than contradicting the original terminology of surface-stratum and deep-stratum earths, the data Lepper presents reinforce the appropriateness of this terminology, thereby expanding our knowledge claims. In short, the maps are valuable data to be used as empirical evidence enhancing the claim that there was a correlation of surface-stratum and deep-stratum earths with Circle and Octagon, respectively, and the further claim that this manifests the upper/lower vertical structural relation. This also empirically grounds the earlier claim that the Circle represents and participates in the Heavens, and the Octagon represents the Underworld, rather than the reverse as favored by Romain.[9]

These data can also extend our knowledge in this area. Given the nature of the surface-stratum earth procurement mode, evidence of it would be difficult to discern. The pattern that is presented by the Salisbury 1862 map was not noted by Whittlesey when he did his survey in the 1840s for Squier and Davis, and, apparently, they were no longer discernible when later archaeologists, such as Fowke, reex-amined the region, although some of the deep borrow pits spatially proximal to the Octagon were still in existence.[10] Therefore, these newly reintroduced data not only reinforce the earlier assessment, namely, that the Circle displays a KS-Profile, or a variant of this, as does the long and low embankment that encircles it to the south, and that the Octagon displays a KPp-Profile, it also suggests just what kinds of limits on earth procurement would count as surface-stratum earth. Since these dispersed depressions are roughly equidistant between the Circle and the outer embankment, it would appear that either these two major features were built simultaneously, or else it was known at the time of construction of the Circle-Octagon that the low embank-ment would quickly follow its completion. In terms of Construction Scenario 1, the change of plan would probably have included the com-mitment to build the low outer embankment. According to this con-struction scenario, when the construction of Feature A initiated the Circle, it was intended that the Octagon be added to the southwest end of this feature. Given the proscription against dismantling em-bankments once built, this would have been an impossible intention if this outer embankment was already constructed. Since this outer em-bankment, in fact, was linked to the long aggregation element leading southwest from the Octagon, and through this element to the aggre-gation element connecting to the Wright Square, and then southeast to encircle the Fairground Circle, and ultimately connecting to the Cherry Valley infix, this suggests that all these elements formed a single com-plex embankment earthwork embodying the Sacred Dual C-R Motif.

What is particularly interesting in this regard is that, quite inde-pendently of this critique, Lepper has established unambiguous empir-ical data indicating systematic selection of different types of earth at Newark. This was the use of contrasting dark brown and light yellow earths in building the Fairground Circle, and smaller clay mounds were built to form the base of this circle. Greber and Ruhl have reported the use of red soils for constructing the Paint Creek Square at the Hopewell site.[11] In short, there is general recognition that soil selection was systematic and rule governed. Apparently, however, I may have been the first to claim that surface and deep stratum soils were relevant emic categories and this suggestion arose simply on the basis of the complementary opposition of Heavens/Underworld that was discussed in the previous chapter.

This highlights the need for confirming Construction Scenario 1 by means of direct empirical verification. A two-pronged, minimally invasive excavation could be carried out both to confirm the construction sequence of the Feature A/Observatory Mound Complex and to illuminate our knowledge of the nature of the earth fill practices and the complexity of the emic categories of soil types. The first goal could be achieved by verifying that Feature A, indeed, extends under the Observatory Mound as a continuous part of the first construction phase of the Circle. The second would be accomplished along the lines reported by Lepper with regard to controlled excavation of the Fairground Circle and by Connolly and Sieg with respect to Fort Ancient, thereby revealing the internal patterning of the soil for both the Octagon and Circle.[12]

INTER-SITE RELATIONS

Earlier it was suggested that the C-R Configuration could be used as a basis for postulating not only intra-site development, as in the case of reconstructing the growth and transformation of the individual sites of Newark, Hopewell, Mound City, and Hopeton, but also as a basis for exploring how this intra-site development can be related to inter-site events. Earlier it was specified that, besides the High Bank Circle-Octagon and the Newark Circle-Octagon, there may be several other sites displaying subvariant expressions of the High Bank C-R Configuration: Circleville, Dunlaps Works and the everted Circle-Rectilinear of Portsmouth, among others. Only the High Bank and Newark Circle-Octagon sites have been demonstrated to have the full complement of the lunar cycle alignments built into them. The others, on first appearances, would not seem to have this, suggesting that they are not High Bank C-R types. However, this is only a problem if belief→action determinism is accepted. If belief→action contingency is accepted, then it follows that the very same purposes and meanings can be served for their respective builders by earthworks displaying different formal properties. Therefore, when two or more spatially separated embankment earthworks are identified that share almost identical formal properties, it becomes important evidence in support of the view that the responsible groups shared ideological commitments. Similarly, the reverse is the case. Should two or more spatially separate earthworks exist that share only some basic formal properties, it can be argued that this marks some shared ideological elements among many differing ones. This suggests that although, or possibly because, these groups shared some ideological protocols but differed in details, they were in considerable ideological disagreement. Each would probably see the other's material warrants as faulty, as flouting one or more of the sacred norms, and, therefore, as only partly felicitous or possibly fully infelicitous media, unfit for fulfilling ideological commitments. Since they shared the same cosmology, and would understand what was at stake, this suggests that such groups would be antagonistic toward each other.

This would not seem to apply to the groups responsible for Newark and High Bank, since these closely emulate each other's ideological know-how. These two sites are almost formally identical, suggesting that they shared ideological postures, and, for this reason, that they had a close alliance relation. As Lepper points out, while the two Circle-Octagon embankment earthworks are oriented differently, this difference would appear to be deliberately complementary.[13] The High Bank axis is oriented NW-SE, and the Newark axis is oriented SW-NE. If these two sites were superpositioned while maintaining their respective orientations, their axes would cross at 90 degrees. It can be reasonably postulated that two separate earthwork groups shared complementary ideological strategies with regard to earthwork ceremonialism, while simultaneously respecting each other's autonomy.

According to the World Renewal Model, the point of earthworks was to ensure the full felicity of world renewal rituals. Therefore, they would be valued resources and alliances would be constituted precisely to promote the construction of these facilities. This is not to deny that other nonritual purposes may have been served by these alliances, such as procuring valued resources. However, these nonritual purposes would be parasitic on the primary symbolic pragmatic meaning of the earthworks. That is, exchanging valuables follows from shared ideology as manifested in the earthwork construction since the latter state of affairs constitutes the alliance that makes this exchange possible.

Lepper has argued along these lines, claiming not only that the complementary patterning of these two earthworks presupposes an alliance, but that this alliance was forged and reinforced through a unique earthwork, what he calls the Great Hopewell Road. He has argued that the third set of parallel aggregation embankments, those leading southward from the Newark Octagon, cut across country in an essentially undeviating line to terminate near Chillicothe at a point about halfway between Hopeton and High Bank. There are

> four locations along that corridor where there is evidence suggestive of road remnants. This evidence consists of parallel linear discolorations in the soil observed within the corridor at the predicted compass bearing. This evidence must be regarded as tentative. Only excavation at these localities will establish whether they are earthwork remnants or some unrelated phenomenon. To date there have been no such excavations. (Lepper 1998, 132)

This evidence makes the Great Hopewell Road thesis very promising and, if it is confirmed in the manner he suggests, then the World Renewal Model and the notion of inter-site ideological cooperation would make accounting for it reasonably easy.

A very significant implication is suggested by treating the Great Hopewell Road thesis and the World Renewal Model as mutually re-lated. This is that the social world of Ohio Hopewell did not recognize exclusive territorialism. This contradicts the view that is dominant in the current archaeological perspective, which argues that, in general, the embankment earthworks were the ceremonial centers of modular communities by which these declared their exclusive control of the territories surrounding the different earthworks. This means that they would also divide those lands between the individual earthworks among themselves. The result that is envisioned is a political geography of autonomous and mutually aggressive modular communities.[14] However, if this were the case, then the region between Newark and Chillicothe would be a patchwork of exclusive territorial modules, each requiring outsiders to negotiate the "right-of-way" to cross its territory. It is unlikely that the complex negotiations that such exclusive territorialism would require for building the Great Hopewell Road would not lead to at least some failures. These would necessitate building deviations in the overall direction of the embankments. In short, in such a world, the probability that even the limited empirical data Lepper reports would even exist becomes minuscule.

This makes these empirical data that he has retrieved even more important for they further ground the World Renewal Model. Under this model, the Ohio Hopewell peoples believed that the world was immanently sacred. In such a world, exclusive control of a delimited territory would have been antithetical and, in effect, unthinkable. Of course, because of the contingency of the cosmology-ideology relation, an immanentist cosmology does not entail the building of a monumental feature like that claimed under the Great Hopewell Road thesis. However, treated as the outcome of an ideological strategy, once built and correctly identified as an iconic warrant mediating human-land interaction in an immanently sacred world, it follows that the land would not and, indeed, could not be divided into mutually exclusive and separably owned territorial modules. These or equivalent conditions are what would be necessary for the existence of the Great Hopewell Road and they are the same conditions that would make the monumental embankment earthworks possible. The implications of these conclusions for understanding the social organizational nature of Ohio Hopewell are considerable, and they make up a major portion of the rest of this book, starting with part 2 in the following chapter.

Table 5.1 Complementary Congruency Hierarchy

Hierarchy	Complementary Opposition	
	X : (is to)	*Y :: ((as)*
1 Cosmos	Heavens :	Underworld ::
2 Vertical Order	Above :	Below ::
3 Natural Elements	Surface Stratum :	Deep Stratum ::
4 Earthwork	Circle :	Octagon

Part 2

AN IMMANENT, SACRED
DEONTIC ECOLOGY

Chapter 6

ECOLOGY, COSMOLOGY, AND SOCIETY

The answer to the question of what caused the "demise" of the Hopewell must rely on nonsubsistence factors or entail a better, more subtle, understanding of the interaction of the subsistence base of human populations with other facets of their culture.

—Dee Anne Wymer, "The Middle Woodland–Late Woodland Interface in Central Ohio"

Dee Anne Wymer makes these insightful concluding remarks in her comparative analysis of the botanical contents of several pit features containing the residue of foraged and cultivated edible seeds from two Middle Woodland and two early Late Woodland sites of central Ohio. Her analysis initially addresses the contradictory claims that, on the one hand, the introduction of maize agriculture in the later Early Woodland instigated the emergence of Ohio Hopewell, and, on the other, its introduction in the later Middle Woodland led to the demise of Ohio Hopewell. By pointing out that no maize was present in any of the pit features, her analysis dispenses with both claims. In fact, it was only in the later Late Woodland, ca. A.D. 800, that maize became an important staple crop, well after the termination of the Ohio Hopewell episode. It is now a widely accepted position that, from as early as the Late Archaic, throughout the Early and Middle Woodland, and into the first half of the Late Woodland, the primary plants used for staple subsistence were indigenous, both wild and cultivated, making up what is generally called the Eastern Agricultural Complex.

In parallel with this indigenous mixed foraging and horticultural system, settlement of the Middle Woodland was bifurcated into two contrasting types of locales: (1) the ceremonial locales such as the earthworks outlined in part 1, and (2) small, almost archaeologically invisible dispersed domestic habitational locales, probably small hamlets and seasonal gardening base camps. These were typically located on the valley floodplains, terraces, and hill slopes. While being dispersed up and down the valleys, they were often near one or another of the Complex G-Form sites, which also occupied primarily the middle and higher terraces of the valley bottoms. In contrast, because of their ridge-top location, the T-Form earthworks appear to be generally isolated from these habitational locales. This bifurcated earthwork/habitation settlement pattern was not only largely the extension of the Early Woodland pattern, it probably was initiated as early as the Middle Archaic with the widespread emergence of collective burial locales, often termed "cemeteries," separate from floodplain camps throughout much of the greater Midwestern region. A radical change in settlement pattern marks the emergence of the Late Woodland in Ohio. Starting ca. A.D. 400, this was characterized by the apparent abandoning of the earthwork locales and the "implosion" of the dispersed habitation locales into nucleated settlements in the middle and upper terrace zone, forming what can be appropriately termed a nucleated settlement pattern that apparently had ceremonial zones embedded in the individual settlement.[1]

If Wymer's archaeobotanical findings are an accurate reflection of the subsistence practices of the Middle and early Late Woodland, it follows that, while the Middle Woodland to Late Woodland transition was marked by this rather abrupt bifurcated to nucleated transformation in settlement patterning, the associated subsistence practices remained largely unchanged.[2] As she rightly puts it, this raises a dilemma for the ecological approaches currently used in accounting for the Ohio

Hopewell, which effectively tie settlement patterns to subsistence practices. She suggests that archaeologists either abandon ecological approaches to account for this transformation and, instead, "rely on nonsubsistence factors," or else develop "a better, more subtle, understanding of the interaction of the subsistence base of human population with other facets of their culture." It is the second proposal that this book will pursue.

This means claiming that the problem archaeologists have in accounting for the Middle Woodland-Late Woodland transition arises largely from within the objectivist ecological approach that has prevailed in North American archaeology and this problem arises from the commitment of this approach to treating the material and the symbolic spheres as, in principle, mutually exclusive. This principled view does not deny that the material and symbolic spheres can be integrated. But this view usually privileges the material over the symbolic by claiming that it is the needs of the material sphere that motivate the people to co-opt the symbolic sphere to serve needs that their ecological practices require to be efficient. The way that this view has interpreted the earthworks can serve as an excellent example of this privileging of the material over the symbolic sphere. In this case, the construction of these monuments is deemed to result from the need to ensure resource control and sustain community integration. Thus the earthwork built by a group works as the symbolic community center by housing the remains of the ancestors and this fact simultaneously reinforces kin-group solidarity and legitimizes its claims to exclusive control of the surrounding territory.

This view of the mutual exclusivity of the material and the symbolic spheres is, at heart, the source of the function/style dualism that plagues our understanding of material culture. In contrast, of course, the symbolic pragmatic approach claims that the material and symbolic spheres are irreducibly integrated since the action nature of material behaviors must be symbolically constituted by means of the stylistics of the material culture. It becomes quite logical to apply the symbolic pragmatic perspective to the ecological system in order to account for the Ohio Hopewell episode. This symbolic-ecological perspective is based on the premise that, while the pursuit of survival is a fundamental aspect of social life, because it is a social and not simply an organic life that is being pursued, the ecological strategies are inescapably structured by the cognitive-normative principles and symbolic pragmatic protocols. Therefore, subsistence and domestic practices are no less symbolically constituted than are earthwork-mediated ritual practices.

THE ESSENTIAL CONTRADICTION

In terms of the core cosmological Sacred Earth Principle, the Ohio Hopewell populations pursued organic survival and reproduction in a natural world conceived of and experienced as immanently sacred. All ecological practices necessitate regularly intervening in the natural order and modifying it to one degree or another. But within this immanentist cosmological frame, the pursuit of survival by means of the basic ecological subsistence and settlement practices would be characterized by a profound dilemma: this pursuit simultaneously diminished or endangered the sacred powers of the land and its resources that made organic survival and reproduction possible. This core dilemma can be properly labeled the essential contradiction of human existence.

The essential contradiction immediately makes the emic norms a central aspect of a culture's ecological system. This view will be termed deontic ecology. Deontics are social rights and duties, values, and principles that make up the ethical, moral, and legal dimensions of social life. This means that ecological relations are irreducibly social relations.[3] A deontic ecological perspective would claim that objective factors, such as population density, carrying capacity, and so on, as critical as they are, cannot fully account for settlement and subsistence patterns, a claim that is partly substantiated by Wymer's findings, in particular, that effectively the same intensive gardening/extensive foraging regime was in place under both the Middle Woodland bifurcated settlement system and the early Late Woodland nucleated settlement system. If ecological practices have an irreducibly deontic aspect, and this also means an irreducibly symbolic pragmatic aspect, it follows that significant variation in these practices and the material cultural assemblages that mediated them can arise from ideological variation. This variation would manifest alternative deontic ecological strategies for resolving the essential contradiction of human existence while sustaining largely unchanged the primary deontic principles implicated by the cosmological perspective.

Since the essential contradiction arises out of the dilemma of pursuing survival in an immanently sacred cosmos, the deontic core of ecological strategies would be practical rules and canons appropriate to this type of sacred world. Anthropological study has firmly grounded the view that, in general, the subsistence and settlement practices of preindustrial societies are heavily weighted with both sacred taboos and sacred imperatives that govern how settlement and subsistence must be pursued. Taboos generally are "negative" directives based on proscriptions, what must not be done; while imperatives can be treated as "positive" directives based on prescriptions, what must be done. In an immanently sacred world where the essential contradiction would be felt acutely, subsistence and settlement practices would be weighted with proscriptions that promoted the avoidance of practical behaviors that would disorder nature, or, when this was practically impossible, that promoted minimizing such behaviors. Prescriptive rules would also underwrite ecological practices, these promoting subsistence interventions that were perceived as enhancing the sacred order of nature. Often, the subsistence prescription would motivate a ritual designed to rectify or reverse the sacred polluting effect that an ecological practice that was required to ensure organic survival would necessarily produce. Examples of this abound in the anthropological literature, such as the Native American practice of offering a gift of tobacco to the custodial spirit of the animal just killed. Failing to ensure that subsistence practices realized the relevant proscriptions *and* prescriptions would be serious transgressions that would necessarily diminish the reproductive capacity of the exploited species, whether animal or plant.

Strategies for resolving the essential contradiction would largely dissolve any separation of subsistence and settlement practices, on the one hand, and ceremonial practices, on the other, constituting a single complex deontic ecological strategy in which the instrumental aspects of both spheres would be woven together into complex and ever modifying ideological structures of proscriptive and prescriptive rules, and, of course, the expressive manifesting of these rules would be mapped by the reproduction and variation of complementary material cultural stylistics.

This complex assertion forms the heart of what will be called the Proscriptive/Prescriptive Ecological Strategy Model. Admittedly, this is a mouthful. The justification for this terminology is that it concisely highlights the double duality of proscriptions/prescriptions and material ecological/symbolic ceremonial that will be treated as central to socioecological practices in a world experienced as immanently sacred. This model will argue that, because it was cosmos-sanctifying, public ceremonialism tended to be prescriptive in nature and, therefore, realized through obligatory world renewal acts performed in specialized sacred locales largely according to sacred, celestially determined schedules. As an integrated system of subsistence and settlement practices that had primarily world polluting impact, everyday domestic life tended to be heavily weighted with proscriptions and rectifying prescriptions so as to minimize the level and degree of the pollution/sanctification that the "mundane" pursuit of survival entailed. Such a bifurcated set of interrelated strategic ecological/ceremonial practices can be described as a *proscriptive* subsistence-settlement/*prescriptive* ceremonial system.

While the Proscriptive/Prescriptive Ecological Strategy Model has particular relevance for the Ohio Hopewell/Middle Woodland period, with appropriate modification, it will also be used more generally to examine the predecessors of the Ohio Hopewell. This historic perspective is necessary in order to analyze its roots, effectively treating the earthwork/habitation dichotomy referred to above as the culmination in Ohio of a particular historical mode of resolving the essential contradiction. It will be argued that this deontic strategic mode was initiated in the Middle or even Early Archaic. To support this claim and to fully elucidate this model for application to Ohio Hopewell, it will be necessary to give an overview of the prehistoric record of Archaic and Woodland subsistence, settlement, and ceremonial systems in the Midcontinent. This will be done following the more general deontic ecological theory that informs this model. This general theory will be presented as the Inclusive Territorial/Custodial Domain Paradigm.

DEONTIC ECOLOGY—A PRELIMINARY SKETCH

The deontic aspect of social life rarely figures in standard human ecological approaches. These speak of subsistence and settlement practices as the ongoing realization of practical strategies. Economizing premises are used to model the practices on the grounds that humans are rational calculators and will attempt to satisfy material needs in the least risky, least costly, and most gainful manner.[4] Avoidances and preferences certainly figure in these calculations. However, these are not deontic in nature since they are treated as the desired survival outcomes achieved by rational calculation.

A deontic ecological approach can assimilate this practical rationalism because it works from the same premise that humans are rational beings and will use practical reasoning to calculate their costs, gains, and risks. However, a deontic ecology requires the addition of another pursuit that is no less crucial than subsistence, this being the pursuit of social survival. Social survival entails deontic structures as these rules and protocols are the basis of social relations. Indeed, most ecological perspectives now recognize that a human community must accommodate to both natural and social environments. But material/symbolic dualism again instrumentalizes the deontics of inter-community relations. Hence, the symbolics of mortuary practices are treated as being co-opted as a political medium to achieve the ecological needs of preserving exclusive control of territories and forcing neighbors to recognize the deontic rights to this land inherited from the buried ancestors.

A deontic ecology takes a more balanced view, recognizing that the pursuit of group reputation is as important to community survival as is the pursuit of material goods. The reputation of a group hinges on the assessment that other groups make of the moral qualities and character of the target group.

> (The) pursuit of reputation in the eyes of others is the overriding pre-occupation of human life, though the means by which reputation is to be achieved are extraordinarily various. Though men compete individually for honour, reputation is a corporate matter and its acquisition a co-operative achievement. It is the production of the recognition of one's worth by others. (Harré 1979, 3–4)[5]

Should the target group fail to sustain a viable social reputation in the eyes of its neighbors, this is bound to lead to confrontations that deteriorate relations and undermine the pursuit of organic survival. Reputation will be treated as the focal concern of social relations, and, of course, it is based on the ongoing monitoring and assessment that groups make of each other. All reputation hinges on presenting the proper appearance. Appearance means manifesting through the expressive dimension the appropriate intentions, motives, beliefs, and so on, in short exercising mutually comprehensive symbolic pragmatics. This requires "universal" standards, both conventions and the beliefs and motives that these conventions can manifest. The standards on which the central worth of others is assessed derive from core collective knowledge about the world, cosmology. Therefore, only through being seen to exercise the appropriate intentions in hunting or gathering will these ecological behaviors be constituted as proper, warranted appropriation of valued resources. When constituted in this manner, the doers are marked as worthy and trustworthy, as "true" hunters and gatherers. If they fail to manifest the appropriate intentionality, say through using tools displaying unrecognized styles, their subsistence behaviors will count as "poaching" and "pilfering," constituting the character of the doers as disreputable and untrustworthy.

The notion of environmental stress plays an important role in ecological accounts. This is also the case in deontic ecology, except that

it is the perceived ecological stress that is important. This means assessing the nature and level of the stress that objective changes make
from within the emic cosmological/ideological framework, in this case,
an immanentist framework that is grappling with the essential contradiction of life. Taking an emic perspective means that communities
having the same overall objective needs for survival but essentially
different cosmologies (e.g., one being immanentist and the other transcendentalist) would have different standards by which to assess environmental events or processes as counting as environmentally stressful
or polluting. What would count as polluting for the community with
the immanentist cosmological/ideological perspective might be perceived by the community with transcendentalist perspective as enhancing. For example, early settlers in North America perceived clear-
cutting the forest as enhancing the productive capacity of the land, an
assessment that, in general, was not in accord with that of the local
Native Americans.

It is not difficult, then, to assimilate the practical rationalist perspective to a deontic ecology. However, this assimilation raises the
point that, for example, in objective circumstances of demographic
and/or environmental stress, survival is not simply a matter of expanding labor to maximize gains and minimize risk. As pointed out
above, in a world experienced as immanently sacred, a community
would draw on their traditional proscriptive and prescriptive deontics
in making the necessary accommodation to these changing material
conditions. It follows logically that a community would avoid labor
that would transgress the sacred boundaries and if transgression was
impossible, they would innovate the symbolic ceremonial dimension
to rectify the unavoidable disorder that the pursuit of survival required,
even if this meant intensifying labor in this area. These greater ceremonial labor costs would be treated as sacrifice, and such sacrifices
would count as reputation enhancement.

Hence, their innovation is both material and ideological. The
ideological changes would be two-fold: additions and changes made to
the proscriptions and prescriptions of everyday subsistence and settle-

ment routines, and additions and changes to the regime of collective ritual and ceremony. The former will be termed midwifery ritual; the latter renewal ritual.

A. Midwifery Ritual

If an immanently sacred world entails that all material interventions, including those forms necessary for survival, are irreducibly polluting, then the essential contradiction would be impossible to resolve. Therefore, the working of practical reason in such a world would promote construing the destructive moments of subsistence and settlement as having a sacrificial as well as an economic purpose. Subsistence and settlement interventions would be experienced not only as the destructive means of the practical pursuit of survival but also as part of the everyday means of participating in the reproduction of the very species being exploited. The success of the latter would be dependent on the incorporation of midwifery proscriptions and prescriptions. The term "midwifery" seems appropriate here since the point of this ritual would be to ensure the rebirth of the exploited species. Usually seamlessly built into the subsistence and settlement practices, these rituals would ensure that the destructive moment of exploiting animals and plants was an intrinsic part of the reproduction of these species through exercising practices that had reproductive consequences, thereby implicating such beliefs as spirit release, human-guardian spirit reciprocity, sacrifice, and so on. Effectively, practicing subsistence and settlement midwifery ritual in pursuing human survival transforms the essential contradiction of human existence into a moral imperative by enshrining warranted exploitation as part of the human way to fulfill sacred ecological duties. Performed as warranted killing, hunting would be the mode by which humans released the spirit of the animal so as to allow it to be reborn. Similar perspectives would underwrite the exploitation of plants. Without humans regularly exploiting these species, the species themselves would not be reborn. Immanentist cosmologies, since they ground the essential contradiction of human existence, therefore, tend to promote the view that human subsistence practices are not the mode of ending the reproductive cycle of the animals and plants but are an intrinsic and sacred part of the reproductive cycles of these very species.

There is considerable anthropological support for this view. With respect to circumboreal hunting and gathering societies of both Eurasia and North America, Ingold has demonstrated that hunting entailed having reproductive midwifery ritual built into the exploitative processes. He has argued that in both continents hunting ritual largely identifies the hunting gear, particularly the weaponry, as having a strong sexual-reproductive symbolic pragmatic meaning.[6] Robert Hall makes a similar claim for historical Native American peoples, arguing that subsistence and war weaponry was viewed as being endowed with sacred reproductive powers.

> The tie between hunters' and warriors' penetrating weapons, the sun father, agricultural fertility, increase of human populations, and animal increase or success in the hunt was both metonym and metaphor in conception but was more than poetic imagery. (Hall 1975, 510)

It is notable that, while most circumboreal peoples share the same type of immanentist cosmology, they do not share the same subsistence practices (many ideological details of subsistence midwifery practices in different societies differ from each other, but the cosmological principles are common). North American circumboreal peoples were, and many still are, hunters and gatherers. Many Eurasian boreal peoples were and still are reindeer herders, supplementing their domestic herding with hunting and gathering. Therefore, immanentist cosmologies can be held by both hunter/gatherers and domestic producers. Finally, Ingold suggests that this notion that the destructive moment of exploitation is simultaneously a reproductive moment can be generalized to characterize much of the attitude of all foragers toward the plant and animal species they exploit.[7]

Sustained by the empirical demonstration of the Sacred Earth Principle, proscriptive and prescriptive midwifery ritual will be postulated as basic to the prehistoric Eastern Woodlands. Midwifery ritual

hinges on the view that humans and the spirit guardians or spirit custodians of the animals and plants have a reciprocal relation in which
hunting and gathering mediate mutual gifting. This means that subsistence has a built-in sacrificial moment. The body of the animal is a gift
from the spirit custodian to humans, who use the resources of this
body according to proper prescriptive and proscriptive etiquette. One
possibly important area of subsistence midwifery ritual would involve
the disposal of "waste." The use of selected bones of the animal might
be proscribed, but also, their disposal in a manner so that they will be
available to be reanimated and thereby reborn is prescribed. Disposal,
then, becomes a form of "burial." This suggests that while humans are
concerned that others perceive their behavior as the type and quality
of act they intended, they are also concerned that the spirit custodians
"see" it this way. Therefore, among the most important midwifery
ritual elements of these routines would be the use of the appropriate
tools, those that display the proper styles that constitute them as warrants in the full richness of this term.

The emic perception that subsistence and settlement practices
were intensifying would entail innovating midwifery ritual. One manner of doing this would be to innovate and/or emulate a new stylistic
dimension, adding it to traditional tools and prescribing this usage. This
might hinge on acquiring exotic cherts perceived in a vision as having
special iconic associations with the spirit custodian of the animals or
plants being exploited, thereby enhancing the sacrificial moment of
subsistence. Tools would come to bear new, more potent warranting
styles. To further enhance this warranting moment, associated with
these new styles would be new proscriptions related to their use (e.g.,
only by males of a certain age, or only by married females, and so on).
Even modes of midden might be innovated to enhance the spirit release
and recycling effect. Clearly, if subsistence practices are intensifying,
this would have an impact on settlement practices. Thus, compensatory
avoidances or proscriptions might be innovated that would encourage
minimizing the concentration of settlement numbers in any given area,
even while subsistence intensification increased. This might lead to

promoting the dispersing of the traditional seasonal aggregations to smaller temporary campsites within a short walk of each other, thereby spreading over a larger area the increasing pollution that more extensive aggregations would cause, and reducing the pollution load per unit area.

B. World Renewal Ritual

Midwifery ritual is hardly the whole ritual story. Although holding to traditional practices, and innovating them when required, can ensure that the midwifery ritual role is effectively discharged, there is the inevitable cumulative disordering of the sacred world that arises by the collective nature of human settlement and subsistence. Collective ritual of a world renewal nature would be linked to subsistence and settlement practices as part of reversing the pollution produced by the social pursuit of organic survival. Therefore, if changing objective conditions required intensifying the degree of exploitation and reducing the degree of mobility, then the logical tactical compensatory move would be to intensify world renewal ritual. Thus, world renewal ceremonialism would escalate in proportion to the intensifying of subsistence and settlement. Just as the prescriptive use of new warranting tools and the proliferation of proscriptions concerning the concentration of population in different areas would be embedded in the subsistence and settlement routines, minimizing the degree of material intervention required in public ritual innovations would also apply. Such minimizing tactics would encourage "doubling up" of ritual practices. Important preexisting forms of communal ritual, such as mortuary practices, could be co-opted to serve both mortuary goals and world renewal rites. Piggybacking one on the other would minimize the overall degree of material disturbance of the natural order that these two ritual forms would require if carried out separately while having an overall consequence of intensifying world renewal. Integrating these two would constitute a prescriptive *mortuary/world renewal ritual regime*.

The spiritual powers of the deceased—immanent in and differentiated across the different components of their material remains:

blood, skin and flesh, skull, long bones, thorax bones, and so on—
would be akin to the power of the multiplicity of the spirit custodians
that collectively animated the cosmos. Mortuary practices characterized
by the manipulation of the deceased through a series of incremental
rites would count as compound mortuary-social-world renewal rites.
This means that the deceased become critically important symbolic
capital of world renewal rites. Since world renewal ritual had to be
regularly performed, this meant that the bank of deceased, as symbolic
capital, would be continually drawn upon and transformed through
rites performed by means of post-mortem manipulation. The deceased
would come to be subject to multiple and incrementally staged rites
mediated by post-mortem manipulation, each step "unfolding" a different range of immanent powers, the spirit of the personal name, the
spirit of the flesh, the spirit of the skull, then the upper bones, lower
bones, and so on, each serving as symbolic warranting devices of prescribed mortuary/world renewal rites. Since the spirits of the deceased
and their different components are central elements of ritual gifting to
the cosmos, then mortuary rites become sacrificial rites. I will characterize this sequential staging of post-mortem mortuary events post-
mortem sacrifice.

Thus, while the initial motive for integrating world renewal and
mortuary practices would have been to reduce the degree of intervention into the natural order that both rituals would involve if done
separately, in time the use of the deceased as potent post-mortem sacrificial media would lead to this integration being perceived as both a
necessary mode of world enhancement and a positive mode of reputation enhancement. Therefore, the expansion of the world renewal
aspect of mortuary practices can be characterized as a prescriptive ceremonial tactic by which to redress the cosmic balance that the intensifying of collective subsistence and settlement practices was upsetting.
The effect on the archaeological record would be the emergence and
escalating development and elaboration of collective burial locales
(CBLs).

DEONTIC ECOLOGY AND TERRITORY

A core aspect of any ecological approach is territory. Territorialism is usually defined as the processes by which a group sustains access to a defined range or "country," usually referred to in the literature as its territory. Territorialism implicates two critically important requirements: informational and social. To have effective practical access, ongoing information about the variety and availability of the resources is important, particularly for foraging groups since many of these resources are mobile and seasonal. Implicated in this information flow, both in terms of gaining it and of being able to act on it, are the social dynamics that allow this information to be distributed and acted upon. In deontic terms, social structures related to territory constitute tenure, these being the rights, duties, and obligations one has towards the land.

There are two faces to tenure: internal and external. The internal face addresses the usage rights and obligations of the resident group that habitually exploits a range or a territory. The external addresses the relations among similar territorial groups. What deontic principles of tenure would characterize a group's assured rights of usage of a given range? The current North American archaeological literature generally postulates the notion of exclusive rights, or the exclusive territorial tenure view. This claims that a group sustains access to its habitual range and the resources that it contains by claiming rights to exclude nonmembers.[8] This has important implications. In terms of the inner face, an exclusive territorial group secures commitment from its members by ensuring access only to members in good standing. In terms of the outer face, this means that there is collective duty to cooperate in excluding nonmembers, strangers, from the use of the group's range. Also, members are aware that, unless special relations with other groups have been established through alliances, they have no right of access to the resources of neighboring ranges, even if their subsistence needs are pressing. To one degree or another, nonmembers are "strangers," "foreigners," often marked by using improper tools and related artifacts, and, therefore, they are potential or actual competitors. This view logically leads to characterizing the external relations as being negatively charged, particularly under perceived conditions of increasing stress on the resources.

This exclusive territorial perspective assumes that, for an indeterminate period of prehistory, human populations were organized as small and highly dispersed groups. Without the need to compete, humans were largely or totally unfettered by deontics of tenure. The need to establish tenurial claims emerged, therefore, only when either environmental deterioration or demographic packing, or both, developed. Resources become scarce or are perceived as being scarce. This means that tenure as a deontic phenomenon is treated as essentially a "natural" adaptive response. Humans are "naturally" aggressive and will compete in conditions of scarcity. This also means that tenurial deontics come to be largely identified as nothing else but the rights and duties of exclusion. Hence, the group becomes a collectivity that owns a range of land deemed its exclusive territory. This is often referred to in the literature as a proprietorial corporation and since the basic social structure that binds the group is deemed to be kinship, it is also often referred to as a corporate kinship group.

This naturalizing of tenure is unconvincing. It distorts the cultural and social reality of Native North American prehistory, in particular, and of foraging societies in many other world regions, and it does so by universalizing the Western notions of property and imposing them on the deep past. The view promoted here expands the picture. First, it argues that the deontic principles of tenure governing human-environment relations are an intrinsic part of the human-human relations of any symbol-using human population, and this would certainly include any human populations from at least the emergence of the Upper Palaeolithic. In these terms, human populations would be constrained "from the beginning" by the deontics of tenure and these would be manifested in the symbolic pragmatics of material culture realized as style.[9] Second, since the deontics of tenure are part of the deontics of human interaction, the access to resources would be based on the same principles as those organizing the distribution of resources in foraging societies, these being sharing and reciprocity. In particular, sharing is widely recognized in the anthropological literature to be the defining characteristic of foraging cultures. As portrayed here, then, prehistoric foraging was based on overlapping and shared territories "from the beginning," and the limits of use on these territories would be largely determined not by any exclusionary deontics but by the straightforward practical needs and capacities of the cooperating resident groups.

This practical limitation does not manifest the absence of tenure but its realization. In this case, tenure is based on the principle of territorial inclusion, which is here taken to be the core of sharing and reciprocity. In this perspective, it is logical to assume that, even should conditions of environmental stress and instability of resource availability emerge, the most likely social response would be to extend the principle of sharing by rationally sharing the resources. This would entail modifying traditional subsistence and settlement practices in order to accommodate to these new constraints while sustaining commitments to traditional inclusive tenure.

This is not to deny that exclusive territorialism is a historically known deontic posture. Rather, it claims that human history constitutes a continuum in this regard. In some historic epochs, the principle of exclusion prevails over that of inclusion, applying equally to real estate and personal property. In others the two principles of exclusion and inclusion might be more or less balanced, with different territorial and personal property regimes operating simultaneously, sorting into communal and private resource zones and portable objects. In others, possibly most characteristic of those epochs when foraging was dominant, the principle of inclusion prevails, with exclusive control largely limited to certain portable artifacts. The total continuum is deontic and, therefore, different positions on it are equally cultural constructs. This book will focus on the two extremes. It will promote the view that the immanentist cosmology claimed here entails that the Ohio Hopewell earthworks be seen as correlating with inclusive territorial tenure.

It will actively counter the other extreme, a position that largely prevails in the current archaeological literature, this being that exclusive territorial tenure emerged in the Middle Archaic and became progressively dominant as mobility became reduced and cultivation became more prevalent.

PROPRIETORIAL AND CUSTODIAL DOMAINS

The principle of exclusive tenure constitutes a territory as a proprietorial domain and the group that is responsible for this domain as a proprietorial corporation. Proprietors may, of course, allow nonproprietors access. However, this simply reinforces proprietorship since warranted access by outsiders to the domains of others involves negotiated exchanges and all parties simultaneously recognize each others' proprietorial rights. Collectively, the living members of the proprietorial corporation make up the resident community. If kinship is the basis of membership, they inherit their rights from those who came before, the ancestors. The living generations are collectively the stewards of the corporate property. As descendants of the ancestors, the stewards are under obligation to pass on the domain to the stewardship of the next generation, and so on. Proprietorial corporate structure is symbolically constituted through collective activities, such as mortuary practices, whereby the reputation of a group is recognized, its legitimacy of tenure is reconstituted through recognizing the ancestors, and the control of its territory is maintained. Because of the scope and generality of this perspective it seems appropriate to call it the Exclusive Territorial/Proprietorial Domain Paradigm.[10]

The principle of inclusive tenure constitutes a territory as a custodial domain and the group that is responsible for this domain is a custodial corporation. The notion of foraging societies having territories based on custodial tenure is thoroughly theorized by Ingold.[11] As he puts it, there are

> numerous instances from the ethnographic record of hunters and gatherers in which there is public recognition of demarcated territories but no conception at all of trespass in the form of taking resources from another's area. There can be no such offence as poaching when intruders

have the same rights of access as anyone else. If a transgression occurs, it lies in not having asked before taking, not in the taking itself. Though territorial boundaries are generally open to movement across them, such movement should be publicized and not concealed; one does not arrive or leave without advertising the fact. . . . The intentions of persons who do attempt to conceal their movements are bound to be suspect: the suspicion, however, is not that they are out to raid the territory's food resources but that they are planning an attack on members of the resident group. This suspicion, in turn, motivates sometimes violent retribution meted out to such intruders. (Ingold 1987, 143–44)

Having territories based on inclusive rights seems to be a contradiction in terms. Such territories would be used much as common land. However, this does not deny territorialism as habitual ranges. That is, these ranges are defined largely on the basis of practical considerations, and the resident group is the group that is familiar with a given territory because of its habitual use of it. Since under this regime no party can own environmental resources, what this usage amounts to is simply the right to distribute the procured resources to the members of the community in terms of traditional rules of sharing. Such rights presuppose familiarity with these rules and protocols that determine what will count as an equitable distribution. Calling this the Inclusive Territorial/Custodial Domain Paradigm seems quite apt, and, because the premises articulated by this paradigm form the core of the Proscriptive/Prescriptive Strategic Ecological Model of Ohio Hopewell, the latter will be treated as a global model for applying the paradigm to the archaeological record of the Early and Middle Woodland periods of the Central Ohio Valley.

However, as indicated in the above quotation, it is clear that Ingold does not realize that an inclusive territorial system could also have its own brand of "poaching and pilfering." This may be partly because he does not address the implications of material cultural style for constituting territorialism. He is not an archaeologist and is not attuned to this aspect of the archaeological record. For this reason, he appears not to have considered the possibility that, even though under inclusive territorialism access to resources is largely a matter of practical distances and information, it also can be constrained by the style of the tools. Indeed, in a world of proprietorial domains, tool styles are as much "birth certificates" as they are hunting "licenses." Knowing the right styles endows the users with the "birth right" to occupy the territory and, therefore, to exploit the resources. In a world of custodial domains, however, where inclusive access to resources prevails, the role of style would be to constitute subsistence tools as midwifery warrants. In a regime of inclusive territorialism and custodial domains, then, *contra* Ingold, stylistic variation that is perceived as outside the norm would be a legitimate basis for denying these others the requisite information to exploit the resources and even the (often) sacred knowledge needed to use the paths and locales used by the local group. The primary grounds would be that those entering the territory must demonstrate that they can conduct themselves properly. If they are neighbors, their styles will tend to correspond to those of the resident group and recognition of the legitimacy of their subsistence intentions will be largely automatic. If the styles of visitors are patently different from those used by local populations, until they can demonstrate that these express the same intentionality as those used locally, their actions could be legitimately denied recognition as fully felicitous forms of resource appropriation.

In her research among the San of Southwest Africa, Wiessner showed separate groups of !Kung, G/wi, and !Xo adult males the arrows used by the others. "A discussion ensued from one small group (of !Kung San) about what they would do if they found a dead animal with such an arrow (G/wi or !Xo styled) embedded in it in their own area, saying that they would be worried about the possibility that a stranger was nearby about which they knew nothing at all."[12]

What is particularly interesting is that it was the possible conduct of the person responsible for an unrecognized arrow that most concerned them, this person being assumed to be a stranger. "Although afraid of !Kung strangers as well, they said that if a man makes arrows in the same way, one could be fairly sure that he shares similar values around hunting, landrights, and general conduct."[13] From this Wiessner concluded that style served to delineate ethnicity. However, ethnicity is tightly tied to exclusivity. Stylistics under an inclusive regime would figure primarily as expressing legitimate appropriative intentions by which to communicate not only that the users of the styles know how to conduct themselves properly toward the resources that they exploit, but that they also know how to conduct themselves toward their hosts: they know what would count as sharing.

Accepting custodianship means accepting the obligation to ensure that the land and its resources benefit rather than suffer from exploitation. Ensuring that visitors conduct themselves appropriately is a heavy responsibility, and the styles that these visitors bear would be the expressive signs by which local people could make initial character assessments. Nor does this mean a standardization so that only one style counts. Rather, it means that different style bearers must demonstrate to each other the equivalence of their particular styles by participating and sharing in collective tasks. When it becomes clear that those bearing strange styles nevertheless know—or have learned—the proper rules of conduct toward the resources they are exploiting and toward their hosts, then the styles will be recognized as alternative warrants. Out of this might come a sharing of styles and, finally, mutual emulation and, possibly, transformation into new stylistic variants.

THE DIMENSIONS OF TERRITORIALISM

Ingold analyzes territory along two-dimensional space, one-dimensional path, and zero-dimensional place. He then argues that, because farmers and pastoralists must have predictable use over a given area of land, they combine all three dimensions to constitute their territories. These manifest differential values, and are structured or ranked in accordance to ecological needs. Two-dimensional space becomes primary and "fixed." Places and paths are important and must be guarded. This structuring of space constitutes exclusive territorial modules. In contrast, largely because foragers typically move from place to place seasonally, path and place become the focus. Hence, the habitually used network of one-dimensional paths and zero-dimensional locales becomes important and structured to constitute custodial domain, while two-dimensional space remains non-exclusive, open to everybody.

The network of custodial paths weaves across the landscape linking custodial places into a ramifying network of custodial domains. The paths are often taken to be routes that the original ancestors or the creator deities walked, and the places are where they stopped and performed acts of creation. They could also be locales where a powerful visionary encountered a guardian spirit, and so on.[14] Specific networks of paths and places are identified as the domains of the groups that claim special relations to the powerful entities that formed them. These groups collectively bear the particular knowledge required for the appropriate use of the paths and locales, and the practical information needed to effectively exploit the resources distributed across the land and to which proper use of the domain paths and locales gives warranted access.

Since these groups are collectively responsible for the care of the custodial domain, the members constitute custodial corporations and one of the primary duties of the custodial groups would be to carry out the appropriate renewal rites at their sacred locales. These would be experienced as being part of their cooperative interaction with neighboring custodial groups, who would also carry out similar performances in their sacred locales. The effects would ramify through the network of paths and locales to all parts of the cosmos. Such a collective experiencing of land, path, and locale would promote and be sustained by inter-domain cooperation at many levels.

ALTERNATIVE SOCIAL SYSTEMS

As pointed out earlier, the Middle Woodland settlement pattern was bifurcated into earthwork ceremonial locales and dispersed domestic locales. This tendency has a deep history in the Eastern Woodlands, showing up by the Middle Archaic with the formation of mortuary locales separated from domestic locales. The currently dominant account of Ohio Hopewell, generally termed the Dispersed Hamlet/

Vacant Center Model, treats this bifurcation as the manifestation of proprietorial kinship groups having the Ohio Hopewell earthworks as their specialized funerary centers, thereby clearly situating this account within the Exclusive Territorial/Proprietorial Domain Paradigm. The region around each earthwork center is treated as an exclusive territory forming a proprietorial corporate module.

In accordance with classic central place theory, the boundaries of a territorial module are postulated by drawing lines to connect up a set of neighboring centers, these representing the linking routes and paths, and then by drawing a second set of imaginary lines vertically crossing these paths at the halfway points. Connecting these halfway points generates contiguous polygonal modular territories.[15] While the lines connecting central ceremonial locales schematically represent real paths, the lines that cut vertically across the latter at the halfway point represent culturally constituted and sanctified perimeter boundaries of the postulated proprietorial domains of the region. In this view, these boundaries count as critical deontic claims that are symbolically contested by the placement of the ceremonial locale in the "epicenter," constituting these earthworks as the social "centers of gravity" of the proprietorial corporate kin groups. The members of these corporations are postulated as living in the dispersed habitational locales clustered loosely around them. The premise of exclusive control of land logically predicates to each the status of a self-governing, autonomous territorial module constituting a proprietorial domain. The sanctity of the boundaries of a corporate module is symbolically constituted via ancestral funerary practices performed regularly in the central ceremonial locales. The modules of the corporate domains of a local region are postulated as interacting with each other as autonomous peer polities, constituting what has been called a peer polity system.[16]

It is notable that, even though habitational and ceremonial locales are bifurcated, the social structure of these modules is treated as unitary or monistic. The classic type of monistic social system would be one based on a single dominant social structure, typically unilineal kinship. Other structural dimensions, such as gender, generation, artisan spe-

cialization, religious specialization, and so on, are subsumed and embedded as subdimensions of the dominant unilineal kinship structure.
The articulation of the social settlement typically takes on the character
of a nesting of incrementally higher kinship units. The domestic habitation group, usually one or more nuclear families linked by close
kinship, forms the largely day-to-day self-supporting economic unit. Its
access to land, resources, and the sacred center would be dependent
on its position within the nested kinship structure of the corporate
domain which makes up the encompassing, territorial module, e.g.,
members of lineage A→clan B→moiety C→tribe D. While usually
taken for granted, it is important that this monistic aspect be made
explicit since it is the core assumption that is shared by the various
and sometimes contradictory ways that the bifurcated settlement pattern of the Ohio Hopewell is interpreted in much of the current literature.[17]

It will be argued here that a monistic social system based on
exclusive territorial modules is inimical to inclusive territorial tenure.
Although in the latter case the basic domestic groups can consist of
nested kinship components, as in the monistic modular system, and
the habitual range of any given nested set of components would tend
to be polygonal in pattern, these polygonal patterns would result from
the practical constraints. This means, therefore, that under inclusive
territorialism the habitual foraging ranges of neighboring kinship-based
custodial groups would have significant overlap, marking shared use.
Furthermore, given inclusive territorialism, kinship would not necessarily subsume or monopolize the social structural axes that articulate
to constitute the total social system. The freeing-up of these multiple
structural axes would allow the formation of companionship-based
groups socially constituted along one or more of these social structural
axes. Hence, groupings of companions based on gender, generation,
specialization, to suggest the most likely axes, might typically be articulated as separate autonomous groupings within the same social system, constituted as military sodalities, religious cults, initiation groups,
and so on. Each grouping would have its own "constitution" of pro-

tocols expressed in its origin myth, its own history, and its own raison d'être. These alternative and parallel groups would also constitute custodial organizations, but ones in which access to each others' sacred locales through sacred paths rather than to two-dimensional foraging zones would be critical. In short, inclusive territorialism and custodial domain would allow for the emergence of rather simple social organizations articulated into complicated patterns of domestic and nondomestic sacred locales linked by networks of sacred paths.

It is postulated that, while proprietorial domains tend to correlate with *monistic modular* social systems, custodial domains tend to correlate with *polyistic locale-centric* social systems. To call this social system "polyistic" is to emphasize the disembedded nature of the social structural axes making it up, and to call it "locale-centric" is to note that all the places or locales of the custodial groups of a particular structural type in a region will be knitted together into a largely seamless social network as seen from the perspective of each respective locale. Each locale would have its particular sacred nature and it would "enfold" the totality of the network of similar locales. A regional social system based on an immanentist cosmology and inclusive territorialism, therefore, has the *potential* of being constituted as a complicated articulation of multiple parallel but autonomous networks. This means that while each group and its locale in a regional system is autonomous, no social system is a discrete, boundaried entity. Rather, its extension is the sum total of the extensions of the different types of social networks that make it up.

I will qualify the above by invoking the contingent nature of the cosmology-ideology relation. Cosmology does not determine social organization. A social system manifesting an immanentist cosmology could be constituted of monistic locale-centric communities. However, the principle of inclusive territorial tenure, reinforced by an immanentist cosmology, would effectively preclude boundaried, modular social systems, and, under the conditions promoting subsistence and settlement intensification, polyistic systems would be likely, typically realized as different custodial groups occupying different types of locales according to the type of ritual and material responsibilities traditionally claimed by each type. These locales could be occupied according to the particular schedules structured by the requirements of each type of group, subsistence domestic, world renewal cult, shamanic cult, military cult, and so on. Indeed, the Proscriptive/Prescriptive Ecological Strategy Model postulates that the Early and Middle Woodland period social system of the Central Ohio Valley can be understood as largely the culmination in this particular region of this type of social system having its roots in the Middle and, possibly, Early Archaic. A major auxiliary model of the Proscriptive/Prescriptive Ecological Strategy Model will be presented shortly as the contrast to the current Dispersed Hamlet/Vacant Center Model of Ohio Hopewell.

CUSTODIAL DOMAIN IN GARDEN-BASED SUBSISTENCE SYSTEMS

There is a modification that must be made to Ingold's territorial perspective. He correlates custodial tenure with hunting and gathering societies and proprietorial tenure with farming and herding societies.[18] This would suggest that locale-centric and modular social systems, respectively, would be correlated with these ecological differences. Certainly, when applied to the Old World societies, there is considerable empirical support for this view. However, this correlation may be overly deterministic in ecological and economic terms. While this is not to deny Ingold's view that farming and herding will promote exclusive territorial tenure, it does not mean that this type of tenure is entailed by farming and/or pastoral systems. To say that effectively ignores the cosmologically informed deontics that give economic and ecological practices an ideological dimension. Of course, it is also the case that cosmology does not determine particular ideological practices, but any ideological practices will be constrained by the nature of the cosmology that makes these practices possible. The essential contradiction of human existence that is claimed here to be endemic with radical immanentist cosmologies suggests that strong proscriptions and prescriptions will tend to be sustained despite a shift to a less mobile way of life based on small-scale gardening.

How could a regime of custodial domain contend with the "exclusive" imperatives of gardening? That is, despite the immanentist cosmology, would a shift to gardening (or agriculture) necessarily transform a preexisting custodial domain regime into a proprietorial domain regime, or could the former be modified to sustain its traditional commitments? I will argue that the latter is possible—indeed, it is very likely. Even proprietorial domain allows for qualified or quasiexclusivity that is characterized by strong proscriptions and prescriptions imposed on the quasi-owners, although the source of this deontic burden is the proprietorial power of the owners, the landlords. Renting is a strongly qualified use of land, thereby constituting the renters as tenants who have a qualified set of exclusive rights termed usufruct. It is useful to be explicit here and specify this usufruct as tenancy. The deontics of tenancy go beyond land usufruct to reinforce social and political subordination. However, as long as tenants fulfill their economic, social, and political obligations to the proprietors, they have first rights of use of the land. In some cases, of course, this qualified exclusivity can be enhanced by first rights (and obligations) to tenancy being inherited by descendants of the tenants.

If proprietorship can be accommodated by tenancy to the needs of landlords to subordinate and exploit labor and to the need of the landless to survive by evolving a strongly qualified quasiownership, then a modified form of custodianship accommodating to ecological needs could be anticipated for custodial regimes. Therefore, it is quite conceivable that an emerging farming and/or herding regime could generate a form of qualified inclusiveness within a basically custodial regime. This might be appropriately termed custodial usufruct. Custodial usufruct would only mimic tenancy since it would be less likely to allow for the political and economic subordination that seems inherent in the latter. In an immanently sacred world, custodial usufruct as a qualified inclusiveness is perfectly consistent for farmers and/or herders. They cannot exclude others. However, those who habitually use a local range of land for gardening have the first rights to its use as long as they have greater need for it than others. In keeping with the principle of inclusiveness, nonresidents would have rights of gardening use to land outside their habitual ranges. However, they could gain these only as long as the land was not being needfully exploited by those already exercising custodial usufruct. Should increasing demands for land usage emerge (e.g., through immigration), given the principle of inclusive territorialism, the resident custodial gardening group could not simply deny this immigrant group use of its paths to gain access to the gardening resources. It would be pressed to act justly by sharing, thereby negotiating and reallocating garden usufruct. This would probably result in more intensive exploitation of resources, of course, and it would promote compensating innovation in ritual, as will be discussed shortly.

In a recent discussion of Native American agriculture in a historic context, Hurt (below) implicitly acknowledges the existence of custodial usufruct. Since he is speaking from within the western cultural proprietorial framework, however, he claims that different levels of social categories "owned" different gardens: for example, families "owned" gardens within their lineage fields, lineages "owned" fields within the village's (proprietorial) territory, and so on. But he then states that this "ownership" was qualified by need. Only as long as a family needed the gardening land would it have its rights of use sustained. He even quotes Chief Black Hawk to this effect: "My reason teaches me that land could not be bought or sold. The Great Spirit gave it to his children to live upon, and cultivate as far as necessary for their subsistence, and so long as they occupy and cultivate it, they have the right to the soil but if they voluntarily leave it, then any other people have the right to settle it."[19]

This translation has to be treated carefully, of course. As in any intercultural case, even the apparently clear statement "The Great Spirit gave it to his children" has to be understood in cultural context. It is likely that Chief Black Hawk was not referring to a specific group but to humanity in general. Furthermore, Chief Black Hawk qualified his assertion by saying that "land could not be bought or sold." This terminology of buying and selling land is necessarily situated within the proprietorial perspective. Therefore, the fact that Chief Black Hawk used the negative suggests that he was actually expressing the notions of custodial domain. It is not surprising that Hurt attempts to make sense of this negation within the proprietorial framework to which he is culturally familiar by invoking the principle of trusteeship, another term for stewardship. In his words: "Indeed, Indian land could not be sold because it did not belong to the present generation, which was acting only as trustee of the land for the generations yet unborn. Consequently, the land belonged only temporarily to the generation presently inhabiting it, subject to their good behaviour."[20] This means that no generation owns the land. Hence, they are not stewards or trustees but custodians, and, as agriculturalists, their use of the land would have been in the nature of custodial usufruct, as suggested above.

Thus, given an immanentist cosmology, a form of custodial usufruct could easily emerge that would allow for a first-rights land use while respecting the principle of inclusiveness. This was probably the case for the Ohio Hopewell, and also for the Late Archaic and the Early Woodland. For this reason the notion of custodial usufruct marks an important emic concept by which the analysis of the custodianship of hunters and gatherers by Ingold can be assimilated to the analysis of farming peoples, including the Ohio Hopewell.

Chapter 7

SUBSISTENCE, SETTLEMENT, AND CEREMONY

The Inclusive Territorial/Custodial Domain Paradigm is the broad deontic ecological framework of the Proscriptive/Prescriptive Ecological Strategy Model, and the latter is the global model that applies the basic theoretical premises of this paradigm to the Ohio Hopewell. Under this model it was claimed that Ohio Hopewell can be treated as the culmination of the prehistoric development of subsistence, settlement, and ceremonial practices in the Central Ohio Valley, and, in all probability, it was a local expression of similar processes occurring across the Eastern Woodlands about the same time. In order to enhance the usefulness of this model, it is important to demonstrate the relevance of its theoretical framework, the Inclusive Territorial/ Custodial Domain Paradigm, by anchoring the latter in deep and broad prehistoric time and space.

Relevance-construction begins with a brief descriptive outline of the development of the subsistence, settlement, and ceremonial practices of the Eastern Woodlands from the Archaic to the Woodland periods. This is done using primarily but not exclusively the relevant archaeological records of the Illinois and Central Mississippi regions of the Midwest. True, these regions are somewhat west of the Central Ohio Valley, but during the past several decades they have been the focus of broad research programs that, in general, surpass work done during this same time in other regions of the Eastern Woodlands. Furthermore, despite the distance separating the Central Ohio Valley from this more western Midcontinental region, the overall environmental conditions and their transformations over time have been sufficiently similar to allow generalization across the Midwest. Along with this, the archaeological record of the Greater Midwest strongly supports the view that interaction among the populations in these different regions was largely continuous. Innovative changes and modifications that can be tracked in the archaeological record of one region probably had equivalent patterns in other regions.

Another important reason to focus on the results of these research programs is that they resulted in the most comprehensive and influential theoretical characterizations and explanatory accounts to date of the development of the Eastern Woodlands prehistoric subsistence, settlement, and ceremonial systems.[1] These are treated here as exemplary expressions of the Exclusive Territorial/Proprietorial Domain Paradigm. Therefore, in keeping with the hermeneutic spiral method, the second step in demonstrating the relevance of the approach taken by this book summarizes and then critiques these accounts. The third step, then, applies the basic premises of the Inclusive Territorial/Custodial Domain Paradigm. This demonstrates the relevance of the Inclusive Territorial/Custodial Domain Paradigm and justifies applying the Proscriptive/Prescriptive Ecological Strategy Model and its auxiliary and ancillary models to the Ohio Hopewell data. The rest of the book is devoted to this task.

THE ARCHAIC-WOODLAND TRAJECTORY
IN THE MIDCONTINENT

The major rivers of the Midcontinent have significant floodplains with low stepped terraces. They are bracketed by fairly abrupt to steep valley sides reaching to bluffs marking the beginning of the upland hinterland, which is itself braided by small creeks that promote erosional formation along the bluffs and escarpments. The Early Holocene was initiated by the retreat of the Wisconsin ice sheet starting ca. 8000

B.C. At first the warming climate was accompanied by a rise in overall moisture so that between 8000 and 6000 B.C. the river regimes were subject to violent flooding caused by rapid down-cutting of the natural moraines holding back the last of the great glacial "inland seas."

The subsistence resource richness of the upland region was seriously diminished, ca. 6000–5000 B.C., as the moist environment of the Early Holocene shifted towards a warming and drying period called the Hypsithermal. As the uplands became more desiccated, the rivers went into an extensive period of aggradation, stabilizing into relatively slow-moving streams forming floodplains, backwaters, marshes, and oxbow lakes. The environmental changes reshuffled the range and distribution of resource opportunities for the human populations, reducing the range of resources in the uplands while creating new opportunities in the riverine zones. As the Hypsithermal developed, accommodative trends in the settlement, subsistence, and ceremonial practices occurred that largely define the subsequent trajectory of the prehistory of the Eastern Woodlands.

The social systems of the Midcontinent during the Early Holocene period had a highly mobile settlement regime focused on the upland zones and associated with generalized hunting and gathering. This regime was characterized by the community exploiting the available resource patches in one locale and then moving as a group to a new locale, although it is also likely that the same locales would be seasonally reoccupied. This is often referred to as the residential mobility strategy. Following the emergence of the Hypsithermal, however, the central tendency of human populations was away from the residential mobility strategy of the Early Archaic toward a more constrained or "tethered" mobility strategy. As part of this process the Archaic hunting and gathering communities increasingly incorporated the resources of the terraces and floodplains of the valley zones into their exploitative strategies. The trend toward a reduced mobility regime promoted the aggregation of groups into larger seasonal or semi-seasonal floodplain bands. The reach of the floodplain base camps was extended by developing the strategy of sending special-purpose task groups to exploit more distant locations in lesser valleys and in the uplands and return to the base camps with food and other resources. This reduced mobility regime of base camps and specialized task groups has been termed the logistical strategy.[2]

By the latter half of the Middle Archaic, ca. 4000 B.C., the logistical strategy was sufficiently developed to promote a considerable degree of territorial "permanency," not in the form of full sedentism but in the sense of regular seasonal reuse of the same locales from year to year and generation to generation. The Middle Archaic base camps were probably reoccupied every summer and fall and, possibly, often during the spring, acting as the domestic centers of the logistical strategy. By the Late Archaic (ca. 3000 B.C.), full seasonal base camps and even, possibly, some year-round settlements had developed.[3] This trajectory, from the residential mobility pattern of the Early Archaic, to the increasingly tethered logistical mobility pattern of the Middle Archaic, to the semisedentary/sedentary settlement pattern of the Late and Terminal Archaic, occurred over about 4,000 to 5,000 years (ca. 6000/5000 B.C. to ca. 2000/1000 B.C.).

The second significant accommodation, in parallel with the above, was an ongoing modification of subsistence practices mentioned earlier, bringing about the domestication of a series of wild seed-bearing plants. During the first half of the Early Woodland, ca. 1000 B.C. to 500 B.C., garden-based domestication became fairly well entrenched across the Eastern Woodlands as an important supplement to the traditional Archaic hunting and gathering strategy.[4]

The third significant and ongoing social transformation was marked by the appearance of the material media of ceremonialism, in particular, collective burial locales, or what are often referred to and characterized as communal "cemeteries." Two major related expressions of ceremonialism are recognized in this process: mortuary and exchange locales.[5] This emergence of major material expressions of ceremonialism was correlated with the shift from largely unfettered to tethered mobility and the associated subsistence changes described above. Indeed, as pointed out earlier, the Ohio Hopewell episode is

largely the culmination in the Central Ohio Valley of those developments in public ceremonialism that were initiated in the Middle Archaic
(or earlier). It is because the intensification of the ceremonial sphere
accompanied the Middle and Late Archaic modifications in subsistence
and settlement patterns that this correlation is so important in understanding Hopewell.

Though it has not done full justice to the detailed work that has
gone into the construction of this developmental history of the subsistence, settlement, and ceremonial trajectories of the Archaic and Early
Woodland periods of the Eastern Woodlands of the Midcontinent, this
brief outline has covered the major points that are germane to the
immediate purposes. It serves as the basis for presenting and critically
assessing the main auxiliary models of the Exclusive Territorial/Propri-
etorial Domain Paradigm evolutionary account of this historical development. The following chapter examines the same historical processes
in terms of the Inclusive Territorial/Custodial Domain Paradigm.

ARCHAIC PERIOD SETTLEMENT TRAJECTORY

The two outstanding theoretical elucidations of the development
of the subsistence and settlement practices in terms of the Exclusive
Territorial/Proprietorial Domain Paradigm have been the subsistence
risk-management strategy approach presented by James Brown and the
evolutionary model of the emergence of domestication as presented in
a series of publications by Bruce Smith.[6] These two approaches address
the same central puzzle: the long, drawn-out nature of the emergence
of domesticated subsistence and semi- to full sedentary settlement. This
problem has been approached from a general systems perspective,
treating human ecological systems as being in natural equilibrium unless disturbed by modifications in objective demographic and/or environmental conditions. However, Brown and Smith both emphasize that
any major shifts in subsistence or settlement practices as outlined above
seem to have been out of phase with any significant changes in these
demographic and/or environmental changes. For example, Brown argues that the Hypsithermal enrichment of the riverine zones reached

maximum levels well before the human populations reduced their mobility and shifted to cultivation as modes of accommodating their settlement and subsistence practices to take advantage of these opportunities.

> The aquatic resource base required evidently acquired most of its present-day productivity by 4000 B.C., thousands of years before the commitment to sedentism was completed. Thus despite the growth in aquatic productivity between 5000 and 4000 B.C., a semi-mobile system was retained for a long time afterwards. *Instead of total commitment to sedentism shortly after aquatic productivity made such commitment possible,* the shift was slow and gradual. Thus the conservatism with which the older mobility pattern was retained and the caution with which greater residential stability was adopted indicates a lagged response to the inducements inherent in the superior productivity of a highly circumscribed zone. (Brown 1985, 220–21, emphases added)

He has also noted that the evidence indicates that demographic expansion did not generate sufficient pressure on resource exploitation to account for the settlement practice changes that occurred. "Hard evidence of (population) pressure is manifest too late in the record to carry decisive weight in explaining the beginnings of sedentism"(222).

Bruce Smith also noted that there appears to have been a significant lag between the initial cultivation of the indigenous wild plants that were traditionally exploited and their domestication.[7] Both Brown and Smith have concluded that, since both subsistence and settlement changes were out of phase with demographic and environmental changes, there must have been some factor or factors, up to then overlooked by archaeologists, that systematically intervened to prevent reduction of mobility and greater reliance on cultivation. Once the intervening variable or variables are established, then, this/these must be the chronic condition(s) that prevented what objectivist ecological accounts claim should have happened: the rapid working out of the adaptive process to its optimal conclusion, this being fully domesticated subsistence integrated with fully tethered settlement, that is, sedentary farming regimes.

Brown has argued that it is the great benefit of mobility that was responsible for the time lag. In general, shifting to a more sedentary life is riskier and more costly than maintaining high mobility.[8]

> The costs of mobility are relatively low when balanced against natural risks. In this respect, I look upon the "mobility" adaptation as essentially a gaming strategy against nature, in which the group's collective memory and its rapid, flexible means of communication are biological attainments of modern man necessary to the success of this strategy. Hunter-gatherer mobility makes sense primarily as a means for balancing foraging success against natural risks with a minimum cost. (Brown 1986, 317)

Since it is claimed here that Early Archaic cultures of the Midwest had an unfettered way of life, in terms of the absence of others and, of course, of *deontic* constraints, reducing mobility would occur only when changing circumstances made maintaining high mobility riskier than reducing it. Brown argues that mobility became riskier when the resource patches of the upland became scarcer, largely as a result of the advancing of the Hypsithermal. However, it was not the failure of the resources but the increasing tendency for these mobile and highly dispersed communities to converge on the same resource patches that changed the balance of risks. In his view, convergences would necessarily instigate intergroup aggression as the converging groups competed for the same resources.

> In the case of autonomous hunting groups, their interests would be most optimally achieved by inter-group co-operation, but since the institutional means for assuring such collaboration are usually lacking, the lesser optimality is withdrawal from the resource rather than to risk conflict. The important feature of this game theory model of *hunter-gatherer decision making in a competitive social environment* is that the risks of conflict can rise to the point where resource intensification will prove to be an attractive alternative to residential mobility. (Brown 1986, 318, emphases added)

Therefore, he concludes that the most effective risk-management tactic would be for individual foraging groups to anticipate and avoid such situations by curtailing their mobility.

[Mobility] becomes curtailed in response to a rise in risks that continual pursuit of this strategy sets up. The risks are not nature's but the by-product of the mobility strategy itself when two or more groups pursuing their own mobility strategies independently attempt to converge on the same resource patch. (Brown 1986, 318)

Notable in the above is Brown's claim that these isolated foraging groups would have preferred to cooperate, but that this strategy was precluded because of the social autonomy and wide settlement dispersal engendered by the original unfettered foraging way of life. In support of this claim, he draws upon one anthropological view that claims foraging societies typically have no institutional mechanisms by which to mediate intergroup cooperation. Also notable is his claim that risk avoidance tactics bring about reduced mobility without making any change to the normative value of mobility. Untethered mobility was a "natural" value because it reduced risks. Since all groups would pursue the same strategy, reducing mobility would reduce the demographic density in the upland regions and, therefore, permit a return to a less tethered mobility until the probability of confrontation again grew, thereby promoting reduced mobility again, and so on. Brown claims that this fluctuating process of mobility tethering and untethering accounts for the long, drawn-out nature of the shift to full sedentism. Thus, the avoidance strategy combined with the lack of "the institutional means" for cooperative intergroup "collaboration" and the continual flip-flops between tethering and untethering mobility produced the long, drawn-out nature of this shift to a domestic-sedentary subsistence/settlement system.

Critique

For the sake of convenience, Brown's risk-management views will be termed the Resource Patch Convergence Avoidance Model. This model could be characterized in proscriptive terms since it postulates strategies of resource patch avoidance as the critical factor in tethering mobility. However, as construed, this avoidance strategy would not be deontic in nature. This requires that the prehistoric populations had a sense of mutual obligation and an understanding that avoiding resource patch convergence was rightful and reputation enhancing to do, rather than simply being an expedient mode of risk reduction. Reconstruing the convergence avoidance scenario in deontic terms is not a trivial matter. While the naturalistic premises of the Resource Patch Convergence Avoidance Model are not accepted here, its central claim, namely, that the reduction of mobility was a practical tactic directed to avoiding resource patch convergence, is accepted as valid. If this can be reconstrued in deontic terms, then it neatly fits into the deontic ecological perspective.

Three critiques of this model follow. In the first case, the Resource Patch Convergence Avoidance Model claims that the lack of institutionalized modes of intercommunal interaction by Early Archaic foraging societies forced groups to reduce their mobility. This claim hinges on the view that, at best, foraging societies have only individualized dispute settlement modes. Individuals in conflict resolve their disputes directly, usually by one or both parties moving out. However, if conflict resolution hinges on individuals *leaving* groups, then it is likely these individuals will do so because they have the viable alternative of easily moving to neighboring groups. This is supported by ethnographic research.[9] Mobile foraging groups can use this personal conflict resolution tactic because, in fact, they have ongoing cooperative interaction, and such mutuality largely presupposes inclusive territorialism. It is not credible to claim that Early and Middle Archaic foraging groups of the Midwest would be hostile, while, however, allowing that individuals would be free to transfer from one local group to another in order to avoid individual disputes.

This claim of the absence of sustained intergroup cooperation can be tested empirically. If Early to Middle Archaic prehistoric foraging groups deliberately avoided each other then there should be little or no indication in the archaeological record of intergroup cooperation when high mobility was dominant in preceding periods, the late and terminal Paleo-Indian. This is implied by the Resource Patch Convergence Avoidance Model when Brown argues that signs of intercommunity cooperation emerge only about 4000 B.C., with the advent of Archaic "cemeteries."

This line of reasoning predicts the early use of tokens of intergroup exchange. A predictable response to giving up security and the risk-averting features associated with mobility is a shift in behavioral strategies toward investment in institutions of *intergroup cooperation*. In the Midwest, tokens of these responses are present by 4000 B.C. in the native copper, galena, exotic flints, and other rocks and minerals. These items circulate well before differences in social status appear in the burials and before burial mounds become prominent territorial markers. As much as a thousand years separate the two developments, leaving little doubt that objects circulating in a balanced system of intergroup reciprocity arose prior to other indicators of cultural complexity. (Brown 1985, 223, emphases added)

This argument clearly demonstrates the commitment Brown has to the exclusive territorial view and confirms that his is a naturalist deontic perspective claiming that the rights in territory emerge only after demographic and environmental stress sets in and, of course, that these are exclusionary deontics resulting from humans being "naturally" competitive. In short, according to the Resource Patch Convergence Avoidance Model there should be little or no evidence of Early Archaic intergroup cooperation, in the form of various exotic materials and styles outlined above, that would serve as evidence of "objects circulating in a balanced system of intergroup reciprocity."

Since Brown presented his argument, evidence has accumulated that does not support this model. Walthall and Koldehoff argue the view that terminal Paleo-Indian or Early Archaic hunters and gatherers of the Central Mississippi Valley (CMV) would have had extensive intergroup reciprocal exchange relations.[10] In particular, they speak of the "Long Blade" Cult of the CMV, ca. 8500 B.C. and 8000 B.C. The identifying artifact of this cult is called the Sloan point, and it fits the requirements of being an expressive medium of a reciprocal exchange network of the type that Brown postulated only emerged ca. 4000 B.C. Furthermore, the form, context, and distribution of the Sloan points are particularly relevant as evidence against Brown's account. Following Daniel Morse and Phyllis Morse in this regard, Walthall and Koldehoff characterize the Sloan point as a "ceremonial" item because of its large lanceolate form, the fineness of its production, the greater length of the blade in comparison with the Dalton point, and the use of two particularly distinct varieties of the Crescent Quarries Burlington chert, High Ridge and Wood Grain varieties.[11] In particular, they emphasize the archaeological context, which is most reasonably interpreted as mortuary deposits and ceremonial caches.[12] Thus, there is persuasive evidence that a complex of cooperatively interacting, mobile hunting and gathering communities existed well before any tendency toward the emergence of the type of tethered mobility that the Resource Patch Convergence Avoidance postulates. The shift to a more tethered mobility would have arisen *within* preexisting social networks of reciprocal exchange and sharing.

Finally, the implications of the theory presupposed by the Resource Patch Convergence Avoidance Model make clear that it fails to explain what it set out to do, namely, the long, drawn-out nature of this subsistence-settlement shift. The model is premised on optimal foraging theory and the notion of the zero-sum game plan. In these terms, humans act rationally by calculating objective economic costs and survival risks in terms of net benefits. The model also works within the assumption that the deontics of territory emerge only with the settling down process and these would clearly be based on the notion of exclusive territorial tenure so that intergroup competition and aggression are the default tactics. Accordingly, the tactical flip-flop between tethered and untethered mobility Brown argues for seems implausible. Certainly some of the foraging groups would choose to return to older levels of mobility to lead a less risky and less costly existence. Indeed, Brown stresses the continuity of the high value attached to mobility during the whole of the Archaic.

However, here is where the notion of exclusive tenure comes into play. Since mobility is simply one tactical move in a complex zero-sum game, different groups would likely choose differently in similar circumstances, some opting to pursue the value of mobility over exclusive territory by reversing their tethering while others would choose to entrench exclusive territory. In terms of the competitive nature of human communities, as presumed by the Resource Patch Convergence Avoidance Model, the former communities would find it impossible to reduce mobility when crowding once more enhanced the risk of resource patch convergence since they would find themselves competitively excluded from the best riverine territories by those who had opted to maintain reduced mobility and exclusive territories. The result would be a rather historically brief period in which the division of "winners and losers" occurred, with those who remained in the valleys entrenching their proprietorial domain modules.

This would not occur "overnight." In human terms, a considerable time lag would probably still be involved. However, in the combined terms of (1) the risk management model and (2) the exclusive territorial tenure premise, an adaptive selection process would *not* lag by up to a millennium behind the ecological changes that promoted such opportunities, and, of course, it should not take several millennia overall to make the transition to a stable, settled, albeit highly competitive lifeway. In short, it ought to take only a matter of several hundred years.

ARCHAIC PERIOD SUBSISTENCE TRAJECTORY

As pointed out above, Bruce Smith has also noted a temporal lag between the emergence of new plant resources and their efficient exploitation by Archaic peoples, suggesting a reluctance to shift from traditional foraging practices to cultivation of wild plants and to their full domestication, a position that reinforces Brown's claim that the permanent tethering of mobility would also be delayed by the people preferring to maintain the flexibility of mobility. The Floodplain Model, as Smith calls it, argues for a coevolutionary selective process leading to domestication. The process started in the Middle Archaic, slowly progressed through the Late Archaic, and only became entrenched in the middle Early Woodland periods of the Eastern Woodland, a period from ca. 5000 B.C. to ca. 500 B.C. The model argues that one of the resources that would be exploited by the Middle Archaic peoples would be the indigenous plants of the floodplains. The annual inundation of the floodplains prevented the forest from invading the habitat of these plants. However, with reduced mobility, humans would start to reoccupy the same lowland forest sites for more and more extended periods, thereby removing the natural forest coverage and transforming these locales into close replicas of the natural floodplain conditions that were ideal for these seed-bearing plants. Although the sites would, in general, be above the annual floods, by regularly bringing the harvested wild seeds of these nutritionally valuable plants to their base camps, inevitably some seeds would be lost and scattered, possibly tossed into midden, where they would proliferate in these open conditions.

In this manner, human populations unwittingly expanded the habitat of these potentially useful wild plants. The model goes on to argue that, following a rather long period of casually exploiting these proximal base camp resources, domestic groups finally started to deliberately cultivate them. This would have led to deliberately storing some of the harvested seeds for later sowing. At first, these stored seeds may have been simply casually scattered in and around the campsites. Although this was a rather casual form of cultivating and harvesting wild resources, by ca. 2000 B.C. at least four of these floodplain plant species had gained domestic status: goosefoot, sumpweed, sunflower, and squash. Supplementing these were other plants that, though still cultivated, did not attain domestic status.

This model argues that these plants were used primarily as a fallback food supply. This casual, small-scale gardening persisted for about a thousand years. Rather abruptly, however, gardening expanded across the Midwest, became entrenched as a source of staple food, and even possibly evolved into field farming regimes. This occurred in some cases 1,500 years *after* the major cultigens had become domesticated. Smith characterizes the emergence of the Woodland period (ca. 1000 B.C.) primarily as this rather sudden intensification of gardening. Certainly, it seems valid to conclude that, during the Terminal Archaic to middle Early Woodland (ca. 1500 B.C.–A.D. 500), while the peoples of the Midwest maintained a broad hunting and gathering economy, they also typically carried out some gardening of wild, semiwild to fully domesticated seed crops.

Critique

Smith characterizes the Floodplain Model as a "minimalist" explanation. By this he means that it relies only minimally on the role of human design and intention as realized in "conscious and directed innovation."

> The explanation is minimalist in that it requires only a quite limited degree of conscious and directed innovation on the part of prehistoric human populations. . . . The seventh millennium B.P. development of anthropogenically disturbed habitat patches is identified as a starting point of the coevolutionary process leading to domestication, while the fourth millennium B.P. appearance of morphological changes indicating domestication in sumpweed, chenopod, and sunflower marks an end point. *In between, given the contexts of domestilocalities, the interrelated and largely automatic set of selective pressures outlined above define the developmental process.* (Smith 1987, 37, emphases added)

He goes on to admit that "Such a minimalist explanation . . . casts human groups in a largely unconscious and automatic role," but quickly backtracks to claim that this does not mean that Archaic populations were automatons, since this explanation "does not preclude attempts at explanatory expansion through the addition of supplemental layers of interpretation, including transformational or social supplementation."[13] In this case he makes direct reference to Brown's risk-management model in which settlement strategies *were* innovated. Settlement strategies were the medium of alliances, the exchange of exotic goods, mortuary symbolism, and so on, and were the outcome of innovations in the social and symbolic spheres. In contrast, subsistence changes were largely the unwitting outcome of the application of practical knowledge about the availability and distribution of local resources and the practical know-how for routinely procuring, preparing, and storing them. In his view, then, Archaic peoples were actively innovative in terms of settlement and ceremonial strategies, but largely non-innovative in terms of subsistence strategies.

The commitment of the Floodplain Model to the innovative nature of settlement and ceremonial practices does not sit well with the claimed noninnovative nature of the subsistence practices. It seems hardly credible that the two contrasting attitudes could coexist, particularly since changes in one strongly rebound on the others.[14] Furthermore, if a calculating, innovative, and practical rationality characterized the modification in the settlement strategies of competing Archaic populations, then it seems implausible that the same calculating, innovative, and practical rationality would not have characterized subsistence strategies. The model claims that extending the period of occupation in base camps would generate significant microenvironmental changes. I can fully concur with this claim. However, these effects would be precisely the type of objective resource opportunities that astute, calculating rationalists would seize upon and promote, quickly embracing the economic benefits that their expanding occupations brought about, particularly if these benefits did not cost them anything in extra labor.

In short, as in the case for the Resource Patch Convergence Avoidance Model, the Floodplain Model actually establishes good reasons to expect a fairly close fit between anthropogenically induced changes and their rational exploitation so that, overall, a rather rapid shift from foraging to cultivation would occur. It is not denied that there would be some time lag. However, the subsistence strategy would be adjusted to the new opportunities in a matter of several generations—not several millennia—and, in keeping with the exclusive territorial premise to which this model is also committed, those populations that did not exercise such rationality, or in doing so, chose the opposite, would be quickly excluded by those who exploited these emerging opportunities.

ARCHAIC PERIOD CEREMONIAL TRAJECTORY

While the Sloan point distribution may have marked a rather early form of "cemeteries," these became common and widespread in the Midwest from the Middle Archaic on. The question that arises is:

"What conditions would promote the emergence of these collective locales for depositing the dead?" Drawing on the body of mortuary theory developed in the 1970s, Douglas Charles and Jane Buikstra explain their emergence as the rational co-opting of mortuary practices to serve the need to control exclusive access to territories, or, at the least, to control exclusive access to the critical resources within such territories.[15] They posit that this emerged as a result of the two tendencies outlined above, namely, the time-lagged, but continual tethering of mobility and the correlated time-lagged emergence of cultivation and domestication.[16]

In the Illinois Valley and the Central Mississippi Valley regions, from the Middle Archaic to the terminal Late Archaic, the place of choice for human burial was characteristically along the bluff tops overlooking the river bottoms. Drawing on Lynne Goldstein's work, Charles and Buikstra argue that should an archaeologist find what appears to be a specialized space for collective burials, then it is highly probable that this marks proprietorial corporate entities. They further argue that these would have been generated as symbols that "reference" the corporate body that claimed control of the catchment area of which the "cemetery" was the spiritual, civic, kinship, political, but not domestic, center.

> From the existence of formal cemeteries we may infer that the populations interring their dead in these facilities were sedentary, in the sense that they exploited specific resource territories, either from permanent base camps or from a series of locations within their respective territories. . . . Furthermore, the use of "monumental" funeral facilities—in this case, bluff top prominences—indicates the size and location of territories were constrained by the presence of other groups; there was overt competition for resources among economic units within the region. The occurrence of single cemeteries at most sites, the general spacing of the sites and the lack of internal differentiation within the cemeteries suggests that these burial facilities were utilized by single villages and, concomitantly, that the village functioned as the basic corporate economic unit. (Charles et al. 1986, 458–59)

This bluff top prominence, they claimed, ensured that these locales would be noted, and, as they were used exclusively for collective burial, they fulfilled the specialization criterion that they claim marks them as formal proprietorial corporate cemeteries. Hence, the argument is that the construction of "cemeteries" was the primary, although not the exclusive, symbolic means by which individual foraging communities could stake out exclusive territorial modules in competition with neighboring peer communities for the limited lands and their resources distributed along the riverine corridors of the Midcontinent. This position is also a deontic ecological claim. Indeed, it is an integrated prescriptive view in that the emergence of Archaic ceremonialism is characterized here as a *prescriptive* response to a progressively *prescriptive* subsistence-settlement regime. In this case, however, the deontics are exclusive tenure grounding proprietorial domain. For convenience, this comprehensive interpretation can be termed the Cemetery Model of Archaic Ceremonialism. Its basic premises, namely, that the cemeteries mark a monistic modular social system, have been largely unproblematically extended by many archaeologists to the mortuary locales of the Woodland periods.[17]

Critique

In the normal sense of the English term, cemeteries are locales where the aggregation of burials is the result of a series of terminal burial rites constituting individual funerals. In terms of the Cemetery Model, the point of the cemetery as a collective burial locale (CBL) was to aggregate the deceased of a proprietorial corporate kin group and transform them into ancestors. For this reason, the burial rite had to be the termination of the funeral of a deceased member, rather than, say, a member of some other group. The tying of funerals to these CBLs is critically important in this model since it is only by being cemeteries embodying the remains of the ancestors that they could serve as collective symbols of the corporate proprietorial domains. In symbolic pragmatic terms, the cemetery would be experienced by the corporate members as embodying their ancestors, thereby literally manifesting the ancestral power and authority. This authority would have endowed the living descendants as the stewards of the corporate territory. Of course, this declaration had to be public since it was constituting this social fact in the moment of communicating it to other groups of the same order, thereby eliciting from them signs of recognition of this claim. These signs of recognition, in fact, would be the construction of their own cemeteries.

It is important to note that while this model treats the cemeteries as monumental symbols, it does not treat the deceased that make up these mortuary aggregations in symbolic terms. That is, it is the "cemeteries" that have symbolic import, and the burials are simply the practical gathering of the ancestral deceased. It is their being deceased that counts in making up the "cemetery," not the treatments that these deceased received. In fact, the Archaic cemeteries are notable for the broad range of post-mortem manipulation of the individual burials, from extended to flexed inhumation, bone bundle burials, burial of body parts, and cremations. This variation is explained not in symbolic terms but largely in practical terms, as rational responses to the material circumstances of death. Extended burial marks the individual as having died near the group's home "cemetery." Flexed inhumation, bone bundle burials, or cremations would indicate that death occurred at a distance or, for example, in the winter, requiring different degrees of post-mortem manipulation: temporary burial, exhumation, removal of residual flesh from the bones, bone bundling, and finally "proper" burial in the group's cemetery.

The major mortuary symbolism internal to the "cemetery" itself is claimed to be largely mediated by two dimensions: the material items associated with the deceased, and the relative labor investment in the mortuary feature. These artifactual and labor-cost aspects are treated as reflecting the social standing that the deceased had in the corporate group at the time of her/his death. In terms of this status-at-death criterion, mortuary symbolism, then, is mediated not by the deceased, as such, but (1) by the variety, quality, and relative quantity of mortuary artifacts, and (2) by the amount of labor and resources invested in the burial facilities. Furthermore, these material elements are taken not as symbolic of death but as symbolic of the social standing of the deceased when living. In effect, the symbolic relevance of the post-mortem treatments of the deceased, the material associations, and the mortuary facilities are largely incidental to the mortuary event itself.

From this it follows that there are *no autonomous symbolic pragmatics of death being manifested in the burial patternings.* This is troubling since it means that archaeologists are characterizing the action nature of prehistoric mortuary behaviors by default, and the default position, of course, is that type of mortuary behavior within the cultural experience of most archaeologists, namely, a funeral. This default position will be termed the funerary view, and it informs much of the archaeological interpretation of prehistoric mortuary data in the Eastern Woodlands. In effect, the funerary view collapses the symbolic pragmatics of the mortuary sphere into the symbolic pragmatics of funerals. It then effectively dissolves even these by using the presence of the deceased, in whatever condition of post-mortem treatment, as a nonsymbolic prima facie indication both that a mortuary behavior had occurred and, simultaneously, that this deceased person, or part thereof, marks the termination of an event that "could only have been a funeral."

This scarcely recognizes that in many societies funerary practices make up only a small part of the mortuary sphere and that the latter may entail a complex set of symbolic pragmatic rules and protocols that structure the mortuary sphere into a differentiated range of emically-recognized mortuary practices, as postulated in the previous chapter. This criticism is not to deny that funerals were part of the mortuary sphere of the Archaic period. Rather, it is to highlight the point that mortuary events, even funerals, are not "objectively" constituted. Just as hunting and poaching are symbolically constituted actions requiring what are objectively equivalent behaviors, so funerals, along with possibly a spectrum of different mortuary events, are also symbolically constituted. Therefore, a society will have a differentiated set of symbolic pragmatic rules and protocols constituting mortuary strategies by which to constitute the total range of its mortuary activities, and, as with all strategies, these rules will have ideological implications. Hence, these would not go uncontested in a society, implicating ongoing debate and negotiation over what forms mortuary behaviors and their material outcome must or should display to count as the intended mortuary event. From this it follows that funerary rules must be only a subset of the range of ideological rules that make up the mortuary sphere of that society. These rules would exercise an autonomous force that could constrain even practical circumstances and possibly override status-at-death circumstances.

Indeed, anthropological studies inform us that in many non-Western societies the mortuary sphere is very complex. For example, many pre-Columbian Mesoamerican societies are known to have practiced human sacrifice that terminated in different forms of burial treatment. The well-known instance of the matched groups of human burials at the base of the Temple of Quetzalcoatl in Teotihuacan, one set on the north side and the other on the south side, is generally interpreted as "en masse" human sacrifice, possibly a foundation sacrifice for the pyramid platform. They were displayed in full military accouterments, having their hands tied behind their backs at death. As such, they would not have been participants in their own funerals, but participants in their own deaths, constituting the mortuary behaviors that generated these deaths as lethal human sacrifice and the terminal burials as sacrificial offerings. Thus, their burials can hardly be characterized as funerary, although they certainly implicate quite a complex incrementally structured set of mortuary behaviors.[18]

Speaking of the mortuary sphere as a complex set of different types of mortuary activities structured by relatively autonomous symbolic pragmatics opens up the possibility that the variation in the Middle and Late Archaic burial forms is only partly accountable for in terms governed by the practical and social circumstances of the deceased at the time of death. Indeed, these may be the least important determinants of the terminal burial behavior. If this is the case, then these CBLs may not be properly characterized as "cemeteries," or alternatively stated, they may have been much more than cemeteries. From this it would follow that the collective burial locales may not be the monumental symbols of proprietorial corporations that they are claimed under this model to be!

The question this critique raises then becomes quite straightforward. Do the Archaic CBLs picked out by the Cemetery Model fit what could be expected of the funerary burials of proprietorial corporate groups? The answer given in the next chapter is negative, and therefore, it is part of the critique of the Cemetery Model. However, it is first necessary to elucidate a mortuary model that situates the funerary perspective in a less privileged position than the one it now holds. Because this model will be particularly apt for analyzing mortuary practices as having a heavy post-funerary or even nonfunerary component, initiated largely by collective mourning rites, it can be termed the Mourning/World Renewal Mortuary Model. It is an important auxiliary of the Proscriptive/Prescriptive Ecological Strategy Model, although it will be appropriately modified for application to the Ohio Hopewell mortuary record. The following chapter, then, will be devoted first to sketching out the basic theoretical framework of the model, and then to demonstrating it by showing how it can more coherently account for the Archaic mortuary data than can the Cemetery Model.

Chapter 8

WOODLAND WORLD RENEWAL MORTUARY CEREMONY

Sketching out the Mourning/World Renewal Mortuary Model necessitates articulating how symbolic pragmatics apply to the mortuary patterning and its variability. In symbolic pragmatic terms, material cultural style constitutes an assemblage as a complex body of action warrants. These warrants, in turn, manifest and constitute the associated social positions, which are deontic complexes of rights and duties that make up the action powers and capacities of those who legitimately occupy these positions. Therefore, through the display and use of the appropriate range of material artifactual warrants, the occupants become active agents whose behavior is productive of the events in which they participate.

The term focal positions is used here to refer to the social positions that are primary in defining and constituting the nature of a social activity or event. The persons or parties occupying these positions can be termed the focal participants. This means that it is also appropriate to speak of focal warrants, these being the symbolic pragmatic devices, both artifacts and facilities, that mediate the social powers of the focal positions. Focal warrants implicate auxiliary, supporting warrants, these being devices used to complement and reinforce the focal warrants. While the use of focal warrants is critical to the successful production of the events, the use of supporting warrants would not be essential, although their absence would probably diminish the reputation of those who failed to conform since this failure, unless justified, would imply that the perpetrators lacked respect for the social positions, their occupants, and the type of events that all participants are attempting to bring about.

A mortuary event is like any other event in that it is constituted by the public display and use of the appropriate symbolic warrants, both focal and supporting. However, it is also unique in that it entails both living and deceased participants. This means that two warranting conditions must be met in order to constitute a mortuary event, one making it a mortuary event per se and the other making it the particular type of mortuary event intended. The first condition is satisfied by the presence of human remains in some state, or, if no body or bodily component is available, in some symbolic substitute form, such as an iconic representation of the deceased. The second warranting condition is satisfied by the material treatment of the deceased, and this hinges on two dimensions: (1) the nature of the post-mortem manipulatory treatment accorded to the deceased (e.g., full extended bodies, flexed bodies, body parts, bone bundles, cremated ashes, ashes that are scattered or clumped, and so on); and, of course, (2) the range of mortuary facilities and artifacts serving as focal and supporting mortuary warrants of the event.

For this reason, the deceased have a unique symbolic standing. On the one hand, they are material things much like the artifacts with which they are often associated and, therefore, can be used as symbolic warrants. On the other, they are not just like these other things since, in fact, it is because of the deceased that the living take themselves to be participating in a mortuary event. The question this raises then is: Should the deceased be treated as participants occupying mortuary social positions or as symbolic warrants mediating mortuary events? This ambiguity does not arise with respect to living participants. Their role is marked and constituted by means of the material warrants they wear, carry, and manipulate, including the behavioral interaction they are expected to have with the deceased (an exception to this is the status of the victim in an event we might term lethal human sacrifice). That is, the living are extrinsically endowed by their warrants with the social powers and their mortuary responsibilities arise from the relationship they hold with the deceased. The deceased, however, are intrinsically endowed in virtue of being in a state of death—in whatever way this state is recognized in different societies. Consequently, the deceased make up a highly valued symbolic pragmatic resource in their own physical persons, and this resource can be manipulated, quite literally, to produce different types of mortuary participation/symbolic warranting of mortuary events.

This focal participant/symbolic warrant duality probably arises from the recognition in probably all human societies that while death brings about changes in the social standing of deceased persons, it does not eliminate their action capacity. Of course, the deceased cannot behave in the same way as when they were living. However, they typically are still recognized as having action capacities, and in many cases, very special capacities, depending on the cultural beliefs (cosmology) and the mortuary protocols (ideology) of the society. Famously, the death of The Inca, ruler of Tawantinsuyu seated at Cuzco, which was the primary axis mundi of the Incan cosmos, was not seen as eliminating his powerful action capacities. If anything these were enhanced. The position and power of the ruler in his deceased state were embodied in his mummy, which was endowed with all the clothing, furniture, palaces, and estates that the ruler possessed when living. The senior son became the new Inca, while the rest of his siblings became the collective custodians of their dead father, who continued to exercise his will in governing his estates through his living sons. They performed regular consultative ritual with the royal mummy and translated the signs they received as the will of the deceased Inca.[1]

This simultaneity of the deceased being both mortuary participants and mortuary warrants will be termed the focal participant/symbolic warrant mortuary duality. Its importance is that it implies a particular symbolic pragmatic meaning can be attributed to the systematic corporeal modification of the deceased. This modification can be used as evidence of deliberate manipulation by which their intrinsic warranting capacities were progressively "unfolded," being used to constitute the incremental series of mortuary events that they mediated, from funerary to rebirth to social renewal to world renewal rites. To illustrate this, assume that a given mortuary program has a broad range of post-mortem manipulatory practices, as marked empirically by the range of burial forms. In those cases when the deceased are presented at burial in a manner that largely replicates their self-presentation when living, at least as closely as practicably possible, this would constitute them as occupying the primary focal position of the event, in this case, being the virtual "hosts" of these mortuary activities. That is, it makes sense to accompany the deceased with the range of artifacts and facilities that would define their social standing in life in order to ensure that they occupy the social position of "hosts." This type of mortuary event would be appropriately termed a funeral. Following the funeral, however, the positions that they subsequently occupy are less focal and more symbolic warranting in nature, as marked by each incremental step of post-mortem manipulation to which they are subjected, flexed burial to bundle burial, to body-parts burial, and so on. In short, each progressive step marks a dilution of the mortuary participant role of the deceased and a strengthening of their symbolic warranting role. These post-mortem reductions do not mean that the deceased are being progressively dehumanized since it is their humanness that makes sense out of the formal modes of treatment. Rather, the incremental stages of modification progressively expose and release the multiple spiritually motivated action or pragmatic aspects that are embodied in the deceased: the name spirit, the spirit of the flesh, the spirit of bones, and so on. Thus, each incremental step counts as a mortuary ritual; but it also generates a new post-mortem state that counts as another type of symbolic warranting capital that can be drawn upon to mediate the next stage, and so on.

This transformational process, marked by stages of post-mortem manipulation, logically grounds what can be termed the focal participant↔symbolic warranting mortuary continuum. This continuum rests on two operating assumptions:

1) The less the post-mortem manipulation, the greater the expression of the participant moment over the symbolic warranting moment.

The corollary:

2) The greater the post-mortem manipulation, the greater the expression of the symbolic warranting moment over the participant moment.

These two assumptions constitute the continuum as an interpretive tool for structuring a social system's mortuary sphere in terms of the multiple mortuary events postulated under the Mourning/World Renewal Model. The two polar extremes of the mortuary continuum, measured in terms of minimal and maximal degrees of post-mortem manipulation, define mutually "opposing" mortuary practices: funerary and world renewal, respectively. The range between is "filled" by the rest of the mortuary practices recognized in a given culture, and, again, these are differentiated in terms of the incremental degree of post-mortem manipulation.

In these terms, the Mourning/World Renewal Mortuary Model postulates that the Archaic mortuary sphere was organized as an incremental series of dual mortuary/world renewal rites. They can be characterized as funerary, mourning, spirit-release, spirit adoption rituals, with each step marked by an incremental stage of post-mortem manipulation marking a decrease of funerary ritual value and an increase of renewal ritual value, from renewing the name, to renewing the group, to renewing aspects of the cosmos, to renewing the totality of the cosmos. This postulated series of mortuary rites will be termed the funerary→mourning→spirit-release→world renewal ritual process.

This ritual process is simply the formalization of Robert Hall's insight relating mortuary and world renewal practices.[2] Using the historically known Native American mortuary/world renewal ceremony as an analogy, he has already noted that many historically known Native American cultures have had complex mortuary practices involving different, incrementally staged rites. These are the death rites, the rites of the adoption of the name of the dead, various spirit release rites, the mourning rites, the burial rites, and so on. The symbolic warranting role of the deceased is implied in this. For example, among the historic Oglala-Lakota Sioux of the Western Dakotas, the "soul keepers" were responsible for bone bundles and were their custodians prior to reburial. While bearing these bones they worked under a number of proscriptive rules. As soul keepers they were considered bearers of the special powers of the bones by which hunting success was enhanced.

> The soul keeper was free to ride with the hunting parties, but it was his role to sit by himself on a nearby hill with his pipe, praying for the welfare of his people and the success of the hunt. If a buffalo were killed near him he could claim it. He could not himself butcher it because he was restricted from touching knives or blood. (Hall 1997, 27)

It is also notable that renewal ceremonies do not have to relate directly to the renewal of the cosmos, as mentioned above. Name adoption rites address the rebirth of the human spirit taken to be immanent in the name. Social renewal is also treated in mortuary terms. The custodial group can come to see itself as having a group spirit, and elements of the group's mortuary process can be co-opted as media for renewing the spirit of the custodial group, thereby reincarnating the spirit of the organization conducting the rite.

> The Feast of the Dead of the Great Lakes Algonquian was . . . an occasion not only for consolidating intertribal relations but also for symbolically reincarnating deceased tribesmen by transferring their names to others, much as the dead founders of the Iroquois League were symbolically reincarnated by transferring their names to their successors during Condolence councils. (Hall 1997, 40)

Hall uses the historic Southeastern societies as his linchpin connecting back to the Middle and Early Woodland.

> The analogy may be extended by considering that world renewal ceremonialism may have been an integral part of some, perhaps much,

Woodland mound burial, as suggested for the Midwest, just as world renewal ritual and the curation of dead Suns were an integral part of mound-top Mississippian ceremonialism in the Southeast. (Hall 1984, 274)

POST-MORTEM HUMAN SACRIFICE

Since the Mourning/World Renewal Mortuary Model postulates that the deceased or their bodily components served the critical dual role as participants and symbolic warranting devices to mediate this progressive process of mortuary/renewal rituals, their use in rituals emphasizing the renewal aspect would suggest that they served as ritual offerings. This notion of an offering immediately implicates a religious sacrificial paradigm. This is precisely what this model intends it to implicate. Therefore, it is postulated that the post-mortem manipulatory behavior constitutes the renewal aspect of these incremental mortuary events as post-mortem human sacrifice. This term has been coined to contrast with lethal human sacrifice. Lethal human sacrifice is a mortuary event in which, done under the appropriate symbolic warranting conditions, the intentional killing of a participant is a necessary part of constituting the behavioral process as a mortuary ritual of the type intended, that is, a world renewal ritual mediated by lethal human sacrifice of the type practiced, for instance, by most, if not all, Classic and Post-Classic Mesoamerican societies.[3] A post-mortem human sacrifice, however, is not lethal. It is a mortuary ritual in which deliberate human killing does not figure as part of constituting the event as the type of mortuary action intended. Rather, the death is natural or caused by some agency extrinsic to the mortuary event itself. Once death occurs, however, it is "cultivated" by treating the deceased as highly valued symbolic pragmatic ritual capital, as discussed above.

This distinction between post-mortem and lethal human sacrifice is also implicated in Hall's comparative overview of the sacrificial dimension of the mortuary practices as marked by the Dickson Mounds site, Mound 72 of the Cahokia site, and the Aztecan and Toltecan sacrificial fertility practices: "It is my belief that in Aztec practices we are witnessing the final phase of a shift from the ritual use of unscheduled deaths, such as those over the Dickson Mounds four, to the use of the scheduled deaths that we have called sacrifice."[4] In characterizing the "Dickson Mound four" as evidence of "the ritual use of unscheduled deaths," while Hall does not use the term "post-mortem sacrifice," he would appear to be expressing the same sense.

Since the theoretical background of the Mourning/World Renewal Mortuary Model postulates a polyistic locale-centric social system, the funerary→mourning→spirit-release→world renewal ritual process would entail a network of autonomous custodial CBLs (Collective Burial Locales) linked to each other via cooperative mortuary and other forms of ritual interaction. This network would be actively realized by the deceased being reciprocally *shared* among the participating groups as symbolic warrants of different forms of post-mortem world renewal sacrifice. Thus, a broad range of post-funerary mortuary manipulation would be generated, as discussed above. In these terms, bundle burials might best be understood not only as, or possibly, not at all as the result of the practical circumstances of death, but the result of intergroup sharing of deceased as symbolic warrants of different post-mortem world renewal sacrificial rites. Similarly, rather than the range of mortuary artifacts, the mortuary facilities, and the mortuary locale itself being a result of the social status of the deceased, these would manifest the symbolic pragmatics required for the felicity of the range of rites typically performed there. Finally, as discussed earlier, remembering Harré's claim that the pursuit of reputation is both collective and highly variable in form across human cultures, the reputation of custodial groups cooperating in the common endeavor of enhancing the cosmos would partly hinge on the effectiveness with which they could share in the rites of neighboring CBL groups. The bearing of their deceased members' bones (not necessarily ancestors or relatives) in the form of sacred bundles would be a major material symbolic warranting medium of such cooperation. All this would count as reputation enhancing since the sharing of these would be a symbolic pragmatic resource by which world renewal rites were enhanced.

Drawing on the notion of multiple social structural axes that was discussed earlier, this mortuary model further postulates the existence of more than one type of CBL in a region, each type displaying a characteristic range of mortuary patterns corresponding to the type of mortuary rituals performed by that type of custodial group. For example, there would be the domestic CBL, which would tend to be devoted to mortuary rites that focus on the funerary aspect integrated with proscriptive subsistence and settlement midwifery renewal rites. Another might be the locales of dispersed clan-based groups cooperating among themselves in public world renewal rites mediated by shared mourning rites. Another type of CBL could be the locale of a non-kinship-based custodial group that specializes in renewal rites directed to cleansing and rectifying the pollution caused by members regularly going on long expeditions in the pursuit of exotic materials to serve as particularly potent symbolic warrants of world renewal, such as copper, mica, exotic cherts, and so on.

In these terms, two obvious major dimensions of the mortuary record can be used as empirical evidence to adjudicate between the Cemetery Model and the Mourning/World Renewal Mortuary Model: the complexity of the mortuary skeletal and associated artifactual and mortuary facility data, on the one hand, and the number of types of CBL locales, on the other. If the former gives clear indication of a focus on proscriptive or minimizing post-mortem manipulation, the Cemetery Model is supported. Indications of prescriptive or maximizing post-mortem manipulation support the Mourning/World Renewal Mortuary Model. If a single type of CBL prevails, particularly if it is kinship-based, this supports the Cemetery Model. If multiple types of CBLs exist, particularly if they display variability that suggests disembedded social structural axes characteristic of polyistic locale-centric social systems, the Mourning/World Renewal Mortuary Model is favored.

A. The Complexity of Post-Mortem Treatment Data

The Archaic component of the Elizabeth Mounds site of the Lower Illinois Valley is not atypical of the Archaic CBLs of this region.

It is a compound CBL on the western bluff of the valley overlooking the important Middle Woodland Napoleon Hollow site. Its most prominent components are seven Middle Woodland CBL Mounds that were superimposed over the earlier Archaic CBL component, which the excavators characterize as a cemetery.[5] This analysis focuses on the latter component. Feature 4, under Mound 1, is a set of burials demarcating the Middle Archaic (Helton) period. This feature was a wide and shallow mortuary facility dug into a natural knoll of the bluff. Burial 32, was a group of five in-flesh extended deceased laid out side by side. There were initially four and a fifth was added before the final earth covering was placed. Burials 24, 31, 33, and 34, as discussed below, were other subsequent mortuary deposits.

Three of the five deceased of Burial 32 were judged by the excavators to be adolescents and two were adult males, between 28 and 40. The sex of the adolescents could not be determined. The bodies were dressed at burial with waist sashes of anculosa shell beads and most had one or two bracelets of the same beads at their wrists. At least four of the five bodies had been subjected to extended exposure prior to burial such that the soft tissues had deteriorated sufficiently to allow rather easy "inferior" insertion of unhafted projectile points into their rib cages.

> The projectile points were placed inside the chest cavities after death; the careful arrangement suggests that they were inserted after the decomposition of most of the soft tissue. In particular, the inferiorly directed tips of the points within Skeleton 8 indicate that they were not hafted and were intentionally placed. There was no damage to the surrounding ribs; the points were inserted base first from below the rib cage. (Albertson and Charles 1988, 34–36)

Burial 31 was a single extended individual covered with a limestone slab, but its skull was missing. There were no associated artifacts. Burial 24 was a compound burial of at least five disarticulated skeletons: three males, and two of indeterminate sex. These were accompanied by one Osceola point, one point of indeterminate type, and two Godar plummets, one limestone and one hematite. Also a ¾ groove ground stone axe and an antler hafting device were included. This burial is reported as having been deposited on top of Burial 32, the initial collective burial, and Burial 31.

It is clear that, in terms of post-mortem manipulation, Feature 4 and its mortuary deposits favor supporting the Mourning/World Renewal Mortuary Model rather than the Cemetery Model. Its overall burial content has modified primary extended burials, secondary and bundle burials. Although the extended burials of Burial 32 would appear to indicate strong funerary treatment, the extensive exposure to which these bodies were subjected suggests otherwise. The projectile points were manually inserted into the chest cavities, clearly indicating considerable post-mortem exposure. This extended exposure suggests that the funerary rites had been performed early during this period, followed by a series of rites, as postulated by the Mourning/World Renewal Mortuary Model. In particular, the insertion of the projectile points into the thorax region suggests spirit-release/world renewal rites in which they were the primary post-mortem sacrificial offerings.

Reinforcing this post-mortem sacrificial claim is the interesting fact that four were laid together, along with *anculosa* shell sashes and wrist bands, indicating that they were equivalently "dressed for the occasion," despite disparities in age. This reinforces the likelihood that the occasion was not funerary in nature since, first, despite the differences in age, this equal treatment suppressed their individual and probably differential social standings at death, thereby reducing their participant role, and emphasized their symbolic warranting commonality. Second, since the usage of shells did not demarcate differential social standing, it would have constituted these deceased as a unitary post-mortem sacrificial set. Third, confirming this interpretation of its being more a collective post-mortem sacrificial world renewal rite than a funerary rite is the fact that the fifth burial was added to the first four. This addition works against the view that the five deceased were focal participants in their own collective funeral, otherwise the late fifth addition would have diminished the status of the initial four as "hosts," thereby diminishing the felicity of the mortuary event as a funeral.

Although tardy, the addition would have been perceived as enhancing rather than diminishing the felicity of this mortuary event. In sum, the extensive exposure, the commonality of artifact warrants, the differences in age, the addition of a fifth deceased, and the deliberate insertion of the projectile points suggests a spirit-release/world renewal rite in which they played the role of primary post-mortem sacrificial offerings.

In line with this interpretation is the possibility that the missing skull of Burial 31 was deliberate. The burial was covered by a large stone, apparently after the skull was removed. This certainly hints that the skull might have been curated for further mortuary rituals. The post-mortem sacrificial treatment is even more apparent in the set of (at least) five disarticulated bodies making up Burial 24. In this case, it is clear that the deceased had already been subjected to extensive mortuary rites involving initial burial, disinterment, bone cleaning, and bundling. Therefore, it is reasonable to interpret the bones of this deposit as another form of post-mortem sacrificial offering.

When these different mortuary deposits in the same feature are examined in their context, they collectively display possibly the total range of incremental mortuary behaviors of this social system. It is not being suggested that this total feature was the result of a single incremental series. Rather, each separate deposit suggests the cumulation of different steps along the total funerary→mourning→spirit-release→ world renewal mortuary process. First, it can be assumed that the deceased in each burial had been given initial funerary rites, probably in some other CBL context. Burial 32 then may be seen as combining post-funerary mourning and spirit-release rites. The deceased making up the large bone bundle of Burial 24 would have been individually and/or collectively subjected to post-funerary mourning and spirit-release rites, and then, possibly, disinterred again and bundled as warrants of another post-mortem sacrificial world renewal rite involving releasing the spirits of the bones. The incremental rites effectively recycled the deceased, each ritual event involving a different spiritual aspect.

There is some artifactual evidence that supports this interpretation. In terms of the Mourning/World Renewal Mortuary Model, CBLs generated as the result of the complex funerary→mourning→spirit-release→world renewal ritual process should display indications of collective rites shared among a regional set of custodial CBLs. Thus, indication should be found of relatively distant groups participating with the local host groups. Burial 24, again, gives evidence to this effect. First, this set of bundled bones was found on top of but physically separate from Burial 32. As a bundle burial deposit it indicates just about the maximal degree of possible post-mortem manipulation without actual bone breaking or cremation. The accompanying artifacts were two Godar plummets (one limestone and one hematite), an Osceola point, and a grooved axe. There were two other burials unrelated to Burial 24 with Osceola points.

> All three (Osceola points) are beautifully manufactured and were discovered in association with three different burials. Discovery of this highly stylized type only in burial association at Elizabeth indicates that the type may have had ritual or social connotations. The type is very similar in shape to a Graham Cave point, both types having been found in Early-Middle Archaic context at Graham Cave and in Koster Horizon 8c and 11. (Odell 1988, 166)

With the exception of the five extended burials of Burial 32, all of the other burials accompanied by projectile points, Burials 7, 19, and 24, had Osceola points, and the bones of the seven deceased making up Burials 19 and 24 were disarticulated. The Osceola point contrasts with the rather utilitarian points inserted in the rib cages of the deceased of Burial 32. As these were extended burials, this suggests that the deceased were local. In contrast, the bone bundles could have been brought from a distance. Since visitors coming from different regions would use working styles that would probably have varied from the local styles, the use of a standardized pan-regional style is understandable. The Osceola point as a symbolic warrant for certain mortuary purposes is reminiscent of the Sloan Point of the Dalton CMV. It would ensure that groups quite distantly separated could cooperate in sharing

the use of deceased in performances that everyone recognized as felicitous world renewal rites.

This means not only that the funerary→mourning→spirit-
release→world renewal process constituted world renewal rites at each
stage, using the deceased as post-mortem sacrificial media, but that the
transformed deceased became the warranting media for the next stage.
Therefore, this would promote CBLs as a means of using and producing this symbolic pragmatic capital that, quite literally speaking, would
be regularly drawn on for the purpose of participating in wider regional
mourning/world renewal rites, thereby reproducing alliances and enhancing the reputation of participating CBL groups.

B. CBL Site Types

Notably, Charles and Buikstra have stressed that there were three
different types of Archaic CBL sites in the lower Illinois Valley, only
one of which they characterize as a "cemetery," the bluff-top CBL. The
two "non-cemetery" CBL types were situated in the bottom lands. One
they refer to as a base camp midden burial place, and the other as a
"local transaction center" having mortuary usage as a secondary function. They treat all three types in funerary terms. Unfortunately, there
is rather scant empirical knowledge of these "local transaction centers."
One of the best known is the Bullseye site.

> While the bluff top cemeteries do seem to have been just that—ceme-
> teries—it is not clear that the floodplain locations were primarily sites
> for the disposal of the dead. Rather, the diversity of artifact types and
> styles, the presence of debitage and isolated tools, and the curated nature
> óf the [bundled] burials themselves suggest that these sites, such as Bulls-
> eye, were periodically and seasonally visited by more than one residential
> community. (Charles and Buikstra 1999, 2)

The other type of floodplain CBL was in base camp midden mortuary
deposits. These are not treated as cemeteries by Charles and Buikstra
since, according to the Cemetery Model, cemeteries are specialized
burial locales constitutive of proprietorial domain. Since these base

camp burials were deposited in midden, they were buried in a place that this model would deem was not specialized for mortuary purposes, and, therefore, these locales did not count as cemeteries. This has important social structural implications. Based on the nonspecialized nature of midden burials, Charles and Buikstra conclude that these deceased would have had little or no standing in the proprietorial corporate group, almost as if they were "outcasts." In this regard, Charles et al. point out that a "major proportion of the population was excluded from the bluff top burial: those adults unable to perform the normal range of subsistence activities. This population segment characterized skeletally by a high incidence of severe bone pathology, was interred in the midden areas of habitation sites."[6]

For comparative purposes, Charles and Buikstra reviewed a number of Archaic midden CBLs and the Gibson blufftop CBL.

> A number of burials have been recovered from Archaic habitation sites in the region, specifically Modoc Rock Shelter, Koster, and Napoleon Hollow. Buikstra has compared the skeletal material from Modoc and Koster with the skeletons from the bluff-top Gibson Archaic component. The burials from Modoc and Koster contain a remarkably high incidence of severe pathologies that would affect the abilities of the individuals to perform normal activities, whereas the mound interments from Gibson are not characterized by a high frequency of incapacitating conditions. This suggests, within our framework, that those individuals unable to contribute to the economic functions of the corporate group were excluded from its bluff-top burial facility. (Charles and Buikstra 1983, 135)

In effect, the physiological attributes of the midden burials indicated that the deceased were suffering from physical disabilities and/or pathologies, probably old and, therefore, they were, in general, economically valueless. Thus, the logic of the Cemetery Model concludes that these Archaic communities were structured into two economic classes based on relative health standing, the healthy and productive at death being buried in the privileged symbolic spot on the bluff-tops; the sick, often old, and unproductive at death being buried (or cast off) in the local midden.

The Mourning/World Renewal Mortuary Model can give quite a different reading of these three CBL types. First, the floodplain "transaction center" CBLs such as Bullseye and Godar will be examined. Charles and Buikstra treat the burials as deceased from different communities, primarily because they display considerable post-mortem manipulation (bundle burials), but also because of the rich artifactual content of these sites, much of it not being directly in burial contexts. They argue that these were primarily transaction centers and funerary ceremonies were used to construct, expand, and/or sustain alliances among highly mobile and dispersed kinship-based communities.

> These floodplain sites served as gathering places for the dispersed households or lineages under regimes of high mobility. Those matters that necessitated inter-community interaction took place here: exchange of mates, valuables, and information; performance of rituals; and status competition between individuals and lineages. Caches of artifacts were deposited both with burials and in isolation, suggesting that status competition was mediated at least in part through the ability to acquire and dispose of portable wealth. (Charles and Buikstra 1999, 2)

For them the bluff-top and floodplain CBLs served domestic communities that were practicing different types of settlement strategies out of chronological phase with each other. The floodplain CBLs such as Bullseye, they claim, were produced by intercommunity cooperative efforts during periods of high mobility. During periods of more tethered mobility, the bluff-top CBLs were used to stake out proprietorial domain over surrounding territories. Therefore, according to their account, the bluff-top CBLs correlate with periods when competition for land among relatively tethered communities prevailed, while the floodplain CBLs, such as Bullseye, correlate with periods when competition for "portable wealth" among relatively untethered communities prevailed.

> Thus, the bluff top cemeteries reflect integration of the community and horizontal (segmental) differentiation among communities at times when the roots of status and power emanate from control of fixed resources,

> while the floodplain sites reflect integration among communities and vertical (hierarchical) differentiation among individuals and communities at times when status and power reside in control of portable wealth. (Charles and Buikstra 1999, 3)

That they see this model as having general and broad explanatory power is clear when they argue that much of the Midcontinental Archaic, Early and Middle Woodland prehistory can be explained in these terms.

> These associations of integration and differentiation with particular types of mortuary locations, that is, the meaning attached to these sites, formed the baseline for the evolution of mortuary behaviour in this region and they provide a key for our interpretation of these events and processes. This phenomenon is not specific to the lower Illinois River valley. Rather, this region is a well-documented example of what appears to be the general pattern over most of the Eastern Woodlands of North America. Thus, what we term Glacial Kame in the Great Lakes region, Poverty Point in Louisiana, and the Maritime Archaic in Newfoundland manifest aspects of this same spatial repertoire. *Similarly, Adena in Ohio and Hopewell in its various guises throughout the Eastern Woodlands are later Woodland developments of this Archaic tradition.* (Charles and Buikstra 1999, 4, emphases added)

This pulsating of tethered and untethered mobility invokes Brown's risk management model, of course, and it is subject to the same criticism in that the postulated cyclic abandonment and resettlement would be antithetical to the principles of proprietorship and ancestral legitimization, and would relatively quickly lead to permanent occupation by those who chose to stand up for proprietorship by entrenching their tethered mobility and, simultaneously, excluding those who were so incautious as to abandon their territory for a less tethered existence. This points to the Cemetery Model as being internally inconsistent. In environmental and demographic conditions promoting competition, proprietorial corporate groups should aggressively defend their territories against others and/or exclude weaker subgroups. Those who abandoned their territory should have done so only under the pressure of overwhelming aggression. Thus, we should expect data supporting significant periods of aggression: high incidence of fractures in skeletal data, defensive settlements, and so on, none of which is supported in the Archaic empirical data.

Notably, the cycles of tethering and untethering that they postulate would, in fact, be quite amenable to the Mourning/World Renewal Model. A reasonable tactic of people practicing a proscriptive subsistence-settlement/prescriptive ceremonial strategy would be to *encourage* periodic abandonment of territory to minimize pollution caused by their ongoing occupation and exploitation. This process, however, would be cooperatively negotiated in order to ensure continuity of reproduction of the resources in the valley regions while avoiding overexploitation of the more fragile resources in the upland regions, and so on. In these terms, there would be no evidence of aggression as required under the Cemetery Model.

As the excavators, what Hassen and Farnsworth have to say about the contents of the Bullseye site is relevant. They noted both differences and parallels between this site and the bluff-top CBL sites. In terms of similarity, the modes of burial treatment in both mark prescriptive post-mortem manipulation regimes. For example, Bullseye burials were predominantly bundle and flexed burials. Where the data from the two CBL site types particularly differ is in the artifactual associations. Bannerstones and axes were abundant in the floodplain CBL sites, while they were almost absent from the bluff-top CBL sites. However, the few bannerstones and axes that have been found in association with bluff-top CBLs, as well as the range of projectile points, display the same styles as those in the floodplain sites.[7] This suggests considerable contemporaneity. Since the base camp midden and blufftop CBLs are recognized by Charles and Buikstra as contemporary, this suggests that all three types were contemporary or else had significant temporal overlap. This still makes it possible that there was some fluctuating in the settlement pattern, as postulated by Charles and Buikstra.[8] However, whether this fluctuation can be tied to objective changes, as they argue, or, as preferred here, to tactical shifts as part of the overall deontic strategic posture, is another question.

Floodplain CBLs

According to the Proscriptive/Prescriptive Ecological Strategy Model, the Illinois River valley social arrangements of the Archaic period would constitute a polyistic locale-centric social system. Since the Mourning/World Renewal Mortuary Model is constructed as an auxiliary of the above model, it postulates that these different CBL types index at least two different social structural axes. In these terms, the Archaic period can be characterized as the initial emergence and elaboration of a dual (possibly a multiple) complementary set of custodial corporate groups, one realized as domestic communities based on principles of consanguineal and affinal kinship (i.e., descent and marriage), the other as custodial groups based on same-age and same-gender companionship. These two possibilities will be the focus here, although it is not impossible for CBLs to exist based on other types of structural axes, e.g., shamanic or female fertility cults, and so on.

Largely overlooked in North American prehistoric archaeology is the fact that in preindustrial social systems, companionship complements kinship. Although mortuary deposits were found in the Bullseye site, the excavators state that, because of the rich assemblage of axes and bannerstones, plus bifaces, drills, copper awls, plummets, pipes, and manos (grindstones), it would be "incorrect to interpret the site as only a cemetery."[9] Much of this material was not in direct association with the mortuary data. Thus, the data suggest that this site was less than a "special purpose" burial locale (i.e., it was not a "cemetery" CBL), and it was certainly not a domestic base camp locale. However, it was also more than a temporary task group locale. As quoted above, Charles and Buikstra opt for the notion of "transaction center" where "exchange of mates, valuables, and information" occurred, as well as the "performance of rituals," and so on, all related to "status competition between individuals and lineages".[10] The complex patterning of the floodplain CBL gives good reason to support their view of the non-domestic nature of this and similar sites. Certainly, exchange would have been among the activities that occurred here. However, it is postulated here that the social structures that undergirded and were manifested in the patterning of this type of center would be primarily same age-based and same gender-based companionship. Kinship relations in this context would be secondary, or even largely irrelevant to constituting the nature of the gatherings at this type of locale. The archaeological data, then, would nicely fit the type of patterning that would be produced by active adult male companions and their dependents regularly congregating to carry out specialized and primarily ritual activities.

The logistical settlement pattern that Brown correlates with the more tethered mobility pattern of the Middle Archaic would entail special purpose task groups acting largely independently but on behalf of the domestic base camps. It is usually assumed that recruitment to these task groups would be from within the given base camp community. This is possible. However, in terms of the Proscriptive/Prescriptive Ecological Strategy Model, access to territory was open. Therefore, it is postulated that typically small groups of older boys and youths from different and possibly distantly spaced base camps would regularly meet up with each other in the uplands and would commonly cooperate in collective hunting and other related tasks. It is further postulated that out of these practices, the youths would forge friendships, constituting themselves as an age cohort of "lifelong" companions. Out of this shared experience, local floodplain CBL groups would have emerged, made up of companions from many different base camp domestic communities. This dual structure of kinship-based domestic and companionship-based age cohort sodalities will be elaborated and further grounded for the Ohio Hopewell.

Midden CBLs

The Mourning/World Renewal Mortuary Model can give a quite different reading of these midden burials from that given by the Cemetery Model. The first step is to reflect on the archaeological notion of "midden." This term is no more culturally neutral than is the term "cemetery." Midden in Western culture is garbage, the unwanted, the unvalued, the dirt and detritus of everyday life, something that must be excluded and expelled from the community. Hence, those found buried in garbage must also be the detritus of society. However, in societies based on immanentist cosmologies, subsistence midwifery ritual would entail a rather complex set of proscriptions and prescriptions stipulating how to treat both the usable and unusable parts of animals and plants. The humanly unusable components might easily be identified as having spiritual essences and, therefore, be treated as apt for mediating interaction with the corresponding spiritual powers of the cosmos.

Therefore, the production of the "midden" may be part of what was involved in constituting a material context where the spirits of the animals or plants can be routinely released so that they could return to the spirit custodians and be reborn. Indeed, this suggests that the real problem facing domestic groups in such a sociocosmological world would be to distinguish between "storage" and "midden." In a transparent and transitive, immanently sacred world, resource storage must be expressively distinguished from midden and both must be distinguished from "waste." Storage of foods must be routinely marked and warranted, possibly by specially shaped pits or pits that were not used for cooking, or by using appropriately styled portable containers, such as baskets, stone vessels, or, when it emerged, pottery, that bear tangible designs that mark and constitute them as storage vessels, and even as vessels for particular types of foods. Unused food to be stored or other curated resources, then, when kept in such containers are manifestly constituted as being stored or curated for later use and not as wasted resources. In these terms, it may not mischaracterize some forms of Archaic mortuary burial as a type of "storage" of the dead. The patterning of the initial burial may be partly the result of anticipating that the dead may be disinterred, the bones cleaned and used as post-mortem sacrificial media, as evidenced by bone bundle burial, and so on.

Therefore, instead of assessing the midden burials from the Western value of midden and concluding that the deceased are merely the social detritus and waste, the midden must be assessed from the probable perspective that the deceased are critical symbolic pragmatic capital. If the deceased had important warranting powers as sacrificial media of world renewal, as argued by the Mourning/World Renewal Mortuary Model, then the midden which contains them was probably a sacredly charged matrix associated with the spirit custodians of the species that were also regularly deposited in it. Therefore, the association of human burial in such a context would logically implicate the burial behavior as midwifery/mortuary ritual directed to a specific range of species on which the community depended, in particular, reincarnating the range of species directly associated with the floodplain midden. If so, it would not be surprising that the deceased display the signs of physical deterioration caused by aging, as pointed out by Charles and Buikstra. Many of these floodplain resources may have been exploited, or their authorization to be exploited may have been given, by the senior generation, and possibly by females.[11] This interpretation suggests that the base camp midden CBLs may have been experienced as mediating a liminal sacred land/water boundary. The midden deceased, associated in their lifetimes with the valley bottom/water resources, would be participating in the dynamic relations that such a sacred natural boundary implicated.

This interpretation suggests that the explanation given by the Cemetery Model for the siting of CBLs on the bluff tops should also be reassessed. It claims that these sites were chosen because of their prominence for display purposes, thereby enhancing claims of proprietorial domain. As with the midden sites, however, the bluff-top zone may have been chosen because of its being the sacred liminal "ecotone" boundary mediating between uplands and lowlands, and the deceased found in these bluff-top CBLs, belonging to the more active groups at death, may have been spiritually as well as economically associated with the exploitation of these mixed upland/lowland resources.

DISCUSSION

In critiquing the Resource Patch Convergence Avoidance Model, it was pointed out that its central claim that mobility reduction was a deliberate strategy to avoid resource patch convergence must be preserved. However, the rationale for this strategy must be modified in terms relevant to the principle of inclusive territorialism that is critical to the Proscriptive/Prescriptive Ecological Strategy Model. In these terms, autonomous community groups would come to see that the increasing tendency for separate groups to converge on the same resources was imbalancing the sacred reproductive capacity of the world. This awareness would promote cooperative modifying of mobility schedules, leading to the increasing use of the underused floodplain resources and the corresponding reduction in the overusing of the upland resources.

This proscriptive avoidance tactic would include careful monitoring to avoid ignoring these traditional upland resources as well as to avoid too much concentration on the floodplain resources. Both overuse and underuse were to be avoided. In the former case, it would amount to a dereliction of sacred duties to the upland species since the midwifery aspect of the exploitation practices was directed to ensuring their reproduction. In the latter case, reduced mobility and the increased focus on floodplain occupation would bring about the problems as postulated by Bruce Smith's Floodplain Model, the tendency for the "domestilocalities" to open up the forest, allowing the floodplain weed plants to invade. Smith claimed that this was only belatedly perceived by Archaic foragers. Under the Proscriptive/Prescriptive Ecological Strategy Model, it is claimed that this would be quite quickly noted and compensated for by proscriptive tactics, for example, carefully limiting the period of stay in any one locale and avoiding reoccupying a locale until the natural coverage had regrown.

Nevertheless, over time the invasion of these plants would be unavoidable and recognized as such. However, the expanding range of these plants would probably come to be treated as a natural part of the environment, and the midwifery protocols would start to be applied to them. Even so, finding that increasing occupation and exploitation of the riverine zone was encouraging one set of plants at the expense of preexisting ones would be taken as an irreversible imbalancing of the sacred order. This would promote both the intensification of traditional forms of collective ritual and the innovation of new ritual in order to rectify the imbalance. Hence, a systematic co-opting and intensifying of mortuary ritual as the medium of post-mortem world renewal sacrifice would occur. By redirecting human spiritual energy back into the cosmos, this emerging prescriptive ritual strategy would be seen as reversing some of the pollution that intensifying settlement was unavoidably generating.

While such renewal ritual might justify cultivating and harvesting these invading weeds, even regular burning so as to enhance their growth, it would not necessarily extend to justifying storing seeds for future planting.[12] It might be feared that such "storage" could be mistaken as resource "wastage" by the spirit custodians and that the deliberate intervention into the natural order that planting entailed might very well initially count as an unwarranted form of polluting—unless further symbolic pragmatics were innovated. Thus, as discussed earlier, storage of seeds not for later consumption but for later planting would itself entail significant ritual innovation, both midwifery and world renewal ritual. In this regard, Watson and Kennedy have cogently argued that the role of women in the processes of domestication should be treated as active rather than as passive and, therefore, as a dynamic force in prehistory.[13] They correctly point out that the transformation of subsistence from foraging of plants to cultivation to domestication required significant innovation. They do not emphasize the sacred and symbolic pragmatic aspect, however. A deontic ecology highlights this as a crucial aspect of effective subsistence innovation in an immanently sacred world. It is suggested here that women of the Archaic and Woodland periods would be centrally involved in the innovation and diffusion of subsistence midwifery ritual, and they would play an important role, as well, in innovating world renewal ritual. The former would include appropriate ritual forms of seed–plant storage, planting, harvesting and cooking, as well as appropriate forms of midden construction, and, of course, the innovation of collective "midden" mortuary practices, such as the well-known shell mound burial practices of the Shell Mound Archaic of Tennessee and Kentucky.[14] Exploring the types of practical materials that could be anticipated in this regard cannot be done now, although some of these have been suggested with regard to storage pits, basketry, and, for the Woodland, pottery.

All this would generate a special conservatism that would easily account for the drawn-out history of the shift from mobile foraging to sedentary cultivating. Furthermore, because material warranting of new exploitative techniques always entailed collective material cultural expressivity, this would involve extensive and ongoing negotiations and experimentation across the region as part of the collective strategy of regional custodial groups resolving among themselves what material forms in the way of tool styles, midwifery taboos, and, of course, mortuary/world renewal ritual would count as warranting the escalating exploitation of new resources. Thus, along with the proscriptive, minimizing tactical adjustments of subsistence and settlement practices, there would be a correlated prescriptive intensification and specialization of mortuary practices as media of world renewal rites. Since this complex was carried out within a polyistic locale-centric social system characteristic of inclusive territorialism, the mortuary patterning of the CBLs of this region—and this can be generalized across the Eastern Woodlands—would be mediated by different types of CBLs, as suggested above (further elucidation of the latter claim to be presented later). With this demonstration of the relevance of the Proscriptive/ Prescriptive Ecological Strategy Model, it will now be directly applied to the Early and Middle Woodland periods of the Central Ohio Valley. This will require appropriate elaborations and modifications, since it is claimed here that the Ohio Hopewell episode is the climactic culmination of the subsistence, settlement, and ceremonial trajectory initiated in the Middle Archaic.

Chapter 9

EARLY/MIDDLE WOODLAND DEONTIC ECOLOGICAL STRATEGIES

Bruce Smith has made probably the most comprehensive comparative analysis of the available data of the subsistence and settlement practices of the Middle Woodland of the Eastern Woodlands. He focuses on those regions that have received the most archaeological attention in this regard: west-central Illinois, particularly the lower and central Illinois Valley; southern Illinois; and central and western Tennessee in the Upper Duck River Valley. He also assesses the data published for western Kentucky and Ohio. At the time of his publication there was very little recent material published on the Central Ohio Valley.[1] However, major volumes and articles since then have been published that summarize the most recent research in this region, and importantly, it largely confirms Smith's descriptive characterization of subsistence and settlement data as being, with some modification, broadly applicable to the Central Ohio Valley region.[2]

Probably the most interesting finding for immediate purposes is that the survey has established the invisibility, or, better stated, the nonexistence of the Hopewell nucleated "village," which had been the postulated "cornerstone" of earlier Hopewellian studies. For example, prior to 1980, a number of sites in west-central and southern Illinois that were over 2 ha in surface scatter had seemed to be candidates for village status. However, as Smith argues, subsequent excavations have established that these were typically not domestic habitation sites. Instead they displayed the specialized characteristics to be expected of sites attached to or involved in the activities associated with nearby CBLs. In short, they were predominantly related to the ceremonial sphere.

Typically, in each of the zones that he summarized, the domestic habitation sites turned out to be small, usually under 2 ha, relatively shallow, and spaced apart usually about 300–800 m and sometimes more. The archaeological contents of these sites were simple and "redundant," to use Smith's terms. When post holes were found they usually defined a small, rectilinear single-pole construction with a relative paucity of such domestic features as storage pits, fire pits, hearths, and ovens. In the Upper Duck River, he noted that usually 50 to 100 m away from the rectilinear single-pole structure there was a free-standing C-shaped structure of posts, often spoken of as a "windbreak." Faulkner has suggested that this was a "summer house" built to complement the nearby rectangular or "winter house."[3] However, the fact that the C-shaped construction has no pits or other features that could identify it as a domestic locale suggests that it might have been used for local ritual purposes.[4]

Although the precise patterning of the domestic features varies from one region to another, in any given region the prevailing pattern indicates a dispersed set of small domestic habitational sites, which Smith often refers to as households or farmsteads. They were simple in plan with minimal construction, and repeated with very little variation. He claims that there is no evidence that domestic locales in a given region varied in size and complexity, and, therefore, there is no evidence that would indicate a settlement hierarchy of hamlet→village→town, and so on. In short, a settlement system of dispersed domestic habitation locales used for seasonal and/or year-round occupation.

According to the current published data, a similar settlement patterning prevailed in the Central Ohio Valley. However, while this dispersed domestic habitational pattern appears to be common for both the Early and Middle Woodland periods, there does seem to be a significant difference. The Early Woodland domestic habitation locales were dispersed over both upland and lowland regions, while the Middle Woodland gardening-based domestic habitation locales seem to have been situated primarily on the lower hills, and on terraces and floodplains, particularly in major tributaries.[5] Particular to Middle Woodland archaeology, however, is a current debate over whether the major Ohio Hopewell earthwork locales (mounds, embankments, or combinations) were occupied in a permanent, residential sense or if they were only used periodically for ritual purposes. Furthermore, caught up in this debate is the precise occupational status of the habitation locales, the focus being on whether they were year-round sedentary hamlets or seasonal gardening base camps. These issues are central to any understanding of the Ohio Hopewell.[6]

THE PROSCRIPTIVE/PRESCRIPTIVE ECOLOGICAL STRATEGY MODEL

Smith's Floodplain Model, critically reviewed in chapter 7, argued that although they were rather underused in the Terminal Archaic of the Eastern Woodlands, both wild cultigens and domesticated varieties of starchy and oily seed-bearing plants were promoted fairly rapidly as important food staples in the Early Woodland and as nutritional staples by the Middle Woodland. The result was a mixed extensive foraging and intensive cultivating subsistence regime. This has been largely confirmed for the Central Ohio Valley by Wymer's archaeobotanical analysis of the organic content in Early, Middle, and Late Woodland domestic sites. In order to address this transformation, the Proscriptive/Prescriptive Ecological Strategy Model can be logically extended by postulating that this shift to a lowland, dispersed domestic habitation pattern based on a mixed extensive foraging and intensive cultivating regime was the outcome of the reversal of causal dominance between the subsistence-settlement and the ceremonial spheres. While the mortuary ritual innovations of the Archaic period that generated the multiple types of CBLs were largely inspired by the perceived need to compensate for the escalating sacred pollution that the intensifying subsistence and settlement practices generated, it is postulated that, starting in the Early Woodland and escalating in the Middle Woodland, the demands of the ceremonial sphere, as manifested in the growth of mound and embankment earthwork construction, intensified subsistence gardening and further tethered mobility.

Although the model postulates that subsistence-settlement intensification was driven by the escalating ceremonial sphere, the model also claims that local domestic clusters were largely unrelated to local earthworks, as such. This is because the polyistic locale-centric social systems that emerged in the Middle and Late Archaic continued into the Early Woodland period. The ideological rationale of building ceremonial CBL earthworks, combined with an escalation in the competitive pursuit of reputation among the non-kinship-based groups responsible for them, promoted an overall shift of domestic populations toward the bottom lands. This enhanced the availability of labor required to feed the escalating earthwork construction. The traditional proscriptive subsistence-settlement stance would be continued in these more constrained conditions by the emergence of local networks of dispersed domestic locales that sustained as wide a dispersal as practically possible. In short, the Proscriptive/Prescriptive Ecological Strategy Model postulates that, by the Early Woodland period of the Central Ohio Valley, the competitive and escalating pursuit of reputation among the autonomous world renewal ritual groups generated a moderately bifurcated ceremonial/settlement system that culminated in the Middle Woodland with a radically bifurcated system characterized by major monumental ceremonial earthworks and almost archaeologically invisible small, widely dispersed domestic locales.[7]

A. The Subsistence/Settlement Sphere

It has been strongly argued by William Dancey and Paul Pacheco that the dispersed domestic locales forming the core of the Ohio Middle Woodland riverine-based settlement system were sedentary year-round hamlets. As pointed out in note 6, others have argued that these dispersed habitation locales were simply a continuity of the Archaic pattern of base camp settlements occupied during the gardening season. Rather than opting for either extreme, the Proscriptive/Prescriptive Ecological Strategy Model postulates these as actual alternatives evolving in parallel with the escalating of earthwork construction programs. This conforms to the logic of a polyistic locale-centric system based on a proscriptive subsistence-settlement/prescriptive ceremonial strategic posture. The seasonal gardening base camp, on the one hand, and the sedentary domestic hamlet, on the other, would be polar limits of a proscriptive settlement system constrained by the labor demands generated by escalating earthwork construction programs. In these terms, the seasonal gardening base camp would be more affirmatively proscriptive in nature while the sedentary hamlet pattern would be more moderately proscriptive.

A seasonal gardening base camp regime would be more affirmatively proscriptive than a sedentary gardening hamlet regime. The former would minimize the concentration of sacred pollution brought about by domestic occupation and subsistence gardening by minimizing the degree of tethering of community mobility while meeting the practical requirements for more intensive gardening arising from escalating ceremonial construction. Such an affirmative proscriptive tactic might also include tactical cycling of the base camps among a proximal set. A domestic unit could return seasonally to the gardening zone to which it had custodial usufruct, but it would use a different occupation locus from one gardening season to another, cycling among three, four, or more.

In diachronic terms, the model postulates that these dispersed seasonal base-camp locales would have been initiated in the Early Woodland and continued into the Middle Woodland, with the latter being characterized by a general shift of settlement toward the valley bottom lands. This shift would likely maintain a seasonal gardening base camp regime. There would be a tendency in certain areas, however, for domestic communities to shift toward the less affirmative, more moderate pole of the proscriptive continuum by moving toward a semisedentary or even fully sedentary hamlet system while sustaining the dispersed settlement pattern. Therefore, in those areas where sedentary hamlets appeared, there should also be indications of prior seasonal base camps. Even when the dispersed hamlet system emerged in a region, a pattern of regularly shifting from one domestic locus to another proximal locus would be typical, again with the goal to allow the disturbance of the natural order that occupation would generate to be reversed.

Correlated with the affirmative↔moderate proscriptive settlement continuum would be a variation in the range of tools and features that would characterize each orientation. Typically, the seasonal gardening base camp would display a minimum range of tool types, those limited to the particular needs of such a seasonal gardening camp. The camps would be used for only the requisite May to October gardening period. During this time, they might be fully occupied when only focused subsistence tasks were required, such as clearing the land for gardens, planting, and harvesting, while a minimum occupation might be sustained to ensure periodic cultivation. Much of the harvested produce would be carried away to fall and winter camps. All this would be consistent with the affirmative proscriptive posture and would suggest a minimum range of such features as post holes, storage pits, and hearths. These would tend to be shallow, and midden would tend to be thin.

Therefore, little indication of permanent shelters would exist, and even the range of domestic activities, such as hide preparation and tool making, would be minimal since the practical reasoning of an affirmative proscriptive nature would encourage carrying out these activities in nearby special task locales. Among these practical constraints, however, might be midwifery proscriptions against mixing the material tasks and tools involved in hunting and fishing with those involved in gardening and gathering. Therefore, even though the seasonal gardening camp would be provided with game by special mobile task groups, the range of butchered and "castoff" bones and the set of tools required for this purpose would be minimal since much of the butchering would probably occur at the kill locales. The requisite tools would also be made in these locales, and, of course, since the camp would be abandoned during the non-gardening seasons, these heavier bifacial tools would probably be taken along by the inhabitants.

Those regions that practiced the more moderate proscriptive orientation would generate sedentary gardening hamlets. Because tactical minimizing of sacred pollution would still prevail, the hamlets or farmsteads would retain many of the above traits. Domestic hamlet locales would cluster to form a domestic residential tract consisting of dispersed hamlets, each being no closer than 100 m or further than about 700 m from one another.[8] Interspersed between these hamlets would be special-purpose activity locales, such as butchering stations, gardening stations, and ritual preparatory locales. Pacheco archaeologically identified the individual hamlets making up a tract by claiming they displayed uninodal clusterings of artifacts and debitage having "crisp" boundaries.[9] He defines these as "formal" refuse deposits or refuse dumps in contrast to the associated special-purpose locales, which he characterized by their having low densities of artifacts and debitage with dispersed or non-uninodal distribution having diffuse boundaries, largely indicative, he claims, of casual or informal refuse deposit. These diffusely formed clusters could be either the result of singular events or a series of temporally staggered but spatially overlapping events, that is, activity palimpsests.

Based on these criteria, and separated by areas that had only a very thin scattering of debitage (and, even more rarely, artifacts), Pacheco recognizes five uninodal clusterings and five (possibly six) diffuse clusterings, making a total of ten (or possibly eleven) separate clusterings constituting the Murphy Tract. These were distributed across the local terrace roughly following the course of a palaeostream, with the five uninodal clusterings, deemed as single-family hamlets, linearly located close to this palaeostream, while the five (or six) low-density, diffuse clusterings were set somewhat away from it.[10]

Discussion

Pacheco takes his analysis as empirically grounding the Dispersed Hamlet/Vacant Center Model, or, in his terms, the Vacant Center

Model. This model characterizes the dispersed hamlets as sedentary. But, in fact, his finding demonstrates only that the Ohio Middle Woodland settlement was not nucleated but dispersed. This is an important finding, but it does not settle the question of whether it was sedentary or seasonal. Nor does it settle the equally important dual claims of the same model, these being that (1) the cluster of dispersed "hamlets" and the proximal earthworks were produced and used by the same social group, and (2), in most cases, this social group was a proprietorial corporate community based on the descent kinship structure. I consider this dual set of claims, one could almost say, taken for granted assumptions, to be the heart of the Dispersed Hamlet/Vacant Center Model. At this point, however, it is appropriate to assess only the claims it makes about Middle Woodland domestic settlement and subsistence patterns in the Central Ohio Valley. The claims it makes about the nature of the social organizations that these patterns manifest are to be fully addressed shortly.

The Proscriptive/Prescriptive Ecological Strategy Model takes a more flexible approach to this question. As pointed out above, it postulates an affirmative↔moderate proscriptive settlement continuum, with the historical tendency for settlement to move from the affirmative toward the moderate pole. Pacheco clearly recognizes that the field method of collecting material from the surface of the plough zone can collapse a series of unrelated activities and events into a single assemblage.[11] However, he explicitly claims that sorting out the assemblage into separate prehistoric units can be effected by using stylistic variations, and by avoiding specific "microfunctional" interpretations. He claims that the clusters that his technique generates are indicative of real settlement similarities and differences. Nevertheless, if these clusters are treated as palimpsests, as Pacheco specifically recognizes them to be, then the same data are equally or even more coherently interpreted as marking a transforming history of settlement, initially based on seasonal gardening base camps and then shifting to the more permanent, sedentary hamlet pattern.

In short, the Murphy Tract site could be taken as evidence not for a historically stable sedentary hamlet system but for the shift from the affirmative to the moderate proscriptive settlement orientation. Certainly the five or six clusterings that have somewhat amorphous patterning would not be inconsistent with seasonal gardening base camps serially revisited over several seasons. The five major uninodal clusters of relatively high density and "crisp" artifactual and debitage boundaries probably do index a real "settling down." But they also probably mask earlier components of the same order as the diffuse clusters. Furthermore, even if they indicate a settling down process, it does not follow that these more permanent domestic locales oriented to gardening activities were continually occupied for a generation or so. They could as easily be the result of a shifting sedentism, as the domestic group cyclically occupied and reoccupied the same spatially close "nodal" locales.

Because the patterning of Murphy I reveals structural differentiation into living area, food preparation area, general working area, and specialized depositional zone, and because of the almost complete lack of observable overlap in such features as post holes and hearths, Dancey has argued that this site was permanently occupied for the duration of use, starting as early as ca. A.D. 100 and being continuously occupied to ca. A.D. 200–250. He also recognizes, however, and pointed out that these dates can be used to support two separate occupations, one around A.D. 100 and the other around A.D. 200.[12] Furthermore, as both Dancey and Pacheco point out, the plough zone is about 30 cm deep so that the many years of modern farming practices have done a great deal of damage. Therefore, even though Dancey notes that the features identified display little overlap, these may represent only the deepest part of the deepest features. These lowest parts would probably be the least subject to modification resulting from reuse. Also, since proscriptive settlement tactics would result in minimizing occupational modification, little overlap could be expected. Further research involving sub-plough zone excavation might help resolve these questions. In any case, as the current status of these data seem to indicate, they are not inconsistent with a historical development from an affirmative to a moderate proscriptive settlement system, as outlined above, thereby giving some important support to the view that settlement decisions were as much ideological as practical.

B. Ceremonial Modes as World Renewal Strategies

Up to this point, I have focused on the subsistence and settlement practices in terms of the essential contradiction. However, it is clear that the same ambivalence toward material intervention would apply to the ceremonial sphere. Even though the destructive moment of earthwork construction could be justified on the grounds that the earthwork was needed to serve as the monumental warranting context by which to constitute ritual practices that enhanced the sacred state of the world, it also entailed deliberate, destructive intervention into the natural order. Therefore, even its builders would have to concede that it also had its intrinsic sacred polluting moment. Therefore, just as hunting and gardening had both destructive and reproductive moments, so also would earthwork construction. The problem facing prehistoric populations, then, would be one of balancing the destructive polluting costs of constructing the sacred embankment with the reproductive and sanctifying gains made by the world renewal ceremonies this construction enabled. As argued under the contingency of the cosmology→ideology relationship, there is no neat recipe inherent in the belief structure itself to resolve this problem. Indeed, experiencing the pursuit of life as an essential contradiction is the logical outcome of the core belief of the cosmology, the Sacred Earth Principle, making that pursuit possible. Whatever would count as a properly balanced ceremonial world renewal regime, that is, what particular ideological posture will prevail, will necessarily be the outcome of negotiations and dispute, and the latter presuppose competition and ideological factionalism, or so it will be argued in detail later.

In these terms, while earthworks present the archaeologist with a map by which to reconstruct the ideological debate that generated their construction and usage, they also represent a methodological constraint. While a major embankment earthwork site clearly manifests a developmental history of a prescriptive ceremonial program that mobilized a major labor force, and therefore involved a relatively large population, it cannot be assumed that a minor site entailed a small support population. It could simply be that the minor site was the work of an equally large group that ideologically opted for a less intrusive construction program. This also means that it was not automatic that the reputation of a group was a function of the magnitude of its earthwork. If a population opted for a less intrusive construction mode of world renewal than its neighbors, they would have perceived the works of these latter as worthy not of admiration but of scorn.

Furthermore, earthwork magnitude would not be the only ideologically meaningful variable at work. Two groups could carry out prescriptive ceremonial programs with earthworks of equivalent magnitude but these could be radically different in form, for instance, the T-Form of the Miami Fort Tradition and the Complex G-Form of the Chillicothe Tradition. In these terms, at the regional and interregional level, a "patchy" distribution of world renewal earthworks could be expected, patchy in terms of magnitude, form, and presence/absence. Within a region such as south-central Ohio, the patchiness might be partly the result of the labor costs involved. Thus a large labor pool across the region might "feed" a few rather distantly spaced but large and complex embankment earthwork locales. In terms of interregional variation, those regions with little or no earthwork construction might be the result of opposition to the prescriptive character of neighboring regions. Finally, in neighboring regions that display monumental earthworks that entail equivalent labor but highly different earthwork forms, commitment to the same intensity of prescriptive construction is indicated but contrasts in forms might display some serious, ongoing ideological or even social structural differences.

Even in cases where a common prescriptive ceremonial regime might prevail, such as in the immediate Chillicothe region, there would be a lack of full ideological agreement in any given earthwork group. The debate over size, form, and aggregation of an earthwork locale would be carried on through factions split between alternative positions, one pushing to increase and the other to moderate the intensity of the embankment construction and related practices. Ideological debate over alternative tactics would both promote factions and reproduce factionalism. Rapid reversals or escalations of ceremonial programs could be expected to occur. Therefore, in any given region, the construction trajectory was not a foregone conclusion.

Given the above, it is not surprising that the Early and Middle Woodland periods of the Central Ohio Valley form a patchwork of divergent subregions that, while individually focusing on one range of possibilities, overall, display the full range, from a restrained, moderately prescriptive bias that was not too different from that characteristic of the Archaic CBLs to a radically escalating prescriptive bias, as manifested in much of southern and central Ohio. While this variation would have been largely the result of factional negotiation and intraregional dispute over the size and shape earthworks must have to count as warranting devices of world renewal ritual, the model postulates that there would also be debate across the Eastern Woodlands of the Midcontinent in these terms. For example, while a number of regions where Hopewellian stylistics are found would correspond fairly closely to the Central Ohio Valley in terms of mound earthwork construction (e.g., Illinois and western Tennessee), most areas appeared to have opted out of or strongly resisted any major expansion into embankment earthwork construction, and many may have resisted any form of earthwork construction.

While the overall model stresses the normative cultural structuring of subsistence, settlement, and ceremony, it clearly requires postulating a social structural context. Only in this way can it account for the radical contrast between the redundantly patterned and rather austere dispersed domestic habitation locales as described above and the highly variable and baroque earthwork ceremonial locales. To this end, the premise of the disembedded nature of the social structural axes that was outlined for the Archaic archaeological record will be elaborated in order to formulate what will be termed the Dual Clan-Cult Model of Ohio Hopewell. However, before initiating this and further auxiliary models of the Proscriptive/Prescriptive Ecological Strategy Model, the hermeneutic spiral justifies devoting the rest of the chapter to summarizing and critiquing the current and dominant model of this settlement pattern of dispersed hamlets and "vacant centers." Although there are variants of it, these will be treated generically as the Dispersed Hamlet/Vacant Center Model.

The Dispersed Hamlet/Vacant Center Model

Probably the most influential socioecological characterization of Ohio Hopewell was initially presented in the 1960s by Olaf Prufer as the Vacant Center thesis. At that time, Prufer proposed treating the Middle Woodland subsistence and settlement system as being based on small farmsteads in dispersed clusters around the Ohio Hopewell earthworks, which acted as "great ceremonial centers in a manner similar to that of the Vacant Center Dispersed Agricultural Hamlet pattern of Meso-America."[13] In the recent refining and recasting of this thesis by William Dancey and Paul Pacheco, it is clear that the social structure and tenurial posture underwriting this Middle Woodland period Ohio "dispersed hamlet/earthwork center" settlement pattern is treated as monistic and modular in nature.

> Ohio Hopewell communities are fundamentally linked to the vacant center-dispersed hamlet dichotomy. The spectacular aspects of the Hopewell societies stem from the social relationships created by interacting dispersed households at vacant centers. If communities were not dispersed, the earthwork/mound centers would cease to function in the same manner. . . . These [vacant centers] existed for and by the efforts of the dispersed communities. (Pacheco 1997, 59, also 1993, 222 and 224)

In short, the earthworks are treated as centers of kinship-based proprietorial corporate domains of tribal communities. Identifying the occupants of the dispersed clustering of "hamlets" as the proprietors of the nearby earthworks is revealingly expressed by Pacheco as the Vacant Center Model. This implies that the dispersed hamlets are "swallowed" or subsumed by the vacant center. Then, more recently, he and Dancey expressed this view slightly differently as the Dispersed Sedentary Community Model, thereby having the dispersed hamlets "swallow" or subsume the vacant centers.[14] These "swallowings" suggest that the nature of the social relation manifested in the earthwork/habitation dichotomy is unproblematically the fusion of the two with the dispersed hamlets being the domestic face and the earthworks the ceremonial face of the self-same, monistic kinship-based group. This comes through clearly in their review of the current "challenges" to this model.

They classify the "challenges" of their model into four categories, labeled as: (1) the Nucleated Sedentary challenge, (2) the Semi-Permanent Sedentary challenge, (3) the Central Place challenge, and (4) the Seasonal Mobility challenge.[15] Predating Prufer's view, versions of the Nucleated Sedentary challenge claim that the embankment earthworks were the focus of nucleated sedentary communities that occupied the land adjacent to the earthworks. The Semi-Permanent Sedentary challenge versions, they claim, are similar to the former except that the adjacent domestic occupation was only seasonal. The Central Place challenge versions treat the earthworks as occupied by the residential locales of an elite corps of an hierarchical social system, constituting simple chiefdoms. The Seasonal Mobility challenges claim the domestic locales were seasonal gardening base camps.

On the basis of the distribution of occupational debris and cultural features (e.g., post molds and pits in and around the embankment earthworks), they reject both the Nucleated Sedentary and Semi-Permanent Sedentary versions, claiming that the intensity of midden, scattered surface debris, and so on, are insufficient for nucleated habitational settlements, whether year-round or seasonal. They also reject the versions of the Central Place challenge, seeing the timber structures found within many of the embankment earthworks not as the residents of chiefs but as special purpose ritual loci. Furthermore, they do not see any evidence in the size and patterning of domestic habitational locales that would justify claiming an hierarchical order. Interestingly, while they admit that the Seasonal Mobility challenges might come closest to their own Dispersed Sedentary Community Model, they also reject these claims on various grounds, including the claim that the artifactual patterns of the known lowland "hamlet" and upland sites do not support the seasonal occupation claim.

What is clear from this review is that, while differing on the occupational statuses of the two types of locales, all five settlement models characterize the earthwork/habitation dichotomy in monistic modular terms. That is, all five share the same assumptions and basically replicate the same premises at the core of the Cemetery Model of the Archaic: Each earthwork was the local ceremonial center and "cemetery" of the associated habitational cluster. This constitutes the earthwork and nearby hamlet clusters as a monistic polity with rather well defined boundaries that interacts in various ways with its neighboring peer polities. Despite the legitimate debate over the occupational statuses of these two types of locales, their sharing this central monistic modular perspective justifies conveniently lumping them under the global title of the Dispersed Hamlet/Vacant Center Model.

CRITIQUE

The rest of this chapter critically examines the monistic modular aspects of the Dispersed Hamlet/Vacant Center Model. Four critical arguments will be deployed: (1) the Adaptation Argument critique; (2) the "Gravity Well" critique; (3) the Artifact Distribution critique; and (4) the Earthwork Distribution and Morphology critique.

1. The Adaptation Argument Critique

At the core of the Dispersed Hamlet/Vacant Center Model there is a contradictory ecological-adaptational claim. This is endemic to the models under the Exclusive Territorial/Proprietorial Domain Paradigm. The central claim replicates the earlier Cemetery Model argument, namely, that under conditions of objective stress, groups must carve out foraging territories and aggressively defend their boundaries. Thus earthwork construction functioned adaptively by promoting social integration and legitimizing exclusive control of its territory. However, it turns out that the exclusive territorial peer polities that emerge as an adaptation to ensure control of adequate survival resources must come to actively pursue alliances with each other in order to overcome the maladaptive aspects of that very same exclusive territorialism.

> Above the scale of individual communities, functionally similar, contiguous communities may form peer polities anchored in centrally located public works, such as represented by the Hopewell, Newark, Portsmouth, and Turner earthwork groups, to name a few. These polities are located at the intersection of major physiographic provinces and originate out of economic considerations. Thus, while adequate focal resources can be found nearly everywhere within a community's catchment, long-term cycles of abundance and scarcity make for risks in the subsistence strategy. Intercommunity exchange balances out such fluctuations. (Dancey and Pacheco 1997, 9–10)

Surely, a more parsimonious account is to argue for cooperative inclusiveness as an essential method "from the beginning." In fact, this rather ad hoc "patch-up" of models under the Exclusive Territorial / Proprietorial Domain Paradigm is not a unique move. Rather, it is a well-used deus ex machina by which to overcome the contradictions and try to empirically demonstrate its premises. It actually first may have been put forward for the Central Ohio Valley by David Brose, and it recently has been invoked for the Adena by Berle Clay, and Jimmy Railey. It appears to be a case of having one's cake and eating it too.

2. The "Gravity Well" Critique

What empirical evidence can actually be marshalled in support of the monistic modular nature of the social structure as posited by the Dispersed Hamlet / Vacant Center Model? The baroque / austere contrast of the distribution of artifacts, facilities, and features constituting the earthwork / habitation dichotomy certainly does not support this claim, a point which will be discussed in more detail later. However, there is the claim that Ohio Hopewell domestic locales are slightly less dispersed in direct proportion to their spatial relation to ceremonial earthwork locales. To put it figuratively, this treats the earthworks much as if they had a "gravity well" effect on the dispersed hamlets. Thus, the centripetal "force" of the earthworks tended to pull the habitational locales toward it while the practical imperatives of intensive gardening and foraging exerted centrifugal "force" encouraging dispersal. The implication here is that if the earthworks were subtracted from the environment, then the distribution of the domestic locales would be governed only by the imperatives of (nonsymbolic pragmatic) subsistence strategies and the local variation in the topography, soils, water drainage, and so on.

The distribution data of the domestic locales are insufficient to demonstrate this alleged "gravity well" effect, particularly since the temporal ordering of these domestic habitational locales is shaky at best.[16] Even if the empirical data were adequate, this is very circumstantial evidence to support the claim, particularly given the absence of a rich sharing of artifactual materials between these two spheres. An alternative explanation for the claim that the dispersed habitation sites are drawn toward the nearby earthworks, at least for the C-R Configuration embankment earthworks, could be that the valley bottoms with wide terraces were a requirement for the latter and, at the same time, these valley bottoms were desirable as domestic habitation locales for more intensive gardening regimes. The fact that both occupied the same zones could be explained in quite separate terms. Finally, when the distribution of the earthworks of all three traditions is examined, it will be shown that there is more variation within and across regions than can be accounted for by the "gravity well" view.

3. The Artifact Distribution Critique

A major empirical claim is that the mortuary practices at the earthwork locales were exclusively for members of the local kinship-based polity (or allied polities?) and that they were funerary in nature, although it is argued that other types of ritual could also have been enacted at these centers.[17] However, the empirical distribution of these mortuary and ritual data does not give much support to this claim. If the ritual assemblages that are so prominent on the floors of the mortuary mounds associated with the embankment earthworks are the responsibility of the local community of dispersed "hamlets," then it is indeed puzzling that there is so little overlap in artifactual material between these two locale types. Of course, there is sufficient artifactual overlap to conclude that these two types of locales were contemporary. In the habitation locales, there is some scattering of waste mica, utilitarian Scioto pottery, actually a rich representation of Hopewell bladelets, and so on, and to confirm this there are overlapping radiocarbon dates.

However, it is what is missing that is important. If the earthwork-habitation relation was monistic in nature, then it would make the earthworks the front stage of major ritual performances and the nearby although dispersed habitational locales would have served as the backstage where much of the preritual preparation would be carried out. Since the model hinges on the notion that the embankment locales would be "vacant" for most of the year, while the "hamlets" would be occupied year-round, or at least seasonally occupied in the case of gardening base camps, then it would be reasonable to find that much of the ritual paraphernalia associated with mortuary deposits on the floors of the embankment earthwork mounds would have been produced and even, possibly, stored at the domestic habitation locales. Not only should there be some lost and misplaced artifacts of the type found at the earthwork sites in the habitation locales, but much of the production debitage of this regalia should be found in the habitations. In fact, neither is the case and, indeed, there is a rich representation of such preparatory debitage in the earthwork locales themselves.

There is one very important exception to this lack of overlap, however. It has been long recognized that the Ohio Hopewell bladelet is a major artifact category that is found in both locales. There was no apparent specialization of production and distribution of the bladelet, as was the case for the complex range of Ohio Hopewell styled materials. On initial appearances, then, the bladelet distribution would seem to support the Dispersed Hamlet/Vacant Center Model. This question, therefore, deserves a much more detailed discussion and it will be returned to in the following chapter and serve as an extension of this critique of the Dispersed Hamlet/Vacant Center Model.

4. Earthwork Distribution and Morphology Critique

In terms of the "gravity well" effect, the social commitment generated by the embankment construction and use is assumed to provide the centripetal force to counter the centrifugal tendencies of an intensive mixed gardening, extensive foraging subsistence strategy. This claim relies only on the logic presupposed by the functional ecological perspective. All that is needed of an earthwork, then, is that it should display sufficient magnitude to be a convincingly strong "gravity well," one that will be impressive and account for the postulated centripetal pull.

However, the variation in the siting of Middle Woodland earthworks suggests factors that were surplus to the effect postulated by this "gravity well" notion. For example, in Kentucky and in the Hocking Valley of southeastern Ohio, there are Late Adena mound and sacred circle earthwork locales that probably overlap in time with the early and middle Middle Woodland, and they are characteristically in the upland zone. While in the Scioto drainage the C-R Configuration locales are almost exclusively on the terraces in the river bottoms of both major and lesser tributaries, the few T-Form embankment works of this region, such as Fort Hill and Spruce Hill, are on ridge tops, replicating the siting of those found in southwestern Ohio. The "gravity well" notion also does not account for variation in distribution among the different traditions. Whole tributary drainages lack any significant Ohio Hopewell works (e.g., the Hocking drainage), or have only Mt. Horeb Tradition earthworks that are largely contemporary with the Ohio Hopewell in neighboring zones (e.g., the Big Sandy and the Hocking drainages), or have both Mt. Horeb Tradition and Hopewell-like embankment earthworks that are uniquely different from both Miami Fort and Chillicothe Traditions (e.g., east-central Indiana).[18] In contrast, neighboring tributary drainages are richly endowed with both Mt. Horeb Tradition and Chillicothe Tradition embankment earthworks (e.g., the Upper Muskingum and the Central and Lower Scioto drainages).

All this morphological and siting variation, earlier postulated as expected under the Proscriptive/Prescriptive Ecological Strategy Model, hardly suggests a simple "gravity well" effect promoted by monistic modular social systems. Indeed, the modular perspective assumes the replication of similarly structured communities across the landscape. Could any community lacking an embankment earthwork, even though it had mortuary mounds, be recognized as an autonomous polity, a peer of those in the neighboring drainage where the whole range of Early and Middle Woodland embankment earthworks is richly represented? Or for that matter, would neighboring earthworks displaying different morphology count as communities that recognized each other as peers? Not slightly addressed by this view is the internal dual variation of these traditions. To be discussed shortly, the Mt. Horeb Tradition had free-standing paired-pole circles and simple sacred circles. The Miami Fort Tradition had the T-Form and C-Form embankment earthworks, as described earlier. Finally, of course, the Chillicothe Tradition had the Complex G-Form embankment earthworks, best exemplified by the Sacred Dual C-R Motif. Furthermore, in the same region there are expressions of the Mt. Horeb sacred circle and the T-Form and C-Form embankment earthworks (e.g., Fort Hill, fig. P. 5, and Cedar Bank, fig. 2.13, respectively). None of this variation is even remotely addressed by the Dispersed Hamlet/Vacant Center Model. It effectively lumps these different earthwork forms together and treats them as equivalent, whether mound or embankment; whether T-Form or C-Form; whether SR-Profile, SL-Profile, or K-Profile; whether High Bank or Paint Creek C-R, and so on.

However, there are some exceptions to this lack of nuanced treatment. By focusing on Adena ceremonial locales, Clay has recognized that variation in the Early Woodland forms may be ritually meaningful. Similarly, Greber and Connolly have recognized that the patterning of the Ohio Hopewellian embankment earthworks may hold considerable social organizational significance. However, possibly representing the majority position, Dancey and Pacheco, while recognizing that an earthwork locale might develop over time, do not give a coherent account of the sequential growth pattern, which is to them a simple aggregation of mounds, the addition of a possible circle around a mound, the addition of one or more "geometricals," and the termination of the site with an extended embankment patched on. Thus, it seems that any form of earthwork would do as the index of a peer polity modular community, as long as it had sufficient magnitude to be impressive.

Chapter 10

CULT, CLAN, AND RITUAL SPHERES

The Dual Clan-Cult Model postulates that the earthwork/habitation dichotomy manifests the simultaneous existence of at least two "virtual" and mutually autonomous locale-centric social networks, differentiated in terms of social structural axes. Specifically, it postulates that one network was based on principles of consanguineal and affinal kinship (i.e., descent and marriage), and the other was based on the principle of same-gender/same-generation companionship. Each local kinship network, constituting one or more clans, was caught up in a network of alliances of mutually autonomous peer clans, making up the clan network of a region. The custodial groups of same-gender/same-generation companionship—lifelong male companions belonging to the same age-grade—constituted world renewal cults. Each cult was an autonomous sodality, autonomous from both the local clan network and from each other, and, typically, each was responsible for building and using one or more earthwork locales.

Because of the inclusive nature of the custodial system, and because the cult system was based on companionship rather than kinship, local cults would draw membership from both proximal and distal clan networks. Furthermore, because of its specialized ritual nature, the social reach of a world renewal cult would extend indefinitely through ramifying contacts with similar cults, allowing for the development of transregional networks, as manifested in widespread stylistic patterns. The point of these alliances would be both to enhance the effects of world renewal ritual and, very importantly, to gain knowledge of and access to widely distributed raw materials that were perceived as critical in the production of ritual warrants, such as obsidian, selected cherts, ocean shells, copper, silver, mica, and so on. Undergirding the dynamics of this interaction would be the pursuit of the cult's reputation. In contrast to the far-flung relations of the cults, the geographical scope of clans would be largely limited to their own and neighboring drainages, and mediated through marriage alliances.

It is worth repeating here that the age-grade structure is an ancient institution tightly tied into Native American cultures, structuring religious, military, subsistence, and artisanal specialization. As Hall argues, among Siouian-speaking peoples, age-grade structures having initiation and world renewal ritual duties may reach back well into the Middle Woodland and, if the association of bannerstones with the central constitutive warrant of initiation in the Hidatsa and Crow Stone Hammer Societies is valid, probably into the Middle Archaic.[1]

Following the Dispersed Hamlet/Vacant Center Model in this one regard, the Dual Clan-Cult Model postulates that, in keeping with the small size, redundant patterning, and dispersed distribution of habitation locales, the domestic units would be nuclear and small extended families. The heads of the units in a local network cluster such as the Murphy Tract site would be lineage and clan siblings. Therefore, egalitarian principles would have prevailed in terms of interhabitational leadership relations. However, since the principle of unilineal descent is based on subordination of the junior to the senior generation, the extended family of the domestic unit would be hierarchically structured by generations. In contrast to the hierarchical nature of ascribed kinship positions, because the world renewal cults of male companions were organized as age cohorts, cult membership would probably be voluntary, and internal relations would have been structured on equality. As such, it is an excellent context for the promotion of communitas.[2]

Much the same ecological conditions that promoted age cohort peer interaction that were postulated for the Archaic would prevail in the Early and Middle Woodland periods. Systems of dispersed domestic habitation locales may have had certain benefits with regard to gardening and extensive wild plant foraging, but as small extended families, organizing cooperative hunting by the younger males within the same dispersed kin network would be difficult. Therefore, logistical special task groups would be made up of small groups of young males who would regularly meet in the upland zones to cooperate in hunting and other long-distance tasks.

Pacheco, in fact, addresses the same point when assessing the role of cooperation in hunting and other collective activities of the Middle Woodland. He concludes quite differently, however, claiming that collective cooperation of the above type was not necessary. While hunting deer played a primary role in subsistence foraging, the most efficient hunting tactic, he claims, would be individual stalking.[3] He may be right to argue that individual stalking is the most efficient means of deer hunting. But stalking is a collective activity, since it requires cooperative communication among individual stalkers to ensure proper spacing across the landscape, and preventing scattering the herd. Furthermore, unlike the Late Woodland use of the bow and arrow, which may have afforded more efficient individual stalking, the Middle Woodland technique oriented around the spear and spear thrower complex, inherited from the Archaic period, may have required a combination of collective stalking, driving, and entrapment. Finally, butchering and carrying the carcasses would call for significant cooperation in labor and general sharing. These cooperative tasks would be the medium by which lifelong bonds of companionship were forged.

Finally, when these young companions approached middle adolescence, they would gravitate to one or another cult organization to go into training as pre-initiates. The cults, regularly regenerated by each new generation of youths, would be characterized as age-grade sodalities based on self-selected, voluntaristic companionship. Their primary, although not exclusive raison d'être, would be to perform world renewal rites to sustain the balance in human-cosmic relations. The over-all result would be, minimally, a dualistic locale-centric social system as postulated under the Dual Clan-Cult Model.

A concrete analogical example of how a polyistic locale-centric social system operates can be very useful here. The Nyakyusa of East Africa were a mixed cattle husbandry and horticultural people of East Africa based on patrilineal descent groups possessing the cattle and male-specific age-grades making up the domestic villages. Each age-grade had a generational span of 30–35 years. The savanna was common land under inclusive tenure and so it was open to cattle grazing by all. Individual villages were based on sacred zero-dimensional custodial tenure and each were occupied by the core age cohort of companions, along with their wives and children. The land immediately surrounding the village was used for gardens cultivated by the women. On retiring from active village life, the senior males became elders, broke up as an age cohort, and took on lineage cult duties by occupying sacred forest locales where the lineage ancestors dwelt and there they carried out rituals directed to encouraging the ancestors to sustain the lineage's cattle herds.

The conditions generating the Nyakyusa system were different but equivalent to those postulated for Ohio Hopewell. Boys from the different age cohort villages cared for their fathers' cattle. From the ages of 7 to 15, they were encouraged to tend the cattle in the savanna, well away from the gardens surrounding their parents' villages. During their initial savanna sojourn, young boys of the same approximate age from the different, widely dispersed villages would chance to meet up with each other and they would become an age cohort cooperating and sharing the care of their cattle. Because a group of companions would all be within the same 5 to 8 year age spread, this meant that male siblings would often end up in different groups of companions. A group of young companions would be encouraged by their parents to form semi-permanent camps, finally, as they matured, forming their own age cohort village.

> Each age-grade, or generation, covers a span of 30 to 35 years, and each village comprises an age-set with a span of 5 to 8 years. . . . The average age of members in each age-grade varies with the date of the last "coming out" ceremony. Just before the ceremony elders with ritual functions are those over 65 or 70; the mature men of the ruling generation are those between 35 and 65; and the "boys" who have not yet "come out," those between 10 and 35. Just after the ceremony the ruling generation consists of those between about 10 and 35, anyone over 35 being an elder. (Wilson 1963, 32)

In structural terms, the Nyakyusa dual, open-networked social system of mutually autonomous villages and forest-based lineage ancestor cults is much along the lines postulated by the Dual Clan-Cult Model postulates for the Ohio Early and Middle Woodland, except in reverse. While the Nyakyusa village was based on companionship, the domestic domain of the Early and Middle Woodland would have been based on descent. The Nyakyusa sacred forest cults were ancestor-based and managed by lineage elders, while the Early and Middle Woodland cults would have been responsible for world renewal ritual and based on age-grade companionship.

DEMONSTRATING THE DUAL CLAN-CULT MODEL

While the mutually exclusive social networks postulated by the Dual Clan-Cult Model would differ in terms of interests, the clan having domestic and kinship alliance concerns, and the cult having world renewal concerns, this does not mean that only the cults and not the clans had a cosmological and ideological basis. Domestic subsistence and settlement practices constituted a ritual as well as an economic sphere; and the religious sphere of the cults had ecological interests as part of its core world renewal concerns. Therefore, these two organizational structures would also constitute two mutually autonomous ritual spheres, in accordance with the autonomy of clan and cult.

Typically, autonomous ritual spheres maintain their separate iden-
tifications through mutual symbolic contrast. Separation of socio-sacred
locales is maintained by explicit proscriptions that safeguard the sanctity
of the two groups. This is not to claim that one ritual sphere had
greater or lesser sanctity than the other. It simply argues that the operation of two distinct ritual spheres based on different structural axes
leads to maintaining their respective purity by a general avoidance of
mixing the resources, objects, and activities across the spheres.
Therefore, it is logical to postulate that the earthwork/habitation dichotomy not only manifested a companionship/kinship contrast, but
also grounded a duality of ritual spheres with the sanctity of these two
spheres being maintained by strong proscriptions by which to avoid
the mixing of ritual resources, objects, and activities across the spheres.
This will be termed the Dual Ritual Spheres Model and it will be
treated as a key ancillary to the Dual Clan-Cult Model—that is, demonstration of the Dual Ritual Spheres Model is considered here to be
direct support of the Dual Clan-Cult Model.

Because the Dispersed Hamlet/Vacant Center Model treats the
earthwork/habitation dichotomy as manifesting a monistic modular
polity in which the earthwork locale is the ancestral mortuary center
of the dispersed domestic habitations, it follows that only one ritual
sphere exists. While it would be expected that some family ritual would
be held in the habitation locale, since the earthwork was the community center of the corporate kin group, the collective ceremonial
activity would be performed there. This can be called the Unitary
Ritual Sphere Model and it can be treated as ancillary to the Dispersed
Hamlet/Vacant Center Model.

Under the Dual Ritual Spheres Model, while there are strong
proscriptions separating resources, materials, and rituals of these two
spheres, the membership would not be mutually exclusive. Individuals
making up the polyistic locale-centric social system are members of
both a clan and, typically, at least one cult. Thus, individual X is a
member of Clan Y and a member of Cult Z and is expected to move
regularly and freely from the Clan Y locale to the Cult Z locale in

accordance with his duties in each. Under the Dual Ritual Spheres Model, a basic practical problem would be ensuring that, in moving between these two ritually contrasting locales, individuals would not mix the ritual materials of one sphere with those of the other. This problem would not exist in a modular system since there would be no proscription between moving ritual resources from habitation to earthwork, or vice versa. Instead, the proscriptions would apply between different ritual locales since, of course, each earthwork locale defines an exclusive corporate polity. Individuals would be full members only of their natal ritual earthwork and "adopted" members of their affinal ritual earthwork. Therefore, they would not have free access to the earthworks of neighboring modular polities to which they had no descent or affinal affiliations.

Thus, the Dual Ritual Spheres Model and the Unitary Ritual Sphere Model can be used to generate different expectations of the distribution of ritual paraphernalia between earthwork and habitation locales. The range of material artifacts, facilities, and features associated with the Ohio Hopewell earthworks (and Mt. Horeb) generally identify what would count as the major ritual paraphernalia of the Central Ohio Valley of these periods. Under the Dual Ritual Spheres Model it is clear that essentially none of these material resources should be found in the dispersed habitation locales. Of course, the reverse applies. Essentially none of the ritual paraphernalia of the habitation ritual sphere should be found in the earthwork locales. Since the proscription applies to ritual materials, this would not apply to "ordinary" materials, those associated with mundane activities, such as preparing food for meals that were not part of the earthwork or habitation ritual practices. In contrast, under the Unitary Ritual Sphere Model, even though major and even lesser rituals would occur in the earthworks, there would be no reason for some lesser rituals not to be held in the habitation locales. Moreover, under the notion that the earthworks were "vacant," it could be anticipated that major ritual preparatory production and even storage would occur in the habitation locales.

DISCUSSION

It is not necessary to repeat the summary of the distribution pattern of the Ohio Hopewell ritual paraphernalia made in the last chapter. Under the Artifact Distribution Critique it was clearly noted that little in the way of Ohio Hopewellian ritual paraphernalia is found in the habitation zone. Even the production of ritual paraphernalia apparently occurred primarily in the earthwork locales. Also, it has long been noted that most of the major embankment earthworks have associated timber structures, some quite large, and there seems to have been a specialization, some being used for the production and possible storage of ritual paraphernalia, and the others being used as the focus of ritual, in particular mortuary.[4] This all suggests that the test implications of the Dual Ritual Spheres Model are favored and, therefore, that this model gives the more coherent account of this distribution.

However, one very important artifact, the Hopewell bladelet, in fact, would appear to work against this conclusion and in support of the Unitary Ritual Sphere Model. In the previous chapter, it was pointed out that the bladelet is ubiquitous and prolific across the earthwork/habitation dichotomy.[5] Furthermore, wear analyses have demonstrated that it displays the same range of wear and damage across both spheres, suggesting that objectively it was used in an equivalent manner no matter in which locale. This clearly would seem to demonstrate that the proscriptions postulated under the Dual Ritual Spheres Model on mixing resources and artifacts across ritual spheres do not apply to this one major category. This suggests that it may mark the earthwork/habitation dichotomy as a unitary ritual sphere, unless it can be argued that the bladelet was simply an expedient tool and, therefore, was not specialized as a ritual tool.[6]

However, as humble and utilitarian in appearance as it is, the bladelet is generally recognized as no less important in demarcating the Hopewell phenomenon in the Eastern Woodlands than the well-known and very elaborate items constituting the Hopewellian artifactual assemblage (fig. P. 8). Because of its rather sudden emergence and ubiquitous distribution across much of the Eastern Woodland, it would be strange if it turned out to be simply an expedient tool. Indeed, George Odell has recently argued that the bladelet was a critical Hopewell ritual tool, and reiterated the widespread view that its appearance, use, and disappearance marked the Hopewell/Middle Woodland period. "It is probable . . . that prismatic-blade [bladelet] technology developed with Hopewell ritual. When these practices died out at the end of the Middle Woodland period, the blade technology died with them."[7]

In short, for Odell the bladelet had special Hopewellian ritual meaning and function. However, he also claims that it was a standard, expedient tool. His dual characterization of the bladelet as both an expedient tool and a ritual tool is made on the notion of "context of use." When the bladelet was used in the earthwork and related ceremonial locales, it was as a ritual or a ritual-related tool. When it was used in the habitation locales, it was as an expedient, utilitarian tool, no different from the expedient flake tools that are, in fact, much more prolific in the habitation zone than are the bladelet tools. "[W]hen blades were either manufactured or introduced in a residential situation, they lost all ritual connotations and became, in the minds of the tool users, indistinguishable from ordinary flakes. In a very real sense, then, blades served different functions in different contexts"(117).

Odell's claim for the dual ritual/utilitarian nature of the bladelet is based on his comparative analysis of wear and usage patterns of bladelets found in four Illinois Middle Woodland sites: the Smiling Dan habitation site, the Napoleon Hollow floodplain site, the Napoleon Hollow hill-slope site, and the Elizabeth Mounds Group site. While the bladelets from each of these sites display the same range of usage, the proportional distribution of type of usage varies. The bladelets found in the Elizabeth Mounds Group site and the hill-slope site reflect a focus on slicing and cutting of soft materials, as in hide preparing, flesh cutting, and some wood working, while the bladelet collection on the Napoleon Hollow floodplain site has a more varied mix with a greater tendency toward cutting, scraping, and engraving of medium-hard materials, notably "wood working, graving, projectile use, and probably some . . . butchery and manipulation of soft materials"(116). He accounts for this difference by interpreting the Napoleon Hollow floodplain site as probably being occupied by elite during times when they were conducting Hopewell ceremonies at the Elizabeth Mounds Group site. As described earlier, this mound site is on the bluff top immediately above the Napoleon Hollow site, and the hill-slope site is between these two. This latter site, in contrast, may have resulted from the deposition of ritual wastage from the higher-up Elizabeth Mounds Group site or from use of the location preparatory to ritual to be held at the latter site.

The bladelet usage at the Smiling Dan habitation site, which Odell points out would be broadly similar to the Murphy Tract site near Newark, along with the very prolific flake tools of this site, also displayed the same broad range of tasks as found in the other three sites. However, the range of tasks that the Smiling Dan site bladelets were used for was the most varied of the total set of sites and, moreover, favored tasks requiring cutting, drilling, and engraving medium and hard materials, including wood, bone, and shell. He also points out that the same range of usage applied to the flake tools, which also were twice as common as bladelet tools. This range of variation across the set of sites, he argues, supports the conclusion that, when found in ritual or ritual-related sites, the bladelet was primarily used as a ritual tool, and when found in a residential locale, it was used as an ordinary practical implement in much the same way as the even more prolific flake implement. "When the mortuary influence became negligible, as at Smiling Dan, it is no longer possible to detect functional specificity within the blade component. When brought back to ordinary residences, blades became dissociated from their raison d'être within the society and were used for the same mundane functions as flakes or bifaces" (114). He admits finding it puzzling, however, that both flake implements and bladelets, equally "expedient," would be used in the same habitation locales to mediate the same interventions. As he puts it, "However, it is difficult to understand why tool users would employ blades at all in a residential situation, when flake technologies had sufficed for these purposes for millennia" (117).

CRITIQUE

Indeed, finding both bladelets and flake tools in the same site bearing the same usage wear is a critical puzzle. As Odell points out, why should the older tradition of expedient flake tools continue to be used in the habitation zone along with the Hopewell bladelet if both could be used for the same range of purposes? This is particularly puzzling since the two implements are produced using different techniques. The flake implement is tied to the bifacial tool tradition and its discoid core technique and, as Odell points out, this technique had been used "for millennia." This technique is also flexible since the same core can be used to produce the whole range of bifacial tools along with flake tools. In contrast, the bladelet was made using the prismatic core technique and this is largely restricted to bladelet production since once initiated, the core cannot be modified to produce bifacial blanks.[8]

While puzzled by the continued use of the flake tool along with the bladelet in the habitation locales, Odell leaves it unresolved. If his conclusion is that the bladelet becomes simply another expedient tool similar in meaning and use to the flake tools, then, this would undermine the Dual Ritual Spheres Model and reinforce the Unitary Ritual Sphere Model. However, by not resolving the puzzle of the dual use of bladelet and flake in the same site, he also undermines his claim that the Hopewell bladelet was critically tied to Hopewell ritual. After all, since the "context" determines the meaning and function of a tool, then, the traditional flake implement could be simply used in the Hopewell earthwork context. This "context" would ensure that its use was ritual in nature. There would be no ritual reason to innovate the bladelet. As his analysis has demonstrated, (1) the flake tool mediated the same range of tasks as easily and as effectively as the bladelet; (2) it was certainly as easily and expediently produced; and (3) its production did not require a new core technique. In short, the continuity of the bladelet and flake tools in the habitation locales is the real puzzle that Odell and others do not address or resolve.

This puzzle can be resolved by addressing it in symbolic pragmatic terms. First, both the bladelet and the flake tools would be warrants. Second, it is assumed here that, indeed, the bladelet was a critical ritual tool. Since the objective usage of both tools was the same, it follows that the tangible differences that mark the bladelet/flake duality manifest a critical symbolic pragmatic contrast. This contrast ensures that the flake tools were seen as not being ritual tools, or as not being tools appropriate for the ritual mediated by the Hopewell bladelets. In this way, the range of material tasks that they mediated was taken as not counting as Hopewellian ritual, despite the two sets of scraping and slicing behaviors and their outcomes being objectively similar—or even identical. The prolific use of the flake tools in the same domestic locale as the Hopewell bladelet suggests that the bladelet was introduced, along with its contrasting core technology, specifically because, whatever the symbolic pragmatic meaning of the flake tools was, it was incompatible with the action nature of the behaviors that the bladelets were intended to mediate.

This conclusion requires reexamining Odell's claim that "context" has primary action meaning. In fact, he is quite correct. This is a basic symbolic pragmatic premise. Pragmatics is very much a study of how meaning is determined by mutual recognition of the meaning of place or situation.[9] When the term is applied to a place or locale, it is being used as a symbolic pragmatic term since it implicates the meaning that this place or locale holds for the people who use it. In short, *just as stylistics of artifacts conventionally constitute their symbolic warranting or pragmatic nature, so the stylistics of locales also conventionally constitute their pragmatic or context warranting nature.* In these terms, if Odell's claim that the bladelet was intimately tied to Hopewell ritual is accepted, then the bladelet, as humble as it is, was part of the symbolic pragmatics by which the ritual context nature of earthwork locale was constituted. This conclusion, however, only partly supports Odell's attribution of the bladelet as a Hopewellian ritual tool. That is, this claim must be modified by deleting "Hopewellian" as the operative adjective. This leaves in place the rest of the claim, namely, that the bladelet was a ritual tool set up to constitute the behaviors it mediated as ritual, thereby having a different nature from the behaviors that the functionally equivalent flake tools mediated. This modification now allows extending the claim to include habitation locales, where blade-lets are also commonly found. The bladelet, in short, would be part of constituting the habitation, or parts of it at certain times, as also having a ritual context nature, one that would be of the same order as the earthwork context, but contrasting in its ritual type. That this ritual nature would contrast with the ritual nature of the Hopewellian earthworks is supported by the habitation locales effectively lacking all the other Hopewellian materials that are associated with the earthwork locales.

Thus, although the introduction of the bladelet marked the emergence of Hopewell ritual practices, it was not used exclusively for these practices. This is clear since, if the bladelet constituted ritual in both earthwork and habitation locales by being a constituent part of the ritual contexts of these two locales, and if the absence of standard Hopewell stylistics from the habitation locales marks these two locales as ritually distinct, then the bladelet was not serving exclusively to constitute Hopewell ritual. Instead it will be postulated that the bladelet is the outcome of an innovation responding to the need to prevent the ritual performed in one locale from polluting ritual performed in the other by the unwitting mixing of ritual paraphernalia. Of course, from this it follows that the regular use of both the bladelet and the flake tool in the habitation locales suggests an internal structuring of these into mundane/ritual contextual spheres.

This means that the two notable properties of the bladelet, its "expedient" nature and its prismatic form, were what made this tool invaluable as a ritual medium. Its formal prismatic appearance clearly contrasted with the non-prismatic nature of the flake tool. This contrast ensured that the same range of material interventions mediated by bladelet and flake tools would systematically count as different types of social activities. Therefore, its expedient nature served as an ideal tool by which the proscription against mixing of ritual resources and artifacts across ritual spheres would be most expeditiously achieved. Two special conditions would make inadvertently breaking the separation of ritual spheres very easy: the use of a single or standard set of curated ritual tools; and the free and regular flow of individuals between ritually different locales. The bladelet resolved both problems. It was easily made as required; it was directly identifiable as different from flake tools and curated ritual tools; and it was easily tossed, thereby ensuring it would not be carried outside the ritual locale.

CRITICAL ANALOGY

To illustrate how "purely" practical tools can serve to constitute ritual spheres, a particularly apt example would be the symbolic pragmatics of everyday food preparation and consumption in Orthodox and Conservative Jewish households. The context nature of the residential locale of such a family entails the possession and use of two easily distinguishable, parallel and complementary sets of kitchen tools and facilities for everyday food storage, preparation, and consumption, one set exclusively used with meat-based meals, the other, exclusively used with meals based on milk and related products. This separation is not to privilege one type of meal over the other. Both "meat meals" and "milk meals" are equally valued. However, mixing "milk" and "meat," or any related components, is strictly proscribed by the religious Kosher protocols and to allow this to happen, even if done inadvertently, is to commit a serious transgression requiring ritual cleansing of both the implements and those who transgressed, and, possibly, the total household (depending on the degree of orthodoxy, disposal of the polluted implements may be required). Above all, the spiritual status of the whole household is critically constituted through fulfilling these rules in the patterning of the kitchen(s). Therefore, this stylistic distinction not only manifests two complementary food spheres, it is partly constitutive of these as ritual spheres. By sustaining their purity, the spiritual harmony and balance of the household is sustained.

The Ohio Hopewell bladelet complex, of course, implicates a proscriptive ritual strategy that is different from that outlined above for the Orthodox Jewish home. This is probably because the conditions of the two were different. In the case of the Orthodox Jewish household, a single, permanent locale is occupied, and avoiding mixing is accomplished by installing permanent but separate facilities and using hardware that is tangibly different and kept separate. In contrast, the Ohio Hopewell strategy was adapted to mobile participants moving from one to the other ritual locale. Thus, as pointed out above, the "expedient" nature of the bladelet tool, as well as its distinctiveness with regard to both curated and (equally) expedient flake tools, would make up its unique value as a ritual tool, an essential constituent medium constituting two ritual spheres having two separate locales being constituted as two separate and different ritual contexts.

However, as indicated in table 10.1, there may be another patterning of the bladelet complex that can be used to develop the Dual Ritual Spheres Model. This is the apparent duality of the chert types, thereby suggesting that the ritual sphere of each locale was structured into two complementary sets of rituals. Characteristically, across both habitation and earthwork locales, the bladelet is predominantly made of two select and contrasting cherts. In Ohio Hopewell, the preferred cherts were often Wyandotte chert and Flint Ridge (Van Port) chert. A similar pattern has been noted for other regions, for instance, Cobden-Dongola and Burlington Crescent Quarries cherts in the Illinois Valley. However, the symbolic pragmatic force of the chert duality would have been different from that of the bladelet, per se. The latter, as noted above, would have been proscriptive in nature, serving to prevent the inadvertent reuse of a ritual tool. Since the Wyandotte/Flint Ridge duality is realized in the bladelet form in both habitation and earthwork locales, the chert duality would not be needed to serve this proscriptive purpose. Therefore, the consequences of transgressing the Wyandotte/Flint Ridge duality (or its equivalent) would not have the same implications that transgressing the reuse prohibition would have. This would give some flexibility to the emergence and sustaining of this chert dichotomy. Circumstances, such as temporary unavailability of one of the preferred cherts, could serve to justify using some "at hand" substitute. The passing of circumstantial constraints would be marked by the rapid return to the preferred cherts for mediating the complementary rituals.[10] Experimentation would also be more permissible. In sum, as indicated in table 10.1, the Dual Ritual Spheres Model postulates a double ritual duality.

It is suggested that the type of ritual activity that would call for the bladelet would be the variable treatment of the deceased as postulated by the Mourning/World Renewal Mortuary Model. In this case, the mortuary practices occurring at the habitation locales would focus on the funerary and ancestor-related ritual, while the mortuary practices at the earthwork locales would focus on the post-funerary or the mourning→spirit-release→world renewal post-mortem sacrificial ritual. Thus, mortuary rites are suggested as the major common factor that led to the duality of ritual bladelet tools in both locales, and the differentiation of types of mortuary rites would be the factor that led to the prescription stipulating that these tools be produced and disposed of as a constituent part of the two ritual spheres, thereby avoiding polluting the sacred natures of the two locales through unwittingly reusing the same tools in both locales.

Even though the Dual Ritual Spheres Model can be used to argue that it would be highly improbable for the bladelet complex to emerge in the monistic modular social conditions postulated by the Dispersed Hamlet/Vacant Center Model, it does not follow that a polyistic locale-centric social system would *entail* the bladelet complex. The Adena archaeological record also probably manifests a polyistic locale-centric social system, as will be discussed shortly. However, it did not manifest the bladelet complex, although some regional Late Adena manifestations apparently did pick up this complex, for example, those in the Hocking and middle Muskingum drainages, suggesting that these may have been emulating neighboring Ohio Hopewell.[11] It is also possible, of course, that the Adena had some preexisting, currently unrecognized, symbolic pragmatic equivalent to the Ohio Hopewell bladelet complex. What the Dual Ritual Spheres Model does claim is that for a phenomenon like the bladelet complex to emerge, a necessary but not a sufficient condition for it would be a polyistic locale-centric social system.

By resolving a number of puzzles (e.g., the context/locale duality, the bladelet/flake duality, and the effective absence of standard Hopewell ritual material from the habitation locales), the Dual Ritual Spheres Model has demonstrated that it can more coherently account for the distribution of Hopewellian materials, including the distribution of the Hopewell bladelet, than can the Unitary Ritual Sphere Model. Since this is an essential ancillary of the Dual Clan-Cult Model, this establishes further reasons for accepting the latter model over the Dispersed Hamlet/Vacant Center Model. Moreover, the Dual Clan-Cult Model allows recognizing that considerable variation in earthwork magnitude, form, and siting can and will occur independently of any equivalent variation in the habitation zones. This is because it argues for the uncoupling of ritual spheres resulting from the autonomous nature of the clan and cult groups based on custodial domain. It also justifies exploring the cultic dimension in its own right, thereby expanding our understanding of Ohio Hopewell. To this end, the following chapter elucidates a complementary ancillary model of the Dual Clan-Cult Model addressing the cultic social organization alone. This is termed the Autonomous Cult Model.

Table 10.1 Dual Habitation/Earthwork Ritual Spheres of Ohio Hopewell

	Types of Locales	*Habitation*
Main Chert Types	Earthwork	Habitation
Flint Ridge (Van Port)	*Ritual Sphere E-FR*	*Ritual Sphere H-FR*
Wyandotte	*Ritual Sphere E-W*	*Ritual Sphere H-W*

Sphere E = Earthwork Ritual Sphere
Sphere H = Habitation Ritual Sphere
FR = Flint Ridge (Van Port)
W = Wyandotte

Chapter 11

AUTONOMOUS WORLD RENEWAL CULT SYSTEMS

The Autonomous Cult Model elucidates the cultic dimension of the Dual Clan-Cult Model with particular reference to the social structure of the postulated world renewal cults of the Central Ohio Valley earthwork traditions: the Mt. Horeb (Adena) Tradition and the two co-traditions of the Ohio Hopewell, the Fort Miami Tradition and the Chillicothe Tradition. To accommodate this wide scope, the model will be presented in two variants, one addressing each major tradition. The primary structural base of these cults, of course, was gender-specific age-grades. Whereas the age-grade span of the Nyakyusa is about 30 to 35 years, with the youngest members of the junior age-grade being about 10, a more conservative initial age for the Adena and Ohio Hopewell might be 15. This would also reduce the spread between the oldest and the youngest of an age-grade to between 25 and 30 years. An age-grade will be internally divided into age-sets or age cohorts. The term "cohort" is preferred here since it has the sense that the individuals making it up would be lifelong companions. Again, using the Nyakyusa as a guideline, an age-cohort would have a spread of 5 to 8 years between the youngest and oldest companion. Therefore, four or five age cohorts would make up a single age-grade.

Two active age-grades would be typical in a given region, one occupying the status of senior age-grade and the other, the junior age-grade. The senior and junior age-grades together would incorporate an age spread from about 15, for the youngest participants of the junior age-grade, to about 65–75, for the oldest participants of senior age-grade. This means that the active cultic life of a given age-grade would be sectored into two periods, first a period of about 25 to 30 years in the junior age-grade status, and then another 25 to 30 years in the senior age-grade status, thereby making up a complete, two-generational cycle of 50 to 60 years.

As age-grade cult sodalities, the different ways the senior and junior cults would relate to each other would characterize the age-grade cult system of a region. On the one hand, the two could constitute autonomous senior and junior age-grade cults. This means they would relate to each other in an arm's-length manner, each responsible for different but complementary world renewal locales and, of course, different but complementary suites of world renewal rites. On the other hand, rather than arm's-length junior and senior age-grade cults, there is another, although less likely possibility. A senior age-grade cult and a junior age-grade cult could join together into a complex cult. A complex cult would be internally structured into two complementary sectors, one composed of a senior age-grade, the other, a junior age-grade. Of these two types, the system of separate, arm's-length senior and junior age-grade cults would be more likely than the system of complex, integrated cults. This is because the combining of senior and junior age-grades into the same complex cult would tend to subvert the egalitarianism of same-generation companionship by the nonegalitarianism of ranked generations. However, this does not make the emergence of a complex integrated cult impossible—if a prerequisite condition existed, this being a system of simple, arm's-length cults. That is, if the latter type of cult system was already in place, rather than inventing a totally new organization, it would be "simply" a matter of integrating established, generationally distinct and autonomous companionship groups while buffering the subverting effects of seniority by maintaining separate senior and junior sectors. Each complex cult would retain its individual autonomy from neighboring cults, whether simple or complex, and from the network of autonomous clans.

The system of complex autonomous cults could be the outcome of negotiations between established autonomous senior and junior age-grade simple cults. These negotiations would constitute temporary arrangements, for example, to coordinate joint efforts in fulfilling their shared raison d'être of world renewal. Such cooperative efforts would not need to have a permanent arrangement as a goal. Rather, initial arrangements would be viewed as expedient, simply a rational way of enhancing ritual performances through sharing expertise and labor. This process, however, has the potential to develop into a more permanent state of affairs, largely to the degree to which it was successful in terms of joint achievements while buffering the clashing effects by maintaining separate senior and junior sectors that would recognize and respect each other's different and largely complementary cultic responsibilities. Of course, even under the best of circumstances, the subverting effects arising from the contradictions inherent in conjoining generational seniority and age-grade equality would not be immediately apparent. But once they started to emerge, resolving them would become part of the negotiating process. Out of this would emerge factional disputes, of course. However, while often contentious, these disputes could as often have the effect of entrenching than of weakening the structural integration.

Therefore, while a complex cult could emerge in the context of a system of simple cults, they would be inherently factious and unstable. Initially, many negotiated arrangements might be short-termed, collapsing before the appropriate protocols and practices were innovated that would buffer and minimize the effects of these structural contradictions. Once innovative modes were in place to buffer the strains, however, they would constitute an organization that would be extremely dynamic as the ongoing structural contradictions intrinsic to intergenerational relations fed ongoing factional competition and cooperation. The emergence of a system of complex autonomous cults is precisely what will be postulated under the Autonomous Cult Model. We know it as the *Ohio Hopewell* and, both logically and chronologically, the *Adena* is postulated as the preexisting system of simple autonomous cults that made the emergence of the system of complex autonomous cults possible.

The modifications to the Proscriptive/Prescriptive Ecological Strategy Model that were made in order to apply it to the Early and Middle Woodland periods postulated an escalation of construction practices promoted by the competitive pursuit of cult reputation as bringing about the emergence of the Early Woodland. This competition promoted further ideological innovations in construction, thereby escalating the magnitude and complexity of earthworks, motivating new strategies of labor pooling. One obvious tactic would be to integrate the expertise and labor capacities of the autonomous senior and junior age-grade cults in a given region. The integrated construction programs that were successful would enhance the reputations of the cults responsible, and this would motivate competitive emulation by others, thereby reinforcing the trend toward "complexification" across a region.

In accordance with the World Renewal Model, the pursuit of reputation through successful collective integration of ritual would promote the rational development of the iconic congruency of the earthworks, motivating the innovation of the C-R contrast and, over time, its development from asymmetry to symmetry, the complexification of the geometrical relations tying the constituent C-R components together, the elaboration of soil selection and procurement as emically perceived to embody sacred elements of the cosmos, such as surface scraped and deep stratum soils, contrasting colored soils, the elaboration of modes of embodying the astronomical alignments, and so on, as well as escalating the overall magnitude of the monumental construction. Cults that avoided such cooperative integration and the symbolic elaboration it enabled would experience a reduction in their ability to recruit new age cohorts, thereby reducing their own growth and reproduction.

This competitive pursuit might promote interdrainage pooling of a number of complex autonomous cults, constituting alliances. It is unlikely, however, that this would occur until after the emergence of successful complex cults. Prior to this, or as an alternative possibility, would be the interdrainage pooling of the labor of the same-generation simple cults, such as two or more junior age-grade cults from different regions cooperating to enhance each others' earthworks. Sustaining the pooling of labor from these different junior age-grade sodalities in order to expand a given earthwork, such as Fort Ancient, would entail the formation of an interregional system of same-generation simple auton-omous cults. While this latter possibility will be explored later, it will be assumed that the historical break defining the Ohio Hopewell epi-sode was the emergence of a system of complex autonomous cults out of preexisting systems of simple autonomous cults. The Autonomous Cult Model postulates that the best known archaeological example of a system of simple autonomous cults is the Adena, and the best known example of a system of complex autonomous cults is the Ohio Hope-well, particularly as manifested in the Chillicothe Tradition. This model may also be applicable more widely across the Eastern Woodlands, but in different variations.

Because of the complexity of the Ohio Hopewell earthworks in contrast to the simplicity of the habitational locales, a question of the role of inequality must be addressed before the empirical grounding of the Autonomous Cult Model can be initiated. This is particularly press-ing given the claim that the social system was based on two separate structural principles, the egalitarianism of age-cohort companionship in contrast to the generational hierarchy that characterizes kinship de-scent. Therefore, the critical question of how to treat the Early and Middle Woodland earthwork groups in terms of egalitarianism and social hierarchy is addressed below.

EMBANKMENT EARTHWORKS, COMMUNITAS, AND SOCIAL HIERARCHY

In his study of Cahokia and the social structure of the Middle Mississippian of the Central Mississippi Valley, Emerson develops a very interesting interpretation of large-scale earthworks that is amenable to the symbolic pragmatic approach. He points out that many scholars identify "feasting, public ceremonials, giveaways, construction of monumental architecture, and other expressions of group power" as the manifestation of expropriating elites. He neatly switches this, however, to suggest that these practices "are best characterized as symbolic expressions of communitas and group solidarity rather than of elite coercive power."[1] He characterizes communitas as a collective action in which the participants experience solidarity through emphasizing commonality and suppressing differences. Although communitas is not limited to religious expression, it is most effectively generated through such participatory ritual. Therefore, communitas rests on a shared conceptual and perceptual experiencing of the group as "unsegmented, homogeneous and unstructured"(13).

Emerson goes on to claim that there is a fundamental contradiction between communitas and its conditions of realization. "But the public expressions of communitas that promote society also carry the seeds of elite alienation" (14). Monumental locales of public ritual and the mobilization of labor that this requires promote hierarchy by the collectivity's mandating selected members to take on directive responsibility for these tasks. The latter emerge as an elite and their specialized associations with the symbols of sacred communitas set them apart from the rest of the community and the community actively recognizes and consents to their special status. Through the regularization or institutionalization of this process, such as by inheritance, a permanent institutionalized ranking or hierarchy of inequality is generated. "Consequently, it is more appropriate to speak of hierarchies as resulting from command activities and the elite as being the recipients of a consensual power rather than expropriators of such power" (14).

DISCUSSION

This is an important perspective, and because the Central Ohio Valley monumentalism of the Early and Middle Woodland periods very clearly fits the parameters that define the conditions of communitas and hierarchy, it must be addressed directly. Emerson initiates his discussion by arguing that the nature of power relations in societies is intrinsically asymmetrical. Therefore, no society can exist without disparities of social power, and, of course, noting Giddens's comments on social power, he points out that it is not to be treated only in negative, suppressive terms but also in positive, enabling terms. Thus, the contradiction between communitas and its conditions of realization in public ritual that results in institutionalized inequality is not to be considered only negatively. The emergence of social relations based on differential power capacities makes the realization of the discursive goals of communitas—in particular the construction of monumental earthworks as symbolic warrants—critical elements by which the behavioral streams performed can count and be experienced as true communitas interaction.

Now Emerson's position in this regard would effectively mean that we cannot assume that there was some original period in history when human society existed in a natural state of social equality and out of which emerged institutionalized structures of power. I can agree with him in this regard, and it relates to my earlier antinaturalistic discussion of territorialism. However, this still raises an important issue that requires clarification. Emerson initiated his theoretical elucidation on the premise that "Personal and institutional inequality is the reality of human society" (4), and that, therefore, the egalitarian/nonegalitarian dichotomy is a "false dichotomy." He seems to distance himself a bit from this when he points out that "the fascinating question then becomes one that attempts to recognize the critical emergence of 'institutionalized inequality' (i.e., stratification), and seeks the processes that created those institutions" (4).

One implication here could be that prior to the emergence of "institutionalized inequality" there was simple, natural equality. However, Emerson's claim that all social interaction is structured on asymmetries of power, both personal and institutional, would suggest that this is not what he meant. Rather, prior to the emergence of institutionalized inequality there would have to have been systems of *insti-*

tutionalized equality. This seems to be his position since he goes on to point out that the emergence of "institutionalized inequality" would result from overcoming social "levelling mechanisms . . . [by] the development of more complex and permanently organized hierarchical societies" (12–13). If this is the case, then these "levelling mechanisms," such as public feasting, must be considered both as forms of institutionalized equality, and possibly as prevailing until the "development of more complex and permanently organized hierarchical societies." If these "levelling mechanisms" prevail (e.g., as embedded in the stylistics of material assemblages, such as hunting gear, thereby manifesting the user's commitment to the deontics of sharing any game he kills), then we can speak of an egalitarian society, one in which the institutions of equality override both individual inequalities, based on attributes of physical strength, youth, senility, and so on, and the institutions of inequality, such as gender and generation. This means that, in fact, these two notions, egalitarian and nonegalitarian, are useful. However, rather than speaking of an egalitarian/nonegalitarian dichotomy, polarizing these as two contrasting states, we can speak of the egalitarian↔nonegalitarian continuum. This would not be a "false dichotomy" but a dynamic continuum. This continuum would, of course, be a "summation" of a multidimensional analysis in which different aspects of a community or social system would be assessed in terms of the interaction of both egalitarian and nonegalitarian institutional practices and principles. From this the overall position of the social system on the egalitarian↔nonegalitarian continuum could be postulated. Of course, any such assessment could be contested. However, this would not detract from the method or its claim. Therefore, as a continuum, the egalitarian/nonegalitarian distinction can be a useful mode of assessing and accounting for the archaeological record.

However, when it comes to extending Emerson's excellent analysis to the Middle Woodland, there is one further problem. His conceptual scheme is critically dependent on the monistic modular view. Although he seems to recognize the possibility of some autonomy between religious groups, such as fertility and military cults, this is not the autonomy of a polyistic locale-centric social system that is being conceptualized in the Dual Clan-Cult Model and the Autonomous Cult Model. For Emerson, the Middle Mississippian is an hierarchical social system embedded in a kinship-based monistic structure dominated by chiefly authority that, apparently, claims exclusive territorial control, hence constituting the chiefdom as a modular unit. Therefore, while there can be competing cults and even competing elites, this competition occurs within a singular, overarching monistic modular framework. His characterization of the Middle Mississippian social system is not the concern in this book, however. But I am concerned with the possibility that his approach might be taken to necessarily link monumental architecture with just such an hierarchical monistic modular system.

While this is definitely one possibility, it is not the only one. It is also clear that levelling mechanisms, or institutions of equality, are the conditions in which *communitas* is most effectively realized, and these conditions can systematically subvert and largely neutralize the alienation of the emergent elite that Emerson argues as being intrinsic to monumental construction programs. This is not to claim that elite hierarchies will not emerge in these circumstances. Rather, it recognizes his point that monumentalism encourages the emergence of elites. However, elite structures are not built on personal capacities. They are social positions and people occupy them. It is how this selection process works that determines whether alienation and class formations emerge and are reproduced. Alienation requires the development of self-reproducing recruitment institutions, institutional practices that systematically exclude the majority from consideration. The classic method is preferential recruiting from among the offspring of the retiring elite. This necessarily separates these from the majority, and, in rather short order, preferentially treated kin groups emerge as lineages or clans having exclusive right to fill certain elite positions. As the ranking groups, they are alienated from those they lead.

However, this is not the only possibility for elite positions, once created, to be filled. The Autonomous Cult Model will incorporate

Emerson's claim that monumental architecture and the related ritual that it mediated would promote the formation of formal elite positions within the cult system. However, it further claims that the companionship principle as embedded in the postulated age-grade structuring, combined with the autonomy of clan and cult as postulated under the Dual Clan-Cult Model, would systematically subvert the formation of groups that would be self-reproducing elite. Thus, while the cult would be hierarchically structured, the autonomy of cults and their voluntaristic nature would make the formation of a self-perpetuating elite class unlikely if not impossible.

Hence, it is postulated that both the simple and complex autonomous cults (Adena and Ohio Hopewell) were structured in a manner that might be best characterized as a form of ranked or stratified egalitarianism. As pointed out earlier, age-grades are internally structured into age cohorts. The principle of same-age equality would operate largely unimpeded at the cohort level, while the age differences separating each cohort would promote ranking that would structure the age-grade hierarchically. This logically leads to characterizing a cult as an hierarchically structured set of junior to senior age cohorts. Since each age cohort was an egalitarian unit, each member would have the same opportunity to pursue reputation by seeking positions of leadership within the cult appropriate to the traditional ritual tasks of the cohort. Typically, the companions of each cohort would select from among themselves those who were to receive the mandates of leadership. Those selected would be recognized cult-wide as the leaders of that cohort and, therefore, the leaders of the ritual tasks traditional to that set and essential to the overall success of the cult.

It is further postulated that this hierarchical structure of equal-opportunity age cohorts would largely govern the ritual tasks each cohort was traditionally allotted and this allocation would be according to received notions of what would count as menial and what would count as esteemed tasks. The proscriptive/prescriptive deontics of an immanentist cosmology would suggest that the menial↔esteemed continuum would be determined according to the degree to which the

material labor that was required for the task was deemed more polluting than sanctifying or more sanctifying than polluting to the cosmos. The former would mean that the task would count as lower in
esteem and the latter as higher in esteem.

In these terms, it follows that the least senior of the age cohorts
would perform the least esteemed tasks. These would probably involve
such tasks as cutting and clearing land for construction of earthworks,
digging and moving the earth, and even the incremental steps involved
in the manipulating of the deceased in mortuary ritual, all of these
probably deemed personally polluting since they would entail primary
material intervention into the natural order. On being promoted, a
cohort would take on a different range of tasks, less personally polluting
and more personally sanctifying than the previous range, and therefore
more esteemed. For example, more senior ranking cohorts of the junior
age-grade would be expected to procure exotic materials through long-
distance travel. This would have a polluting moment in that it involved
crossing sacred natural boundaries. However, such expeditions might
be deemed less polluting than earthwork construction itself since, first,
they did not involve permanent disordering of nature in the way that
earthwork and timber construction did, and second, they would be
highly esteemed in that they ensured the supply of the raw materials
that were necessary for the production of material warrants that would
be critical to the felicity and success of ritual practice: mica, copper,
ocean shell, select cherts, and so on. Thus, procurement and exchange
expeditions would be highly valued tasks, probably performed as rites
of passage that would initially pollute the participants but then be
terminated through ritual cleansing and rebirth, constituting a promotion.

In sum, while the leadership positions would be allocated across
the age-grade, they would be tied to each traditional age-cohort position in the age-grade hierarchy. Those selected would occupy these
positions by being granted custodial care of the focal symbolic warrants
required for leadership participation in the associated ritual, copper
plates, celts, headdresses, and so on. Since leadership was a matter of

taking on custodial responsibility for these focal warrants, they would be accumulated by the custodian over a lifetime. Accumulating esteemed warrants was the measure and essential constituent of the overall reputation and standing of the individual in a cult, and, tied to the age-cohort promotional process, it would be time-dependent. Since the age-cohort hierarchy would be permanent while the occupation of the positions would be time-dependent, and since these were companionship- and not kinship-based sodalities, no system of exclusive class control and self-reproducing selection is expected to emerge.

HISTORICAL PERSPECTIVE

As Hall has pointed out, world renewal ritual is often the performing of a creation myth which, in its performance, recreates or at least reenergizes the object of the myth.[2] If this was the performance of a world renewal myth, then it would be the world or those aspects of it that figured in the myth that were deemed to be the targets of renewal effort. Part of a world renewal ritual is often the telling of the story of the cosmic creation, and only those who are warranted to recite it can do so. To become warranted or authorized often involves a layperson approaching a ritual specialist, e.g., a shaman or a cult priest, to ask to be taught the esoteric knowledge and know-how craft. The shaman or priest may require a fee-gift from the layperson, thereby accepting her/him as an "apprentice." She/he then instructs and guides the apprentice or novice in learning the esoteric knowledge and know-how. This goes beyond learning the story itself, which, because of its public recital, would be collective knowledge. It involves the detailed esoteric know-how of ritual preparation, questing, cleansing prayer, and so on. This would be the critical part of the know-how that is gained and often kept secret.

Hall points out that many historically known cults, such as the Menominee Medicine Society, incorporate individuals into graded ranks in relation to the different degrees of know-how they have been taught.[3] Each rank adds its know-how to that gained when the apprentice was in the lower ranks, thereby endowing members of that rank with the capacity and authority to perform more specialized ritual. Therefore, the learning and transmission of ritual knowledge for purposes of public performance creates a largely age-sensitive hierarchy of degrees of responsibility, and membership is a matter of achievement. These degrees are marked and constituted by material warrants that are often taken by their users to be participating in the sacred powers of the cosmos that are invoked and presenced by them in the ritual.

However, it is important not to identify such known historical groups as the Menominee Medicine Society, specifically a shamanic order, with the cult groups being modeled here. Rather, these are analogically useful to suggest the possible ritual and organizational framework of Early and Middle Woodland period cults. Also, unlike historically known shamanic societies, these cults would be open voluntary sodalities. The degree of openness would have varied over the history of a particular cult and depending on whether it was a senior or a junior age-grade cult. Because the primary structural constraint to membership would be generational, a senior cult would tend to be closed, except to lateral recruitment of members, discussed below. The junior age-grade cult, however, would be more open since the primary recruitment into the cult system would involve a cohort of adolescent companions approaching a junior age-grade cult to be initiated as a novice unit. However, it is quite possible that a given junior age-grade cult, anticipating taking on the responsibilities of the senior age-grade cult, would close itself to new companion age cohorts, and instead encourage these to participate in the cult rituals as age-cohorts "in-waiting." With the shift or promotion, that is, generational promotion, these pre-initiate age-cohorts "in-waiting" would be initiated and recognized as the new junior age-grade cult.

The nature of a locale-centric social system as an open network would enable individuals to belong to more than one cult simultaneously. Both senior and junior age-grade cults would probably be open to lateral recruitment, although the junior age-grade cults would be more affected than the senior by this. Since the retirement of a senior cult probably was instigated when the number of active survivors was reduced by natural death, the cults of neighboring regions could be out of phase. Because of this an individual could be a member of a junior cult in one region and a member of the senior cult in the neighboring region. A complex cross-cutting of membership would be part of the building of intercult alliances. As voluntary sodalities, then, the general tendency would be to encourage membership by as many individuals as circumstances made possible.

In postulating this autonomous cult system, in both simple and complex versions, it is important to consider structural constraints other than generational, in particular, gender. For example, in such a system it is possible that: (1) a cult might be open equally to males and females of the same generation; or (2) a cult might be open equally to both but with a ranked differentiation according to gender, females having more sanctifying tasks than males, who might tend to have more menial tasks, or vice versa; or (3) a cult might be internally structured into two sectors based on gender; or (4) the cults might be exclusively male, or there might be two parallel but separate gender-based cults. All these possibilities should be kept in mind for accounting for variation in the archaeological record. However, given earlier claims that companionship would be forged through cooperative extradomestic tasks such as hunting, and since the tasks that the junior age-grade cults in particular would perform would usually be long-distance expeditions entailing extended periods of absence from the domestic sphere, it will be assumed that these cults were male-specific. Along complementary lines, if females were the core of the gardening domestic habitation locales, then the ritual sphere of the habitation zone that is postulated by the Dual Ritual Spheres Model could easily accommodate the claim that women, as lineal and/or affinal clan members, would have had a strong presence in these clan-based ritual activities. Of course, should relevant data not appear to support this claim, then the exclusive gender assumption is one possible source of misfit, and alternative possibilities would have to be examined.

In sum, as postulated by the Autonomous Cult Model, the cults would be male-gendered, generationally structured age-grade sodalities, and the majority of members would be "ordinary" lay individuals, rather than shamans or their equivalents. In fact, later it will be postulated that the basic distinction between simple and complex cults largely hinged on the shaman/priest distinction. As stated above, the affairs of each autonomous cult would be run by the peer-selected elite of the different age cohorts. Their occupying elite positions would probably be achieved via a ritual recognizing the individual as a custodian of focal symbolic warrants, also possibly involving a name-adopting ceremony whereby the new occupant would take on the name of a prior custodian or a founding companion who, possibly through a visionary experience, was taught by a spirit companion(s) the rituals that became the basis of the ceremonies of the cult, and so on.

ADENA RITUAL ORGANIZATION

The demonstration of the Autonomous Cult Model by applying it to the Ohio Hopewell requires first examining the Adena archaeological record in order to demonstrate that this is most coherently accounted for as manifesting a system of simple autonomous cults that, as predicted, preexisted the complex type, the Ohio Hopewell. For this purpose Clay's interpretation of the Adena as a ritual organizational type will be critically used, and the process he refers to as sacred locale "replacement" will be interpreted as the result of the junior and senior age-grade cycling of simple cults, as postulated by the Autonomous Cult Model. Because of the importance of the generational factor, a core part of its demonstration will be to account for the patterning of the Adena CBL (the "classic" Adena mound) as marking the operation of the dual senior and junior age-grade structure.[4]

William Webb and his associates were responsible for the excavation of many of the important Kentucky Adena CBL moundings during the 1930s.[5] It was Webb who noted the circular paired-pole pattern found under many of these CBL mounds. He interpreted these as the wall remains of chiefly houses, arguing that these had probably been surrounded by the other houses of the local Adena village. Berle

Clay, however, has challenged this claim by arguing that there probably never were any Adena villages. In his interpretation, the paired-pole circles define free-standing, roofless, and relatively isolated circular timber structures that were not domestic in nature. He made several empirical points to support his claim. First, the average diameter of these paired-pole circles was between 15 m and 20 m, making these very large for houses, particularly when few were less than 10 m and a number were almost 30 m in diameter. He noted the absence of internal post holes, indicating the lack of any roofing supports that would allow for spanning these areas. The poles of the paired-pole circles were typically angled outwards approximately 15 degrees, making them unstable as roof supports. Confirming that these were not dwellings, he noted the absence of domestic facilities, such as earth ovens, hearths, storage pits, and so on. In fact they were remarkably clean of any domestic debris. Clay concluded that these open paired-pole circular structures served as Adena ritual locales.[6]

Clay argues that while a paired-pole circle was built to serve some nonmortuary ritual purpose, at a certain time in its use-history it would start being used for burials, most often initially cremations, thereby transforming the nonmortuary nature of the locale into a mortuary ritual context. Through the ongoing addition and accumulation of extended, cremated, and bundle burials, the "classic" Adena burial mound (here termed the Adena CBL) was created. He calls this nonmortuary→mortuary cycling of ritual function ritual replacement, and it is the core characteristic of what he refers to as the Adena ritual organizational type. Sometimes, instead of the process being a dual staged transformation (i.e., the paired-pole nonmortuary circular structure to the mortuary Adena CBL), it involved a triple-staged transformation, the Adena sacred circle to the paired-pole circular structure to the Adena CBL. Clay argues that, although the precise nature of this replacement of the ritual function of sacred spaces is still puzzling, it was tied to the particular developmental history of the alliances of communities that were responsible for these sacred locales. However, he admits that:

I find it difficult to adequately explain the significance of replacement beyond the suggestion that it points again and again to a basic instability in the Adena or Hopewell ritual centers of the central Ohio Valley. It suggests to me that the structure of inter-group cooperation may have gone through cycles during which ritual cooperation in mortuary was differentially expressed. I assume that such cycles, if they existed, were linked to the life cycles of cooperating groups. I am, however, not prepared at this point to detail this fascinating relationship between site type and small group social . . . evolution. (Clay 1998, 19)

Clay situated this Adena ritual organizational type in the context of his settlement system model. This latter is an important characterization of the Early Woodland social system that will be critically contrasted to other Early and Middle Woodland subsistence/settlement-ceremonial models. Railey, for example, has given a settlement system account of the Early Woodland Adena that effectively replicates the Dispersed Hamlet/Vacant Center Model of the Ohio Hopewell, making the Adena earthwork/habitation settlement dichotomy simply an earlier but more dispersed version of Ohio Hopewell.[7] In fact, specifically arguing that Adena earthwork construction practices acted to reverse the strong centrifugal tendencies of an extensive foraging and gardening system, Railey gives a thoroughly functional account of them as exclusive territorial markers, social integrators, and proprietorial domain legitimizers.

Clay speaks of Railey's treatment as being a version of the "bull's-eye" view.[8] However, his own is only a partial contrast since it is also firmly rooted in the monistic modular perspective. Nevertheless, he denies that individual Early Woodland monistic modular communities were responsible for the Adena ritual organization. Instead, he argues that, starting around 500 B.C., these dispersed communities began building peer polity alliances, and the material expression of these was the Adena sacred circles, paired-pole circles, and Adena CBLs. That is, an Adena earthwork, whether CBL or sacred circle, did not demarcate the ritual center of a monistic modular community. Rather, it was the ritual center of an alliance of these peer polity communities and they chose to build their alliance center on the boundary edges separating the territories of the allied polities.

> [It] is clear that the early burial mound functioned ritually as an expression of society on a *different level of organization* from that of the local group. I interpret this to mean that the burial mound probably served multiple groups and not an isolated polity. However directed, mortuary ritual probably represented the negotiated outcome of interaction *between* allied local groups. Reconstructed, the Adena mound occupied an "edge" location with respect to the groups that used it and, because of this placement, it is difficult to interpret from the mound and its burial population what groups used it. (Clay 1991, 32, emphases in original)

He goes on to argue that the central function served by the Adena ritual organization was not, as argued by Railey, to ensure the solidarity of the kinship–based domestic community, but to minimize long-term survival risks. That is, he specifically draws on Brose's view that alliances served to reduce survival risk[9] and interprets the Adena ritual organization type as functioning to forge alliances so as to ensure access to crucial resources *outside* the territorial modules of the individual peer polity groups that made up these alliances.[10]

This seems mistaken. To the degree to which Clay rejects the "bull's-eye" view and locates Adena ritual locales in a "no-man's" land, his view would be consistent with the Dual Clan-Cult Model and its premise of inclusive territories. For this reason, there would have been only practical constraints to gaining access to resources outside the habitual ranges of exploitation, making his risk avoidance claim irrelevant. In short, his "territorial edge" view implicitly presupposes custodial tenure. Despite his commitment to a monistic modular view, it seems appropriate, then, to draw on Clay's notion of ritual replacement, which is central to his modeling of the Adena ritual organization, to see how the Autonomous Cult Model would apply it to the archaeological record.

THE CYCLING OF THE AGE-GRADES

Ritual replacement will be spoken of as the cycling of the symbolic pragmatic meaning of ritual locales and, in terms of the Autonomous Cult Model, it will be claimed that it largely manifests the cycling of junior and senior age-grade cults. Therefore, the junior and senior cults of a given region would be responsible for their own traditional locales and repertoires of world renewal rites and these two repertoires would be complementary, thereby embracing a total world renewal ritual cycle for the region, for instance, one repertoire oriented to world renewal ritual under the lunar aspect of the cosmos, the other, under the solar aspect. The model postulates that these locales should display this organizational duality so that at least two formally different but ritually complementary sets would typically exist in any given region.

In fact, there are at least three formally different locale types: (1) the paired-pole circle feature, (2) the Adena CBL, and (3) the Mt. Horeb sacred circle. Excluding the (rare) polygon earthworks, such as Peter Village, this set defines the monumental facilities that have been labeled in this book as the Adena of the Mt. Horeb Tradition. As outlined above, the second locale, the Adena CBL, is usually built over the paired-pole circle and, as Clay argues, these actually represent two stages of a complex history of ritual usage.[11] It is postulated here that, first, the sacred circle (Simple G-Form/SR-Profile) and the paired-pole circle constitute mutually contrasting sacred locales, what will be termed here the paired-pole circle/sacred circle dichotomy. This dichotomy is claimed to manifest the arm's-length relation of the senior and junior age-grade cult sodalities in a region. It will be further postulated that the Adena CBL mortuary mounding, usually found to be covering one or more of the paired-pole circles, constitutes a dual CBL shared by the senior and junior age-grades. Also, the radical claim will be made that, in general, the paired-pole circles were locales under the custodianship of the senior age-grade cults, and the sacred circles were locales under the custodianship of the junior age-grade cults. The Adena CBLs, as the outcome of terminal mortuary practices, then, would be under joint custodianship of at least two cooperating senior and junior cults. The complex arm's-length relation was realized in the formation and cycling of this set of sacred locales.

In addition, it is claimed that when in active use, the paired-pole circle feature was a symbolic iconic warrant of world renewal ritual under the solar aspect and the sacred circle was built and used to mediate the equivalent lunar aspect world renewal ritual. When speaking in symbolic warranting terms, then, and based on Clay's initial characterization of the paired-pole circles as "rising-sun structures," it might be appropriate to refer to them as solar circles.[12] This suggests that the sacred circles would, in effect, be the complementary sacred locales under the lunar aspect, thereby extending to the Mt. Horeb Adena the notion of a sacred duality already established in the World Renewal Model for the Ohio Hopewell.

Calling the paired-pole circle the solar circle is contrary to Clay's most recent position in this regard. He has retracted his earlier 1986 suggestion that this feature was a "rising-sun" structure. He now calls that position a "fanciful astronomic speculation."[13] When the symbolic pragmatic view of the meaning of material culture is applied to major material features, however, we should treat any patterning that appears to manifest alignments or orientations toward astronomical horizon events as probably a form of iconic congruency. The earlier discussion of Hively and Horn's work at Newark and Chillicothe, as well as Greber's work at Seip and Liberty Works, and Romain's work confirming the systematic nature of the forms and alignments, certainly can be used as circumstantial supporting evidence, if it is accepted that the Ohio Hopewell is simply a developed expression of the earlier Adena ritual organizational type. In these terms, then, the material variation contrasting Mt. Horeb and Ohio Hopewell can be taken as largely resulting from variation of symbolic pragmatic protocols at the ideological level, while continuity was sustained at the cosmological belief level. Thus Clay's original attribution of the term "rising-sun structures" will be treated here as quite appropriate and a legitimate aspect of the Autonomous Cult Model.[14] Of course, rather than the senior/junior cults correlating with the Solar/Lunar Motif duality, it could be the other way around. The validity of the Autonomous Cult Model does not hinge on correctly matching cosmic aspects to senior and junior age-grade cults, as such. Rather, it is the duality that is important.

When the Ohio Hopewell situation is examined, further support of this postulated matching will be presented.

If this complex structuring of Adena sacred locales was realized in the cycling of ritual space, then the promotion of the junior age-grade to senior age-grade status would logically entail the retiring seniors transferring their responsibilities to them. This would be materially marked in one of several possible ways. First, there could be a formal transfer of locales so that the newly promoted senior age-grade cult would become the custodial occupants of the solar circle locale, and, of course, it would in turn hand over its sacred circle (lunar) to the new junior age-grade cult. Or it might involve a formal closing of one or both locales and the building of new locales.

Certainly, the closing of the solar circle seems indicated at some point since its floor typically became the base for a new Adena CBL. As postulated above, this new Adena CBL would be under the joint custodianship of the new senior and junior age-grade cults. With its closing (i.e., transformation into a CBL), the newly promoted senior age-grade cult would need to build a new solar circle, probably near the old one. The rite of the "closing" of the old solar circle or "rising-sun structure" and the founding of a new CBL may have entailed the post-mortem sacrificial deposition of human cremations that had been produced and/or curated for this purpose, and/or the building of an initial crypt in the center of this circle.[15] This "closing" rite would constitute the conversion of the old solar circle into a new CBL. Since it occupied the space of the solar circle of the retired seniors, it is postulated that it would be used jointly by both the new senior and the new junior age-grade cults for terminal mortuary ritual purposes.

Even though the Adena CBL was the locale of crypt-related mortuary rites and the ritual repository of the terminal burials, in terms of the Mourning/World Renewal Mortuary Model, the post-funerary world renewal rites leading up to these terminal practices would have been held in either a nearby solar circle or a nearby sacred circle, depending on the cult in which the deceased held membership at the time of death. (The deceased could also be a dependent of a living member.) That is, the Mourning/World Renewal Mortuary Model can be used to postulate that both senior and junior age-grade cults would use mortuary practices as the post-mortem sacrificial medium of their world renewal rites. Therefore, only the terminal stages of the funerary→mourning→spirit-release→world renewal ritual process would be performed at the Adena CBL mounding. Therefore, contra Clay, who claims that the solar circle was not involved in mortuary practices, it is suggested here that it was, and critically so in that it was a sacred locale for the custodial cult to perform the initial set of incremental mourning/world renewal rites leading to terminal burial in a nearby CBL. Furthermore, the sacred circle would be similarly involved in mortuary practices, the latter possibly mediating lunar aspect world renewal rites.[16] Penultimate rites might occur in the central crypt of the Adena CBL mounding. Ritual manipulation involving the deceased in the crypt would be followed by her/his removal, possible cremation, and terminal burial in a tomb or small prepared platform built laterally to the crypt.

A possible alternative would be for the new senior age-grade to convert its old sacred circle to serve as a solar circle by building a paired-pole circle inside it, as at Mt. Horeb. With their retirement after about twenty-five to thirty years, this solar circle, originally converted from a sacred circle, would again be converted, this time into an Adena CBL, thereby initiating the accumulation of burials as outlined above. The growth of this CBL mounding would come to fill the original ditch and cover the original sacred circle embankment, as may have been the case at the Adena site itself. This possibility is reported by Clay for the Adena Mound. Greber also comments that the Adena Mound was encircled by a Simple G-Form/SR-Profile embankment.[17] There is another possibility, however. Surface-scraping was part of the procedure to procure earth to cover the Adena burials. This could start from the edge of the paired-post circle and work outward. This procurement pattern may have generated a trench without an "external" embankment. The excavators, however, may have thought that the embankment had become obscured by the addition of earth in the final formation of the Adena CBL mound.

Indeed, although Thomas Hemmings found no evidence of a paired-pole circle, his report on the Grave Creek Mound (Adena CBL) argues that the "circle" around its base is a "moat" and that it was probably the result of the last stage of construction.[18] In this case, the builders apparently removed the soil from around the mounding, much as described above, and placed it around the base. Hemmings claims that there was no accompanying embankment. Another way of looking at this, however, is to see the terminal construction as a variant SL-Profile earthwork. The typical C-Form embankment has the ditch ("moat") on the "outside" with the embankment on the "inside." In this case, if Hemmings's analysis is correct, the "moat" is on the "out-side" but the earth that was procured from it, instead of being used to build a free-standing embankment, was placed around the base of the preexisting mounding, completing this latter construction. In a sense, the "moat" and the perimeter base of the Grave Creek mound consti-tute an SL-Profile "embankment earthwork."

DEMONSTRATING THE MODEL

As stated above, the Adena CBL mounding was postulated as the shared locale of the terminal mourning/world renewal rites of both Solar and Lunar age-grade cults.[19] Working from this premise, there are several lines of evidence that could be pursued to test the claim that the Adena mortuary data index a system of simple junior and senior auton-omous cults. These would include the contrasting sizes of the known Adena mortuary locales, the nature of the rules as displayed in the pro-curement and use of soil types for burial, the possible duality of cult and clan burials, the patterning of the Adena CBL burials, and indica-tions that patterned age variation was a major source of structuring the mortuary pattern. Since this analysis of the Adena serves as back-grounding of the Ohio Hopewell, focus will be on only the burial pat-terning of the Adena CBL moundings. This will be an incomplete empirical analysis but it will be the opportunity to introduce an important notion, this being the idea that over its active cult period, an age-grade would generate at least two separate burial populations, one population consisting of those members, possibly including their dependents, who died while the age-grade was in its junior phase, and the other consisting of those members, also possibly including their deceased dependents, who survived into but died during the senior phase.

In terms of the Autonomous Cult Model, assuming that the classic Adena CBL mounding type was shared by the mutually autonomous senior and junior age-grade cults, and assuming that in death the arm's-length relation manifesting the separation of the generations was respected, then there should be a separation of the terminal mortuary rites according to age-grade, all those associated with the senior cult at death being in one sector of the CBL and those associated with the junior cult at death in a complementary opposite sector. Therefore, a junior age-grade cult and a senior age-grade cult sharing the same CBL would tend to generate a dual sectoring of the mounding into two contiguous but separate cumulative mortuary populations. It can be expected that contemporary senior and junior age-grades should have differential death rates, the latter having a lower death rate than the former during the same period of time. This means that from the initial formation of an Adena CBL, the senior sector should grow faster than the junior sector, and then, with the reduction in the number of surviving senior age-grade members and with the aging of the junior age-grade members, this should reverse. Thus, as the mound grows, two sectors should emerge and vary in growth over the usage period in the above terms, one sector growing faster than the other, and then the reverse occurring.

In terms of the null hypothesis, since it is generally recognized that the Adena CBL moundings are the result of the cumulative addition of burials, unless some systematic factor is intervening, they ought to grow laterally and vertically in a more or less symmetrical manner. However, if this is not the case, and, instead, there is marked skewing, then it indicates some intervening factor or factors.

In fact, one type of skewing that has been broadly noted is the shifting of the epicenter of the typical mounding toward one side of the solar circle over which it was being built. However, correlated with this lateral drift, there is also a skewing of the vertical growth, one sector of the mounding growing more rapidly than the other, and then the reverse occurring. This dual skewed lateral and vertical growth is clearly indicated by the history of cumulative additions of the Robbins Mound (fig. 11.1).[20] Using this feature as illustrative, it would appear that at any given stage in the developmental history of an Adena mounding, or at least of the type conforming to the Robbins Mound, it was asymmetrically bimodal. It may have taken on a symmetrical appearance only when the terminal mantle was added.

There are complications that could be expected. As stated above, the promotion from junior to senior cult position would entail abandoning old locales and building new ones, and this would mean initiating a new CBL. If a new CBL was built with each transition, it would mean that the deceased of an age-grade would be split between two Adena CBLs, the companions who died during the junior phase being in one Adena CBL mounding and those who died in the senior phase being in a later, probably nearby mounding. As was also suggested, reusing a CBL for several cycles could be done, and then abandoning it and building a new one. Both alternatives are possible and these can be called single-cycle and multi-cycle Adena CBL moundings, respectively. Since a single-cycle CBL would contain only "half" the deceased of one senior age-grade cult and "half" the deceased of one junior age-grade cult, it would represent a single generational spread of about 25–30 years. Another way of putting this is to point out that a new CBL would be started only after the promotion of a junior age-grade to the senior status, and it would be terminated with their retirement and/or decease. This would be the span of one generation, between 25–30 years.

A multi-cycle CBL would probably entail terminating a cycle by covering the burials with a mantle (a recognized pattern in Adena moundings), constructing one or two "central" crypts on this mantle, and initiating a new cycle. In this case, the new senior age-grade cult members would be the surviving companions of those buried in the junior sector of the first cycle, below the mantle. As these seniors died, they would be added to the senior sector side but above the first mantle. Of course, the deceased of the second-cycle junior sector would also be buried above the first mantle but in the sector where the previous first-cycle junior age-grade deceased were buried, and so on with each cycle. Because of the lateral skewing described above, the burials of sequential cycles would actually not be neatly aligned vertically. Those Adena CBLs, then, having two or more mantles, each mantle being postulated as defining a cycle having a single-generational time span, would embody multiple generations, a dual-mantled mounding covering a two-generational spread (i.e., 50–60 years), and so on. With respect to the Adena mortuary program, then, the senior/junior age-grade structural factor would appear to be the simplest explanation accounting for this dual asymmetry, as illustrated in the computer-simulated reconstruction of the Robbins Mound (fig. 11.1).

Because of the very limited data, it can be only tentatively concluded that the Adena burial mounding is structured by having two burial populations, one being senior in generation to the other, along with some of their dependents who had died during the time of usage of that mounding.[21] Before concluding, however, it is relevant to explore other possible factors that could account for a dual, asymmetrical modal growth of this nature. Under Clay's model, there should also be two or more burial populations since two or more allied communities share the same CBL. Assuming that community unity would tend to be expressed by burying members of the same community in the same sector of the mounding, then at least two burial populations could be expected, one from each modular community. However, if these are peer polities, then they should have the same age-structure and, therefore, there would be no reason for a differential accumulation of deaths as described above. Another possibility is gender-specific burials. If two cohorts of opposite genders were to share opposite segments of the same mound, differential death rates between males and females of the same generation might be registered. William Webb and Charles Snow argue that male:female burial proportion was 2 to 1 in favor of males. The problem with this, of course, is that Adena burial mounds are not segregated by gender in this manner. Furthermore, reexamination of the extant skeletal evidence has led to reassessing the male:female proportion, making it much closer to the norm of 50:50.[22]

Another factor that could influence the rate of growth of Adena moundings would be the "retention rate," this being the number of actual deceased buried in the CBL as compared to the total number of deceased of the cult or cults involved. This assumes that not all the deceased of a cult ended up in the "classic" Adena CBL mounding. This notion of retention rate makes sense in the context of the negotiations between clan and cult as postulated above under the Dual Clan-Cult Model. Many cult deceased would end up in their clan CBLs, which may have been small mounds along the lines noted by Clay with regard to the Auvergne mound site.[23] It will be argued later that, for the Ohio Hopewell, the retention rate did vary for senior and junior age-grades. At this point it is not possible to make a valid estimate of the retention rates for Adena and, therefore, this will not be attempted. The retention rate will be addressed when these ideas are applied to the Ohio Hopewell mortuary data.

Before turning to the next step in applying the Autonomous Cult Model by focusing on the Ohio Hopewell system of complex autonomous cults, it should be pointed out that no definitive conclusions concerning the Adena mortuary moundings can be made at this time. The above needs further empirical verification. Also more expectations could be postulated and tested against the empirical data (e.g., age differentiation correlated with vertical height of burials—the higher the burial in a single cycle unit, the greater the tendency for the age at death to be greater, and so on). Space prevents exploring these possibilities. The most important conclusion that can be made, then, is that the dual skewing of the Adena CBL is more coherently accounted for in terms of the Autonomous Cult Model than in terms of a model treating Adena from a monistic modular perspective.

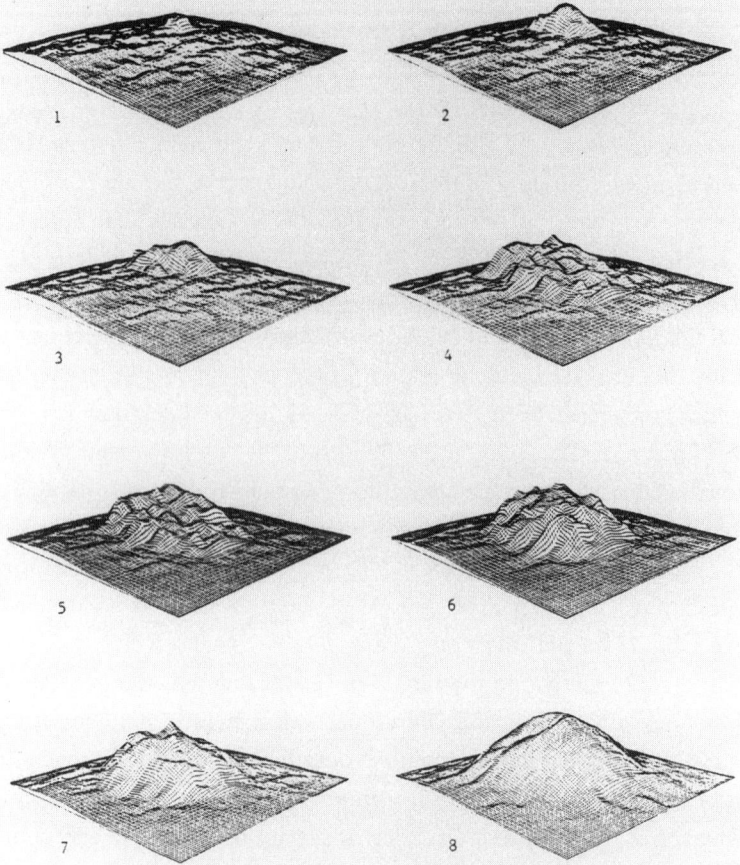

Figure 3. Eight episodes of mound construction, from the earliest (Stage 1) to the latest (Stage 8).

38

FIG. 11.1. The Robbins Mound Cumulative Stages (Milner and Jeffries 1987, 38, figure 3)

Chapter 12

THE SEIP AND HARNESS
GREAT HOUSE CBLS

The Ecclesiastic-Communal Cult Model is an ancillary of the Dual Autonomous Cult Model. It is designed to apply to two of the three Middle Woodland earthwork traditions, the Miami Fort and the Chillicothe Traditions. It claims that these were characterized generally as two separate networks of complex autonomous cults that, as postulated under the Autonomous Cult Model, emerged from the preexisting Adena system of simple autonomous cults. These complex organizations will be termed ecclesiastic-communal cults. As discussed earlier, the Adena may have maintained systems of simple autonomous senior and junior age-grade cults into the middle Middle Woodland in Kentucky, West Virginia, the Hocking Valley of southeastern Ohio, and possibly in east-central Indiana.[1] However, this claim may have to be modified in the light of further research since it is possible, particularly with regard to east-central Indiana, that local expressions of the Mt. Horeb Tradition emerged as variant forms of ecclesiastic-communal cults.[2]

A STRUCTURAL THEORY OF RELIGION

Before laying out the Ecclesiastic-Communal Cult Model, it is necessary to summarize Wallace's cultic view of religion since this will serve as the theoretical framework of this cult model. However, since his theory is generally applicable to monistic modular social systems, some significant modifications will need to be made. Wallace characterizes cults in terms of their social organization and suggests a fourfold typology: individualistic, shamanic, communal, and ecclesiastical.[3] Individualistic and shamanic cult institutions are situation-dependent in organization. The individualistic cult is ephemeral, realized as an individual periodically and directly interacts ritually with her/his companion spirit(s). In Native American religious practices, either party can instigate this interaction.[4] A powerful spirit can appear before a subject and become her/his companion spirit. Often this event occurs as part of a vision quest in which a young person undertakes a spiritual journey during which she/he invokes or is confronted by a spirit and they enter into a reciprocal relation involving mutual assistance and sacrifice.

The shamanic cult is characterized by the special nature of the shaman. The shaman is usually treated in the literature as an individual who has one or more particularly powerful companion spirits. This does not mean that the power of the shaman is totally dependent on the spirits that she/he can call upon and/or invoke, since among the reasons these spirits have become the shaman's companions is that she/he has personal powers or charisma. In any case, a shamanic cult gathering is also usually situation-dependent, emerging when an individual or group requests a shaman to conduct a seance, usually in response to some critical situation, such as illness, poor hunting, and so on. The shaman is expected to achieve some direct resolution, such as curing the sick by driving out hostile spirits or improving the spiritual power of the hunters. With the resolution of the crisis situation, the shamanic seance and cultic posture would dissolve.

In contrast to the above cases, the communal cult is much more situation-constitutive. However, in Wallace's theory the communal cult does not have an independent or autonomous existence and, therefore, it does not have a fully realizable congregational structure. Rather, the congregation is embedded in a relatively autonomous communal organization, such as a lineage, clan, military sodality, and so on, and these are treated by Wallace as different aspects of an overarching monistic social system. The communal group often works on a schedule and plans in advance to shift into its communal cultic posture. The concerns of the communal cult are linked to human and natural cycles on which much of the scheduling is based and this means its concerns are collective and even universal: reproduction of the group, and enhancement of the sacred powers of the local region or of the cosmos. Since the cult is the communal group in its religious posture, its structure remains the same under the religious aspect. If it is a clan, then the clan leaders simply shift into the religious aspect of their leadership position. If they require religious expertise in conducting certain rituals that is beyond their own know-how, they usually call on local shamans while maintaining their presiding authority.[5]

The ecclesiastical cult is, in one sense, simply the communal cult written large in that, as with the communal cult, it is situation-constitutive, and it has a strong schedule orientation. In Wallace's estimate, however, it is also very different. First, it is a "true" congregation in the sense that its primary raison d'être is in being an autonomous organization defined by its religious nature. It is also a congregation in that it is based on the laity/clergy structure. The clergy occupy the specialized position of priests, that is, religious leaders of the congregation, and the laity are the "ordinary" members. The laity/clergy structure constitutes the two positions as complementary, each requiring the other for its social existence.[6]

Wallace targets the laity/clergy structure as the source of structural contradictions that generate the dynamism of this organization. Although laity and clergy share a common interest in that they derive their social identities as members of the same congregation, each also perceives the other as a competitor for power in organizing the overall activities of the cult. Using Western religious types as his template, Wallace classifies ecclesiastical cults by the different ways the laity/clergy power distribution is worked out. Directly using the Western notion of "church," he speaks of two complementary extremes, episcopalian and congregationalist, and a third, presbyterian, as the fluctuating middle.

In cultic terms this would mean that an episcopalian type ecclesiastical cult is one ruled by a hierarchy of "bishops," and the local clergy, answerable to the hierarchy, would largely dominate in the affairs of the congregation. The congregationalist type of ecclesiastical cult is the opposite, an autonomous religious group in which the laity, or laity members selected by their peers, dominate the affairs of the congregation. Presbyterian type ecclesiastical cults fluctuate between these two extremes, characterized by different sectoring of power between the laity and clergy.[7] It is postulated here that if a rapprochement or stable "draw" is achieved, it is because the laity leadership and the clergy leadership arrive at a "balancing of responsibilities." For example, the clergy might be recognized as necessary for constituting the rituals, and they will largely control their own recruitment, training, and promotion. The laity, however, control the material infrastructure required for the ritual and also may have considerable say in the way the ritual is performed.

It is clear that the history of western Christianity and its structural transformations are strongly reflected in this scheme. However, Wallace has abstracted social organizational variables that have broad application. Still, there is a major modification that must be made. When speaking of a society, Wallace has a modular social system in mind, and, of course, as pointed out above, with respect to individualistic, shamanic, and communal cults, in his scheme these are not autonomous organizations but religious postures of the social components making up a social system of modular communities. He construes a cult institution as "a set of rituals all having the same general goal, all explicitly rationalized by a set of similar beliefs, and all supported by the same social group" (73). Therefore, excluding the ecclesiastical cult, the type of social system in which these three cults prevail is not only modular, it is also monistic. It is only when complex monistic modular societies emerge that Wallace sees autonomous ecclesiastical congregations emerging.

As it stands, Wallace's model may be applied appropriately to Native American social systems only if they fit the monistic modular

social system premises, and these would have only individualist, sha-
manistic, and communal type cults. Treated structurally, the Adena
would constitute a system of simple and the Ohio Hopewell, a system
of complex "communal" cults, respectively. However, Wallace's model
precludes treating communal cults as autonomous in nature, and, of
course, it precludes the possibility of the laity/clergy structure as existing in pre-state or pre-complex chiefdom social systems.[8] The Dual
Clan-Cult Model explicitly claims otherwise. In its terms, it is possible
in pre-state times not only for embedded individualistic, shamanic, and
communal cults to exist, but also for autonomous communal and even
autonomous ecclesiastic-like cults to exist. This is because the model
characterizes these Archaic, Early Woodland, and Middle Woodland
social systems as polyistic and locale-centric in nature. *Another way of
putting this is to postulate that, in conditions of inclusive territorialism,
autonomous cults can emerge independently of and prior to the emergence of
complex chiefdoms and/or states.* These can manifest simple communal
cult structures or, under special conditions as outlined by the Autonomous Cult Model, they can manifest complex ecclesiastic-like structures. These cults, simple or complex, will have their own sacred locales linked through networks of sacred paths.

This means that a major modification of Wallace's approach is
required. It entails claiming that in polyistic locale-centric social systems, autonomous communal and ecclesiastic-like cults can exist "side-
by-side" with individualistic, shamanic, and communal cult postures.
Therefore, as a general theoretical claim, it is postulated that a complex
modular social system is not a necessary condition promoting the
emergence of autonomous cults. Indeed, if the characterization of Eastern Woodland prehistory presented in this book is adequate, it means
that conditions equivalent to those outlined under the Inclusive Terri-
torial/Custodial Domain Paradigm would promote autonomous communal cults. Once these emerge, then at least two types of autonomous
social organizations can come to exist simultaneously, cults and clans.
Each type will have its own sphere of concerns, as postulated earlier.
With this type of social system in place, the possibility is opened for

some of these autonomous communal cults to evolve laity/clergy structuring, forming into ecclesiastic-like congregations. Of course, as an autonomous organization, an ecclesiastic-like cult could develop major forms of material expression that would seem to be incongruent with the network of kin groups which surrounds it. This is precisely the state of affairs the Dual Clan-Cult Model postulates for the Early and Middle Woodland periods of the Central Ohio Valley.

THE ECCLESIASTIC-COMMUNAL CULT MODEL

The term "ecclesiastic-communal cult" seems apt to refer to the Ohio Hopewell complex cults because these are postulated as having the laity/clergy ecclesiastical structure while retaining strong communal cult tendencies from their Early Woodland Adena simple communal cult roots, particularly with regard to the dual senior and junior age-grade sodality structure. Therefore, the Ecclesiastic-Communal Cult Model postulates that, typically, the members, both laity and clergy, would have gathered as a congregation in their sacred locale(s) in order to conduct rites. Traditionally these gatherings would be scheduled according to the seasonal cycle as mapped by astronomical movements. These astronomical events would be seen as marking particularly propitious periods when renewal ritual was most required. As argued earlier, even for a simple autonomous age-grade communal cult, such as the Adena type, it was necessary to overcome the effects of social inequality arising from the wide scope of 25–30 years in a single generation. This was achieved by the formation of an hierarchy of age cohorts identified with differential degrees and mixes of menial to esteemed labor tasks. A similar problem would arise with the integration of junior and senior cult types into a single cult. This would necessarily reduce the arm's-length relation characteristic of the system of simple autonomous cults by enhancing the implicit expectation on the part of the senior age-grade members, given their greater collective experience, that they could rightfully dominate in the overall affairs of the congregation. The junior age-grade would persistently resist this and attempt to retain its traditional autonomy in leadership and ritual. Clearly, in such circumstances, mutual constraint and formality between the two laity age-grade sectors would be endemic, possibly at times breaking into conflict and the rupture of a given cult organization.

The contradictions inherent in the senior/junior structure would be aggravated by the integration of the shamanic role into the cult. In communal cults, shamans play a part-time specialist role, as needed. With the integration of the senior and junior age-grade cults, the shamans, no longer peripatetic among simple communal cults in a region, would tend to become identified as religious specialists of a particular complex cult or set of complex cults. The fixed nature of this membership transformed the shaman into a shamanic priest and the age-grade sectors into a laity, thereby generating the laity/clergy structure. As characterized for the ecclesiastical cult type under Wallace's theory, this would lead to an ongoing struggle for control of the cult between the dual hierarchy of co-leadership, one hierarchy made up of the laity leaders of both the senior and junior sectors. At the same time, however, the clergy, as a small group of "resident" shamanic priests, would also be called upon by the leaders of the two laity sectors to act as mediators in disputes arising over the contradictions resulting from the integration of the senior and junior age-grades. The combination of these two new structures, senior/junior age-grades and laity/clergy, each with its intrinsic contradictions, would lead to some very interesting debate and negotiation.

However, the Ecclesiastic-Communal Cult Model postulates further that the emergence of the complex cult would exacerbate a third social structural source of tension, this time inherent between the mutually autonomous clans and cults. As postulated under the Dual Clan-Cult Model, the central source of strain would be the differential mortuary focus that clan and cult had on the treatment of their shared deceased members. The interests of the cults were best served by prescriptive post-mortem manipulation maximizing the impact of post-mortem sacrifice. The interests of the clans were best served by proscriptive post-mortem manipulation mediating primarily funerary and related ritual by which to constitute their deceased as honored ancestors.

It was postulated under the Autonomous Cult Model that the Adena CBL was under the co-responsibility of autonomous senior and junior age-grade communal cults. Because the leaders of the senior age-grade cult would also probably have an important presence in their respective clans, there would be a strong tendency toward a rapprochement that would satisfy the diverging mortuary interests. This would be manifested by the cults, in general, fluctuating between moderate and more pronounced post-mortem manipulation. The norm would be extended mourning periods mediating memorial, naming, and re-incarnation rites with constrained bodily manipulation that simultaneously served as renewal rites. Full prescriptive post-mortem manipulation (e.g., bodily dismemberment and cremation) would occur only under exceptional conditions, such as the formation of a new CBL, and it wold be largely limited to the junior age-grade cults.

With the emergence of the ecclesiastic-communal cult system, this rapprochement of senior and junior age-grade and cult and clan would be seriously strained. The primary ideological lever for the junior age-grade sector to counter the pressure from the senior age-grade leadership for dominance would be to promote highly prescriptive post-mortem mortuary manipulation as the preferred form of world renewal ritual. Ongoing and sometimes heated negotiations would occur, and, of course, these would implicate clan and cult relations. Again, having leadership responsibilities within the clan hierarchies, the senior age-grade sector leaders would tend to promote a more moderate approach. In this type of complex senior/junior age-grade and clan/cult conflict of interest, the role of the clergy as mediators would be critical, thereby enhancing the importance of the emerging shamanic-priest position.

This complex of structural contradictions arising from the conjunction of senior/junior age-grade, laity/clergy, and clan/cult structures would be particularly expressed in factional dispute between clergy and laity for cult power. Wallace's theory suggests that this competition for power in ecclesiastical cults tends to generate three general orientations. These grounded his tripartite classification of ecclesiastical cult types: congregationalist, presbyterian, and episcopalian. There is great potential in this tripartite classification for exploring the variation of the Ohio Hopewell archaeological record, particularly when this classification is seen within the context of the range of structural contradictions that is postulated above. However, rather than speaking in terms that articulate separate types of cults, these would be more effectively treated as tendencies resulting from power dynamics and, therefore, as forming a continuum. This will be called the Congregational↔Presbyterian↔Episcopalian continuum or the C↔P↔E continuum. In terms discussed above, a cult in which the laity leaders dominate over the clergy in the affairs would place it toward the congregationalist extreme on the continuum; the opposite would place it toward the episcopalian extreme. A balancing of power between laity and clergy leadership would situate the cult in the presbyterian middle. This position would vary over time as factional disputes shifted the cult to different positions on the continuum. In short, by establishing the correlates in the material culture of these different positions, the C↔P↔E continuum can be a very flexible tool for interpreting the variation and transformation of the Ohio Hopewell record. However, its full utility requires the elucidation of the Ideological Cult Faction Model, an ancillary of the Ecclesiastic-Communal Cult Model. This can only be done after empirical confirmation of the Ecclesiastic-Communal Cult Model. Therefore, I now turn to this latter task.

DEMONSTRATION

In conformity with the hermeneutic spiral methodology, the empirical confirmation of the Ecclesiastic-Communal Cult Model begins by focusing on the remains of three major timber building constructions and their associated mortuary facilities, burials, and artifacts. These will be described and then, in conformity with the hermeneutic spiral method, the dominant account of this archaeological record will be presented and critiqued. This account is termed here the Civic-Ceremonial Center Model. It is an exemplary expression of the monistic modular perspective and is largely an expression of the same premises as the Cemetery Model, only modified for application to the Ohio Hopewell. This will be followed by the reanalysis of these same data in terms of the Ecclesiastic-Communal Cult Model, demonstrating that this model gives a more coherent account of these important data than does the alternative Civic-Ceremonial Center Model. Since both the Civic-Ceremonial Center Model and the Dispersed Hamlet/Vacant Center Model share the same general monistic modular perspective, this conclusion reinforces undermining the global Exclusive Territorial/Proprietorial Domain Paradigm.

THE TRIPARTITE GREAT HOUSE/EARTHWORK PATTERN

The Edwin Harness mound, situated in the infix component of Liberty Works (fig. P.3), covered the remains of one of the three timber buildings, and the remains of the other two were found under the major mounds of Seip, also in the infix component of this earthwork (fig. P.4). Using the terminology devised by Greber, these latter two will be termed the Seip-Pricer mound and the Seip-Conjoined mound.[9] Although the Seip-Conjoined mound is the smallest of the three, they were all large, between 60–70 m (220–240 feet) in length. This is only slightly less than half the length of Mound 25 at the Hopewell site, the largest known Ohio Hopewell mound at about 140–150 m in length.

All three mounds have been excavated. N'omi Greber has supervised the most recent excavation of the residue of the Edwin Harness mound, particularly the undamaged elements of the floor and the peripheral elements of the mound strata. The work she and her colleagues have carried out has recovered significant data undestroyed by earlier excavations, these latter having been conducted before excavation and curating techniques were developed to current standards. In particular, she has revealed a complex but definable patterning of post holes that marks the remains of a major timber building, as shown in the schematic diagram of fig. 12.1.[10]

Shetrone and Greenman did not note a similar pattern in their report of the excavation of the Seip-Pricer mound and there were insufficient post holes for Greber to reconstruct an equivalent pattern there.[11] However, Greber's 1976 and 1983 reconstructions of the patternings of the burial facilities on the floors of the Seip-Pricer and Edwin Harness mounds replicate each other, as illustrated in fig. 12.2. Combining this finding with her noting a similar overall size of the two mounds, a similar internal patterning of their construction strata, a similar stratified construction of their floors, and a similar patterning of ritual features and facilities in the form of burnt areas, clay basins, ritual nonmortuary deposits, etc., she concludes that Seip-Pricer covered a timber building similar to the one covered by the Edwin Harness mound. She also concluded that the mortuary residue on the floor of Seip-Conjoined indicates a similar timber construction (fig. 12.3). She concludes reasonably that all three mounds covered the residue of timber buildings conforming to the same pattern, which she generically terms the Great House. Because of the mortuary features, facilities, and post-mortem remains found on their floors, they will be termed here Great House CBLs.

Probably the most striking property of these structures is that they were two-storied with major mortuary facilities built on the ground floor. The upper floors of these three Great House CBLs were divided into three large rooms, two rectangular and one circular (figs. 12.1 and 12.2). The rooms are aligned axially forming a Rectangle-Rectangle-Circle pattern, or what will be termed the R-R-C pattern. The two major rectangular rooms, joined by a short passageway, display the "classic" Ohio Hopewell form of rounded corners and double side walls. They are slightly different in size. The circular room is the smallest and is situated at one end. There is a small rectangular section that is located to one side of the corridor connecting the two rectangular rooms. However, this may be only on the ground-floor level.

The axis of the Edwin Harness Great House CBL was oriented about 15 degrees west of north with the larger of the rectangular rooms at the northern end. Greber refers to this as the North Structure or North Section.[12] The Middle Structure or Middle Section is the smaller rectangular room, located between the larger North Structure and the smaller South Structure or South Section, the southern circular room. The small "side" room associated with the Edwin Harness Great House is the East Structure or East Section in Greber's terminology. Apparently, the directional orientation of the axes of the Great House CBLs could vary since the Seip-Pricer and Seip-Conjoined Great House CBLs were oriented east-west (actually skewed slightly north of west). In both cases, the larger rectangular room, which in this case Greber refers to as Room 1, is at the west end, and immediately to its east is the smaller rectangular Room 2, connected by the short corridor.[13] Room 3 is the circular room conjoined directly to the east side of Room 2. Except for these different directional orientations, the Seip-Pricer Great House is largely a replication of the Edwin Harness Great House.

As stated above, these rooms made up the upper floor of the Great House CBL, being raised on major support posts above the ground-level prepared floor.

> The size and placement of these (48 timber) supports strongly suggest that there had been an upper floor or platform area. There was at least one post from the North Section which was just over 10 feet (3 m) high . . . the major construction timbers were young hickory trees . . . [and] for similar environments, 58 foot (17 m) heights are recorded for 18 cm diameters. (Greber 1983, 27)

At Edwin Harness each of the two rectangular rooms had forty-eight support posts. The support timbers averaged 24 cm in diameter. Therefore, the young hickory trees that were used would have been of substantial height when cut down. The supporting posts were placed in seven columns and seven rows. Both rooms lacked a center support post. The posts of the Middle Structure were spaced about 1.6 m apart, east-west and north-south. The posts of the North Structure were also spaced east-west about 1.6 m apart, but the north-south spacing was about 1.9 m, making the floor of North Structure the more spacious of the two. It was about 14 m north-south by 11.5 m east-west. The Middle Structure was about 12 m x 10 m and the circular South Structure had a floor 6 m in diameter. If the small connecting corridor is added, the total length of this Great House would be about 35 m. The wall of the circular South Structure at Edwin Harness was made of posts with a smaller average dimension (actually three different sizes) than those of the Middle and North Structures. It had only four support posts, two in the northern sector of the wall and two in the southern sector. The posts of the small ground-floor East Structure, adjacent to the corridor connecting the North and Middle Structures, were small in diameter.

The ground floor under each of these three main rooms was devoted to ritual performances, in many cases terminating with human cremated burial. Most of these ground-floor burials were within the confines of the pattern demarcated by the support posts. As stated above, the central space in each rectangular room lacked a support post, creating an open area on the ground level. If the term "charnel house" is understood in the broad sense of the Mourning/World Renewal Mortuary Model, then these double-leveled Great House CBLs may have been that way partly in order to handle the whole range of mortuary events postulated by this model, from the mourning period through to the cremation and the final burial.

Greber has used different terms to refer to the rooms and the burial spaces: rooms, sections, structures, lobes. The term "room" will be used to refer to the major upper-floor spaces and the term "section" will be used to refer to the major ground-floor spaces. Following Greber's usage in her discussion of Seip, these will be numbered Rooms 1 and 2, for the two rectangular rooms, and the circular room will be termed Room 3. Sections 1, 2, and 3 are the corresponding ground-floor spaces. The East Section of Edwin Harness and the equivalent section on the north side of Seip-Pricer will both be termed Section 4.[14] There were other post holes and features that could be described, but this summary should be sufficient for immediate purposes.

THE CIVIC-CEREMONIAL CENTER MODEL OF THE GREAT HOUSE

Greber characterizes these major Ohio Hopewell embankment earthwork and mound locales as the locales for activities that might be most adequately characterized as combining civic leadership and ceremonial tasks.[15] Therefore, it seems appropriate to label her account as the Civic-Ceremonial Center Model. She considers the Great House to have been used by a local community as its center for conducting rituals. Although she does not use the term "clan," she does claim that the individual components would have been based on lineal kinship and they would have been related to each other by common apical ancestor descent that was, however, ranked hierarchically and cross-cut by ritual specialization. For convenience, the components will here be termed clans. She interprets the R-R-C patterning as correlating with these ranked and ritually specialized clan components. In the terms developed in this book, this ranked clan-based community would be a hierarchical monistic modular social system.

Initially she noted that both the embankment earthwork and the Great House CBL displayed a tripartite pattern: the R-R-C pattern of the Great House and the small circle/large circle/square pattern of the earthworks (in terms of this book, the Paint Creek C-R Configuration). This double tripartite patterning was partly the reason for her claiming the community was based on three major social components, each occupying one of the three major rooms of the Great House CBL, each using the corresponding floor sections, and each assigned to using one of the three major spaces that made up the tripartite patterning of the embankment earthwork.

However, her subsequent excavations at Edwin Harness, published in 1983, revealed the East Section of this structure. Therefore, she recharacterized the patterning from being tripartite to quadripartite.[16] To maintain the formal embankment/Great House parallelism, she suggested that the East Section corresponded to the "lobed circle" found between the C and R components. Since this fourth element was a small side section, she viewed the group associated with it as different from the groups associated with the three large sections. This separation would still leave the community, as such, with three main or clan-like components. In any case, these social characterizations allow classifying the Great House CBLs as the civic-ceremonial centers of monistic modular communities based on proprietorial domain.

Importantly, Greber does not claim that this model can be generalized to all the Ohio Hopewell. Based on her comparative analyses, she concludes that the variation in the forms of the Ohio Hopewell buildings and the variation in the patterning of the mortuary data are such that, while the general principles of this model may apply to Ohio Hopewell, only a few of the major Ohio Hopewell sites fit the tripartite (or, possibly, quadripartite) patterning displayed at Seip and Liberty Works. For example, the Raymond Ater Great House CBL, she concludes, was based on a dual-ranked component community.

THE MORTUARY DATA AND THE FUNERARY PARADIGM

In common with the Cemetery Model, the Civic-Ceremonial Center Model gives a funerary view interpretation, therefore treating the burials as the termination of funerals. This means that the mortuary patterning is taken to rather directly reflect the social structure of the living community. Accordingly, the Civic-Ceremonial Center Model argues that any statistically significant variation displayed in the mortuary data of the three funerary burial sections would reflect the organizational complexity and differential ranking of the groups making up the community. This variation is then used rather directly to map onto the R-R-C patterning of the Great House, claiming that this patterning manifests a community based on a compound form of functional and rank-based differentiation. Indeed, it was precisely argued that rank constrained the range of valued functions that a component group could perform, or vice versa.

In these funerary view terms, the artifactual associations of the individual funerary burials mark the relative social ranks of the deceased at the time of death. This served as the basis for establishing a rank order for each individual in each burial section, and, in the aggregate, the rank order in the community of the social component associated with each section.[17] Eleven major categories of artifacts were established as having statistically significant distribution variation across the total set of burials. The number of categories and the number of items of each of these categories associated with each burial were counted and these two measures established the burial's ranking within the total burial population. It is possible that a given burial could rank high in one category and very low in another. In order to make an objective comparison, the ranks that each individual had in each category were summed to generate individual ranksums. Since first rank in a category was numbered 1, then the lower the ranksum, the higher the social rank of the individual burial.

Sections 1, 2, and 3 for Edwin Harness and Seip-Pricer were compared along three variables: (1) The allocation of burial space; (2) The sector profiles of the individual ranksums in each section; (3) The distribution of individual burials in relation to those individuals who ranked among the top ten percentile.[18] It was established that there was a direct relation between the median ranksum of each section and the average amount of space per individual burial so that the lower the median ranksum of the section the greater the overall burial space available for those buried. It was also established that the variation in distribution across the three sections was statistically significant for only three of the eleven artifact categories.

> There is a statistically significant differential distribution of rankings among the 4 groups for 3 classes of grave goods: rectangular copper plates, celts or axes, stone and copper ear spools; and for a fourth attribute, the grave area per individual.[19] All groups have statistically *similar* distributions or rankings for pearls, marine shell containers, flint blades, nonutilitarian worked animal bones and teeth, shell beads, and other rare objects. There are no individuals in Lobe 3 [the circular section] with rectangular copper plates, ear spools, or celts. (Greber 1979b, 42, emphasis in original)

This allowed concluding that the three artifactual categories of copper plates, ear spools, and celts defined leadership in terms of both functional tasks and rank.

> It is relatively easy to assign these three classes of artifacts to a sociotechnic category. These items visually symbolize socially recognized functions performed by the individuals associated with high prestige. It is more difficult to estimate what these functions might be. Because copper plates, ear spools, and celts can occur separately, I assume that each signifies at least some separate social functions. (Greber 1976, 89, 20)[20]

By analyzing the distribution of these three categories it was concluded that celts marked the possessor as a leader who was "specialized in nature and of importance within the society" (89), while the ear spools marked the wearers as "leaders of possibly kin-related subgroups, but with wider influence [than celt bearers] which enables concentration of these individuals at burial, one concentration in each of Lobes 1 and 2" (89), and the possessors of copper plates were leaders "within activities which involve external societies. Because of the nature of the distribution of the plates, this would appear to possibly involve only high ranking individuals of Lobe 2 but a more varied range of ranked individuals from Lobe 1" (90).

It was also concluded that copper plates, ear spools, and copper celts marked ascribed leadership positions. There were also achieved leadership positions, best indicated by Individual 82 of Burial 73, the highest-ranking individual in Section 3 of Seip-Pricer, whose burial facility was surrounded by granite boulders. He was the only individual in this section with marine shell gorgets, small bone awls ritually broken, and a bird effigy pipe. He was one of eight burials that had a large ocean shell container, and one of four to have wild feline canines.

A second dimension basic to this analysis was the construction of the ranksum profiles for each burial section. When the individuals in a burial section are ranked relative to each other, this creates the ranksum profile for that section. The argument is that the patterning of these profiles should give a rough measure of the number of ranked orders in each section and the greater the number then the greater the organizational complexity of the section and the greater the rank of the highest-ranking individual.

> The status [of an individual in a ranking order] is the sum of the product of the number of individuals in each rank below this person and the weight for the given rank. This is a measure of complexity, as previously defined. . . . [Thus] for a given number of individuals there is a greater complexity and a corresponding rise in status of the top ranking individual with a strung-out hierarchical arrangement that has one individual in each rank than for one individual heading a group of similarly ranked individuals. (Greber 1976, 75–76)

In these terms, then, the section with the highest number of high-ranking individuals should also be the highest-ranking section overall. This spread (fig. 12.4) is used as a measure of relative organizational complexity of each component. Therefore, it is notable that, despite the fact that Section 1 and Section 2 have different median ranks, the former being the higher of the two, both have a similar profile with no significant gaps splitting the low- and medium-ranking individuals. Even the individuals making up the top 10 percent of Sections 1 and 2 are not separated from the highest of the middle-ranking individuals to any extraordinary degree. The ranksum profile of Section 3, however, quite distinctly differs by showing the majority of the individuals clustering near the lowest rank, a minority strung out as the middle rank, and then a large gap between this group and the highest-ranking individual. Greber notes this difference.

> Lobe 3 [i.e. Section 3] is the only one [of the three sections] which has a statistically significant difference in ranksum spread. This can be seen in the difference, in terms of order spread, between the quartiles of Lobe 3 individuals as compared with those of Lobe 1. Lobe 1 distances between quartiles are relatively equal. Lobe 3 distances are not equal; the large distance between the first and second individuals is even more noticeable in the magnitude of the ranksum. (Greber 1976, 74)

As Greber points out, it is clear that the ranksum profiles of Sections 1 and 2 are more similar to each other than either is to Section 3. This is not to deny that Section 1 has a lower ranksum median (about 775), and therefore has a higher overall ranking than Section 2 (about 790), and that both of these sections outrank Section 3 (about 815). Furthermore, as claimed under this model, because the spread of rank-sum profiles is a rough measure of the number of different ranks in a group, the complexity of the three sections varies directly with their ranking. Finally, despite the fact that the highest-ranking individual in Section 3 also has the highest-individual ranksum of all the burials, because this person is in the group with the least complex organization, it is claimed that he is actually outranked by the highest-ranking individuals of Sections 1 and 2. "Accepting such a model as a theoretical framework, then there is less complexity within Lobe 3 than within either Lobe 1 or Lobe 2" (76). On these grounds it is concluded that the three sections of the Great House CBL were differentially ranked and that this ranking order reflects the rank order of the three components that they represent.

Accordingly, this analysis would give grounds to conclude that the Great House CBL does not mark an "equal opportunity" society based on stratified egalitarianism, as postulated by the Autonomous Cult Model. There are similar opportunities (i.e., leadership positions, functional positions) in each component, but these are claimed to be ascribed to each group, thus preventing competent individuals from lower-ranking groups from competing with those in the higher-ranking groups for esteem-enhancing positions. Furthermore, the quantitative distribution favors the "ranking clan," with the greatest number of esteemed positions, many relating to external affairs (copper plates), available only to those born into the component represented by Section 1, a lesser number available to those born into the component represented by Section 2 (fewer copper plates), and, for those born into the component represented by Section 3, the least number as well as the least esteemed ascribed positions, since artifacts of the three selected categories were not found with any of the burials in this section.

The third dimension deals in further detail with the question of organizational complexity. It was noted that the distribution of burials in each section was different when the patterning of burials was examined in respect to the spatial positioning of individuals in the top 10 percent. The clearest patterning would appear to have been found in Section 2 of Seip-Pricer. Here it was noted that there were five clusters of burials, each associated "with an individual who was high ranking within the total population and may represent kin-based units."[21] Four of these were "linear" clusters, each "headed" by a top-ranking individual. These were distributed linearly around the center of Section 2. The fifth cluster was a roughly circular set of burials around one of the "crematory basins." Since none of the burials in this central cluster was among the top 10 percent, it is suggested that the basin(s) substituted for highly placed leaders in defining the overall ranking of this group. Section 3 has a different patterning, one in which the highest-ranking Individual 82 is in the center and the others are spaced around him on the periphery of the section forming a rough circle.

There are two major "hot spots" of high-ranking individuals, one in Section 4 (the equivalent of the East Section at Edwin Harness) and one at the west end. It was also suggested that, in ranksum terms, there was a "hot" linear cluster on the north side, another on the south side, and another around a large cache of copper plates just south of the center of this section.[22] Based on these spatial distributions of high-ranking burials, it was suggested that Section 1 displays greater complexity than Section 2, and both display greater complexity than Section 3. These conclusions would appear to confirm the interpretation of the sectional ranksum profiles as indicating differential organizational complexity across the three sections, and, therefore, across the three major components of the Great House CBL group.

Although focus has been put on three burial components, as mentioned earlier, four mortuary components were actually identified based on spatial distribution, the three ground-level sections (four, if we treat Section 4 at Edwin Harness and at Seip-Pricer as separate from Sections

1, 2, and 3), and the out-lobe group. This latter group was made up of individuals who were primarily the above-floor burials. These occurred largely in step with the multiple stages in the construction of the Great House CBL mounds, the latter covering the burned and abandoned remains of the Great House CBLs themselves. These burials were interpreted as playing a significant role in the ritual construction of these multiple strata. It must be stressed, however, that even though the Seip-Conjoined Great House CBL had a circular room and an associated ground-level Section 3 (fig. 12.3), it contained no burials, and, also, this figure shows no post holes that might indicate a fourth section equivalent to those found at its near neighbor Seip-Pricer or the Liberty Works Edwin Harness Great House CBLs. Still, the model extended these findings with regard to Sections 1 and 2 of Seip-Pricer to the equivalent sections at Seip-Conjoined as well as to Edwin Harness.

CRITIQUE

While the claim of the Civic-Ceremonial Center Model that the patterns and internal treatments of space correlate with and define ranked social components, it is important to note that the model does not account for the most distinctive aspect of this Great House CBL pattern, this being the formal rectangle/circle contrast embodied in the Room/Section layout. Indeed, given the commitment of this model to the social symbolic meaning of space, it is surprising that it does not target this formal contrast, particularly since the differential patterning of the ranksum profiles neatly correlates with the rectangle/circle contrast. That is, as pointed out above, the ranksum profiles of Section 1 and Section 2 are more like each other than either is like the ranksum profile of Section 3. The fact that the spatial forms and ranksum profiles of the former are iterated and contrast with the spatial form and ranksum profile of the latter immediately suggests that two different types of groups are being compared according to the same artifactual measuring criteria. The absence of copper artifacts from Section 3 and their prolific distribution across both Sections 1 and 2 reinforces the conclusion that, rather than three social components of the same type simply differentially ranked and functionally specialized, there are marked here two different types of social components structured according to two different social characteristics R and C, with the R being represented by two groups of the same type.

Since the parallels in morphology and ranksum profiles of Sections 1 and 2 mark them as probably of the same social type, whatever that was, then it is relevant to compare them in terms of the same ranking criteria. However, since the morphology and ranksum profiles of these two contrast with Section 3, then these criteria cannot be applied to the latter section in the same way. In short, the variation in the ranksum profiles across the three sections correlated with the rectangle/circle morphological contrast of the rooms/sections can be reasonably argued to manifest a social structural contrast. If this is the case, then the fact that Section 3 displays the lowest median ranksum may have no relevance in assessing the standing of that group or of those who are in it. Because of its probable different social nature, this group may have played a specialized function that accounts for its qualitative differences in spatial form, qualitative difference in individual ranksums, quantitative difference in average sectional ranksum, and qualitative difference in ranksum-profile patterning. To speak counterfactually, if the three components were of the same social nature while differing only in rank and ritual specialization, which is the position of the Civic-Ceremonial Center Model, neither the morphological contrast nor the contrast in ranksum-profile patterning should be expected. Instead, there should be three differential ranksum group averages, three similar ranksum-profile patterns, and three rectangular (or three circular) rooms/sections differing only quantitatively in overall area.

The hermeneutic spiral requires that the above critique of the Civic-Ceremonial Center Model be completed by giving an alternative account of the same mortuary data, while eliminating the weaknesses that this critique has raised. The alternative model, of course, is the Ecclesiastic-Communal Cult Model of Ohio Hopewell. In these terms, it will be postulated that the Great House and its associated embankment earthworks constituted a CBL of an ecclesiastic-communal cult. Therefore, the next chapter is devoted to empirically grounding this model in the same data as presented above, but supplemented by further observational material.

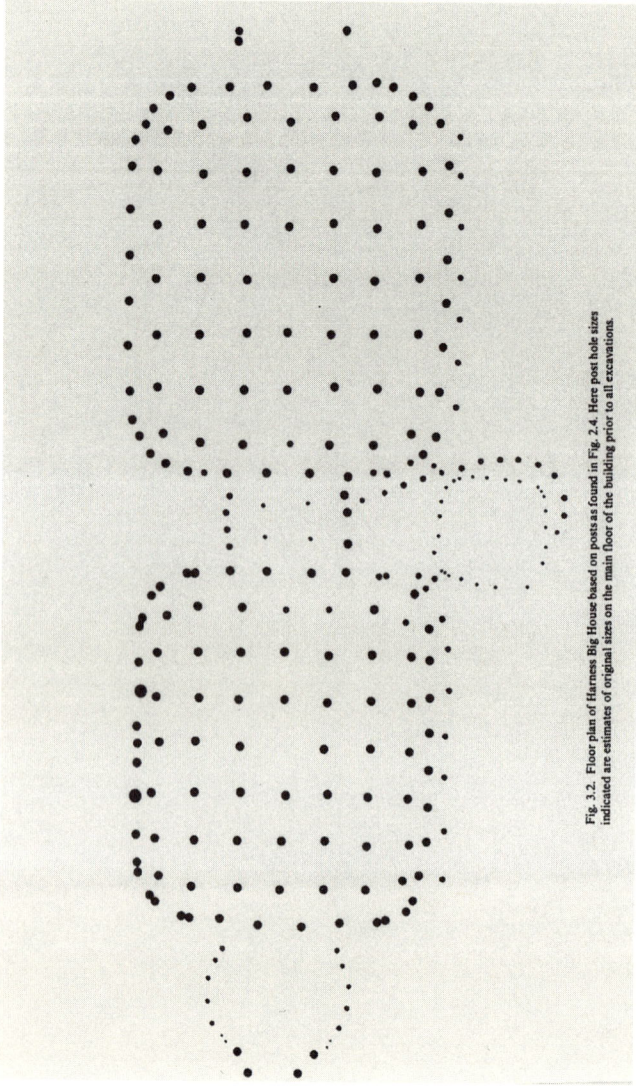

Fig. 3.2. Floor plan of Harness Big House based on posts as found in Fig. 2.4. Here post hole sizes indicated are estimates of original sizes on the main floor of the building prior to all excavations.

FIG. 12.1. The Edwin Harness Floor Plan of Post Features (Greber 1983, 28, figure 3.2)

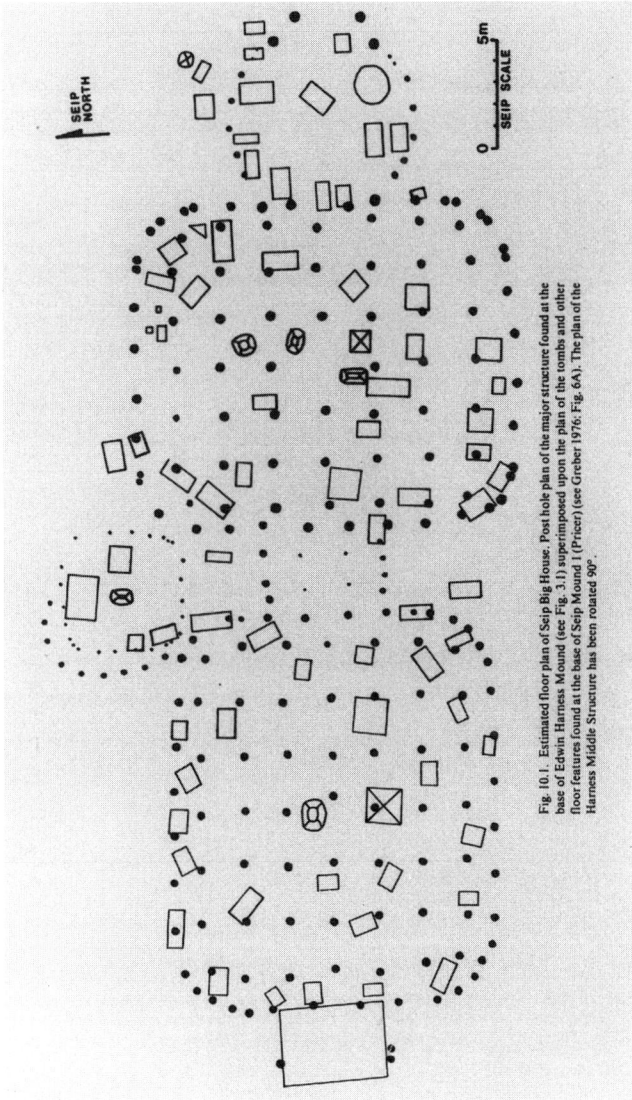

Fig. 10.1. Estimated floor plan of Seip Big House. Post hole plan of the major structure found at the base of Edwin Harness Mound (see Fig. 3.1) superimposed upon the plan of the tombs and other floor features found at the base of Seip Mound 1 (Pricer) (see Greber 1976: Fig. 6A). The plan of the Harness Middle Structure has been rotated 90°.

FIG. 12.2. The Superimposition of the Seip-Pricer and Edwin Harness Floor Plans (Greber 1983, 88, figure 10.1)

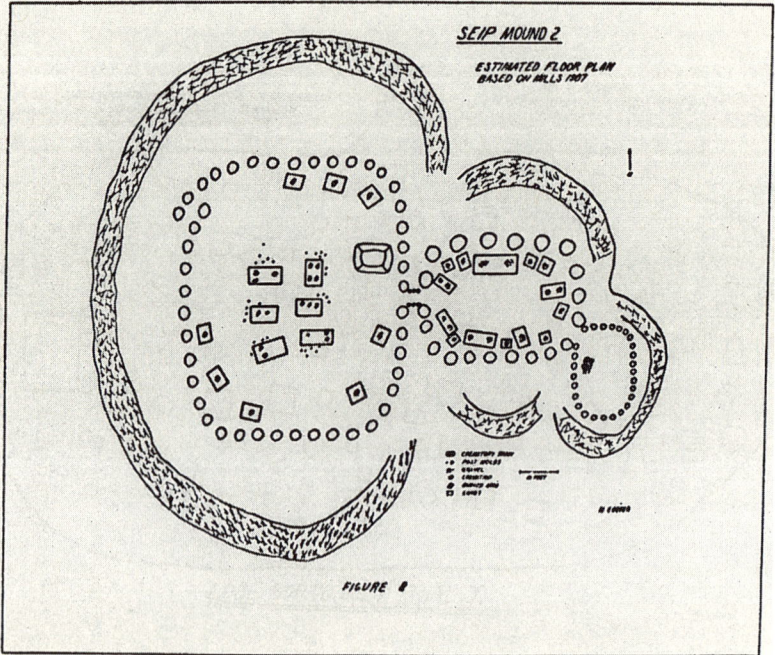

FIG. 12.3. The Seip-Conjoined Floor Plan of Mortuary Features (Greber 1979b, 66, figure 7)

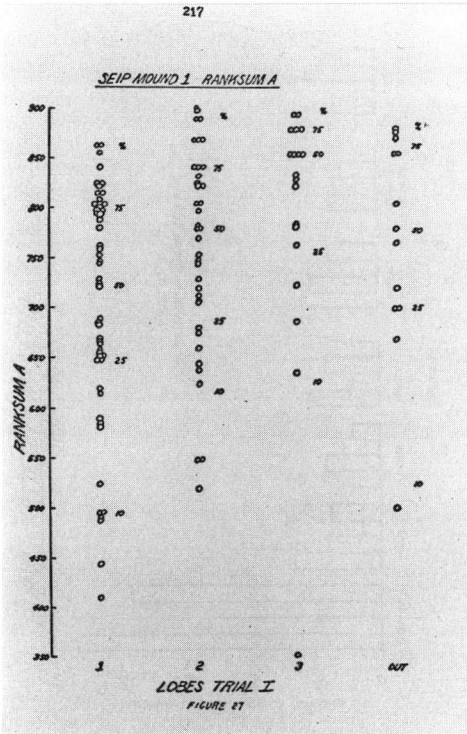

FIG. 12.4. Ranksum Profiles of Seip-Pricer Burials (Greber 1979b, 217, figure 5)

Chapter 13

THE RITUAL CYCLE OF GENERATIONS

The Ecclesiastic-Communal Cult Model of Ohio Hopewell gives an alternative interpretive account of the Great House CBLs and their associated embankment earthworks to that of the Civic-Ceremonial Center Model by claiming that they were the locales of the type of complex autonomous cults I have termed ecclesiastic-communal cults. There are two lines of empirical evidence that can be presented to support this claim. The first concerns the R-R-C pattern and the correlation of this pattern with the mortuary data. The second requires applying the notion of the split age-grade burial population to the distribution of the mortuary deposits of the Great House CBLs. The split age-grade notion was presented in chapter 11 and applied to analyze the Adena CBL mounding. It was argued then that the age-grade structure of the postulated senior and junior cults sharing the same CBL had distinctive effects on the internal patterning of the CBL mounding. Since the ecclesiastic-communal cult sustained the dual senior and junior age-grade structuring, an equivalent impact on the patterning of the mortuary residue of the complex cult system should occur. The Split Age-Grade Mortuary Model of Ohio Hopewell is presented for this purpose and is treated as an ancillary of the Ecclesiastic-Communal Cult Model. Any data that support the operation of a split age-grade mortuary scheme, then, are considered to be significant support for the model.[1]

THE R-R-C AND THE MORTUARY ARTIFACT PATTERNS

In the previous chapter, it was pointed out that the Civic-Ceremonial Center Model does not adequately account for either the patterning of the mortuary residue found in the burial sections of the Great House CBL or the rectangle/circle contrast as embodied in the R-R-C patterning. It should be remembered that this R-R-C patterning of the Great House CBL reflects a two-storied structure with the ground floor R-R-C pattern delineating burial sections and the upper floor R-R-C pattern delineating ritual rooms. The Ecclesiastic-Communal Cult Model postulates that the organizational core of the cult is the conjunction of the laity/clergy structure and the senior/junior age-grade structure. This conjunction would favor the generation of two contrasting spaces, manifesting the laity/clergy structure, and two similar spaces, manifesting the senior/junior age-grade structure. These needs are quite adequately served by the R-R-C pattern. The contrasting aspect of the pattern is highlighted by conceptually collapsing R-R iteration to the compound R' component (prime). This highlights the core R'-C pattern contrast, thereby manifesting the laity/clergy structure. This patterning seems particularly appropriate to express and constitute the social distinction of the clergy and the laity positions. Of course, the R' iteration is quite appropriate to manifest the type of social structural equivalence of two age-grades distinguished by senior/junior generation.

The layout of the post holes of the major support timbers of the Great House CBL reinforces the interpretation that the size differentiation of these two rectangular spaces and single circular space was deliberate. The former are conjoined by a narrow corridor and the latter by direct juncture. The two rectangular rooms display differences of size and detail of construction that are not inconsistent with what could be anticipated when groups of the same social nature are separated according to the principle of seniority, thereby recognizing social similarity while sustaining social distancing from each other.

Consistent with the Ecclesiastic-Communal Cult Model would be finding quantitative and qualitative differences in the mortuary data that could be accounted for by the same factors postulated as causing the complementary iteration and complementary contrast making up the R'-C pattern. Indeed, the patterning of the mortuary data as summarized in the last chapter would nicely fit these criteria. Here there are two important areas: the differential distribution of the three key artifactual categories—copper plates, celts, and ear spools—and the qualitative difference in ranksum profiles of the groups. In the latter case, the Section 1 and Section 2 ranksum profiles do not differ qualitatively from each other while differing equally with the ranksum profile pattern of Section 3. This would be consistent with the premise that the R' pattern manifests groups of the same social nature and the contrasting R'-C pattern manifests groups that are complementary oppositional in their social natures.

Even the differential distribution of the copper plates, celts, and ear spools is consistent with the claim that the R'-C pattern manifests the laity/clergy structure. Since this structure constitutes complementary contrasting social natures, it would be partly constituted by mutually exclusive sets of artifacts having contrasting symbolic pragmatic powers. In these terms, it would appear that these three categories of artifacts were possibly avoided by those associated with Section 3 while being reserved for the leadership associated with Sections 1 and 2. This would be consistent with the laity/clergy structure. Furthermore, assuming that Section 3 represents the clergy, many may have been young acolytes or "apprentices." There was probably only a small core of shamanic priests at any given time and, although these would be custodians of powerful sanctifying warrants, the relatively small number of ranking clergy in any given generation and the relatively large number of "apprentices" would be responsible for both the overall low median ranksum and the wide separation of high and low ranksum individuals in the group's ranksum profile. In sum, the low median ranksum of Section 3, the postulated clergy section, may be attributable to the hierarchical nature of the clergy group, a possible proscription against the use of copper, and the significance of young "apprentices."

As postulated under the Autonomous Cult Model, the distribution of esteem-endowing copper artifacts across the senior and junior age-grade sectors would be strongly time-dependent, largely constrained by the age-cohort hierarchy. Therefore, it would be expected that, in the aggregate, *since the deceased buried in the senior age-grade sector would have lived a longer time, they would have had more ritual opportunities to accumulate a richer set of potent copper and other types of warrants than the deceased in the junior age-grade sector.* Also, as argued earlier, despite both laity sectors being of the same nature, based on male, same-generation companionship, the junior age-grade would have a rather strong hierarchical ranking of the age cohorts, while in the senior age-grade this ranking would be relatively unimportant, the assumption here being that aging correlates with sharing experiences and the latter come to figure more importantly in same-generation interaction than does intra-generational age differences. Thus, the "hot spots" noted in the mortuary data in Section 1 of Seip-Pricer may be related to shared experiences of the senior age-grade custodians of these symbolic warrants more than to differential types of leadership. At the same time, identifying four linear clusters of "low-ranking" deceased in Section 2, Seip-Pricer, each anchored to an "esteemed" burial, and one cluster of low-ranking deceased encircling "esteemed" crematory basins may, in fact, reflect stronger age-cohort hierarchy in the junior age-grade than in the senior age-grade sector.[2]

THE SPLIT AGE-GRADE MORTUARY MODEL OF OHIO HOPEWELL[3]

The central notion of the Split Age-Grade Mortuary Model was discussed earlier and applied to the Adena CBL moundings. In the latter case, it was argued that: (1) the junior/senior age-grade structure would generate two asymmetrical sectors of the mounding; and (2) the cycling of ritual spaces would lead to splitting an age-grade into two separate burial populations. These premises also form the core of the Split Age-Grade Mortuary Model of Ohio Hopewell. The Ecclesiastic-Communal Cult Model postulates that the senior age-grade occupied upper Room

1/ground floor Section 1, and the junior age-grade, upper Room 2/ ground floor Section 2. In terms of the Split Age-Grade Mortuary Model, then, there should be a gap between the age-at-death of the Section 1 and Section 2 mortuary populations that would be consistent with a generational difference.

It is well known that control of the temporal dimension in Ohio Hopewell is a major problem, largely because so much of the data was recovered prior to the development of modern archaeological techniques. Of course, in principle, the relative age-at-death of the deceased in the two sections could quickly be determined simply by analysis of the anatomical age indicators of the mortuary remains. However, almost all the deceased on the floor were cremated. According to Konigsberg, even though many were in-flesh cremations, there was sufficient damage done to the critical age-determining bones that he was able to construct only three general age categories: infants, sub-adults, and adults, the latter being estimated as anyone over 20.

> All adults were simply classified as 20+ years old. This classification was necessary because of the extensive damage from cremation, which obscured or destroyed the macroscopic morphological indicators traditionally used to estimate adult ages-at-death. Although recent work . . . suggests that microradiography could be used to determine ages-at-death for cremated adult remains, this technique has not yet been applied to the Seip materials. (Konigsberg 1985, 125)

Until such analytical techniques can be applied, other evidence must be used to establish whether there is a differential age-at-death between Section 1 and Section 2 consistent with a generational difference.

The Seip-Pricer and Seip-Conjoined Great House CBLs present the best current possibility for demonstrating the Split Age-Grade Mortuary Model. This is because the strongest comparative evidence in support of this model would be two CBLs that are prima facie linked by an unbroken organizational continuity of several generations and that display evidence that would support the above differential age-at-death claim. The Seip-Pricer and Seip-Conjoined Great House CBLs fulfill the first requirement since (1) they are in the same embankment earthwork, (2) they are in the same part of this earthwork, the infix, (3) they are the only two known CBLs there, and (4) based on the stratigraphic evidence, the Seip-Pricer Great House CBL was constructed and used before the Seip-Conjoined Great House CBL was built and occupied. Finally, important mortuary ritual variation between Seip and Liberty Works to be presented below supports the above differential age-at-death claim.

Only Sections 1 and 2 of the three R-R-C sections are relevant for the Split Age-Grade Mortuary Model since it is these two sections that have been postulated as the senior and junior age-grade burial sectors. Section 3 has been demonstrated as being qualitatively different from the other two sections. Section 4 and the above-floor burials can also be set aside for the same reasons. If any temporal markers are available, then they should be found in Sections 1 and 2 such that it can be concluded that Section 1 contains burials of individuals who, on average, were older at the time of their deaths than were the individuals buried in Section 2. If this differential age-at-death is consistent with a generational gap, it follows that Section 1 contained the deceased of the senior age-grade sector, and Section 2 contained the deceased of the junior age-grade sector. This finding would constitute strong support of the Ecclesiastic-Communal Cult Model.

The Split Age-Grade Mortuary Model hinges on the central premise that ritual space was cycled as the result of the promotion of age-grades. The promotion would see the junior age-grade moving to the senior age-grade sector and the adoption of the abandoned sector by the newly initiated junior age-grade. As pointed out above, the Seip-Pricer Great House was built and used first, and when it was abandoned, the Seip-Conjoined Great House was built and used. Assuming that these two were used as single-cycle CBLs, when Seip-Pricer was abandoned, being the survivors of the former junior age-grade, the newly promoted senior age-grade group would have taken occupancy of Room 1 of the newly constructed Seip-Conjoined Great House CBL. The new junior age-grade group would occupy Room 2 of this Great House.

As postulated under the Mourning/World Renewal Mortuary

Model, as members of these two age-grade sectors aged and died, they would be subjected to incremental mourning→spirit-release→world renewal ritual, ending as cremated burials in their respective sections. This means that when all members of the senior age-grade of the Seip-Conjoined Great House CBL had died, those members buried at Seip would have been split into two burial populations, one in Section 1 of Seip-Conjoined and the other in Section 2 of Seip-Pricer. Each generational span is estimated at 25–30 years. Therefore, the burial population of Section 2 of Seip-Pricer, marking the age-grade in its junior 25–30 year phase, and the burial population of Section 1 of Seip-Conjoined, marking the age-grade in its senior 25–30 year phase, would together span a two-generational spread of about 50–60 years. However, if these two Great House CBLs were multi-cycle split age-grade CBLs, then only some of those buried in Section 2 of Seip-Pricer during the junior phase and only some of those buried in Section 1 of Seip-Conjoined during the senior phase would make up the deceased of the same age-grade spanning two generations. Others would belong to the cycling of earlier or later age-grades. This would mean that, in fact, together, these two Great House CBLs would incorporate more than 50–60 years, depending on the total number of split age-grade cycles they incorporate together.

Is there evidence that can be marshaled to support the Split Age-Grade Mortuary Model, and if so, is it sufficient to establish how many cycles they incorporated in total? If this could be done, then an estimate of the total occupation period of Seip covered by the building and use of the two Great House CBLs could be made. A positive answer can be given in both cases, at least sufficiently positive to support the Split Age-Grade Mortuary Model, and, of course, the Ecclesiastic-Communal Cult Model that presupposes it, and to give a reasonable estimate of the occupation period, although the latter estimation will be made later in the book.

EMPIRICAL DATA

William Mills excavated and reported first on the Edwin Harness mound at Liberty Works, and he then excavated and reported on the

Seip-Conjoined Mound.[4] In the latter report he referred to Seip-Conjoined as the Seip Mound. He did not excavate the Seip-Pricer Mound. This was excavated some years later by Shetrone and Greenman. Mills's published comments on the Seip-Conjoined Mound and the Edwin Harness Mound are very illuminating for the above purposes. The most relevant comments indicate that the Seip-Conjoined Great House CBL burials had two different modes of burial termination, while the Edwin Harness burials had only one of these two modes.

> As stated, the single graves in the first section [Section 1, Seip-Conjoined] were similar in construction to those of the Harness Mound, *though the final burial ceremony must have been different*. In the [Edwin] Harness Mound, after the incinerated remains had been placed in the grave, objects of clothing, together with straw, bark, or twigs, were placed over the remains and set on fire: while burning, clay was placed over the grave, thus preserving the cloth, the grass, and twigs in a charred state.[5] This ceremony was *dispensed with* in this section of the Seip Mound [i.e., Section 1, Seip-Conjoined], and the incinerated remains were placed in the prepared grave, and a covering of wood, usually split pieces, was placed on the top and the grave covered with earth to a depth of a few inches. (Mills 1909, 24, emphases added)

He then comments with regard to the burials of Section 2, Seip-Conjoined that "[t]he final ceremony of burning straw, bark, and clothing over the remains, *similar to the burial methods at the [Edwin] Harness Mound*, was in evidence in 9 burials of the 19 found at the base of the [second] section [of Seip-Conjoined]."[6]

These two different burial termination modes will be termed the *fire-clay rite* and the *split-log rite*. It is clear from the above that the fire-clay rite was used exclusively at Edwin Harness, while both rites were used in Seip-Conjoined, the split-log rite being used exclusively in Section 1 of Seip-Conjoined, and both the fire-clay rite and the split-log rite being used in Section 2—and almost equally (nine and ten cases, respectively). The way that Mills interprets this patterning can be used as important chronological evidence. As he puts it, the fire-clay rite ceremony "was *dispensed with* in this section of the Seip Mound" (Section 1 of Seip-Conjoined). This comment implies that, because he had already excavated the Edwin Harness mound, and because the fire-clay rite was universal at this site, he took it as being the norm at Seip-Conjoined and that the fire-clay rite was partially *replaced* by the split-log rite in Section 2 and completely replaced by the split-log rite in Section 1.

From this evidence the sequence of burials in the two sections can be reconstructed. According to Mills's data, it can be argued that, in parallel with the Edwin Harness practice, initially only the fire-clay rite burials occurred at Seip-Conjoined and all nine of these occurred in Section 2. Then, partway through the use of Section 2 of Seip-Conjoined, the split-log rite was introduced, possibly being innovated at this site. It is also possible that, for a while at least, both rites may have been used in parallel, and, of course, some split-log burials could also have been performed in Section 1 while the fire-clay rite was continued in Section 2. This also indicates that, from the time that burials were initiated at Seip-Conjoined, there were at least nine deaths in the group occupying Room 2 while there were no deaths or possibly only a few deaths registered for the group occupying Room 1. This certainly is suggestive of a differential death rate for these two groups. The distribution of these two burial rites, with the split-log rite being in the significant majority, also suggests that the fire-clay rite was completely abandoned well before the termination of the usage of Seip-Conjoined.

In short, the disparity of distribution of these two burial rites across Sections 1 and 2 of Seip-Conjoined suggests that a differential death rate of the two groups would be responsible, and this implicates a differential age-at-death that may very well be consistent with a generational difference. The problem is, however, since the initial deaths marked by the fire-clay rite were all in the group occupying Room 2 and using Section 2, then this would appear to be the senior age-grade group. However, according to the analysis of the mortuary data, the distribution of the esteem-endowing copper artifacts, spatial size differences, differential ranksum profile averages, and so on, it should be the reverse, with Section 1 containing the deceased of the senior age-grade and Section 2 the deceased of the junior age-grade. Therefore, while the distribution of the two terminating burial rites appears to support differential age-at-death between the two burial populations, as predicted by the Split Age-Grade Mortuary Model, it is the reverse of the order postulated by the Ecclesiastic-Communal Cult Model.

The disparity is easily resolved, however. The excavations of Seip by Shetrone and Greenman established that, in fact, the split-log rite that Mills's earlier excavation showed was used throughout Section 1 and partly in Section 2 of Seip-Conjoined was used throughout the floor level of Seip-Pricer, and, as noted above, it was only after this Great House was abandoned that Seip-Conjoined was built and used. Importantly, based on the current radiocarbon dates, Greber gives a median date of A.D. 300 for the Seip-Pricer and Edwin Harness Great House CBLs.[7] Therefore, *the split-log rite was the norm at Seip-Pricer while the fire-clay rite was the norm at Edwin Harness and these two types of rites were largely contemporary.*

Therefore, the split-log rite was used during the full period of Seip-Pricer at the same time that the fire-clay rite was being used at Edwin Harness (Liberty Works). Following abandoning the Seip-Pricer and transferring to the Seip-Conjoined Great House, the split-log rite must have been continued in the latter for a considerable period before the fire-clay rite was introduced, probably from Edwin Harness. The above temporal analysis using Mills's claims, then, must be reversed. Sections 1 and 2 of Seip-Conjoined would have been used for senior and junior sector burials, respectively, thereby fitting the expectations of the Ecclesiastic-Communal Cult Model.

With the correct sequence established, can the differential distribution of these two rites between Sections 1 and 2 be more coherently accounted for by the differential age-at-death claim of the Split Age-Grade Mortuary Model or by the differential clan ranking of the Civic-Ceremonial Center Model? Since the latter model postulates clans or clan-like components, then they would have had comparable age-structures, with individuals of each group having comparable life experiences, even when the posited difference in hierarchical ranking of the groups is taken into account. Therefore, the death rates should have been more or less equal. Furthermore, under this model, since these different sections mark different "clan" burial spaces, then they would have been used simultaneously. Once the fire-clay rite was introduced, it should have been distributed about equally across both sections. This is clearly not the case. It could be argued, however, that since the burials in Section 2 were split between the two rites, while all the burials of Section 1 of Seip-Conjoined were terminated with the split-log rite, that there was some free choice or leeway between these two modes. Therefore, the group responsible for Section 1 might simply have been very conservative. However, if the two groups were of the same social nature with the same age structure, it could be expected that the same range of attitudes toward ritual would be about equally distributed across both groups. Therefore, this conservative/liberal argument is not persuasive.

Not being able to account for this distribution in these terms reinforces the differential age-at-death claim. However, were these tendencies sufficient to indicate a generational difference as postulated by the Split Age-Grade Mortuary Model? What must now be done is show that this disparity in the distribution of the two terminal burial modes across Sections 1 and 2 marks differential death rates consistent with contemporary age-grade groups separated by a generational span, with Section 1 being the locale of the senior age-grade sector burials and Section 2 of the junior age-grade sector burials.

On first appearances, there are three possible scenarios accounting for the use, introduction, and replacement of the two terminal burial modes. These would seem to exhaust the logical possibilities, given the current understanding that (1) Seip-Pricer and Edwin Harness were largely contemporary and (2) Seip-Conjoined succeeded Seip-Pricer. The scenario that most coherently accounts for the patterning will be accepted and the others rejected.

Scenario 1: This postulates that the introduction of the fire-clay rite at Seip-Conjoined terminated the use of the traditional split-log rite.

Scenario 2: This postulates that, following the introduction of the fire-clay rite, a period of co-usage of the two ritual modes occurred with the phasing out and termination of the split-log rite well prior to the final abandonment of the Seip-Conjoined Great House CBL.

Scenario 3: This postulates that, once the fire-clay rite was introduced, there was co-usage during the rest of the occupation period of Seip-Conjoined, the split-log rite being continued in Section 1 while the fire-clay rite was used in Section 2. There are two possible variations of this continuous co-usage scenario. A radical scenario would argue that following the introduction of the fire-clay rite, the split-log rite continued to be exclusively used in Section 1, while the fire-clay rite was exclusively used in Section 2. A more moderate scenario would allow a short period of co-usage of the split-log and fire-clay rites in Section 2, followed by the exclusive fire-clay rite usage in this section in parallel with exclusive split-log usage in Section 1.

To summarize the relevant mortuary data: together there were 48 burials in the Seip-Conjoined Mound, 43 cremated burials on the floor and 5 noncremated burials at different positions in the mound but above the floor. As pointed out above, of the 43 individuals buried on the floor, 24 were in Section 1 and 19 in Section 2. Nine of the latter 19 were subjected to the fire-clay rite. The rest were subjected to the split-log rite.[8] A sequential presentation and critique of the three scenarios follows. It turns out that they must all be rejected. A new Scenario 4 is postulated and critically analyzed. This latter turns out to be the most coherent and is accepted.

Scenario 1

If Scenario 1 were the case, then, when the introduction of the fire-clay rite occurred, all the 34 split-log rites of the 43 floor burials would have been in place, 24 in Section 1 and 10 in Section 2. The 24 would be the total number of burials in Section 1 while the 10 would make up only 52 percent of what would become the total in Section 2. That is, 9 more burials would occur following the introduction of the fire-clay rite and prior to the abandoning of the Seip-Conjoined

Great House. Assuming that the two groups occupying Rooms 1 and 2 were initially about the same size, if all these 34 split-log ritual events occurred following the construction of the Seip-Conjoined Great House CBL, then the 24 mortuary events in Section 1 and the 10 in Section 2 suggest a differential death rate of 24:10 or about 2.4:1. Everything else being equal, death rates will be higher in older than in younger age cohorts. Therefore, the average age-at-death of the two groups at the beginning of the occupation of the Seip-Conjoined Great House CBL, when only the split-log rite was used, was significantly different, possibly sufficient to mark senior and junior generations. If so, Scenario 1 would constitute support of the Split Age-Grade Mortuary Model.

Critique

However, it might not be quite so straightforward. In terms of the Ecclesiastic-Communal Cult Model, the promotion of the incumbent junior age-grade sector to the senior age-grade sector would occur because of the inability of the incumbent senior group to continue the ritual tasks required. This inability may have been simply the inevitable outcome of aging. It follows that the abandonment of Seip-Conjoined and the construction of a new Great House would have occurred shortly after the final split-log burial in Section 1, this containing the remains of the active senior member whose death would have instigated the retirement of the surviving companions of this senior age-grade. However, Scenario 1 postulates that 9 fire-clay rite burial events occurred in Section 2, the junior age-grade sector, following this last split-log burial in Section 1. This does not satisfy the above requirement that the move would have occurred shortly after the final split-log burial in Section 1. Therefore, Scenario 1 would appear highly unlikely and is rejected.

Scenario 2

This suggests that Scenario 2 might be more likely. With the introduction of the fire-clay rite, the split-log rite was continued as an alternative for a short period, and then was phased out. Allowing for a period of simultaneous use of both rites requires computing a reasonable estimate of the number of burials in Sections 1 and 2 of Seip-Conjoined prior to the introduction of the fire-clay rite. With this estimate, the average death rates in both sections can be compared under the same assumptions as Scenario 1. It is suggested that, conservatively, the first 2 of the 9 fire-clay rite burials in Section 2 were performed while split-log burial rites continued to be performed in both Section 1 and Section 2. Since 2 out of 9 fire-clay rites would be 22 percent of this category of burials, then a reasonably conservative estimate under Scenario 2 is that about 22 percent of the 34 split-log burials, or 7.48 burials, would also have occurred during this period of overlap. This figure must be rounded out to 7 burials so that a total of 27 split-log burials occurred prior to the introduction of the fire-clay rite. Of course, these seven must be proportionally distributed in whole numbers between the total of 24 split-log burials of Section 1 and the 10 of Section 2. The 24/34 ratio gives a factor of 0.7 for Section 1 and the 10/34 ratio gives a factor of 0.3 for Section 2. For Section 2, then, $0.3 \times 7 = 2.1$ burials, rounded out to 2 burials, is reasonable; and for Section 1, $0.7 \times 7 = 4.9$ burials, rounded out to 5 burials, is reasonable. This would mean that, during this overlapping period, Section 1 had an estimated 5 split-log and Section 2 had an estimated 2 split-log burial events. This would mean that, of an estimated 27 split-log burials on the floor of Seip-Conjoined immediately prior to the introduction of the fire-clay rite, 19 were in Section 1 $(24 - 5 = 19)$, and 8 in Section 2 $(10 - 2 = 8)$. This would generate a 2.375:1 ratio (19/8). Since this very minimally reduces the Scenario 1 ratio of 2.4, it would not justify changing the conclusion that Section 1 was the burial locale of a group that was at least one generation senior to the group in the Section 2 burial locale.

Critique

However, this still presents a problem. Scenario 2 suggests a rather short period when the two rites were practiced simultaneously, two fire-clay rites and seven split-log rites. Following this short period of overlap, while no burial rites at all would have occurred in Section

1, seven more fire-clay burial rites would have been performed in Section 2. Scenario 2 can be criticized on the same grounds as Scenario 1, namely, a considerable period of time would elapse between the last burial event occurring in Section 1 and the last burial event occurring in Section 2, at which time the Seip-Conjoined Great House CBL would have been abandoned. Therefore, Scenario 2 is rejected.

Scenario 3

Possibly, Scenario 3 can resolve this problem. This postulates that once the fire-clay rite was introduced, both were used until Seip-Conjoined was abandoned. The radical version argues for a sharp bifurcation of usage between the two sections with the split-log rite continuing exclusively in Section 1 while terminating in Section 2 and the fire-clay rite being exclusively used there. The moderate version allows for a short period of dual usage in Section 2.

Critique

However, there is a problem with Scenario 3. This is actually the most probable scenario under the Civic-Ceremonial Center Model and it was earlier rejected on the grounds that it is unlikely that two groups of the same social type and having the same age-structure would display such radical disparity in terminal burial modes while retaining similarity in the rest of the burial treatment. Therefore, it is again rejected. Now, Scenarios 1 and 2, which would clearly not be likely under the Civic-Ceremonial Center Model, have also been rejected under the Ecclesiastic-Communal Cult Model. If these three scenarios effectively exhaust the logical possibilities, then these analyses have created a quandary. There would apparently be no way of accounting for the numbers and distribution of fire-clay rite / split-log rite burials.

RESOLUTION: SCENARIO 4

At this point it is important to reexamine the assumptions of these scenarios. In fact, the central one is that Seip-Conjoined is a single-cycle CBL. Under the Autonomous Cult Model, both single-cycle and multi-cycle burial locales are possible. Therefore, it will now be postulated that the Seip-Conjoined burial pattern marks a *multi-cycle CBL*. This suggests that the split-log/fire-clay contrast in Section 2 marks both the introduction of a new terminal burial rite and an age-grade promotion. This will be called Scenario 4.

Treating Seip-Conjoined as a multi-cycle or, more precisely, a dual-cycle CBL allows setting aside Scenario 3 and reconsidering Scenarios 1 and 2. The timing of Scenario 1 still applies: the introduction of the fire-clay rite generated an exclusive bifurcation of rites, the split-log rite at Seip-Conjoined continuing in Section 1 and the fire-clay rite being introduced and used in parallel with the latter, but exclusively in Section 2. Even the moderate Scenario 2 postulate could apply: that is, the fire-clay rite was introduced and used in Section 2 while the split-log rite was continued in this section for a period, and then was phased out. The important change, however, is the claim that the new fire-clay rite marks an age-grade promotion. This postulates that the second-cycle senior age-grade sector when newly promoted would have shifted its ritual locus from Room 2 to Room 1, where it continued the same split-log rite in the Section 1 burial locale that it had used in Section 2. The second-cycle junior age-grade sector, moving into Section 2, of course, would have introduced the fire-clay rite, which it may have picked up from the Edwin Harness Great House CBL, and this new junior age-grade sector would have used the fire-clay rite in Section 2 until Seip-Pricer was abandoned.

Scenario 4 not only replaces Scenarios 1 and 2, it actually adds to what these latter postulated. In Scenarios 1 and 2, Seip-Conjoined was treated as a single-cycle CBL. This does not work, as demonstrated by the critique of these two scenarios. Scenario 3, which supports the Civic-Ceremonial Center Model, does not work either. If Scenario 4 works, then it not only supports the Split Age-Grade Mortuary Model, it also adds to our understanding of Seip-Conjoined by empirically grounding the view that it was a dual-cycle CBL. This has considerable significance, as will be argued shortly.

CRITICAL ANALYSIS

This dual-cycle CBL reading raises a problem by appearing to introduce a leveling out of the death rate. If two split age-grade cycles are represented in these data, then the twenty-four split-log burials in Section 1 can be treated as two sequential senior age-grades. It is reasonable to postulate that each would have an estimated twelve burial events. The nineteen burials in Section 2 can be neatly divided by the split-log/fire-clay rite distinction into ten and nine, respectively. Assuming that both sections were used consistently during the period Seip-Conjoined was occupied, with the first cycle of split-log burials being twelve in Section 1 and ten in Section 2, this gives a 1.2:1 senior-to-junior death-rate ratio. In the second or combined split-log/fire-clay burial rite cycle, Section 1 had another twelve split-log burials added and Section 2 had nine fire-clay burials, giving a 12:9 or 1.38:1 ratio. The average would be 1.29:1. Alternatively, the overall death rate could be calculated by simply collapsing the two postulated generations for each section, 24:19 or 1.26:1. While these calculations still indicate a higher death rate for Section 1 than for Section 2, the differences are far from the 2.4:1 ratio of Scenario 1, or even of the 2.375:1 for Scenario 2.

How can this be accounted for, or can it? Are there other elements to this analysis that have been overlooked? In fact, there are several. These analyses have been made on the assumption that, everything else being equal, senior and junior age-grades will have significantly different death rates. This is a reasonable assumption, if everything else is equal. Given the earlier characterization of age-spread of each age-grade and the division of labor between senior and junior age-grade sectors under the Autonomous Cult Model, this is probably not a reasonable assumption. A second assumption of these analyses, and possibly more important, has been that the death rate in both sectors would be directly reflected in the number of burials.

In regard to the first assumption, although by definition the senior age-grade members would be older than the junior age-grade members, it does not follow that the junior age-grade members would all be young during their tenure in the junior age-grade phase. Wilson points out that, for the Nyakyusa, a junior age-grade had an age spread in which the youngest would be 10 and the oldest between 40 and 45 years of age.[9] This may be directly applicable to the Ohio Hopewell, but, as suggested earlier, a spread of 25 to 30 years for a generation is more conservative. This would give the junior age-grade a more plausible age-spread with the youngest member of the youngest age cohort being about 15. However, the eldest of the oldest junior age-grade cohort would still be 40–45 years old. By the time the junior age-grade was promoted, at least half of the junior age-grade members would be in their thirties and a number of these would be in their early forties, advanced middle-age or early old-age in a preindustrial society. Furthermore, it would be individuals in this age-grade who would be exposed to the greatest stresses and risks, since most of the tasks that they would perform would include the heaviest labor, the riskiest subsistence tasks, and the longest long-distance travel required for exotic resource procurement and exchange. In short, the risk and stress levels would be higher for junior age-grade members than for senior age-grade members. Now, these factors do not mean that the junior age-grade death rate would approach that of the senior age-grade, but it does suggest that it would still be considerable, particularly as the promotion transition was approaching.

The second assumption, however, is also problematic since it unwittingly assumes the funerary view, in which the number of burials is considered to reflect rather directly the group's death rate and demographic structure. As soon as the analytical framework is shifted to take into account the factors implicated under the Proscriptive/Prescriptive Ecological Strategy Model, a different conceptualizing of the relation between the mortuary population and the living members of the cult is required. First, in terms of the Dual Clan-Cult Model, the burial populations are part of a polyistic locale-centric social system with the embankment earthworks manifesting a system of world renewal cults and the dispersed domestic habitations manifesting clan-based communities.

In terms of the Mourning/World Renewal Mortuary Model, these burial populations realize a complex ritual process of integrating mortuary and world renewal rites. In the context of these two conceptual structures, it should have been anticipated that, in applying the Split Age-Grade Mortuary Model, the mortuary record of the cult system would not directly reflect the death rate or demographic structures of the cults since only some of their deceased members would become symbolic warrants of post-mortem sacrificial world renewal rites, the rest being retained by their respective clans and given funerary type terminal burial. This is because, as pointed out earlier, the clans and cults had different and contradictory interests towards the deceased. The clans would push for mortuary practices that would minimize the post-mortem manipulation of the deceased since the latter would constitute the clans' reproductive symbolic pragmatic capital as ancestors. The cults would push for the opposite, wishing to use their deceased members as symbolic pragmatic capital mediating post-mortem sacrificial rites. In these terms, then, not only would the mortuary data of the Great House CBLs not be representative of the demographics of a domestic community, they would not be directly representative of the demographics of the cult itself. In sum, while it is assumed that the deceased who did end up as burials beneath the room of a particular age-grade were members of the cult, or their dependents, many of the deceased cult members or dependents of cult members would end up in the CBLs of their respective clans. (More will be said later about these clan CBLs.)

In this light, it is appropriate to speak of the post-mortem retention rate of a cult. This argues that for every X number of cult members or their dependents dying, a cult would bury $X - Y = N$ deceased, where Y is some integer smaller than X but greater than o. Y would represent those deceased related to the cult who were actually buried elsewhere, probably in their clan CBLs. N is the number of cult deceased who would actually be buried in the cult CBL. Thus, a 1:3 retention ratio would mean that for every three deceased members, a cult would retain one for the final cult burial rite. However, it is a little more complicated than this. It is likely that the post-mortem retention rates would vary among the laity/clergy and senior/junior age-grade components. For example, if the circular Section 3 is the clergy section, since Section 3 of Seip-Conjoined contained no burials, while Section 3 in both Seip-Pricer and Edwin Harness contained a considerable number of burials, it could be concluded that the clergy retention rate significantly fluctuated between Seip-Pricer and Seip-Conjoined. Also, the life circumstances of members in the senior and junior age-grades would significantly vary, thereby generating different retention rates. The senior age-grade members of the cult would tend to have a greater degree of involvement in the affairs of their clans than the junior age-grade members would with their clans. From the perspective of the clans, it is likely that, in general, deceased senior clan members would be more strongly valued in their role as ancestors than would deceased junior clan members. Therefore, the cult's retention rate of the deceased of its junior age-grade sector would tend to be higher than its retention rate of the senior age-grade sector. Even so, despite the lower cult retention rate of the senior age-grade members, because the death rate of the senior age-grade would be greater overall than the death rate of the junior age-grade, the number of deceased buried in the senior section would tend to be greater than the number of deceased buried in the junior section.

Using this rationale as the theoretical basis, and the mortuary data of the Hopewell site as the empirical basis (these data are described and analyzed in detail in chapters 16 and 17), I have established a means of estimating a reasonable retention rate for the Seip-Pricer and Seip-Conjoined mortuary data. It can only be a rough estimate since the Hopewell site mortuary data can be used at this time to estimate only the overall cult retention rate rather than a retention rate for each postulated age-grade. For the Hopewell site this was roughly a 1:4 ratio. Applying the above assumptions of differential valuing of the senior and junior age-grade deceased, I have estimated a 1:6 retention rate for senior age-grade deceased and a 1:4 retention rate for junior age-grade deceased. This gives an overall retention ration for Seip of 1:5, slightly lower than the estimated retention rate for the Hopewell site. This lower average is justified later when it is argued that the Seip cult was more radical in its orientation than was the cult at the Hopewell site, thereby leading to a lower retention rate for the senior age-grade sector than at the Hopewell site.

In order to be clear, a ratio of 1:6 for the senior sector, and 1:4 for the junior age-grade means that, on average, for every six senior age-grade members to die, and for every four junior age-grade members to die, one each would be given full mourning/world renewal rites and end as a burial in Section 1 and Section 2, respectively, while the other five and three deceased, respectively, would have been given partial mourning/world renewal rites by the cult and final funerary burial elsewhere, usually at their clan CBLs. This 1:4 ration seems reasonable for members of the junior age-grade since, even though not as valued by clans as senior deceased, as pointed out above, many junior age-grade companions would, in fact, be middle-aged and, if they died while still in the junior age-grade sector, they might still be viewed as seniors in their clans.

If the estimate of the differential retention rates is reasonable, then the nineteen burials found in Section 2 would represent about one-quarter of the actual number of deceased members of the two junior age-grade sectors who died during the occupation of Seip-Conjoined, and the twenty-four burials found in Section 1 would represent about one-sixth of the members of the two senior age-grade sectors during this same period. Adjusting these actual burial numbers would mean a total of 220 deaths for the cult: [6 × 24 = 144 for Section 1] + [4 × 19 = 76 for Section 2] = 220. As a dual-cycle CBL, however, the cult during the Seip-Conjoined occupancy period would have had about 110 members die per cycle (220/2). From this, the active membership per cycle could be computed using the estimated total deaths for the senior age-grade as the base. The postulated senior sector is preferable because, according to the Split Age-Grade Mortuary Model, this number, adjusted for retention, would closely fit the size of the senior age-grade when it was promoted from the junior age-grade sector. Then, adding to this figure the estimated number of the deceased of the complementary junior age-grade sector, adjusted for retention, would give the estimated size of the new junior age-grade sector. The addition of these two figures would give a reasonable estimate of the total size of the laity at the time the senior age-grade was promoted from its junior age-grade status. Of course, an estimated number to account for the clergy must be added, on the one hand, and the elders, on the other, these being the few survivors of the retired senior age-grade sector. Given the specialized nature of the clergy, about ten at this critical time would be estimated and about five for the elders.[10]

In these terms, the average size of the cult during the first cycle can be calculated. The twenty-four burials of Section 1 are divided by two and this figure, twelve, is assumed to represent the number of senior laity deceased who were retained. Based on the 1:6 retention ration, this figure multiplied by six would give seventy-two as the estimated size of the senior age-grade when newly promoted. This would also be the first senior sector to occupy Seip-Conjoined. That is, their promotion would mark the abandoning of Seip-Pricer and their deceased companions buried on the floor of that Great House CBL. There were ten split-log burials out of the total nineteen burials in Section 2. These are postulated to be the deceased of the first-cycle junior age-grade of Seip-Conjoined, and, adjusted for a retention rate of 1:4, there would have been forty deceased in this sector. The size of the junior age-grade during the first-cycle occupation of Seip-Conjoined, then, is estimated to be $72 + 40 = 112$.

But it must be remembered that the junior age-grade sector would only reach this maximum level of 112 about four-fifths of the way through its "mandate" before it was promoted to the senior age-grade sector. As mentioned earlier, the junior age-grade sector would be open to new members and would continue accepting age cohorts of companions, the primary constraint being that they fall within the generational span of the junior age-grade. This would probably mean that about five years prior to the end of its "mandate" as the junior sector, it would close itself to new age cohorts. Prospective groups of young companions would be given "associate" or pre-initiate status, many of these possibly becoming apprentices to the clergy. When the cult-wide promotion occurred, this group would be initiated as the new junior sector, replacing the older group, which would itself become the new senior sector, and so on.

The estimated seventy-two senior age-grade sector individuals would be supplemented by the estimated five surviving elders to give seventy-seven as the size of the senior+elder age-grade sector. The maximum estimated number of the pre-retirement laity during a single cycle, then, would be 77 senior and 112 junior sector members, or 189/190. Then the ten clergy would be added to give an estimated total of about 199/200. To allow for some leeway, and the fact that the five above-floor burials have not been figured into this calculation, this could be rounded out to between 210 and 220 as the optimum size of the cult. Even though this number would be reduced by deaths, the total number affiliated with the cult could remain fairly steady as groups of youth would become initiated as cohorts into the junior age-grade sector. As stated above, some pre-initiates might attach themselves to the cult as "apprentices" to the clergy, possibly twenty or thirty. If premature death occurred, some of these pre-initiates might have been buried in Section 3, at least at the Seip-Pricer and Edwin Harness Great House CBLs.

It should be noted that the second cycle would be slightly smaller, based on the nine fire-clay rite burials. These are postulated to be the deceased of the second-cycle junior age-grade of Seip-Conjoined, and, adjusted for retention rate of 1:4, there would have been thirty-six deceased in this sector. Therefore, controlling for the differential retention rates, 108 can be estimated for the new second-cycle junior age-grade sector, based on adding the estimated size of the new senior age-grade, seventy-two, to the estimated number of deaths among their companions during their junior age-grade tenure, thirty-six. However, this also means that, everything else being equal, while thirty-six out of 108 died in the 25–30 years of their junior phase, almost all the survivors, seventy-two, would die in the next 25–30 years of their senior phase. The 72/40 for the first cycle would constitute a 1.8:1 death rate for the first cycle and the 72/36 would constitute a 2:1 death rate for the second cycle, giving a 1.9:1 average. This is effectively a 2:1 differential death rate. This is lower than Scenario 1 (2.4:1) and Scenario 2 (2.375:1), and higher than the estimate under the funerary assumption (the average 1.26:1), while representing a gap that would not be inconsistent with a generational difference, given the constraints discussed above under Scenario 4.

On the basis of this analysis of the data, it can be concluded that Scenario 4 accounts for these mortuary data more coherently than any of the other scenarios. Although further research will be required, it gives sufficient grounds to tentatively claim that Section 1 and Section 2 mortuary records mark a differential death rate that can be accounted for in terms of generational difference.

SUMMARY DISCUSSION

This concludes the first step in grounding the Ecclesiastic-Communal Cult Model. This split age-grade mortuary analysis has highlighted a number of important points that will be drawn upon later. One of these is the fact that, while the Seip-Pricer and Edwin Harness Great House CBLs each had a Section 4, the Seip-Conjoined Great House CBL did not. In each of the former cases, the burials of Section 4 displayed treatment that would make the deceased count as notable custodians. As suggested above, they might represent some of the elders of the cult, those who had survived their companions but continued in the capacity as elders to carry out important ritual duties for the cult and had been buried in this spot during the active life-usage of the Great House CBL. It is notable, then, and will be commented on again later, that this section is absent at Seip-Conjoined. It is equally important to understand the ideological significance of the introduction of the fire-clay rite. In terms of the Mourning/World Renewal Mortuary Model, the symbolic pragmatics involved in modifying part of a rite would mark an ideological change. In effect, it would imply a "dispute" over what forms of behavior would count as proper ritual acts by which to discharge sacred duties. According to Scenario 4, Seip-Conjoined was a dual-cycle CBL, and this means that both split-log and fire-clay rites were performed during the second cycle, each being exclusively held in a separate section. This would seem to indicate an ideological dispute in the second cycle of the Great House CBL between the senior and junior laity sectors of the cult.

Finally, there were no burials in Section 3 of Seip-Conjoined. This section is postulated to be the clergy burial locale. In the presentation of the Ecclesiastic-Communal Cult Model it was argued that the clergy would find its primary influence in the cult as the mediators between the senior and junior age-grade sectors and would use this position to promote a prescriptive post-mortem manipulation regime, since this would be seen by them as maximizing the success of the world renewal mission of the cult.

All this suggests that during the second postulated cycle of Seip-Conjoined, which would also have been the final period of use of the Great House CBL institution at the Seip site, several important changes occurred. First, the fire-clay rite was introduced. This rite and the split-log rite were treated as equivalent symbolic pragmatic forms of burial behavior that were differentially favored by the junior and senior age-grade sectors, respectively, suggesting that these sectors were in some ideological conflict over what forms would properly count as the type of rites intended and required. Second, the absence of Section 4 and the lack of burials in Section 3 suggest that, during the total period of Seip-Conjoined, neither the elders nor the clergy (including the pre-initiates) were participating in the affairs of the cult in a manner similar to the way that the members of these categories had when the Seip-Pricer Great House CBL was occupied. Third, it is notable that the standard method of completing the covering mound with a gravel stratum was not done. Immediate conclusions will not be drawn from the above at this time. However, these points, along with several others, will be returned to later in this book to argue that, in fact, factional dispute split this cult and the above anomalies indicate the development toward this split.

Setting these considerations aside for now, as mentioned earlier, treating the burials of the Seip-Conjoined Great House CBL as marking a dual-cycle CBL is significant in its own right, not only because it allows for some coherent estimate of the demographics of the cult, but also because it can be used as a basis to estimate the period of occupation of Seip. As a dual-cycle CBL, this would mean that the first cycle of Seip-Conjoined would be one full generation, about 25 to 30 years, and the second cycle would be effectively the same, another 25 to 30 years, giving the total span of usage of Seip-Conjoined as two generations or between 50 and 60 years, assuming that the second cycle was completed in the normal manner. This also suggests that the twenty-four burials found in Section 1 and the nineteen burials in Section 2 bracket the expected number per section for a dual-cycle CBL. Using this as a benchmark, an estimate can be calculated of the probable number of split age-grade cycles the Seip-Pricer Great House CBL incorporated, based on its floor burials, thereby estimating the total number of generations this structure was occupied. Adding the number of generations cycled by these two Great House CBLs would result in an estimate for the Seip site in terms of the Great House CBL occupations. This is done in part 4.

Part 3

OHIO HOPEWELL, SACRIFICE, AND WORLD RENEWAL

Chapter 14

FUNERARY CREMATORIES OR SACRIFICIAL ALTARS?

The last several chapters have focused on characterizing the social organizational aspect of Ohio Hopewell as a network of autonomous ecclesiastic-communal cults variably distributed primarily within the Miamis, the Scioto, and the Muskingum drainages of the Central Ohio Valley. Critically important to confirm is the ritual nature of the practices that it is claimed this cultic context made possible and, indeed, demanded. To this point, the nature of the mortuary data of the great earthworks as manifesting world renewal ritual has been assumed, largely by relying on both the associations between the mortuary mounds and the earthworks, which the findings of the World Renewal Model concluded were the context and outcome of world renewal ceremony, and the Mourning/World Renewal Mortuary Model. However, the empirical evidence in support of this latter model was presented in the context of the Archaic period as part of the initial critique of the Exclusive Territorial/Proprietorial Domain Paradigm. This chapter initiates the empirical grounding of the Mourning/World Renewal Mortuary Model in the Ohio Hopewell record, thereby firmly tying the nature of the mortuary practices to the symbolic pragmatic function of the embankment earthworks as world renewal locales. This is done first by demonstrating that the mortuary patterning of Mound City is more coherently accounted for in terms of the mourning→

spirit-release→world renewal process than in terms of the funerary view, and then by extending this interpretation and empirical grounding to the mortuary patterning of the Hopewell and Turner sites.

THE MORTUARY PRACTICES AND
CULTURAL PARADIGMS

The symbolic pragmatic approach of this book argues that material cultural style is a critically important symbolic expressive medium by which humans constitute the action nature of their regular material behavioral interventions. If this is accepted, then it follows that the users of material culture have a vocabulary to express in ordinary conversation the pragmatic meanings that correspond to the stylistics this material culture bears and by which it is conventionally endowed with action constitutive force. Recalling the earlier discussion of the bladelet/flake dichotomy, it can be imagined that, in conversation, Ohio Hopewell participants would unproblematically speak of an elongated or prismatic blade made of Flint Ridge flint or Wyandotte chert in terms that could be glossed as, say, a "mortuary shaving blade," while speaking about a non-prismatic flake tool made of local river cobble chert in terms that could be glossed as, say, a "hide scraper," even though either could equally mediate the same objective steps required for shaving the dead deer or the deceased human. These terms would implicate overlapping but contrasting terminological paradigms mapping what might be instrumentally equivalent but pragmatically contrasting material assemblages constitutive of contrasting contexts, for instance, cult Renewal Lodges and extended family households. Thus the use of bladelet and flake tools would be part of the production of the context natures of the respective locales, and of the social systems of which the context nature of these locales were partly constitutive.

In short, it becomes very important to ensure that the terminology that is used to refer to the Ohio Hopewell material features, facilities, and artifacts related to the mortuary sphere adequately characterizes these in terms that do not contradict and, preferably, correspond in sense with the terms that the Ohio Hopewell would have used themselves. Unfortunately, this may not currently be the case and, indeed, this terminology not only may be inadequate but may actually distort our understanding of the social reality manifested in the materials that it labels. As pointed out in the presentation of the Mourning/World Renewal Mortuary Model, the clearest example of this mislabeling may be the use of terms that articulate what was called earlier the funerary view. It was argued then that this is effectively the default interpretive posture of archaeology with respect to the mortuary record.

As stated then, the funerary view collapses the symbolic pragmatics of the mortuary sphere of a prehistoric community into the practical and social circumstances of the deceased at death, reducing mortuary variety to the single-dimensionality of funerals, and then effectively dissolving even these symbolic pragmatics by using the presence of deceased as sufficient grounds to mark both that a burial behavior had occurred and that this burial event "could not have been anything else but" the termination of a funeral. Since funerals require the deceased to be treated as the focal participants, then their social standing at the time of their death becomes the critical factor in the mortuary treatment. In this way, the overall mortuary variation is taken to mark somewhat directly the social structure of the living community. This chain of reasoning is most explicitly initiated by the unproblematical use of such Euro-American mortuary terms as "cemetery," or "funeral" mound, "grave," "tomb," "charnel house," "funeral or grave goods," and so on. This paradigm of concepts then harmonizes with the term "enclosure," often used to refer to the embankment earthworks, and this "enclosure-funerary" couplet then presupposes a social world in which mortuary practices are co-opted as symbolic pragmatic modes to legitimize proprietorial or "enclosed" domain.

Of course, if the basic mortuary terminology is indeed adequate in characterizing the data, then the social world that these data manifest must have existed. The corollary, however, is that if it is inadequate, then the social world exists only in the imagination of the archaeologist. Therefore, rather than prescribing that we should refrain from trying to construct an emically adequate vocabulary on the supposed grounds that it is unverifiable, the opposite is the case. There is no way that we can avoid trying to do it. To pretend that we can avoid the problem by being objectivist immediately fills the pragmatic vacuum with the archaeologist's own experience, thereby imposing on the data the type of mortuary practices with which the archaeologist is culturally and socially familiar. Of course, as a scientific endeavor, any claims to establish an adequate emic terminology must be theoretically justified and empirically validated via the hermeneutic spiral method.

With regard to Ohio Hopewell, this is precisely what has been attempted in this book, initially through the embankment earthwork classification and its use in grounding the World Renewal Model in the empirical data, then through the Proscriptive/Prescriptive Ecological Strategy Model and its structured set of auxiliary and ancillary models. This has led to the current position, as articulated by the Dual Clan-Cult Model, the Dual Ritual Spheres Model, the Autonomous Cult Model and the Ecclesiastic-Communal Cult Model, as presented in the previous chapters, this being that these major Ohio Hopewell sites were custodial locales of ecclesiastic-communal cults. Therefore, the treatment of the deceased at the time of burial relates directly to the ritual needs of the cult context and the position of the deceased within it and only indirectly to the ritual needs of the domestic or clan context. The mortuary data, as the end-product of this cult-focused collective behavioral stream, would be constituted as mortuary ritual having only attenuated, if any, funerary relevance or meaning and, instead, it would manifest the constituting of different forms of world renewal ritual.

Interestingly, earlier archaeologists were not so limited to the current default funerary view. Squier and Davis, for example, perceived the heavily burnt clay basin features and prepared floor spaces on which the mortuary remains were placed as sacrificial altars, and, as such, marking the residue of lethal human sacrifice.[1] Of course, the notions of "savages" and "human sacrifice" were central to the default view that many nineteenth-century Westerners had of Native American peoples at that time. Indeed, Squier and Davis were apparently only too willing to impose upon the Ohio Hopewell data the mystique of the Mesoamerican pyramid-platform/temple complex and its association with human blood sacrifice and lethal human sacrifice. Combining these views, they concluded that such Mesoamerican-inspired lethal human sacrifice was at the basis of the mortuary residue found under many of the earthwork mounds.

Although Mills was successful in countering Squier and Davis's claims in both cases, his success was won at considerable cost. Central to his strategy was his insistence that these basins and platforms were simply funerary paraphernalia—crematory basins, burial platforms, graves—in the Western mortuary sense.[2] Thus, he shifted the interpretive framework away from the lethal sacrificial extreme to the funerary, nonsacrificial extreme. Of course, given the scholarly context of his time and the prevalence to mythologize the Ohio Hopewell, Mills's determination to discourage the mystical and mythical interpretations can be understood and appreciated. However, his success may be responsible for what might amount to generating simply another myth in the prevalence of the funerary paradigm in the Ohio Hopewell interpretations.

There are a number archaeologists, however, who have expressed the importance of not identifying the Hopewell ceremonial sphere with the mortuary data in this funerary way. As discussed earlier, Robert Hall has been particularly emphatic in this regard and the earlier presentation of the Mourning/World Renewal Mortuary Model was largely influenced by his argument that Hopewell mortuary practices strongly manifest a complex combination of both funerary and world renewal ritual, including the view that such ritual tends to be sacrificial in nature. The Mourning/World Renewal Mortuary Model is very much in tune with this latter perspective and, therefore, it is not inimical to Squier and Davis's earlier sacrificial view, minus the Mesoamerican connection and minus their insistence on lethal human sacrifice. Therefore, before moving to this analysis it is important to review the position of the Mourning/World Renewal Mortuary Model in regard to lethal sacrifice.

The model postulates post-mortem human sacrifice to be central to Ohio Hopewell ritual, and this constituted a variable range of world renewal rites in which the deceased humans, their physical components, or items of personal use (i.e., personalty) treated as icons of the deceased, figure as post-mortem sacrificial offerings. This does not deny, however, that lethal sacrifice also may have been practiced. However, claims of lethal sacrifice would require empirical demonstration that the conditions that would promote it were in place. Theorizing the lethal human sacrificial conditions is beyond the scope of this book, but it can be briefly suggested that the polyistic locale-centric social system as postulated under the Proscriptive/Prescriptive Ecological Strategy Model would tend not to promote lethal human sacrifice, at least not as a regular aspect of the mortuary ritual process, and especially not under social conditions of the dual clan-cult type outlined earlier. The dual, open networks of companionship-based cults and kinship-based clans would constitute complex overlapping relations that would ramify across regions, making systematic processes of procuring victims for lethal sacrifice very unlikely. Lethal human sacrifice, if it did occur, would have been rather rare, largely because it would have a strong consensual aspect to it. That is, selecting the victims would entail a high degree of consensus building among those who were closest to them, both relatives and companions, thereby suppressing the possibility of vendettas and vengeance feuds. This would most probably entail a form of lethal endo-sacrifice in which the gifting of the victim by relatives or companions would be itself part of the sacrificial process. One possibility, then, would be the use of infants who were gifted by their relatives as victims. Clearly, such endo-sacrifice is strongly self-limiting and, therefore, it would tend to be a rare practice requiring unusual circumstances, such as serving foundation rite imperatives (e.g., the construction of a new ritual locale). Some data that at least hint at this possibility will be presented later. If this is the case, a more expansive regime of lethal sacrificial practices would require a class-structured polyistic locale-centric social system, that would promote exo-sacrifice by the waging of war, thereby procuring "enemies" as world renewal sacrificial victims. The Mississippian might be seen as enabling such lethal exo-sacrificial world renewal practices.

FUNERARY CREMATORIES AND SACRIFICIAL ALTARS

> Squier and Davis . . . persistently used the term "altars" with reference to what are now termed "basins" and "platforms," a usage which was followed by Willoughby in his account of the Turner Group of Earthworks in Hamilton County. (Shetrone and Greenman 1931, 497)

In his discussion of the Edwin Harness mound, Mills pointed out that the rectangular feature in which Squier and Davis found a partial human cremation was not an "altar," as they termed it, but a "burial platform," in his terms.[3] Typically these platforms were log cribs with a prepared floor and completely covered with mud or clay so that they bore a superficial appearance to the burnt clay basins, which Squier and Davis also called altars.[4] Furthermore, as pointed out in the last chapter, the mortuary event was terminated with the fire-clay rite at Edwin Harness, thereby adding to the similarity with the "crematory basin." For Squier and Davis, however, this partial cremation in what they took to be an "altar" was proof that all the other "altars" (in Mills's terms, "crematory basins") that they had exposed at Mound City, the Hopewell site, and so on, in most cases with nothing in them, were, in fact, like the platform "altars" that did have human remains, which they interpreted as the termination of lethal sacrifice.

Mills promoted the funerary view in his report on the excavations of the mortuary mounds of the Mound City by devoting considerable space in his analysis to counter this lethal human sacrificial model, and particularly, as commented above, the diffusionist argument that Ohio Hopewell was connected to Mesoamerican cultures. He summarized what he considered to be Squier and Davis's "principle . . . questionable conclusions" of Ohio Hopewell:

> That the builders of the Mound City group practiced human sacrifice; and that, from this custom, they should be in some way related to the great culture groups of Mexico and Central America; that certain basin-

like receptacles constructed upon the floors of the mounds were altars, upon which human sacrifices were made; that the so-called stratified mounds were not places of sepulture. (Mills 1922, 561)

Starting with the lethal human sacrifice claim, he then rebutted this position, using this as an opportunity to present and support his own funerary view.

[T]he idea of human sacrifice was in no way borne out by our investigations. The sites of the Mound City group were found to be similar in every way to that of the Tremper mound, on the lower Scioto, where the sacred structure, with its crematories and depositories was used *solely for the cremation and burial of the dead, and for the attendant funereal ceremonies*. This present conclusion regarding the surmise as to human sacrifice automatically answers that as to relationship with the southern culture groups. (Mills 1922, 561, emphases added)

To complete his critique of their sacrificial model, he then addressed Squier and Davis's notion of the "altar."

As to the question of "altars," upon which human sacrifice was made, it has been demonstrated once again that these basin-shaped receptacles were merely crematories, used in preparing the dead for burial in what to their builders was the customary manner. . . . Although Squier and Davis declared that the so-called altars in mounds number 8 and 3 served as depositories for artifacts, not a single one of the twenty uncovered by our survey were used as such. All were found to be devoid of contents beyond scattering charred human bones and fragments of artifacts carelessly left within them. (Mills 1922, 562)

He claims here that Squier and Davis were mistaken in saying that the large deposit of burnt artifacts on the floor of Mound 8, Mound City, including the well-known platform pipes (more than 200), was inside an "altar." When Mills re-excavated the remains of this mound, he established that these artifacts had probably been in a large leather bag that had been placed beside the "altar" at its southwest corner. Therefore, he concludes that "the cremation or burning of the deposit found in this basin had occurred elsewhere, possibly in the adjoining mound, number 9, which appears to have been supplemental to number 8 in purpose" (440).

However, this is not the only possible scenario. While Baby and Langlois confirm that some of these pipes had been subjected to fire, the associated clay basin feature had been subjected to heavy fire usage and displayed major cracks.[5] Mills pointed out that it had been repaired with clay which, subsequently, however, had not been fired. It could well be that this basin had actually been used for the burning of the deposit found beside it, then cleaned out, the cracks repaired, and the deposit placed beside it, as Mills described. Of course, either scenario is equally probable. However, more important is that his observation raises a puzzle for the funerary view that he did not address, namely, if this feature was simply a practical instrument for funerary cremation, as he claims, then why would it have been carefully repaired but not used again for that purpose? Given that the burnt deposit was found beside it, either having been burnt in the basin prior to its repair, or burnt in another basin, given that the basin was carefully repaired and then buried, implying that it was not intended to be used again, then the repair that he describes seems pointless. However, if this feature was perceived by its users in a sense closer to Squier and Davis's use of the notion "altar," then such careful repair prior to its usage as the receptacle for the final artifactual deposit becomes quite reasonable. That is, what was being repaired was an iconic warrant by which the final usage would be constituted as the type of ritual intended, even if this usage did not entail a final burning event. If this were the case, then speaking of it as an altar could well express the symbolic pragmatic sense it had for the users.

To reinforce this alternative interpretation, it is interesting to note that Mills comments on the plurality of these heavily burnt facilities by pointing out that most floors at Mound City Proper had two or three substantial, rectangular basins, all apparently being used in parallel. He explains this redundancy in terms of specialization, arguing that in many cases the deposition of the "funerary" cremation occurred in a neighboring and complementary structure. This is quite possible, of course, although it is also the case that both these paired structures had basins. In short, there would appear to be too many of these features for the funerary crematory purposes he claims. The alternative explanation, of course, is to recognize these features as crematory altars, in which case their multiple numbers would not be determined by funerary needs but by symbolic pragmatic imperatives by which, as iconic warrants, similar forms of controlled use of the burning of deceased humans may have been constituted as different types of mortuary rituals involving post-mortem sacrifices to different aspects of the cosmos.

Finally, his main criticism of Squier and Davis is based on his claim that the mortuary patterns that he exposed at Mound City Proper "were found to be similar in every way to that of the Tremper mound, on the lower Scioto, where the sacred structure, with its crematories and depositories was used solely for the cremation and burial of the dead, and for the attendant funereal ceremonies".[6] However, this assertion adds nothing substantive to support his funerary view nor does it really help to dismantle Squier and Davis's sacrificial view. In fact, his invoking Tremper as the "touchstone" of what will count at Mound City as a funerary cremation is counterproductive since there is a distinct difference between the cremated depositions of these two sites, a difference that he noted himself. Those at Tremper were in large collective depositories while Mound City cremations were, for the most part, individual deposits. The finding of two different mortuary depositional modes by two groups who, according to Mills's own argument, had the same funerary practices does not, in fact, support his funerary claim. Indeed, his recognizing that cremation played an important role in mortuary practice can be used to bolster the possibility that the variation in deposition modes indicates a considerable degree of contingency related to this ritual aspect, a contingency that he tried to deny by emphasizing only the cremation aspect. The variability in the deposition of the cremations, however, suggests that much more was going on than simply funerary rites and that this surplus element could be part of an ideologically variable post-mortem sacrificial ritual. If this is possible, then the complex process that Mills was responsible for mapping could easily be redescribed in terms already articulated by the Mourning/World Renewal Mortuary Model.

In short, by eliminating the term "altar" and referring to these two facilities as the "crematory basin" and the "burial platform," Mills imposed the funerary view on the mortuary residue with little or no theoretical and certainly variable empirical support for this characterization. An observation by N'omi Greber and Katharine Ruhl is relevant in this regard, namely, that the terms crematory altar may be closer to the meaning of the mortuary features than the terms crematory basin chosen by Mills.[7] According to the Mourning/World Renewal Mortuary Model, cremation would be situated toward the extreme warranting pole of the focal participant↔symbolic warrant continuum. For this reason, it will be postulated here that Squier and Davis's use of the single term altar in referring to what probably were, in fact, two different mortuary facilities has touched on a symbolic pragmatic meaning that they shared, namely, being symbolic warrants of world renewal rites mediated through post-mortem sacrificial cremation. Therefore, when the data seem more appropriately interpreted in terms of the symbolic warranting pole, the term altar will be used to characterize the associated features—e.g., crematory basin altars and platform altars.

However, it would be a mistake to move to the sacrificial extreme and claim that all Ohio mortuary data mark full world renewal rites or that such rites were necessarily always mediated by human cremation. As stated above, because of the cultic context, the Mourning/World Renewal Mortuary Model incorporates the total range of mortuary treatment from the focal participant extreme to the symbolic warranting extreme. Furthermore, not all world renewal ritual was necessarily mediated directly by human remains. For example, there are good reasons to speak of burnt offering altars as distinct from crematory basin altars, the former displaying little or no evidence that would support claims that they were used for human cremation. As will be argued later, there are examples of burial platforms that may be quite adequately characterized as funerary burial platforms, while others are more adequately interpreted as mourning or laying-in crypt altars. These latter features may have been intended as facilities for carrying out a series of mourning/world renewal rites mediating only the post-funerary stage of the mortuary process, and only under special circumstances would this ritual usage be terminated by burial. These possibilities are theoretically and empirically explored in greater detail later.

THE CUSTODIAL-ARTIFACT RELATION

In interpreting the Great House CBLs, certain categories of the Ohio Hopewell artifacts, such as copper plates, copper ear spools, celts, and other classic Hopewellian artifact types, were noted as having been highly valued as symbolic warrants for constituting important ritual offices and their related activities. At that time, the question of the nature of the deceased-artifact relation was not critically raised. Under the funerary view, the artifacts accompanying the deceased are treated as being "attached" to her/him, usually as personal property of some sort. Of course, this largely presupposes a proprietorial perspective. Since the deceased is deemed to own them, their being buried with her/him does not reduce the community's action capacities to any greater degree than the actual death of the owner has already done since the community has lost the warranting benefits of the artifacts with her/his death. This suggests, however, that for proprietorial regimes, if the deceased used artifacts during her/his life that were critical to the action capacities of the community, and therefore were partly constitutive of its social nature, then these would not be counted as private property, and they would be excluded from the burial itself. In effect, under the proprietorial regime, the deceased would have been the steward of the artifacts, holding and using them on behalf of the corporate group, and this stewardship would be embedded in some critical social position of the group. Such stewardship artifacts would be perceived as embodying some essential property making up the nature of the corporate group or community that owned it. It is likely that at the death of the steward, then, these artifacts would not accompany the deceased but would be withheld and passed on to the next steward. These types of artifacts can be generically termed stewardship regalia.

A custodial regime would also have critical focal symbolic warrants. However, these would tend to be divided into two categories: one set that is withheld from burial and the other that is normally associated with burial. This follows from the claim that custodianship is usually correlated with an immanentist cosmology. As such, focal artifacts would be perceived as agents in their own right, participating in the essential sacred nature of that aspect of the cosmos that they represented. Therefore, they could not be "owned," only used. The group would be their collective custodians, caring for and using them appropriately. In effect, the group had a form of custodial responsibility toward them. As a general class, these critical focal artifacts will be termed custodial regalia. This general class, however, can be divided into two subcategories. One subcategory would be the equivalent of stewardship artifacts. Although not "owned" by the group, these would still be perceived as embodying some essential nature of the group itself, usually conceived as its "spirit." Therefore, the custodian-artifact relation would be realized as in the steward-artifact relation by usually being withheld from burial and passed on to the next custodian(s). These will be termed group custodial regalia. The second category of custodial regalia would constitute a set of agents that "chose" their individual custodians, thereby creating a usufruct custodial relation with them. This category will be termed usufruct custodial regalia. For Ohio Hopewell, one way that responsibility for usufruct custodial regalia artifacts would have been allocated to particular persons has already been postulated under the Autonomous Cult Model, namely, by peer selection for age-cohort leadership. Therefore, allotting usufruct custodianship would be part of a ritual in which the "adoption" of the selected companion by the artifact would be publicly recognized, and thereby constituted, and simultaneously, this "adoption" would be a world renewal ritual. A similar allocation process presided over by leading custodians might apply to the group custodial artifacts. Both sets of custodial artifacts, therefore, would be critical to the success of the operations of the organization.

Since group custodial regalia would be systematically withheld from the terminal mortuary rites of their custodians and passed on to the next custodians, they would normally be absent from the archaeological record. Finding them would mark very special circumstances, as is more fully discussed and exemplified later. However, the usufruct custodial regalia would be very visible because of the usufruct nature of the custodian-artifact relation. In effect, the death of the custodian would entail surrendering custodial usufruct and this would normally be accomplished by returning the artifact to its origin, the cosmos, by means of a world renewal ritual in which the artifact figured centrally. This would constitute the mortuary event as a form of sacrifice for the group. In many, not all, cases, deceased humans would be associated with the usufruct custodial regalia item, and, as discussed more fully below, in some cases these would be as custodians, in other cases, as post-mortem sacrificial offerings.

This justifies postulating that when such potent artifacts are identified in the mortuary residue of a custodial group, it is appropriate to treat these items as powerful agents and, therefore, as "focal participants" in their own right. In principle, this applies to both group and usufruct custodial regalia, even though the former should be rarely found in the archaeological record while the latter should be common. This means that, as focal participants, their presence and usage would play a critical role in constituting the action nature of the ritual events in which they participated. In particular, the usufruct custodial regalia are critical because they are used as the symbolic pragmatic media by which the group performs those ritual activities that discharge its raison d'être. Finding group custodial regalia in mortuary contexts is also critical, but in a different way. Finding them in mortuary contexts would be unexpected, indicating some form of group crisis. For this reason, and in order to simplify discussion, group and usufruct custodial regalia are not routinely distinguished. Instead, the term "custodial regalia" is used to reference only the usufruct category. The distinction between group and usufruct custodial regalia is made only when necessary for clarity.

Of course, this leaves auxiliary and supporting artifacts. These would be more personalized and could very well be treated in the mortuary context much in the way that personal property in proprietorial regimes structures artifact production and usage. Therefore, archaeological terms, such as personalty and mortuary gift, that are used to speak of funerary goods could be appropriate. And, of course, referring to them as grave or funerary goods would have to be justified on the grounds that the mortuary event displays funerary characteristics. Otherwise, it may be more appropriate to label them as mourning or memorial gifts. Furthermore, it must be kept in mind that the semantic value of these terms varies between the two different regimes. For example, under a custodial regime it might be customary for a person receiving a compliment about her/his personalty, for instance, copper ear spools, to immediately remove them and hand them to the complimenter, insisting that she/he now use them. In similar circumstances under a proprietorial regime, the recipient of the compliment might be pressed simply to reciprocate by commenting on the complimenter's own finery.

This theoretical elucidation now requires a methodological discussion specifying logically how custodial regalia might be identified. Given this custodial regime/immanentist cosmological perspective, an important supplement has to be made to the focal participant↔ symbolic warranting continuum initially presented in chapter 8. This latter was presented on the assumption that only deceased humans possess this dual aspect of being both participants and symbolic warrants of the mortuary events which they mediated. This duality grounded the methodological rule that the degree and type of deliberate post-mortem manipulation to which the deceased were subjected could be used as a rough measure of their position on the focal participant↔symbolic warranting continuum: when minimal post-mortem manipulation is indicated, the deceased were being treated more as the focal participants at their own funerals than as a symbolic warrants mediating post-mortem sacrifice, that is, as "hosts" or, at the least, as "honored guests" rather than as rather potent but depersonalized symbolic warrants.

Custodial regalia are powerful agents that are only "on loan" to the custodial groups. In effect, they could, and normally would, be treated as focal participants of the mortuary behaviors in which they participated. Therefore, they will also typically display minimal or no deliberate destructive mortuary manipulation. In these terms, then, it is postulated that Ohio Hopewell mortuary artifacts, or more broadly, those artifacts found in circumstances that can be argued are ritual in nature, will be subjected to a range of treatments that is largely parallel and equivalent to the range of post-mortem treatments to which human deceased were subjected, and for broadly the same reasons. This generates at least two broad categories of artifacts underwriting Ohio Hopewell ceremony. Those mortuary or ritual artifacts that are typically *not* subjected to deliberate destructive mortuary treatment, and, in contrast, those that commonly display signs of modification, either deliberate or as the by-product of post-mortem manipulation of the associated deceased. Custodial regalia fall under the former, while the latter would be supporting or auxiliary warrants of the deceased, serving as personalty, mortuary gifts, and so on.

In short, the focal participant↔symbolic warrant continuum is paralleled by the custodial regalia↔auxiliary symbolic warrant continuum for the treatment of mortuary artifacts. Human-artifact mortuary associations require interpretive judgments based on these criteria. In general, when custodial regalia are identified, they are assumed to determine the primary nature of the ritual events in which they participated, and these will be world renewal rites directed to different aspects of the cosmos and having different degrees of importance. When these regalia are accompanied by deceased humans, the nature of the post-mortem manipulation to which the latter were subjected will determine whether they were co-responsible with the custodial regalia in determining the nature of the ritual event, or if they were being treated as post-mortem sacrificial offerings serving to enhance the sanctifying renewal aspect of the custodial regalia mortuary ritual. Here, again, the focal participant↔symbolic warranting continuum is relevant. Given the mortuary context, if the deceased display minimal post-mortem manipulation similar to that displayed by the custodial regalia that they are accompanying, this mortuary event might be best interpreted as a ritual having the mortuary and world renewal aspects balanced. In this case, it is probable that the deceased is the custodian of the regalia, the expectation being that, through the powerful agency of the custodial regalia, the custodian will continue in death to act as an agent of the custodial group in its ongoing pursuit of enhancing the human-cosmos relation.

In the case in which the custodial regalia are accompanied by deceased who were subjected to significant post-mortem manipulation (e.g., full cremation, including personalty and related personal artifacts), this would constitute the latter as post-mortem sacrificial offerings accompanying the custodial regalia, thereby enhancing the ritual through magnifying the impact of the sacrificial moment. Although this mortuary event would also be an important world renewal rite, it would not be as major as those in which the custodian and the regalia are treated more or less equally, as in the previous case. Then there is the case in which artifacts are not associated with human deceased. Both burnt deposits and unburnt deposits are possible. The burnt deposits of artifacts would be accumulated and curated artifacts of previous mortuary events. As iconic representations of past mortuary events, their usage constituted the event they mediated as a form of memorial/ world renewal mortuary rite. The unburnt deposits suggest custodial regalia that are being used to constitute the events in which they participate as major transformative renewal ritual. Both types will be illustrated and analyzed later in the following chapters.

This is only a rough means of operationalizing the human-artifact mortuary relation. Because of human ingenuity, and because of the contingency of both the belief→action relation and the particular circumstances of its realization in action, it is impossible to predict precisely the full range of expressive symbolic pragmatics of ritual life. As

emphasized previously, the same cosmological beliefs can be shared by
different populations while their material warranting media can vary
as a result of ideological variation. The premise of the focal partici-
pant↔symbolic warrant continuum can be universal when applied
within the cultural scope of the mortuary/world renewal phenomenon.
Nevertheless, its expressive outcome can vary just because of the historical contingency of the cosmology-ideology relation. Therefore, it is
conceivable that intervening factors can modify the outcome of the
focal participant↔symbolic warrant continuum and the custodial re-
galia↔auxiliary symbolic warrant continuum.

THE MORTUARY SET-PIECE

This view of the deceased and the mortuary artifacts as differentially subject to the equivalent range of treatments along two parallel
continua, the focal participant↔symbolic warrant continuum and the
custodial regalia↔auxiliary symbolic warrant continuum, requires developing an interpretive framework that can be applied to the mortuary
data. The patterning of a mortuary deposit will be assessed in terms
of what can be conveniently called a *mortuary set-piece*. This is a heuristic concept and it presumes the context/local duality that was discussed earlier in regard to the warranting role of the Hopewell bladelet.
In these terms, the differential treatments of human remains and material artifacts, as well as the spatial relations among these and the
features that contain them, are meaningful and can be meaningfully
related such that their symbolic pragmatic weight can be assessed in
terms of the two above continua when placed within the same framing
devices. That is, a mortuary set-piece is encapsulated by frames or
framing devices. These are the aspects of the locale and its constituent
features that have symbolic pragmatic relevance, thereby constituting
its context nature as ritualistic. The feature that is most proximal to
the deceased/artifact mortuary deposit will be termed the primary
framing device. This is "contained" within the secondary framing device, which is "nested" within the tertiary frame, and so on, to practical
limits. For example, the feature earlier termed the burial platform altar

might be considered the primary frame of most mortuary set-pieces. (There are mortuary deposits known not to have a platform or any apparent framing device of this sort, but a simple prepared floor.)

Naming the primary framing device in mortuary terms is critically important since it implicates the total symbolic pragmatics of the nested set of framing devices, including the earthworks and their immediate surroundings. For example, if the primary frame of the mortuary set-piece is termed a funerary burial platform, then its mortuary contents would logically be the residue of a terminal funerary burial. Therefore, the mortuary set-piece up to the secondary frame—the mortuary deposit, the funerary platform, and the primary mound covering it—would be appropriately termed a grave. This, of course, is in line with the funerary view. In contrast, if the primary frame were treated as the contextual medium of a mortuary/world renewal event, then it would be appropriately termed a burial platform altar, and the mortuary contents could be termed a post-mortem sacrificial offering. Therefore, up to the secondary frame, the mortuary set-piece would be appropriately described as the residue of a world renewal rite mediated by post-mortem sacrifice. Of course, this is in line with the mourning/world renewal view. Along with this would be the naming of the ancillary facilities, such as naming the clay basins as crematory altars rather than as funerary crematories, and so on.

In either funerary or world renewal terms, the tertiary frame would be the structure and floor on which the mortuary set-piece was constructed, and the quaternary frame would be the mounding covering this structure and floor. If the set-piece is characterized as a grave, then this entails that the tertiary frame, the building and its floor, be termed a charnel house or, at the least, that it be characterized as having just such a funerary function among the ritual activities that took place within it. However, if the set-piece is characterized as the outcome of a world renewal ritual mediated by post-mortem sacrifice, then the funerary function may have been only a minor part of the total range of mortuary activities and, of course, the building would be properly termed a Renewal Lodge.

Importantly, while the descriptive labels of primary, secondary, tertiary, and so on, move outward from the mortuary deposit itself, the constitutive "direction" of the symbolic pragmatic forces described move in the opposite direction. If the tertiary frame is, indeed, a charnel house, then the mortuary deposits are best characterized as graves. However, if it is a Renewal Lodge, then the deposits are best characterized as post-mortem sacrificial burials. That is, the symbolic pragmatic meaning of the quaternary frame endows the tertiary frame with its pragmatic meaning, and the latter endows the secondary frame with its meaning, which then endows the primary frame with its action nature. This chain of pragmatic meanings constitutes the mortuary deposit and its patterning as the type of ritual that it was for those who were responsible for performing it.

The above two readings, in fact, summarize the alternative mortuary approaches, namely, the funerary view for the Civic-Ceremonial Center Model and the post-mortem sacrificial view for the Ecclesiastic-Communal Cult Model. In terms of the latter, these earthworks are treated as the sacred locales of autonomous ecclesiastic-communal world renewal cults. As such, the embankment earthworks served as monumental iconic framing devices of the timber construction(s) they "encircled," constituting the latter as Renewal Lodges. The Civic-Ceremonial Center Model, in contrast, treats the earthworks as the context of the ritual practices of a clan-based community, what has been termed here a monistic modular social system. It interprets the mortuary data as the outcome of the community's funerary ritual. However, as developed by Greber, the Civic-Ceremonial Center Model does not reduce Ohio Hopewell ceremonialism to funerary practices. She has clearly indicated that the tertiary framing devices, the Great Houses, should not be identified as charnel houses but as civic-ceremonial houses. As such, while the claimed funerary practices as mediated by the mortuary remains had an important place in the ritual life of the community, in this view they were only one of a range of ceremonial practices performed in them. Nevertheless, with regard to the ceremonial data identified by the presence of human remains, she clearly does take the funerary view.

In these hermeneutic spiral terms, the question then becomes: Which model more adequately and coherently accounts for these same data? Does the patterning of the mortuary set-pieces manifest primarily funerary rites or mourning/world renewal rites? To answer this question, the following chapter is a detailed interpretation of the contents and make-up of Mound 7 of the Mound City site. While a full interpretation would entail applying both funerary and post-mortem sacrificial views to the same data, in order to expedite matters, the mortuary contents will be primarily treated in terms of the Mourning/World Renewal Mortuary Model, supplemented by the important addition of the custodial regalia↔auxiliary symbolic warrant continuum. When appropriate, the alternative funerary reading of the same data will be mentioned, largely in order to demonstrate why it is inadequate. Although this might seem to be an unbalanced treatment, it should be noted that the funerary view has already been well argued by Mills, and repeating his and subsequent funerary interpretations would be unnecessarily redundant.

Chapter 15

WORLD RENEWAL POST-MORTEM SACRIFICE AT MOUND CITY

While Mound City was reported by Squier and Davis as having 24 mounds, their plan shows only 23 (fig. 2.12).

> The first and most striking feature in connection with this work [Mound City] is the unusual number of mounds which it contains. There are no less than *twenty-four* within its walls. All of these . . . have been excavated, and the principal ones found contain *altars* and other remains, which put it beyond question that they were places of *sacrifice*, or of superstitious origin. (Squier and Davis 1973, 54, emphases in the original)

Mills excavated the residue of those mounds that survived following the serious damage caused to most of them by agricultural modification in the nineteenth century and, in particular, by the construction of Camp Sherman, a First World War U.S. military base.[1] At the urging of museum officials, Mound 7 was preserved more or less intact by the camp commandant's order that it be built around. As Squier and Davis reported the conical mound, it was 90 feet in diameter at the base and about 17 feet high, making it the largest mound in this site.[2] They undertook an exploratory excavation of the mound by digging a vertical shaft from the top center to the floor. They noted that the mound had been built in a series of earth, clay, and sand strata, a form of

construction that was to become recognized as typical of Ohio Hopewell mounds.[3]

Despite its size, Mills was able to establish that there were only
13 separate mortuary deposits, or burials, as he usually called them (fig.
15.1), and all were cremations. Three of these were found individually
deposited above the floor. These will be commented on later. The
other ten were deposited on the floor of the mound. Six of these
mortuary deposits were on relatively small prepared bases and were
"casually" covered with earth and sand, meaning that they did not have
the formal stratification of sand and earth constituting them as primary
mounds. These six were apparently deposits of single cremated individuals. However, Mills comments that at least one, Burial Number 8,
may have combined several individuals. The other four mortuary deposits, Burial Numbers 3, 9, 12, and 13, were placed in large prepared
log-crib facilities and covered with primary mounds. The log cribs were
rectangular with the widths about 5 feet and the lengths about 6 feet,
although precise sizes varied among them. Burial Number 3 appeared
to be the least complex of these four mortuary deposits, having the
cremated human remains deposited in a pile over one half of the platform and associated with a few artifacts, including a "large obsidian
spear, copper 'button' and a shell and pearl bead necklace." Burial
Number 9 may have also involved more than one human cremation.
In Mills's view, this feature was the central mortuary deposit on the
floor. Burial Numbers 9, 12, and 13 will be described and interpreted as
mortuary set-pieces, starting with Burial Number 13.[4]

BURIAL NUMBER 13 MORTUARY SET-PIECE

Treated as a mortuary set-piece, the primary framing device of
Burial Number 13 (fig. 15.1) was a large rectangular pit dug into the
northeastern sector of the floor and lined with a log cribbing (it was
the only sub-floor burial deposit). Mills does not give the lateral dimensions of this pit. However, the floor map of Mound 7 indicates
that the log cribs of Burial Numbers 12 and 13 were about 6½ by 5 feet
and these appear to be about the same size as the log crib of Burial

Number 13. This was within the normal size range of most of the prepared platforms on the floors of the Mound City Proper. The floor of the pit was covered by four inches of clay. Placed in the center was a large copper celt. Sheets of mica were carefully placed over this object and, indeed, covered most of the bottom of the pit. "Upon the mica were scattered the incinerated human bones, with which were the fragments of a large crystal quartz spear, a necklace of shell beads and two bone needles" (496). Seven ocean shell containers (*Fulgar perversum*, probably from the Gulf Coast) were placed on the periphery of the prepared floor, one in each corner, one in the center of each end, and one in the center of one of the sides. He does not comment on the covering of this "burial," but it can be assumed it would have had a primary mound, probably with several strata of earth and sand, since this was characteristic of the complex mortuary deposits.

Mills interprets this elaborate complex as a funeral because of the scattering of "incinerated human bones" over the mica sheeting. It is interesting to note that only after describing these cremated remains as being scattered did he then add that there were seven, carefully positioned, ocean shell containers. His description reads as if he were recapitulating the steps in the production of this deposit. It will be assumed that this is precisely how he intended this to be read since it would correspond with his interest in distancing these data from the sacrificial perspective promoted by Squier and Davis. By first speaking of the large copper celt and the mica, and then describing the scattering of human remains, leaving to the last the placement of the shells, he is suggesting that the first and last steps framed the placement of the cremated remains. Therefore, this descriptive sequence effectively makes the human mortuary deposit the primary focal element, as focal participant. Since this description is being set up so as to make the deceased a focal participant, it is clear that readers are meant to treat cremation as the termination of a funerary event. Thus the copper celt, the mica, and, "finally," the shells are to be interpreted as funerary materials that accompanied the human deceased as personal property, or personalty.

These same data, however, can be given a quite different reading, one that interprets this deposit as having the nature of a post-mortem sacrifice mediating a world renewal rite. Since the copper and shell artifacts were not subjected to cremation, these will be treated as custodial regalia. Cremation indicates the warranting extreme of the human focal participant↔symbolic warrant continuum. The scattering of the ashes and bones, along with "the fragments of a large crystal quartz spear, a necklace of shell beads and two bone needles" (496), can be interpreted as auxiliary symbolic warrants that accompanied the cremated ashes. The copper celt and ocean shells were focal custodial regalia. The mica sheeting can be interpreted as a framing device separating the copper celt from the scattered cremated remains (and possibly the shells, although this is not definitive).

In this alternative reading, then, the constituent elements of this set-piece, given in probable order of placement, would be: (1) the primary frame, this being the sub-floor pit with a log cribbing and prepared base constituting a platform *altar*; (2) the seven ocean shell containers in their particular peripheral placement; (3) the copper celt in the center (2 and 3 could be reversed); (4) the covering layer of mica; (5) the *scattered* cremated remains with the broken point, the shell bead necklace, and two bone needles; and (6) the primary mound covering.[5]

What is particularly interesting here is the extreme post-mortem manipulation of the human deceased, extreme in that not only is this a cremation, it is a "scattered" deposition. This scattered deposition contrasts with the treatment of the shell and copper artifacts and the mica sheets. These were carefully placed and they were not subjected to any modification. As stipulated earlier, this constitutes them as custodial regalia, at least, this would be the case for the copper and ocean shell artifacts. Furthermore, the care in the placement of the latter two, the copper celt in the center and the seven ocean shells deliberately placed around it, indicates that these two were intentionally related. This suggests a copper/shell complementary duality with several subordinate complementary oppositions of copper/mica, mica/human, upper/lower, medial/corner, and so on.

Burial Number 13 as a mortuary set-piece would appear to be expressively replicating and embodying some fundamental powers of the cosmos, and the mode of adding the human cremation would suggest that the latter is being treated as auxiliary to the copper and shell focal custodial regalia. The production of the primary framing device, the pit with log cribbing and the prepared base, was carefully considered, suggesting that it had distinctive iconic meaning as a pit type of platform altar. The copper and shell may represent a land(copper)/water(shell) duality, and the fact that the pit was set into the floor suggests that this land/water duality of the cosmos is subsumed under the Underworld/Underwater aspect. The mica sheets were placed such that they separate the copper celt from the cremated human remains (Middle World). Finally the stratified primary mound is placed on top of the total mortuary set-piece, possibly replicating some aspect of the vertical structuring of the cosmos.

According to the above interpretation, the cremated remains may be the residue of the deceased as the custodian of these custodial regalia artifacts. If so, then the custodian was also being treated as a post-mortem sacrificial offering. Since the residue was scattered across the mica sheets, the cremation occurred separately, probably in the *crematory altar* nearby on the floor of Mound 7, although Mills also argues that cremation may have occurred in the separate but complementary structure on the floor of Mound 3, which is immediately east of Mound 7. In any case, this suggests that the cremation was carried out as a separate step in the total rite and that this step was several times removed from any funerary rites that may have been held earlier for this deceased. Therefore, the scattering of these ashes and bones, along with personalty, across the mica sheets would be a ritual means of completing the return of the copper and the shell custodial regalia to the cosmic powers from which they were derived.

From all this, three conclusions follow. (1) The cremation and the deposition of the copper celt/ocean shells were two different action components of a single but complex rite. (2) The context nature of this complex behavioral process would have been constituted by the copper/shell duality framed in the pit-like platform altar. (3) The scattering of the human cremation was an auxiliary warranting mode of behavior whereby the destructive moment of the cremation would be transformed into a reproductive moment of sacrificial offering.

Confirmation of this interpretation can be suggested by the form of the pit and the positioning of the seven ocean shells. The latter is similar to the positioning of four corner and three medial gates of the nearby High Bank Octagon (where the fourth medial "gate" should have been was filled by embankment earth, while, in fact, there were eight mounds). As argued earlier, Mound City Proper would have preceded the emergence of the High Bank C-R Configuration. According to the World Renewal Model, the rectilinear component of the C-R Configuration, in this case the High Bank Octagon, may simply be a monumental expression of the same sense being expressed in the primary pit-like platform altar framing device of Burial Number 13. If the rectilinear of the High Bank C-R Configuration constitutes the Lunar Motif, as argued earlier, with the seven gates representing critical elements ("portals") mediating between the Underworld and the Heavens, then the seven shells in Burial Number 13 may be an earlier although variant symbolic pragmatic expression of the same notion. In this case, however, the use of ocean shells suggests that they are aquatic "portals" of the Underworld. The fact that copper was traditionally derived from mining pits in the Keweenaw Peninsula and on islands in Lake Superior also suggests a close aquatic relation, but one in which the contrast of land and water figures prominently.

DISCUSSION

Whenever an interpretation of the possible representational content of material items as symbols is given, it can always be challenged in semantic terms. This is not a critical problem, however, since it is not the representational content, as such, that is epistemically critical. Rather, it is the contrasts that are important since these indicate the exercise of symbolic pragmatics. As argued above, these contrasts are embedded in the facilities, as framing devices, and the material items, as symbolic warranting devices. Therefore, in the collective understanding of the agents responsible for their construction, they would have been experienced as part of what they represented, the cosmos, under whatever aspect that might have been.

According to Mills, Burial Number 13 marks the termination of a complex funeral. However, besides the radical reduction of the human parties involved, there are two components of the Burial Number 13 feature that seem to be superfluous to funerary rites. There is the nature of the primary framing device—the large prepared pit—and then there is the differential treatment of the artifacts and the deceased. In the first case, it is clear that the size of the pit is not a function of the need for depositing the cremated remains. In fact, as stated earlier, the majority of the cremated deposits on the floor of Mound 7 have rather minor prepared spaces, not much larger than that required to receive the cremated deposits. For example, Burial Number 4 "occupied a low platform 18 inches in diameter" (486) and was accompanied by copper objects with two large projectile points, one of "hyaline quartz" and the other of obsidian, both of which had "been broken into fragments," as well as a necklace of pearl beads and wooden beads covered in silver. Burial Number 5 was placed in a "short stump-like section of a tree, into the top of which had been excavated a bowl-like cavity to contain the cremated remains . . . [accompanied by] numerous remnants of perishable objects, including cut jaws and teeth, beads, and so forth, practically destroyed by cremation" (486).

Burial Numbers 6 and 7 were placed on "slightly raised platforms, and with each was placed a necklace of shell beads" (487). Burial Number 8 may have been a compound cremation. Mills made no comment on the nature of the area on which it was deposited, and, since he usually commented if a log cribbing was involved, it can be reasonably assumed that it was also small, within the same range as those already discussed. Similarly, he made no comment specifically about the nature of the prepared space of Burial Number 10 on the northwestern side of the floor, except to say that its "content of cremated remains was larger than usual, and with them were placed a necklace of bone beads and several perforated bear canines" (493–94). Therefore, of the ten floor mortuary deposits (all cremated, including Burial Number 13, as a sub-floor mortuary deposit), six were deposited on prepared floor areas commensurate with the practical requirements of depositing cremated remains. Some of these were associated with significant artifacts, such as copper materials and large projectile points, one of which was an obsidian point. However, in general, these had been subjected to ritual destructive manipulation.[6] Seen in terms of the focal participant↔symbolic warrant continuum, these mark the use of the deceased, along with their personalty, as symbolic warrants of post-mortem sacrificial rites.

What is implied by these lesser cremation deposits is that the type of major platform facility that is associated with Burial Number 13 (as well as Burial Numbers 9 and 12, as discussed below) is in excess of any requirements of depositing human cremated remains in Mound 7. Thus, the platform framing device played a role that was surplus to the needs of simple cremation mortuary deposition, as such (or, if the funerary paradigm is applied, of funerals, as such). The above analysis has suggested that these needs would be those that only a platform altar could fulfill, this being an essential warranting facility of a world renewal ritual.

As pointed out above, Burial Number 13 is not the only complex mortuary feature in Mound 7. Indeed, the floor of this mound had some of the most impressive mortuary features at Mound City Proper. Furthermore, as Mills reports, Mound 7 was a dual stratum locale. It initially started as a large oblong semisubterranean pit carefully dug into the gravelly ground (fig. 15.1). The bottom of this pit was covered with a clay floor and the clay was continued up the walls, effectively cementing and preventing the loose subsoil of the walls from collapsing. This whole construction was quite substantial, about six feet below the natural surface, and with a southwest-northeast length of 40 feet and a width of about 30 feet. The northeast entrance was built as a ramp and the walls had a supporting perimeter of posts. Toward the southwest end of this structure there was a large "crematory pit," 6 feet by 4 feet, well used as Mills explains. No mortuary deposits, that is, no "burials," were found. This "basement," as he called it, was a sacred place; he concluded that it

> was used for a long period, but that its purpose was mainly that of cremation. The cremated remains apparently were then removed to adjacent sacred places for deposit and burial. In the end, the site was abandoned, the excavation filled to a level with the corresponding natural surface, and upon this restored surface mound Number 7 was constructed. (Mills 1922, 479)

Burial Number 13 mortuary deposit was built into the floor that filled and covered this "basement," in a position that situated it over the northeast end of this substructure, just south of the old entry ramp. If Burial Number 13 mortuary deposit as a set-piece was the context of an important world renewal rite, then this suggests that we can extend this pragmatic meaning to the "basement" and its "crematory pit," calling this a crematory altar.

BURIAL NUMBER 9 MORTUARY SET-PIECE

Burial Number 9 mortuary deposit (fig. 15.1), treated by Mills as the central burial feature of Mound 7, is one of three major complex mortuary deposits on the same floor as Burial Number 13, the other being Burial Number 12.[7] Note, however, that while the log cribs of Burial Numbers 9 and 12 mortuary deposits were built on the floor that was created by filling in the substructure described above, the log crib of Burial Number 13 mortuary deposit was built into this same floor. Burial Number 12 was on the north side of the floor and Burial Number 9 was on the south side. There is no doubt that both the contents and the primary mound of Burial Number 9 make this feature the most elaborate in Mound 7.

The primary mound is very interesting. The initial layer was loam and the second layer was white sand. On its northwest side, and partially covering the base of the sand stratum, there was an 8-foot by 4-foot set of overlapping mica sheets. The platform was built with a cribbing that was two logs high, and the western side was held in place by outer rocks. Its long axis was oriented about 15 degrees west of north and its floor was built up about six inches, although Mills does not mention if this was puddled clay. In the center was placed a unique copper-covered wooden implement that was

> apparently made to represent a toad-stool of the death-cup variety, and suggesting a wand or baton as its purpose. The object is 13½ inches in length, and is made of wood, covered with thin copper. Directly over and around this peculiar object were placed the cremated remains of the dead. Adjacent to these remains, at the south, was a copper plate, ten inches in length, bearing a striking conventional decoration in repouse, with the eagle-head as the motif. (Mills 1922, 490)

He goes on to point out that there was another copper plate with a similar bird motif adjacent to the "toad-stool" but on the north side. In the southwest sector of the log-crib frame there was an elaborate copper effigy headdress of a horned animal. There were two copper repouse cutouts of a flying raptor, one in each of the southeast and northeast corners and, as Mills writes, "Elsewhere throughout this grave were placed copper pendants, pearl and shell bead necklace, and broken-spear-points of rock crystal" (491). Covering the "entire grave," but not the log cribbing, were overlapping mica plates, and the copper objects at the south end of the "grave" were covered by a "woven coarse matting," the latter apparently being placed before the mica sheeting was laid down.

Again, this can be treated as a mortuary set-piece in virtue of the cremated human remains. Unfortunately, Mills's reference to the quantity of the latter is ambiguous. The natural reading and the range of materials presented would suggest several deceased. However, there is no ambiguity about the order of the "mushroom" artifact and the cremations. The former was placed first and center, and then the cremated human remains were placed directly over and around it.[8] The mica is used in two different instances, in this case. First, unlike the Burial Number 13 mortuary set-piece where the mica sheets separated the copper celt from the cremated remains, here the mica was used to cover the baton and the associated cremations, as well as all the copper artifacts that were laid out across the burial platform altar. Second, it was used to make the large ballast of mica sheeting on the northwest side of the primary mound covering this mortuary set-piece. The ballast of mica sheets was started over the foot of the sand/clay stratum and was extended northwest across the adjacent floor.

This set-piece also had a centrally placed copper (and wood) artifact, the "toad-stool," surrounded by other copper artifacts in a manner akin to the copper celt/ocean shell dual pattern in Burial Number 13. While the human remains are not separated from the central copper artifact, it is debatable whether these remains are accompanied by the artifact. It is more likely the reverse. First, all the artifacts were copper and were not subjected to any destructive manipulation. Furthermore, the placement of the "toad-stool" artifact in the center of the platform altar and the placement of the other copper artifacts symmetrically around it, similar to the Burial Number 13 copper celt/ocean shell duality, suggest that these were being treated as a structured assemblage of focal custodial regalia. By the positioning of the cremated remains over and immediately around the "toad-stool" artifact, it is clear that this latter item was intended to be "privileged" over the other artifacts, thereby making this an hierarchical order.

The deceased, having been subjected to maximum post-mortem manipulation, was/were being treated as symbolic warrants that were auxiliary to and supportive of the pragmatic meaning of the custodial regalia. This conclusion is reinforced by the contrast between the treatment of the artifacts as "individual" pieces carefully placed in relation to the central "toad-stool" artifact and the "collective" manner in which the cremated remains were treated. As suggested above, if more than one individual was involved, their treatment effectively eliminated their individuality. This simply reinforces the implication that they may have already been subjected to funerary, mourning, and various spirit-release rites. This left the residual spiritual content of their remains as appropriate post-mortem sacrificial media by which the world renewal aspect of this rite would be enhanced.

This latter conclusion is reinforced by noting the cosmological implications of the contrasts among the peripheral copper items that surround the "toad-stool." The east side of the platform altar is dominated by artifacts bearing avian motifs, vultures centrally adjacent to the "toad-stool" artifact, and raptors in the northeast and southeast corners. The southwest corner had the elaborate horned-animal headdress. The copper bird motifs in the eastern half of the platform were complemented by the animal-motif headdress in the southwestern sector. Mills claims that this headdress was one of the most elaborate he ever found. It was so poorly preserved that he had difficulty identifying the animal. Because its 6-inch copper horns were straight, he suggests that it may have depicted a "young buffalo" (547). Since the bison is identified by many historic Native American peoples as the form the Underworld spirit custodian of the animals often takes, it might represent and thereby participate in the power of the Underworld. The bird motifs, therefore, would represent and participate in the powers of the Heavens. If the "toad-stool" baton represents certain powers of the Middle World, possibly associated with night, then its central position on the platform altar could also be seen as integrating this Heavens/Underworld duality under the lunar aspect.

It can be further noted that in Burial Number 13 mortuary deposit set-piece, the cremated remains were scattered across the large platform, while in this Burial Number 9 mortuary deposit set-piece, they were heaped on and around the "toad-stool" baton. Since human remains could be heaped or scattered, this further suggests that the size of the platform was not the result of any concern for their deposition. It is more likely that the area *and its form* were needed for the patterned laying down of the artifacts. This reinforces the view that it is the artifacts, here identified as custodial regalia, and not the mortuary remains that are the focal constituents of these set-pieces and that they were the result of a carefully planned series of ritual steps in which the artifacts and features figured more importantly as constituents than did the human remains. This is not to deny pragmatic significance to the human remains. In the terms already argued, these would probably have been the cremated remains either of the original custodians or else of deceased who were deemed appropriate offerings to accompany these custodial regalia artifacts. Prior to their being cremated, it is probable that they would have been subjected to mortuary rituals of a mourning and spirit-release nature.

The complementary dualities that have been discussed also enhance the probability that these mortuary set-pieces mark world renewal rites. The different artifactual and featural patterning of the set-pieces would have evoked different relevant aspects of the cosmos. Furthermore, as the secondary framing device, the construction of the primary mound over this set-piece would itself be part of the world renewal ritual that generated the set-piece. This is supported by the strata of the selectively varied materials, loam, puddled clay, sand, and, of course, in this particular case, the mica ballast. As argued earlier, the primary mound and its complex layering might represent the vertical stratification of the Heavens, implying that the construction of the primary mound also participated in the renewing of the cosmos.

BURIAL NUMBER 12 MORTUARY SET-PIECE

Continuing the hermeneutic spiral, these two latter interpretations can be further confirmed by examining the patterning of Burial Number 12, the third of the major mortuary deposit set-pieces of Mound 7. This was also covered by a carefully constructed primary mound. It is immediately north of Burial Number 9 mortuary set-piece, about halfway between the latter and the set of post holes that defines the northern wall of the mortuary structure. Located between these two set-pieces is the single largest burnt basin facility on the floor of Mound 7, apparently a crematory altar.

The log cribbing of the Burial Number 12 mortuary deposit was similar in form and size to Burial Number 9, but with its north-south axis oriented about 15 degrees east of north, suggesting a complement of the 15 degrees west of north axis orientation of the Burial Number 9 mortuary set-piece. "In its centre were the usual cremated remains and with these and covering all parts of the platforms [sic] were numerous artifacts of copper, obsidian and mica."⁹ To say that the "centre" of this platform was occupied by the "usual cremated remains" is a bit misleading. While it is true that all ten cremated mortuary deposits on the floor were in the "centre," as pointed out earlier, six of these were on small prepared zones having only adequate space if the cremated remains were centered with the artifacts. Of the three major mortuary deposits, however, it should be noted that only two had the cremated remains in the center, this one and the Burial Number 9 set-piece, with the cremated remains of Burial Number 13 being scattered across the platform altar. Furthermore, while the cremated remains were in the center of the platform of the Burial Number 9 set-piece, Mills recognizes that this same position was shared with the copper-covered wooden "toad-stool" artifact.

This sharing of the central position with an important artifact was also the case for the Burial Number 12 mortuary set-piece, since he goes on to comment that

> Near the centre of the grave was a large copper plate, finely made and well preserved. It is covered on one side with leather. In conjunction with this plate were found a pair of spool-shaped ear ornaments, one lobe of each being of copper, and the other of native silver. (Mills 1922, 494–95)

Again, then, the focal point of this set-piece is a major copper artifact deposit. As in the above case, because of the undamaged nature of these artifacts and their central positioning, rather than treating these as accompanying the deceased, it may be more adequate to treat them as being accompanied by the deceased, that is, the cremated human remains. Indeed, when the total set-piece is considered, the cremation and the artifacts can be seen as mutually transformative. At the northwest corner of the platform was a belt-like leather object having attached a series of 18 mounted hollow copper artifacts in the form of turtles with loose pebbles and beads inside. Near it were two large obsidian projectile points, slightly less than 8 inches long. In the central point of the northeastern sector, the platform was covered with more than a dozen star-like figures cut from copper and averaging about 2¼ inches in diameter. Associated with these stars were two conventionalized objects of copper, resembling bats, each 5 inches long and 6½ inches wide. In the same part of the grave was found a copper plate 10 inches in length, representing the hawk or eagle in an upright posture. The eyes, feather markings and body lines are executed in repousse, while at the neck, but on the reverse side, is a large pearl bead. (Mills 1922, 495)

On the east-central side of the platform there was a set of leaf-like copper pendants, ranging from 6 inches to 8 inches long, some with shell and pearl beads attached to the obverse side. A large mica sheet was just south of the cremated remains, suggesting that the copper plate that Mills described as "near the centre of the grave" and, therefore, adjacent to these same remains, would probably have been on their north side. If this is the case, then the cremation was bracketed north and south by copper and mica plates, respectively.

Again, without giving its precise position except to say that it was "in close proximity to the mica sheet," Mills listed a copper effigy of a mountain goat horn. In fact, Mills expressed the opinion that

[f]rom the appearance of the specimen, it is not improbable that it consisted originally of a genuine horn, over which the copper plate was fashioned as a covering. No evidence remained, however, of the original horn. No record exists of the presence in Ohio of mountain goat, either in historic or prehistoric times, hence the inference is that it may easily have been a product of commerce with the aborigines of the Mound City group, along with obsidian, copper and other commodities from distant sources. (Mills 1922, 546)

He interprets this copper horn, 9 or 10 inches long and 2 inches in diameter at the center, as the "central and important part of a head-dress" (495). In the southeast corner were deposited "several necklaces of pearl beads, bear claws and sharks' teeth, and a number of small ornaments of copper" (495–96). Deposited in the southwest sector of the platform, near the corner, was a curved copper headdress in the form of a female human torso, headless and handless. This state would have been deliberate. It is possible that it implicates warfare or feuding. Alternatively, it could implicate sacrifice, thereby being an important warrant of the shamanic priest responsible for the process of post-mortem sacrifice.

The three-dimensionality of the cosmos may be represented here two-dimensionally by the patterned distribution of the artifacts in the Burial Number 12 mortuary deposit set-piece, supporting the view that the total was a critical warranting aspect of a world renewal ceremony. The positioning of the copper/cremation/mica in the center of the platform appears to constitute this as a complex focal custodial regalia deposit with the cremation serving to integrate the copper/mica du-ality. This might represent the Middle World occupied by humanity, symbolizing the role of copper and mica as mediating between humans and the cosmos. The vertical dimension would be represented by the north-south axis creating an east-west division of the north sector of the platform. The belt with the turtle effigies and obsidian points in the northwest sector may have represented the earth and water surface of the Middle World and the "bat" effigies and possible eagle/hawk effigy, along with the copper "stars" in the northeast sector, then, would have represented the Heavens. If the bat-like items were intended as bat effigies, then, accompanied by the eagle/hawk effigy, this duality might represent a temporal structuring of the cosmos into night (bat) and day (eagle/hawk), lunar/solar. The mountain goat headdress was nearest to the obsidian plate, which was itself south of the cremation. Therefore, this headdress, it is assumed, would have been in the south-ern sector and more or less centered. The southwestern corner had the remains of the headless/handless human effigy headdress and the southeastern corner, the necklaces.

Treating the human effigy headdress as a post-mortem sacrificial warranting device could suggest an association with the Underworld and the latter might suggest that the whole western sector, including the northwest and southwest corners of the platform altar, were iden-tified with the Underworld, both in terms of water and earth. The mountain goat headdress in association with the mica/cremation/copper plate axis might contrast, identifying the Middle World. The eagle/ bat effigies on the northeastern quarter would represent the Heavens in the night/day:lunar/solar complementary aspects. The necklaces in the southeast corner may represent both the earth (bear claws) and water (pearl, shell, and sharks' teeth) aspects of the Middle World.

THE OFF-FLOOR MORTUARY DEPOSITS

There are three known off-floor cremated mortuary deposits in Mound 7 that are relevant to this analysis. Mills accounts for them in practical circumstances of death terms by suggesting that these were burials resulting from deaths occurring after the mortuary structure was abandoned and while the mound was being built. This is not an atypical explanation under the funerary view, of course. Indeed, this may be among the early exemplars of this practical approach. However, given other comments that Mills made, in particular, that there were examples of "charnel houses" operating as a complementary pair, Mound 7 and Mound 3 immediately to its east being one example, this is clearly an inadequate explanation. If funerary burial was the primary concern, then it would appear that there was always a mortuary facility available for this purpose. Therefore, the practical explanation actually works against the funerary interpretation since, of course, in these terms there should be no off-floor mortuary deposits. Thus, the question remains: "Why would off-floor mortuary deposits, whether cremated or noncremated, be placed in the mound during construction?"

As mentioned earlier, Greber has noted a similar pattern at Edwin Harness, Seip-Pricer, and Seip-Conjoined, involving both cremated and noncremated mortuary deposits. She claims that these were part of the ritual involved in constructing the mounds. This claim is fully supported here. This also implies, however, that the nature of this ritual was not funerary, even though mortuary elements were involved. The relative positioning and the patterning of the three deposits give further support to the view that this mound and its contents were primarily the locus of world renewal ritual mediated by mortuary practice. If the floor mortuary deposits and their primary mounds resulted from the

performance of post-mortem sacrificial world renewal rites, then building a mound over what will now be called the Renewal Lodge and its
associated plaza would seem to be part of an even more important
world renewal rite. That is, as earlier suggested with respect to the
Great Houses of Seip and Liberty Works, the mound "buried" the
Renewal Lodge and its plaza as a major sacrificial medium and only
circumstantially did this involve "re-burying" the floor mortuary deposits.[10]

In short, this claim requires demonstrating that the mound and
its contents constitute a monumental mortuary set-piece resulting from
the very same type of process that has already been discussed in interpreting Burial Numbers 9, 12, and 13 mortuary deposit set-pieces, only
the process has been expanded to apply to the Renewal Lodge itself.
This requires interpreting the mounding as constituting a monumental
symbolic pragmatic device framing the associated features, in particular
the cremated remains of the Renewal Lodge and its contents. In this
view, then, rather than the off-floor mortuary deposits being the result
of unanticipated deaths during the construction of the mound, they
would have been an essential part of the construction process itself,
contributing to transforming the labor into being part of a major world
renewal rite. As such, timing is very important. Once the process of
"abandoning" the Renewal Lodge is initiated, it cannot be stopped. It
is quite possible, then, that post-mortem sacrifice may not have been
sufficient as the timing may have imposed the need for lethal sacrifice.

Squier and Davis noted that Mound 7 was over 17 feet high and
90 feet in diameter. However, they also noted that the vertical shaft
that they dug through the center reached 19 feet before it touched the
"floor" of the mound.[11] Since the latter citation refers to pages devoted
to describing their "excavations" of this mound, it will be assumed that
19 feet was the more accurate of the two. In any case, once they had
penetrated through the top 20-inch gravel mantle, which they commented was "found to be broken up" (154), they encountered two
copper celts that had been deposited three feet from the top. At seven
feet down, they encountered the first of four sand strata and then the

next three at one-foot intervals. At nineteen feet deep, according to
Mills, what they claimed to be the floor was actually the puddled clay
mantle covered by the thin stratum of sand on which were placed the
sheets of mica that formed the ballast on the northwest side of the
primary mound of the Burial Number 9 mortuary set-piece, as described above.

By the time Mills excavated this mound, it had been reduced
from 19 feet, as reported by Squier and Davis, to only 12 feet in height.
He noted that the circumference of the base of the mound covered a
ninety-foot-diameter circle of single post holes. About 18 inches inside
this circle, he noted a puddled clay stratum of about ½ inch thick and
covered by a layer of sand, about 1 to 2 inches thick. He initially took
this to be the floor of the mound. But as excavation proceeded across
the mound from the south side, he noted that it was a clay and sand
mantle that covered what had been the earth stratum making up the
first vertical seven feet of the mound. He noted another similar feature
about one foot below the first one. Thus the center of these two strata
marked the six-foot and seven-foot vertical height of the mound. As
he then commented, "The true floor of the mound was easily disclosed
and proved to be very marked in character. It had been constructed of
puddled clay, with a light covering of fine sand" (481). He claims that
more sand was added from time to time to create what he characterized as a "peculiar cement-like layer."

Finally, it should be noted that the circle of post holes defining
the base of the mound was not the only post hole pattern. There were
two others. There was the "oval" that defined the semisubterranean
structure in the northeast sector with its entrance facing northeast, and
there was the main mortuary structure or Renewal Lodge built over
it. This was the structure containing all the floor-level mortuary deposits. It was rectangular with rounded corners and also had its entrance to the northeast.[12] This structure took up most of the northern
half of the large circle, leaving the southern half largely devoid of post
holes. Thus, at its height of usage, the prepared floor was defined by
a ninety-foot-diameter timber post circle, and located in the northern

half of this circle was the large southwest-northeast-oriented rectangular structure that was referred to above as the Renewal Lodge (fig. 15.1).

Mills did not hesitate to correct Squier and Davis when their excavation reports were in error. Therefore, it is interesting to note that he found only two sand strata while Squier and Davis noted four. Also, Mills commented on the fact that the covering mantle of gravel that Squier and Davis reported cutting through before encountering the two copper celts was "continuously ploughed from the top toward the base".[13] This suggests that the two clay/sand strata excavated by Mills, separated by one foot of fill, represent the two lowest of the four vertically stratified sand layers reported by Squier and Davis, particularly since, as they commented, these four were separated by only about one foot each. Since Mills comments that the first stratum that he excavated covered the whole mound and topped seven feet, then we can assume that the two highest sand strata reported by Squier and Davis, at eight and nine feet from the base floor, were destroyed by the cultivation process that eliminated the mantle of gravel.

This summary sets the scene for the three off-floor mortuary deposits that Mills found. Two were deposited at the three-foot level, one on the southwest side, which Mills refers to as Burial Number 2, and one on the northeast side, his Burial Number 11.[14] The axis joining these two mortuary features would be nearly parallel to and just slightly south of the southwest-northeast axis of the Renewal Lodge over which the mound was being erected at the time of these depositions. Burial Number 1, the third off-floor mortuary feature, was also in the southwest sector, but at the four-foot vertical level, thus higher than Burial 2. Unfortunately, Mills comments only on the positioning of Burial Number 11, the northeast feature at the three-foot level. Here he states that

> [t]he cremated remains were deposited on what, at that stage, was the surface of the mound, and covered with earth. With the burial were two flint knives, and a necklace of barrel-shaped bone beads. The natural

supposition with regard to burials placed above the floor is that they represented individuals who died during the erection of the mound. (Mill 1922, 484)

He does not comment on the positioning of Burial Numbers 1 and 2 relative to their sand strata. But since they are close to each other on the southwest side, and Burial Number 2, like Burial Number 11, is at the three-foot level, while Burial Number 1 is at the four-foot level, it will be assumed that all were associated with sand/clay strata. Burial Numbers 2 and 11, both at the three-foot level and located at opposite sides of the mound on a line that would be closely parallel to the axis of the substructure, would be on the lowest or first-built sand/clay stratum. Burial Number 1, in the same southwest sector as Burial Number 2 and at the four-foot level, would be on the second-lowest level. It is quite possible, then, that there is a missing second stratum northeast mortuary deposit, one that would have complemented Burial Number 1 in the same way that Burial Numbers 2 and 11 complement each other. It could have been destroyed by the agricultural cultivation. Furthermore, of course, it is also possible that this mortuary pattern was repeated on each of the four sand strata, but we will never know.

What is important, however, is that apparently all three of the known off-floor mortuary deposits were associated with two of the four known sand/clay mantling strata, and that two of them, Burial Numbers 2 and 11, were positioned on the same axis closely or actually parallel to the axis of the Renewal Lodge, suggesting that these mortuary deposits were not simply the result of these two people dying at an inconvenient time. Rather, their remains were available at a time during the constructing of the sand/clay mantles when ritual cremation may have been needed as part of the symbolic pragmatics of the mound construction itself. Burial Number 1, then, would have been part of the same process, only at the next stage. Finally, as reported above, Squier and Davis found two large copper celts deposited three feet below the surface level of the center of the mound, echoing the use of copper artifacts as focal elements in the Burial Numbers 9, 12, and 13 mortuary set-pieces.

The nature of the artifacts accompanying the three cremated mortuary features should be noted. Burial Number 1, the southwest four-foot-level feature, was accompanied by a double-bladed copper celt, quite unusual, as Mills comments, since the typical celt is single-bladed. Burial Number 2 was accompanied by "two interesting copper pendants, spoon-shaped, and a number of bone and shell beads" (483). These were placed in a "pocket-like receptacle in the earth," suggesting that a small hole was dug for their deposition. Burial Number 11, as noted in the above quotation, had two flint knives (probably bladelets), and a bone bead necklace. With the exception of Burial Number 1, there is nothing unusual about these artifacts. Indeed, taken out of the context that has been argued for, these would be quite unassuming mortuary features. However, added to these known features should be the two large copper celts deposited near the apex of the mound. These clearly cannot be explained in the terms that Mills has argued for these mound burials, namely, that someone died at an inconvenient time.

In terms of the World Renewal Model, the construction of the mound over a locale sacredly charged by preceding rituals is most reasonably interpreted as a critical part of a major world renewal rite and that, indeed, the clay/sand strata bearing the cremated mortuary deposits may have served as monumental sacrificial altars. As such, the stratification of earth, sand/clay, earth, etc., may have been designed to embody, replicate, and renew the stratification of the cosmos. The situating of the cremations at these levels along the southwest-northeast axis of the Renewal Lodge would suggest that their place-ment was a significant part of the constitution of this ritual, and the deposition of the copper celts, major custodial regalia, was the termi-nating sacrificial offering. It is postulated that the point, then, was to return the spirit and its powers embodied in the Renewal Lodge to the cosmological powers in which they participated. The construction of the mound with the cremations was part of this process, constituting it as a monumental set-piece that embodied the sacred powers that animated the cosmos. We lack an adequate ritual vocabulary, but such construction practices, mediated by post-mortem, or, in some possible cases, lethal sacrificial rites, might be adequately termed a form of spirit release and renewal. In this case, the spirit that was being released might have been those aspects of the cosmos that were embodied in the Renewal Lodge itself.

CONCLUSION

One major purpose of this analysis of Mound 7 and its contents was to demonstrate that the default funerary view cannot adequately account for the mortuary residue when situated within the complexity of total patterning in which it was caught up. This is not to claim that funerals were irrelevant. Rather, it is to claim that the funerals were only part of the total Ohio Hopewell/Middle Woodland (cult/clan) mortuary sphere(s). The funerary rites may not even have occurred in these locales but rather in the clan locales. In any case, following the funerary rites, mourning and spirit-release rites would have been performed so as to transform the deceased into warranting media of important renewal rites, and some of those deceased would have been retained to become the primary media of post-mortem sacrifice constituting the terminal mortuary aspect of these set-pieces.

The hermeneutic spiral has built up a series of interpretations that serves most reasonably in support of the initial claim of the Mourning/World Renewal Mortuary Model. With the relevance of this model established for Ohio Hopewell, it can be strengthened and expanded by using it to examine the symbolic pragmatics of formal deposits in other Ohio Hopewell sites. This is done in the next several chapters by looking at both burial and nonburial features at the Hopewell and Turner sites, leading to further, incrementally constructed models for interpreting the social dynamics that were mediated by the mortuary practices of these sites.

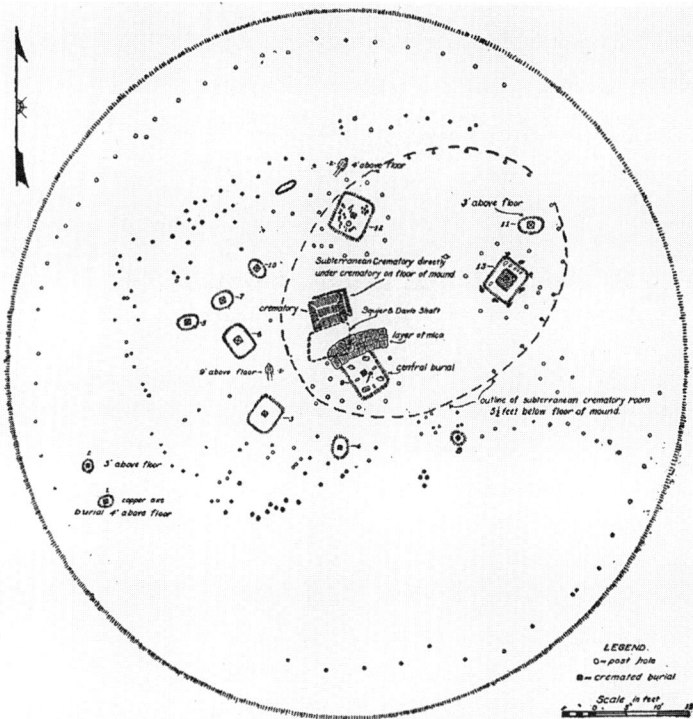

ARCHAEOLOGICAL SURVEY OF MOUND *7
MOUND CITY GROUP
FIELD EXPLORATIONS
BY
W.C. MILLS M.Sc.
AUG 1920 ~ JULY 1921
E. EVANS \ J H JEFFERSON SURVEYORS

FIG. 25. Plan of floor of Mound No. 7.

FIG. 15.1. The Mound 7, Mound City Proper, Floor Plan (Mills 1922, 480, figure 25) (Courtesy Ohio Historical Society)

Chapter 16

THE VAULTED CHAMBER CRYPTS OF OHIO HOPEWELL

The Hopewell site may be the best source of data currently available to support the set of incremental rites that the Mourning/World Renewal Mortuary Model postulates. This is not to deny that Mound City, Seip, and Liberty Works are important grounds for the model. For example, Mound City Proper certainly is significant since its mortuary data, as exemplified in Mound 7, were used to demonstrate that a mourning/world renewal ritual interpretation is a more coherent account than a funerary interpretation of the Ohio Hopewell mortuary practices. However, Mound City Proper seems to mark only the endpoint of this postulated complex mortuary-mediated process. With the exception of two extended inhumations found on the external slope of the southeast embankment wall and two burials outside the northwest and southeast corners, the mortuary data of Mound City Proper cannot be used as direct evidence of the mortuary stages through which the deceased may have passed prior to being cremated and deposited.[1] This situation was largely replicated at Tremper, Liberty Works, and Seip, where cremation also dominated. In contrast, extended burials constituted the majority pattern at the Hopewell site, and many deceased displayed signs that suggest an extensive period of exposure during which they could have been used to mediate the incremental series of mortuary rites postulated under the Mourning/World Renewal Mortuary Model. There are also partial and full cremations, the latter making up only about 25 percent of the known burials, and even several instances of the use of curated human parts: "trophy skulls," drilled human mandibles and maxilla, and, finally, there is even one example of a bundle burial.[2] Therefore, the Hopewell site is a major source of empirical data to support the funerary→mourning→spirit-release→ world renewal process claimed and articulated by the Mourning/World Renewal Mortuary Model.

If all the groups responsible for these major earthwork sites in the Chillicothe region were autonomous ecclesiastic-communal cults and operated the same type of mourning/world renewal ritual program, why would the Hopewell site seem to be the only major one to display elements of possibly the whole range of incremental mortuary rites while the former display pretty well only the end-product cremations? Lloyd (note 2) has also commented on this contrast. He recognizes three forms of apparently terminal mortuary treatment at the Hopewell site. He refers to these as extended, "charred," and cremated inhumations. In terms similar to those that will be explored here, he suggests that these three forms are not three different terminal mortuary treatments but three steps in the total mortuary program and that this patterning was largely the result of a truncated cycle of mortuary ritual.

However, he does not clarify if he means by this that, given sufficient time, all the deceased would have been put through all three steps, with cremated burial being the actual terminal step, or if, even with sufficient time, only some would have been subjected to the total cycle by ending up as cremated burials. Since he pointed out that only one of the forty-nine mortuary deposits found on the floor of Mound 23 was cremated, the rest being extended inhumations, of which 25 percent (twelve) were charred, this appears to indicate a relatively incomplete cyclic development of the ritual program held in the structure under this mound compared to the mix found under Mound 25, in which 25 percent of the 102 burials had received the full cremation treatment and only one of its extended inhumations displayed evidence of charring.[3] If the latter indicates the "normal" cycle, then it would appear that about 25 percent of the Hopewell mortuary deposits would be full cremations. Therefore, the mortuary patterning under Mound 23 would appear to index a rather abrupt rupture of the normal cycle. In fact, it will be argued here that the Hopewell site demarcates two rather abrupt interruptions in the normal mortuary cycle, one marked by the mortuary record on the floor of Mound 23 and the other on the floor of Mound 25. This argument will be termed the Laying-In Crypt Model of the Hopewell site.

THE LAYING-IN CRYPT MODEL

Set into the broader Proscriptive/Prescriptive Ecological Strategy Model, the Laying-In Crypt Model of the Hopewell site can be treated as an ancillary of the Mourning/World Renewal Mortuary Model as applied to the Middle Woodland period of the Central Ohio Valley. It postulates that the range of mortuary treatments at the Hopewell site is best treated as the expression of a program based on incremental stages in the treatment of the deceased and that not all the deceased were given the "full treatment." Therefore, as stated above, the mixed treatment of the deceased on the floors of both Hopewell Mounds 23 and 25 mark two ruptures of the mortuary cycle. As a result of these two major ruptures many of the deceased were "frozen" at different stages of the funerary→mourning→spirit-release→world renewal process, some possibly in the initial stages and others further advanced, and only some had completed the mortuary process to the normal terminal mortuary deposit stage. This means that the mortuary patterning exposed by the excavations of this site should be treated as at least two synchronic "snapshots" of this complex diachronic mortuary/world renewal cyclic process.[4] It will be argued later that the first rupture occurred when the Renewal Lodge Complex under Mound 25 was abandoned and the second, when the Renewal Lodge under Mound 23 was abandoned. It is likely that with this second rupture, the site was abandoned.

However, as important as these postulated ruptures would have been, they cannot be the whole explanation for the differences between the Hopewell site and its near neighbors—Seip, Liberty Works, Mound City, and so on. In fact, these ruptures do not even account for the differences between the mortuary patterning under Mounds 23 and 25. Therefore, while it will be maintained that the truncation of the mortuary program was a crucial factor in determining the mortuary patterning, a number of other factors were involved and these would have been part of the reason for the differences both within the Hopewell site and between the Hopewell site and its near neighbors. These factors relate primarily to different site trajectories in the history of factional disputes and will be elucidated and explored later.

The feature that will figure most prominently as a primary mortuary set-piece framing device is the large vaulted burial chamber that seems unique to the Hopewell site. All known vaulted burial chambers of this site are found on the floors of Mounds 23 and 25 and they contain the majority of the burials, these being extended inhumations. The Laying-In Crypt Model will treat these chambers as mourning/world renewal processing crypts. Before elaborating this model, it will be useful to summarize the data that are found on the floor of Mound 25 (fig. 16.1). Mound 25 is the large "conjoined" mound found within the "halfmoon" SL-Profile embankment earthwork of the Hopewell site (also see fig. P.5). It is the largest known Hopewell mound. Mound 23 is the only other one of the thirty to forty mounds found at this site that approaches it in magnitude, and this was less than half the overall size of Mound 25.

Mound 25 was about 550 feet long (about 180 m) with a base-width of about 190 feet (about 62 m) at its maximum. Greber and Ruhl point out that the floor of the mound was much more extensive than the core or central mound that covered the residue of the complex of structures found on the floor. This complex included two major, probably free-standing, timber structures, and several ancillary ones (fig. 16.2). Greber and Ruhl have labeled the two main features Structure D/E and Structure C. They term the lesser features Structures A1, A2,

B, and F.[5] The small Structure F was near the west end of the mound floor, quite separate from its nearest eastern neighbor, Structure D/E. This latter is a large oblong construction with its axis oriented slightly north of east and parallel to the axis of Mound 25 that covers it. With some imaginative stretching, this structure might be viewed as sub-rectangular. It is divided N/S internally by a row of post holes into the eastern Sector D and the western Sector E. Immediately east of Structure D/E is the smaller Structure C, and it appears to be more rounded in form than the latter.

In the relative size and positioning of these two major structures, Greber perceives a similarity with the patterning of the Seip and Liberty Works Great House Renewal Lodges. In this case, Structure D/E would be equivalent to rectangular Rooms 1 and 2, and Structure C the equivalent of circular Room 3. However, there is no indication that Structure C was linked to Structure D/E to form a single building, as was the case for the Seip and Liberty Works Great House Renewal Lodges. She extends this parallel by suggesting that Structures A1, A2, and B, which were separate structures south of but close to Structure D/E, may be the equivalent of the North Section of the Seip-Pricer, and the East Section of the Edwin Harness, and that a high-ranking set of burials covered by a primary mound over Structure F, the feature isolated at the west end of the Mound 25 floor, may be the equivalent of the high-ranking set of above-floor burials at the west end of Seip-Pricer. There is no indication that these structures were dual-leveled. All activities would have occurred on the ground level. Greber concludes that the total set outlined above were contemporary and were abandoned together. These will be treated here as a Renewal Lodge Complex.

In contrast to the multiple structures making up the Renewal Lodge Complex under Mound 25, only a single Renewal Lodge was constructed on the floor under Mound 23 and it was largely a repli-cation of Structure D/E on the floor of Mound 25 in both size and internal organization.[6] Assuming that these two mounds mark sequen-tially spaced ruptures in the mortuary cycle, these differences imply that significant changes occurred between the periods of use of the two Renewal Lodge complexes. The changes may be akin to those registered between Seip-Pricer and Seip-Conjoined, that is, some major shift in cultic postures; they are discussed in chapter 23.

THE VAULTED CHAMBER

> [The] most common type of tomb recorded by both surveys [of Shetrone and Moorehead] appears to have been constructed of logs placed about a 'platform' and supporting a bark or timber roof. Such tombs . . . are associated with the characteristic easily recognized 'arch' of loosened earth which results from the roof's collapse. None of these typical log tombs were found outside these three burial groups [under Mound 25]. (Greber and Ruhl 1989, 52–54)[7]

Although some of the vaulted chambers may have been only a little over 1 m high, many appear to have been 2 m or more high. As Shetrone comments, a number almost reached to the top of the primary mound and its gravel cover.

> A good illustration of the caving down of earth above a grave following the decay of the enclosing timber structure is shown in [Burial 12]. The arch-like opening in this was 7 feet long and 4½ feet wide, extending 6 feet above the floor level. It was not unusual to find this loose arched condition extending practically to the top of the mound, in some instances, more than 10 feet above the base. (Shetrone 1926, 72)

Consistently, these log-crib vaulted chambers are referred to in the literature as graves or tombs. This terminology is motivated by the funerary view, of course, and it predisposes the excavators to identify the deceased found in them as intended terminal funerary burials, and the whole mortuary process as simply a complex series of rank-sensitive funerary practices. In contrast, the Laying-In Crypt Model will characterize these facilities as *laying-in crypt altars*, and, as such, it will claim that they played a critical symbolic pragmatic warranting role in the funerary→mourning→spirit-release→world renewal ritual process of the Hopewell site. Therefore, the vaulted chamber is postulated as having served as the mortuary facility in which to place the body such that it was easily accessible during this relatively extensive period while being protected from the elements. However, this practical use would have been eclipsed by their symbolic pragmatic meaning, which constituted them as important symbolic warrants of the mortuary rites that they mediated. I first thought to call them mourning crypts. However, this term may lead to an overly narrow characterization of the symbolic pragmatic usage since mourning rites implicate only part of the incremental stages making up the funerary→mourning→spirit-release→world renewal ritual process. Therefore, the terms "laying-in crypt altar," or more simply, "laying-in crypt," seem to serve the broader nature of this postulated process.

Robert Mainfort accounted for what appears to be an equivalent Adena mortuary facility in similar terms.[8] He criticized the claim put forward by Andrew Shryock that, because of the labor involved in the construction of the large Adena "tomb," the individuals buried in these features were probably chiefs and that those burials positioned on the shoulders of the Adena CBL mounding would be commoners.[9] Instead, he argued that these features were more coherently understood as crypts that would have been reused to process a series of deceased. The individuals found in these crypts, then, would have been the last deceased to be cycled through the facility. Instead of being removed and buried on the shoulders of the moundings as previous deceased had been, they were left in the crypts.

The same core insight can be fruitfully applied to interpreting the Hopewell site mortuary data. However, there is a significant difference between the Hopewell and the Adena crypts. In the latter case, only one or possibly two crypts at the same CBL were operating at a time. In the case of the Hopewell site, there were many, and, potentially, most or all of them could have been in use simultaneously. Therefore, at any given time during the usage history of the Renewal Lodge, there would have been a number of deceased individuals, most probably being at different stages of the funerary→mourning→spirit-release →world renewal process. The individual deceased would have been removed at the normal termination of the laying-in period, some (probably the majority) transferred to clan CBLs, as will be discussed later, while the minority would have been retained by the cult and moved into the post-laying-in period to be processed through these rites, either as focal participating custodians or as auxiliary symbolic warranting media of world renewal rites, terminating as burials at the Hopewell site.[10]

In fact, as noted above, not all the burials in the Renewal Lodge Complex on the floor of Mound 25 involved crypts. There were also facilities lacking vaulted roofing but containing cremated human remains, very similar to the minor mortuary deposits at Mound City Proper. Given Shetrone's comment that the "typical" extended burial under Mound 25 was found under a vaulting of loose earth and gravel, it would appear that the crypts contained only extended burials. There were also a number of non-crypt extended burials in lesser mounds, as will be discussed in detail below.

THE INCREMENTAL MORTUARY PROCESS

The Laying-In Crypt Model is complex, and its demonstration requires three steps combining qualitative and quantitative data analyses. The first step is to generate a set of expectations of how the mortuary data ought to be distributed as a result of funerary→mourning →spirit-release→world renewal ritual mortuary process as postulated under the Mourning/World Renewal Mortuary Model. Summary descriptions of selected extended burials will be given, both crypt and non-crypt, in order to give qualitative support of the predictions, followed by an overall quantitative review of the data to demonstrate the degree to which they fit these predictions. The second step will be to summarize the evidence in support of the claim that several of the major "nonmortuary" burnt deposits were, in fact, part of the mortuary process as postulated, in this case, as the outcome of collective memorial/world renewal rites. The third and final step confirms this claim by analyzing the well-known Great Copper Deposit to demonstrate that the Hopewell site is more adequately characterized as a world renewal ritual locale than as a large-scale funerary locale, thereby reinforcing the view that the burnt deposit data, Mound 25 in which it was placed, and the Hopewell embankment earthworks operated as the framing devices by which the range of multiple inhumation types constituted a complex funerary→mourning→spirit-release→world renewal ritual process.

1. The Laying-In Period

As stated above, the process entailed under the Laying-In Crypt Model would require an extensive period of exposure of the deceased to allow for the post-mortem manipulation entailed by the postulated ritual cycle. Therefore, a full demonstration of the model should include marshaling the evidence required to show that, in fact, many extended burials display evidence of marked exposure. However, to limit the length and detail of this demonstration, it should be sufficient to point out that such extensive exposure is already well attested to in Ohio Hopewell literature. For example, Lloyd comments that "Hopewell burials were, in most cases, left exposed for some time before a mound was erected over them." Brown also makes this very clear.[11] However, it is also important to note that while showing that an extensive period of time during which the deceased were exposed was typical, it is not sufficient to demonstrate that the rites postulated under the Laying-In Crypt Model were actually performed. The following analyses are directed to demonstrating this claim.

In terms of this model, then, typically the deceased would have been placed in an appropriate crypt to initiate her/his laying-in period during which an incremental series of public rites would have been performed, possibly initiated by funerary rites, and then, more importantly, the incremental series of mourning, spirit-release, and related rites, such as name adoption/transfer rites. Each rite would be followed by a "mourning" period prior to the next step. Since these mortuary rites had the added value of being post-mortem sacrificial world renewal rites, each engaging a different spiritual aspect of the deceased combining different sacred values, then each incremental rite would probably supplement the number of attending local kin and companions by drawing in an expanding range of participants from more distant clans and cults. These incrementally scheduled gatherings would have involved the cumulative deposition of appropriate artifacts with the deceased in the crypt, appropriate being defined here not only in mortuary but also in world renewal sacrificial terms.

These can be usefully classified into four broad artifact categories. One set would be the laying-in warrants proper. These were artifacts that were specific to and partly constitutive of the laying-in period, such as mortuary garments, blankets, and shrouds, and other materials appropriate to the laying-in period and distinguishing it from the subsequent post-laying-in period. This latter non-crypt period is postulated as focusing heavily on the post-mortem sacrificial moment critical to world renewal ritual. The second categorical set would be both formal and informal artifacts that would have been accumulated by the deceased during her/his active social lifetime. Following archaeological tradition, these will be called the personalty of the deceased. With the termination of the laying-in period they would accompany the deceased to the post-laying-in period. A third category would be artifact types appropriate as mourning gifts/sacrifices. That is, these would be recognized as doing double duty as both sacrificial offerings and mortuary gifts and would be given during the laying-in period by cult companions, kin, and those attending from other world renewal cults. Because these contributors would be differentially linked to the deceased, and because of the double symbolic pragmatic nature of these items, the mourning gift category is postulated as having two subcategories, *formal mourning gifts* and *personalty-mourning gifts*. In general, the formal types may have been contributed by those who held a rather formal or distant relation with the deceased (e.g., the postulated attenders from other cults), while the personalty-mourning gift type may have been given by closer kin and companions. The fourth categorical set would be artifact types making up the custodial regalia, as described earlier for Mound City Proper. Like personalty, these also would have been accumulated during the lifetime of the deceased. However, only *some* deceased would be associated with them since they map the achieved custodial leadership trajectory of the deceased in the cult. The summary list of these four categories of mortuary artifacts is given below.

1. The Laying-In Warrants
2. The Personalty
3. The Mourning Gifts: Formal and Personalty
4. The Custodial Regalia

Concrete examples of these categories and the method used to delineate them will be given shortly. It should be noted that these categories are logical constructs and they are postulated as "fitting" the type of mortuary process specified under the Laying-In Crypt Model. Therefore, these will be treated as four emically real categories.

2. The Post-Laying-In Period

According to the Laying-In Crypt Model, with the termination of the laying-in period, the deceased would be removed from the crypt and transferred to the next ritual stage, in this case, involving rites having incrementally greater world renewal relevance. This removal from the crypt and transfer to the post-laying-in period mortuary locale, such as a small prepared floor area either in the main Renewal Lodge or in a lesser or secondary Renewal Lodge, would also entail a sorting out of the accumulated artifacts. This would be done according to the perceived four categories outlined above, some accompanying the deceased in the post-laying-in stage, and others being removed and curated, to be used for future ritual, as will be discussed below.

3. Artifactual Contexts

This total process in which mortuary artifacts were contributed, processed, passed on and/or accumulated would generate three different artifactual contexts: A. The laying-in crypts with the accumulation of the associated laying-in period artifacts; B. The post-laying-in period burials with their associated artifacts; and C. The ritual context in which the curated mortuary artifacts were "sacrificed" as media of collective memorial/world renewal rites. These will be called:

A. The Crypt Burial Context.
B. The Non-Crypt Burial Context.
C. The Burnt Deposit Context.

With the rather abrupt truncation of the funerary→mourning→spirit-release→world renewal ritual process as postulated under the Laying-In Crypt Model, the distribution of these four artifact categories would be "frozen" across these three major contexts. The Crypt Burial Context A would be made up of those deceased "frozen" partway through the laying-in period when the cycle was truncated and the crypts were co-opted as tombs. The Non-Crypt Burial Context B would be the burials marking the normal set of post-laying-in period world renewal rites, some of these also being "frozen" at different stages, while a number of them would have actually reached the terminal state, some being cremated and some being given extended terminal burial. The Burnt Deposit Context C would result from the use of the curated laying-in warrants and mourning gifts as symbolic pragmatic devices in constituting the sacrificial moment of collective *memorial/world renewal rites*. This is an important context and it is the reason that Greber and Ruhl's analyses of the burnt deposits and the Great Copper Deposit are important.[12]

As mentioned earlier Hall argues that memorial rites formed a significant mortuary event in historic Native American practices. The Laying-In Crypt Model postulates that the accumulated *laying-in warrants* and *mourning gifts* would be excellent candidates for this purpose, ensuring that these artifacts were appropriately used both as symbolic warrants of the deceased mediating collective memorial/world renewal rites and as sacrificial offerings, thereby enhancing the world renewal aspect of the memorial rituals. The period of accumulation would be reasonably terminated with the abandonment of a Renewal Lodge, and their final use as sacrificial media would probably be via massive collective burning in a ritual context.

It immediately follows from the above description that there should be observable and specifiable differences in the distribution of the four categories of artifact types across the three Deposit Contexts. This patterning is postulated in table 16.1 and it is the basis for both delineating the artifact types fitting under each of the four categories and demonstrating the model. Importantly, since in many cases those found buried in these crypts would have been only partway through the process when it was terminated, the crypts with their deceased would have been converted en masse into terminal post-mortem sacrificial altars. Therefore, the Crypt Burial Context should display artifact types from all four categories and the range should be highly varied in quantity and quality. The Non-Crypt Burial Context would have involved only the post-laying-in period deceased. Therefore, it should be characterized by an absence of artifacts belonging to the laying-in warrant and mourning gift categories, since these would have been removed and curated at the termination of the individual laying-in periods. The Burnt Deposit Context should lack both the personalty of the deceased and custodial regalia, while being rich in laying-in warrants and both formal and personalty-mourning gifts, since these would represent an accumulation of these mortuary artifacts deposited during the many individual laying-in periods prior to the abrupt termination of the Renewal Lodge usage.

While this postulated distribution isolates the custodial regalia artifacts, since none of these should appear in the Burnt Deposit Context while being in both the Crypt and Non-Crypt Burial Contexts, it does raise some problems of identification of the artifacts in the other three categories. This is primarily because of the overlap in the content of the personalty and mourning gift categories. As postulated above, items classed under personalty-mourning gifts would have been the actual personalty of the donors given as sacrificial mourning gifts. Therefore, they would be very similar and even identical to the personalty items of the different deceased who were processed. From the point of view of the active participants, of course, this overlap would not be a problem. For them, the personalty and personalty-mourning gift items would be publicly recognized as having distinctly different mortuary value. Therefore, those presiding over any particular mortuary process would be sure that only the actual personalty accompanied the deceased into the post-laying-in period rites while the other observably identical items, being personalty-mourning gifts, were curated along with the formal mourning gifts and the laying-in warrants.

This postulated overlap with requisite sorting has methodological significance since it is the basis for predicting that the laying-in crypts should have a superfluity of personalty-type artifacts compared to the personalty associated with the post-laying-in period deceased, which would be limited only to those actually accumulated by the deceased during their lifetimes. Furthermore, this overlap makes it relatively easy to identify the formal mourning gift types, thereby separating out what artifact types would count under the personalty and personalty-mourning gift categories. This is because, according to the distribution scenario postulated above, the only artifact types that would be common to all three contexts would have to be personalty, some used as personalty-mourning gifts, and some as the personalty of the deceased. This is indicated below in table 16.2, which is a modification of table 16.1, showing the distribution of the two mourning gift subcategories: formal mourning gifts and personalty-mourning gifts. By subtracting the custodial regalia from the Non-Crypt Burial Context, the remaining artifact types in this context would make up the emically identified personalty-type artifacts—Category 2. Because all four categories are represented in the Crypt Burial Context, by subtracting from this context the laying-in warrant artifacts (Category 1), the personalty-type artifacts as identified above (Category 2), and the custodial regalia (Category 4), the remaining types make up the formal mourning gift subcategory of the mourning gift category (Category 3).

However, as both table 16.1 and table 16.2 indicate, there is still a problem. In the above method, in order to identify the formal mourning gift artifacts it is necessary to subtract the custodial regalia, the personalty, and the laying-in warrants from the Crypt Burial Context. Since personalty and custodial regalia are already identified, this simply means subtracting the laying-in warrants. This raises what would appear to be an intractable problem: "What would count as laying-in warrants?" This can only be known by noting that they would be absent from the Non-Crypt Burial Context while present in the other two contexts. However, this would also be the case for the formal mourning gifts. That is, both these categories have the same postulated cross-context distribution, both being absent from the Non-Crypt Burial Context and both being present in the Crypt Burial Context and the Burnt Deposit Context.

To overcome this redundancy problem, it is reasonable to assume that the range of laying-in warrants will be narrower than the range of formal mourning gifts. This is because the former are partly constitutive of the laying-in period and the rituals that make it up while the formal mourning gifts are important but auxiliary warrants serving dual mourning and sacrificial purposes. Therefore, while there would be very little discretion in the former, there would be wider discretion in the latter. As a result it could be anticipated that the formal mourning gift category would be characterized by a much greater range of artifact types than the laying-in warrant category. Also because the laying-in period was made up of incremental rites, each drawing a wider range of participants, a variable qualitative and quantitative distribution of this type among the individual deceased across the Crypt Burial Context would be expected. Therefore, what is called for is a detailed, "up close" contextual analysis of the quantitative and qualitative distribution of the actual artifact types with these two assumptions as guidelines. This will be done later as part of the empirical grounding of this model.

As stated above, since burials making up the Non-Crypt Burial Context would have occurred after the laying-in period, they would have been "stripped" of mourning gifts and laying-in warrants. This suggests a test prediction of the Laying-In Crypt Model. Compared to the Crypt Burial Context, not only should the Non-Crypt Burial Context lack formal mourning gifts and items that can be identified as laying-in warrants, it should display an overall impoverishment of personalty-type artifact association, certainly in terms of quantities of the same type, but also possibly in terms of variety of categories. This impoverishment-claim follows from the above sequential scenario in which the deceased would be "stripped" of these surplus artifact types, the laying-in warrants and the mourning gifts, both formal and personalty-mourning gifts, when the laying-in period was terminated and they were moved into the post-laying-in period.

To control for variation that might be a result of differential prestige among the deceased, only custodial regalia burials in the two burial contexts will be compared, the assumption being that custodial regalia constitute notables in the cult, and, therefore, under the null hypothesis, it would be expected that there should be no significant quantitative or qualitative differences between crypt and non-crypt custodial regalia burials in the distribution of noncustodial regalia artifacts, that is, those artifacts postulated here as falling under the other three categories of laying-in warrant, personalty, and mourning gift. If this is not the case and there is a significant difference, then the null hypothesis can be rejected and the artifact context distribution supports the Laying-In Crypt Model.

DESCRIPTIVE ANALYSIS

Table 16.3 summarizes the distribution of seventeen artifact types across all three artifact contexts. A quadripartite patterning occurred, as table 16.1 and table 16.2 would lead us to anticipate. There were five types of artifacts that were absent from the Burnt Deposit Context while being associated with both the Crypt and Non-Crypt Burial Contexts. These included copper plates, copper headdresses, "trophy skulls," ocean shell containers, and copper celts. It is concluded that these are candidates for the custodial regalia. Equally importantly, table 16.3 has isolated four artifact types that were found in all three contexts: ear spools, shell beads, pearl beads, and bear canines. As discussed above, being the only artifact category predicted to be in all three contexts, these would be candidates for both personalty and personalty-mourning gifts. Eight artifact types were found to be shared by the Crypt Burial Context and the Burnt Deposit Context, while being effectively absent from the Non-Crypt Burial Context. They have been separated under the two categories of Formal Mourning Gifts and Laying-In Warrants. Copper-covered wood, clay, or bone buttons and cones have been chosen as candidates for laying-in warrants, and stone celts, platform pipes, flint and obsidian bifaces, and mica effigies and bone effigies have been chosen as candidates for formal mourning gifts. The reasons for this division will be discussed below.

As stated above, the null hypothesis would claim that all crypt and non-crypt burials associated with custodial regalia should have an equivalent variety and quantity of noncustodial regalia items. If this is the case, then the Laying-In Crypt Model would not be supported. If there is significant difference along the lines postulated above, then the null hypothesis can be rejected and it can be concluded that the Laying-In Crypt Model is supported by the patterning of the data. This will be tested in two steps. First, the data associations across all the Non-Crypt and Crypt extended burials have been summarized below in tables 16.4 and 16.5, respectively. Before addressing these tables, however, a descriptive summary will be given of the typical examples of custodial regalia burials, starting first with the Non-Crypt Burial Context and then focusing on the Crypt Burial Context. This will serve to present a qualitative summary of the patterning of the relevant data in order to confirm precisely what would count as laying-in warrants, personalty, and mourning gifts, including both the personalty-mourning gift and formal mourning gift subcategories. With this established, using tables 16.4 and 16.5, a summary overview will be presented to demonstrate that all the expectations of the Laying-In Crypt Model were fulfilled, and that the null hypothesis can be rejected. This clears the way to complete the demonstration by moving to steps two and three in the following chapter.

NON-CRYPT BURIALS WITH CUSTODIAL REGALIA
MOUND 2

Prime examples of the non-crypt extended burials associated with custodial regalia would be those in Mound 2. The floor of this mound primarily consists of the very large deposit of 8,000-plus Wyandotte chert bifacial blanks (also termed Indiana Hornstone). There were also five extended burials. All these burials were well supplied with esteem-endowing custodial regalia and moderately supplied with formal personalty artifacts (table 16.4). Shetrone reported that Burial Number 1 was a large male, about 25 years of age. In each hand he had a copper ear spool with silver onlay. His skull had been displaced and where it should have been there was only a copper headdress that was

> curved to fit the crown, with perforations at the corners for attachment. This helmet-like plate was partly embedded in the loose underlying soil and upon investigation it was found that a pit, 18 inches in diameter and 4 feet deep, had been excavated directly beneath the skull. This was filled with loose loam and apparently had contained a large post, or had been left unfilled, with some sort of temporary covering. When the perishable content or covering gave way, the skull had fallen into the pit and was found 2½ feet below its normal position, face downward. Scattered between the skull and the surface were found the lower jaw and the cervical vertebrae. (Shetrone 1926, 21)

Clearly, the displacement of the skull is evidence for a considerable period of exposure prior to burial. In terms of the Laying-In Crypt Model, much of this exposure would have occurred in a laying-in crypt and the removal of the body and its placement on the floor of Mound 2 would represent the post-laying-in period ritual. However, this period must also have been quite extensive, sufficient to allow the supporting board to rot. It is likely, therefore, that the skull was still in its proper position when the black "muck" of the primary mound was deposited. It was probably this work that caused the supporting board to fail and, unknowingly to those doing the work, the head was abruptly displaced.

Other than the copper headdress plate and the ear spools, no other artifacts were reported as accompanying this deceased.

Burial 3 was placed in a large burial locale, 11 feet by 3 feet, defined by standing stones. Associated with this deceased was an ocean shell container, several thousand shell beads, two copper plates, one below the left shoulder and the other below the hips, a copper ear spool in each hand, and a small copper axe between its legs. Burial 4, placed on a minimally prepared platform, had a copper plate, an ocean shell container, a copper ear spool in each hand, and many shell beads with a few pearl beads. Burial 5 was the most elaborate of the set. An elderly male, he was associated with an ocean shell container, two copper ear spools in each hand, many shell beads, a shell spoon, a small mica plate, a copper plate under his head, and a "trophy skull"— which also was fitted with a "helmet-like copper plate."[13] The redundancy of shell beads, ocean shell container, copper plates, headdress, and ear spools across these five burials is clear. No doubt this pattern is related to the special nature of the locale, and it might indicate that these were the primary custodians of this major flint deposit. Despite its special nature, however, the range and quantity of formal personalty were not that different from other well-endowed non-crypt custodial regalia burials, and, as discussed below, they were distinctly different in this regard from crypt custodial regalia burials.

MOUND 26

Mound 26 was located south of Mound 25 in the half-moon SL-Profile embankment earthwork that enclosed the latter. It contained seven individual burials, two of which were cremations (Burials 1 and 3). Except for one individual, all the burials, both extended and cremated, contained "standard" artifacts (e.g., ear spools, shell beads, pearl beads) in modest amounts, that is, none had an excess beyond the number that could be worn without unusual display. Burial 6, however, was more richly endowed. It was an extended male burial who had a copper headdress with attached pearls and residual textiles indicating that it had been quite elaborate. Around his neck there was a large pearl necklace. Also he had a copper plate, an ocean shell container, a "trophy skull," a platform pipe (the only one found with the non-crypt burials), along with shell beads and shell disks. Shetrone specifically states that these latter items were directly associated with the "trophy skull," not the extended burial.

These elaborate non-crypt extended burials are a fair representation of the range and quantity of artifacts reported to accompany the extended non-crypt custodial regalia burials. However, there were two cremated non-crypt burials on the floor of Mound 25 that were also associated with custodial regalia, Burial 4, with an elaborate copper headdress, and Burial 39, in section D of Structure D/E, associated with a copper plate, a sheet of mica, four drilled bear canines, and some flint flakes (bladelets?). Burial 4 also had some copper cutouts and two mica discs. These Mound 25 non-crypt cremated burials associated with custodial regalia display the same range and variety of personalty items as the other non-crypt extended burials discussed above associated with custodial regalia.

CRYPT BURIALS WITH CUSTODIAL REGALIA, MOUND

A good number of crypt burials (table 16.5) had the same range of custodial regalia artifacts as described above for non-crypt burials. However, the artifact associations significantly differ from the latter in two dimensions. Some of the artifact types associated with these crypt burials are found in the burnt deposits and are effectively absent from the non-crypt burials, and, in those cases where both the crypt and non-crypt burials share personalty artifacts found in the burnt offerings, there is, as predicted, a larger number and variety associated with the crypt than the non-crypt custodial regalia burials. In particular, the crypt custodial regalia burials were often richly endowed with pearl and shell necklaces and bracelets, often with attached animal canines, as well as "surplus" copper ear spools and other personalty.

Shetrone Burial 22 is situated in the D (eastern) sector of Structure D/E. It contained two extended burials. As Shetrone points out, the two individuals occupied "an earthen platform hardly wider than

that usually prepared for a single individual, and the log-molds enclosing it [were] much smaller than a typical grave" (79). These were young
male and female adults and they were so close that they were touching
at the shoulders. The male was associated with esteem-endowing artifacts in the form of hundreds of pearl and shell beads at his neck and
on his wrists, four grizzly canines set with pearls, two copper plates
(one at his pelvis and one on or under the right side of his chest), a
large projectile point, ear spools, and a large cannel coal celt, among
other items. Also, there was a copper headdress plate placed between
the two skulls on the bark matting of the platform. Interestingly, a long
rectangular cut mica strip "lay across the left humerus of the female
and the right humerus of the male, apparently so placed as a connecting
tie between the two" (79–80).

In terms of custodial regalia artifacts, Burials 34 and 35 in Structure
C were the equivalent of most of the non-crypt custodial regalia burials
described above. These two burials were also side by side, although
they had separate platforms, and the crypt that they shared was quite
large.

> A feature of . . . [Burial 34] was the size of the log structure enclosing it
> and of the individual logs employed. The dome-like aperture above the
> grave, resulting from the decay and falling down of the log structure,
> was strikingly large and bold, reaching almost to the top of the mound
> which at this point was ten feet in height. . . . [T]he gravel strata of the
> primary mound have broken off and dropped down with the loose earth
> filling the cavity. (Shetrone 1926, 86)

Associated with this burial was a "trophy skull," a "shield-shaped" copper plate at its chest, and four copper ear spools. But it also had a wild
cat mandible, and there were numerous split bear canines along with
many shell and pearl beads arranged across the pelvis area from wrist
to wrist. Several hundred pearl and shell beads were also found at the
left humerus and across the neck. There was a human maxillary on its
chest, four split bear canines set with pearls, and a dozen bear canines,
apparently all strung around its neck. There were three large mica
effigy spear points, mimicking the formal style of the well-known ob-

sidian blades found in Burnt Deposit 2 in Structure D/E, one to the right and two to the left of its head. Above its head there was a large (13-inch) headless female torso cut from mica. Burial 35, its immediate neighbor, was no less elaborate.

> *Burial Number 35* so closely adjoined Burial 34 that they might with propriety be considered together as a double burial; the platforms occupied by the remains, however, were separate. The grave was altogether similar to that of its companion burial. (Shetrone 1926, 89–90, emphasis in original)

Again, associated with this burial was a "trophy skull," a copper headdress, and two copper plates. It had two copper ear spools, and "about the neck and chest were several hundred pearl beads of every size and form. . . . At the neck were several pearl-set bear canines and two bear canines set into sockets of bone, the latter decorated with incised designs" (89–90). Shetrone comments on the copper headdress, which appeared to be a compound effigy, the upper body being human and the lower that of an insect. This artifact was considerably damaged by having been placed above and possibly on top of the crypt vault. He suggests that when it rotted, the effigy fell into the crypt and near the head of this deceased.

Burials 260 and 261 were two other crypt burials endowed with custodial regalia artifacts and sharing a common platform. They were found in the E sector (western) of Structure D/E. In fact, the quantity of copper plates and copper axes found with them exceeds the total quantity of these two categories in the rest of the Hopewell site. The simplest way of presenting these burials would be to quote Moorehead in full.[14]

> Two skeletons, Nos. 260 and 261, lay together near the base line, with the heads west. The mass of material deposited with them exceeds that associated with any other burial so far discovered in the United States. The objects were laid so as to form a rectangle 7 feet long and 5 feet wide, and were frequently so closely spaced as to overlap one another. The most remarkable find was a copper celt 22 inches long, which weighed 38 pounds. In spite of its size it was very symmetrical.

The objects covering the two skeletons were as follows: Sixty-six copper celts, ranging in length from 1¼ to 22½ inches; one stone celt 11 inches long; 23 copper plates, mostly fragmentary, and a great number of pieces; one very large jaw; a curious copper head ornament (deer antler plate with bud-like antlers); a broken shell; some very fine pearls, pearl and shell beads and teeth; carved bones and bone fragments; effigies; meteoric iron, partly worked copper, etc.; and colored earth.

Although Moorehead focuses on the large number of copper axes and copper plates, the other artifacts should not be ignored. There was the headdress and a "broken shell," which was probably an ocean shell container. Therefore, these burials were associated with the same core custodial regalia (e.g., headdress, copper plate, ocean shell), as those in Mound 2. They lacked "trophy skulls." No mention is made of ear spools, but there was a jaw present, as in Burials 34 and 35, as well as effigies.[15] He simply mentions pearls, and pearl and shell beads and teeth, without commenting on the relative quantity or their position. In terms of the copper axes and copper plates associated with these two burials, it is almost as if this collection constituted an accumulation of custodial regalia, a "treasury," that required special sacrificial deposition, somewhat equivalent to the large Seip-Pricer Section 1 floor deposit of copper plates with one central copper axe. This was deposited in its own platform altar and primary mound, with no directly associated custodian. As a "treasury," this large deposit may represent group custodial regalia.

Burials 6 and 7 were found in Structure F on the floor of Mound 25. Structure F appears to have been a small mortuary locale on its own and, although it was not a "vaulted chamber" in the sense discussed here, it may have functioned in the same manner as the vaulted chambers found in Structures C and D/E. Burials 6 and 7 are well known archaeologically as the "copper nose burials." They were placed on a platform made of a thick pile of bark sheets set immediately on the floor (in contrast to gravel used for the other extended burials on this floor). Shetrone assessed Burial 6 as a young adult male and Burial 7 as an adult female.[16] Both were accompanied by ear spools at their heads, and the female also had a bark strip running alongside her, almost the length of the platform, on which had been placed a row of about fifty copper ear spools. Each also had a pair of long copper "skewers," as Shetrone refers to them, ⅜ inch in diameter and about one foot long, one placed on each side of each skull, and lying across the collar bones. As mentioned above, each skull had a copper nose-piece. Shetrone describes these copper nosepieces as having funnel-shaped attachments allowing them to be fully inserted into the nostrils.

> These copper noses were firmly imbedded in place in both skulls, and afford striking imitations of the original features. The copper noses doubtless were post-mortem insertions, and if one wished to indulge in surmise it is easily conceivable that the Hopewell builders, obviously familiar with human anatomy and thus realizing that the nasal append-age quickly decomposes after death, supplied this young couple with imperishable noses, that they might not pass into the beyond lacking these useful and ornamental facial features. (Shetrone 1930, 66)

Indulging in a surmise based on the perspective of the Mourning/World Renewal Mortuary Model, these may have been spirit-release items rather than substitute noses. In any case, it is clear from his description that not only were they post-mortem insertions but also, because they were "firmly imbedded in place in both skulls," their insertion occurred considerably late in a post-mortem period of expo-sure. Copper plates had been placed beneath the heads, hips, and knees of both, making a total of 6 plates. Both had copious numbers of beads. As Shetrone describes it, Burial 6, the male, had "hundreds of pearl beads, ranging from very small to exceedingly large fine specimens" at his neck, wrists, and knees. Of particular importance was his comment that along with these "were more than 50 button-shaped ornaments, made variously from stone, clay and wood, with copper coverings."[17] In similar if not even more copious a manner, Burial 7 had hundreds of pearl beads "at the head, neck, knees and hips," and also "encircling her entire skeleton was a line of pearl beads and copper-covered but-tons" (65).

DISCUSSION

The selected custodial regalia crypt and non-crypt burials, along with burial data as summarized in tables 16.4 and 16.5, constitute a fair representation of the range, variety, and quantity of artifacts reported to accompany the custodial regalia extended inhumation category and should be sufficient to support the following discussion. Copper ear spools and pearl and shell beads are found across the Non-Crypt Burial, Crypt Burial, and Burnt Deposit Contexts (table 16.3). However, when large numbers of ear spools and beads are found, they are concentrated in copious quantities in the Burnt Deposit Context and, to a lesser extent, in the Crypt Burial Context. Therefore, since large numbers of these items are associated with crypt burials, it is reasonable to interpret the surplus over one or two pairs of ear spools or necklaces and brace-lets as personalty-mourning gift contributions elicited from kin and companions during the laying-in period. The mode of deposition of these personalty-mourning gifts is particularly important. For example, in the case of the fifty-plus ear spools associated with Burials 6 and 7, these were placed almost as a display, aligned on the left side of Burial 7, the side furthest from Burial 6.

The personalty, formal mourning gifts, and laying-in warrants distribution corresponds closely to the predictions generated by the Laying-In Crypt Model. This claim is particularly supported by the distribution of shell and pearl beads; copper-covered wood, bone, and clay buttons; copper cones; bear claws; and animal teeth. When the distribution of these small items is compared across the three deposit contexts, the complex patterning is illuminating. First, copper-covered wood, clay, and bone buttons were associated with both the Crypt Burial and Burnt Deposit Contexts, but were absent from the Non-Crypt Burial Context or, if some were present, the numbers were so small that they did not warrant mention by the excavators. In contrast, as pointed out in table 16.3, animal teeth, pearl and shell beads, bear claws, and so on, were found in all three contexts, copiously so in the

Burnt Deposit and Crypt Burial Contexts, while much scarcer in the Non-Crypt Burial Context, as predicted. Second, there were two patterns of body-associated distribution of these small items. One is the linear distribution found at necks, waists/hips, and ankles. This linear body-associated distribution will be called Pattern A. The second is the broadcast distribution, with these small items being distributed across and around the body. This broadcast body-associated distribution will be termed Pattern B. Pattern A probably marks necklaces, bracelets, anklets, and other like items. Pattern B probably marks elaborately decorated garments, sashes, shrouds, and other like items. Pattern A distributions were typically made up of bear claws, animal teeth, pearl and shell beads, but rarely, if ever, of copper cones and/or copper-covered wood, clay, and bone buttons. Instead, almost all Pattern B distributions were characterized by copper-covered wood, clay, or bone buttons, and copper cones, along with some pearl and shell beads, and a few animal teeth.

Crypt Burial 34 displays both of these patterns. Along with a "trophy skull," a copper plate, and four copper ear spools, it had a number of bear canines distributed across the neck and upper chest region, and "extending across the pelvis, from wrist to wrist, . . . numerous split bear canines and many shell beads, both regular and barrel-shaped; at the left humerus and at the neck were several hundred beads of shell and pearl" (88). These represent Pattern A and they were found in both Non-Crypt Burial and Crypt Burial Contexts, although much more commonly in the latter. Associated with Burial 34 was a Pattern B example. This appears to be the residue of a shirt or body garment on which many pearl and shell beads, as well as copper-covered clay, wood, and bone buttons, had been attached. Reports conforming to Pattern B were also made by Moorehead for several crypt burials. For example, his Burial 248 was also well supplied with esteem-endowing items, including a headdress, a large platform pipe, and 3 copper plates. He specifically interprets a Pattern B distribution of beads as defining a garment.

The body had apparently been dressed in a cloth garment, extending from the neck to the knees, upon which had been sewn several thousand beads, some of pearls and others of shell. Upon the skirt of the garment had been sewn some of the largest and most beautiful beads found in any of the mounds, together with bears' teeth, etc. The head had been decorated with a remarkable head-dress of wood and copper. The mass of copper in the center was originally in the form of a semi-circle reaching from the lower jaw to the crown of the head. . . . [T]he antler-shaped ornaments were made of wood encased in sheets of copper, one-sixteenth of an inch thick. They originally had four prongs of nearly equal length. (Moorehead 1922, 107)

According to Willoughby, "With many bodies buried in the mounds of the Hopewell group finely woven textiles were placed."[18] Many of these textiles were preserved because of their contact with copper items, particularly copper plates that were associated with the bodies. There are frequent comments that a given copper plate had been wrapped in a textile, or in bark or leather, these being preserved by the copper salts. Many of the plates were also placed on or immediately below different parts of the body (head, chest, hip, knees), and it is possible that some of the preserved textiles included elements of body garments. Finally, both Burials 6 and 7 display Pattern A and Pattern B distributions. Certainly, the bear canines at the neck of Burial 6 display Pattern A. However, "the line of pearl beads and copper-covered buttons" that encircled her body would mark the residue of a shroud-like cover placed over or under her body and these beads and buttons were attached along its edge.

When *non-crypt* extended burials associated with such custodial regalia as headdresses, "trophy skulls," and copper plates are noted, any associated animal canines and beads typically display the Pattern A distribution. This is not to suggest that these non-crypt extended burials went "naked." However, if garmented, they may have been somewhat austerely so since they lacked elaborate decorative shell and pearl beads, copper cones, and copper-covered wood, stone, and clay buttons. There are exceptions, of course. However, when a Pattern B distribution is associated with a non-crypt extended burial, rather than covering the full body or large parts of the body as in the case of Pattern B crypt burials, it was associated with a specific part of the body, such as the head or waist. For example, non-crypt Burials 3 and 4 of Mound 2 were buried with head garments that were elaborately decorated in the Pattern B manner. Shetrone describes Burial 3 as having "several thousand beads of ocean shell . . . strewn over the head, face, neck and shoulders," and Burial 4 had "many shell beads and a few of pearls . . . strewn over the head and neck".[19] However, there is no indication that either of these individuals wore the decorated shirt or skirt-like garment similar to that of crypt Burial 248 described above by Moorehead.

A non-crypt burial that was more akin to the equivalent crypt burials in this regard is Burial 6 of Mound 26. This accompanied a "trophy skull," an ocean shell container, a copper headdress elaborately decorated with pearl, shell, and marginella beads, as well as Pattern A necklaces of beads and bear canines. It also displayed the residue of an elaborate Pattern B loin item.

> At the hips were the remains of an elaborate loin covering, consisting of a rectangular copper plate, upon which, at the perforations, were two very large oblong pearls. This plate had been fastened to a coarsely woven fabric, or loin cloth, which was further decorated by numerous beads. Additional split bear canines were at the hips. (Shetrone 1926, 104)

Again, as in the case of Burials 3 and 4 of Mound 2, this was not the full body garment described by Moorehead or as suggested for Burials 6, 7, and 34 of Mound 25. This does suggest, however, one possible usage of copper plates, particularly those found at the hips—which also were not uncommon in Mound 25—since in this case it was apparently associated with an elaborate loin garment. Besides, similar focused Pattern B distributions were found with several crypt extended burials, indicating head garments or elaborate loin gear. Finally, it should be noted that only canines and pearl and shell beads are associated in all these "constrained" Pattern B cases. What is conspicuously missing are metal-covered buttons of clay, wood, or bone, and the copper cones.

It is possible that the occurrence of buttons, beads, cones, spheres, and rings on deceased individuals in Mound 25 is underreported in the published literature. For example, in discussing buttons under his category of "personal ornaments," Shetrone simply comments, "Numerous button-shaped objects made variously from stone, clay and wood with coverings of copper, silver and meteoric iron, were found with burials *throughout the group*."[20] Thus, there would appear to be a much higher incidence of the Pattern B distribution in crypt burials of Mound 25 than can be discerned from what was actually reported and these instances would be full-body-covering garments. Since equivalent esteem-endowing custodial regalia items are found in both crypt and non-crypt burial cases, this suggests that the difference in garments associated with these two separate burial contexts relates to differential symbolic pragmatics. It is concluded that the elaborate body garments associated with some, and possibly many, crypt burials were, in fact, *laying-in shrouds or garments*, rather than *burial shrouds or garments*. That is, they were laying-in warrants in the sense that they were constitutive of the laying-in period. Not all such laying-in garments, of course, would necessarily have been elaborately constructed with beads and buttons attached.

These postulated laying-in garments definitively link the crypt burials to the burnt deposits since, among other items that were found in the burnt deposits, there were copious quantities of small items in the way of canine teeth, beads, spheres, metal-covered buttons, and pyrite and chlorite cones, as detailed in the following chapter. It is true that canines and beads were also strung as necklaces and bracelets so that many found in the burnt deposits could have derived from the personalty-mourning gifts. However, as pointed out above, necklaces and bracelets, Pattern A, are common in both Crypt Burial and Non-Crypt Burial Contexts, and much more copiously in the former than the latter, and, therefore, many of the canine teeth and beads found in the burnt deposits would have been from both necklaces and bracelets as well as from elaborate laying-in garments. Furthermore, rather than being suspended, spheres, cones, and buttons appear to have been decorative elements that were typically sewn to garments. Willoughby carefully describes how these were produced so as to be attachable to textiles and skins.[21]

This analysis has also demonstrated the reality of the formal mourning gift category. As shown in table 16.3, not only were the decorative materials associated with the laying-in shroud common to both the Burnt Deposit and Crypt Burial Contexts, these two contexts shared mica cutout effigies and symbols, as well as carved bone effigies, all of which were absent from the Non-Crypt Burial Context. Along with these latter items were platform pipes, stone celts, and flint and obsidian bifaces. Since these are found almost exclusively in the Burnt Deposit and Crypt Burial Contexts, and were effectively absent from the Non-Crypt Burial Context, this suggests that they made up the formal mourning gift category. They were normally found in the Crypt Burial Context beside and in close association with the head or upper part of the deceased. However, it should be noted that some mica cutouts may also have been used as elements for laying-in shrouds and garments, since some of these appear to have been caught up in Pattern B distribution as part of the button, cone, and bead decoration.

Shetrone makes a relevant comment concerning the provenance of a human torso effigy cut from mica, shown with a similar, larger mica item associated with Burial 34, along with three cut mica "points".[22] The larger human torso effigy was found beside the head of Burial 34, a typical placement for formal mourning gifts. However, he comments that the second, smaller effigy "was found in a log-mold of Mound 25" (210). Unfortunately, he does not make any specific reference that would place this item more precisely. However, as a formal mourning gift item, and based on the reported positioning of the large mica effigy with Burial 34, it probably had originally been placed on or near the head of its associated deceased. Since he reports it as having been found in a "log-mold," this suggests that the effigy had fallen between several logs when the associated deceased was removed from the crypt. In any case, this adds to the evidence that these structures were reusable crypts.

As pointed out earlier, Shetrone reported that Burials 6 and 7 were laid on a thick stack of bark sheeting, similar to the usage in the "vaulted chambers." This reinforces the earlier claim that this mortuary structure played a role equivalent to that of the crypts in Structures C and D/E. Although associated with copper plates, copper "rods," and the famous "copper noses," these burials were richly accompanied by artifacts also found in the Burnt Deposit Context, especially the copper beads and the button-like items of copper-covered wood, stone, and bone, and so on, and of course, many of these were distributed in the Pattern B manner. Finally, while Structure F lacked the vaulting of the latter crypts, Shetrone stated that there were the signs distinctive of a large log cribbing. When the artifactual patterning and the log cribbing are added to the use of a bark platform, this strongly suggests that Feature F was a major laying-in crypt without the vaulted roofing. Alternatively, when abandoned, it is possible that the vaulting was removed, the bodies were covered with more bark sheets, and the primary mound with its gravel/sand surface stratum was built over them. As suggested by Greber and Ruhl, it is likely that Structure F was abandoned at the same time as Structures C and D/E and that Burials 6 and 7 were caught up in the larger world renewal rite that this abandonment mediated.

As displayed in table 16.4 and table 16.5 and summarized above, considerable descriptive detail has been marshaled indicating that there was a significant difference in the distribution of laying-in warrants, mourning gifts, and personalty between custodial regalia mortuary deposits of the Crypt Burial Context and the Non-Crypt Burial Context, and that this patterning conforms to the expectations of the Laying-In Crypt Model. This means that the null hypothesis can be rejected. Completing the demonstration of the model requires confirming that the burnt deposits are most coherently interpreted as an intrinsic part of the total mortuary program as postulated above. To this end, four of the major burnt deposits are analyzed and interpreted in the next chapter, followed by the analysis of the Great Copper Deposit, the point being to demonstrate that the Hopewell site was the sacred locale of a major world renewal ecclesiastic-communal cult.

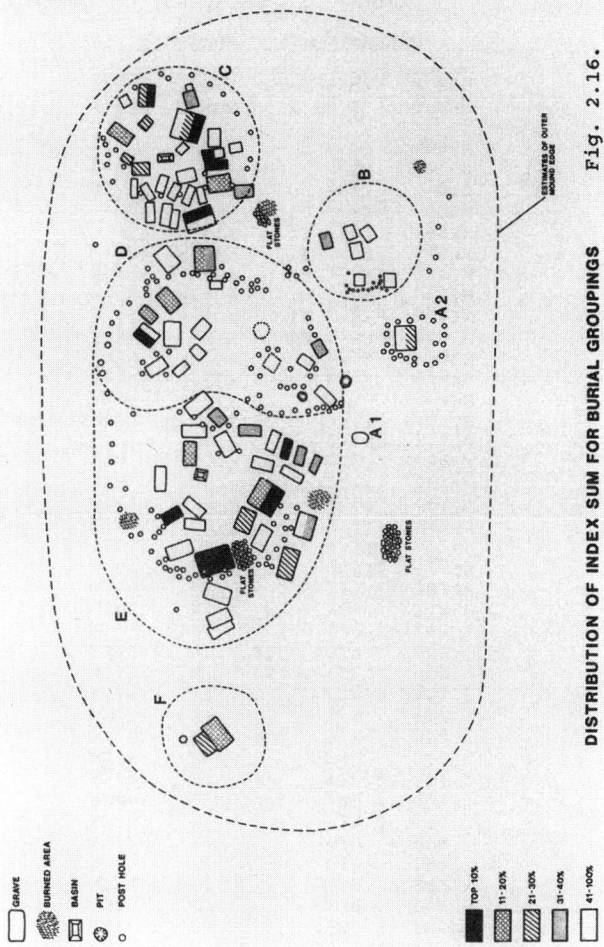

FIG. 16.1. Structures C and D/E and Ancillary Structures (Greber and Ruhl 1989, 50, figure 2.16)

Table 16.1 Artifact Patterning across the Three Artifact Deposit Contexts

Artifact Categories	Deposit Context		
	A. Crypt Burial	B. Non-Crypt Burial	C. Burnt Deposit
1. Laying-In Warrants	Present	Absent	Present
2. Mourning Gifts	Present	Absent	Present
3. Personalty	Present	Present	Absent
4. Custodial Regalia	Present	Present	Absent

Table 16.2 Artifact Patterning across the Three Artifact Deposit Contexts Modified by Mourning Gift Subcategories

Artifact Categories	Deposit Context		
	A. Crypt Burial	B. Non-Crypt Burial	C. Burnt Deposit
1. Laying-In Warrants	Present	Absent	Present
2. Mourning Gifts			
Formal	Present	Absent	Present
'Personalty'	Present	Present	Present
3. Personalty	Present	Present	Absent
4. Custodial Regalia	Present	Present	Absent

Table 16.3 Illustration of Distribution of Selected Artifact Types of Four Categories across the Three Artifact Contexts (R=Rare; P=Present; A=Absent)

Artifact Types	A. Crypt Burial Context	B. Non-Crypt Burial Context	C. Burnt Offering Context
Custodial Regalia			
copper plates	P	P	A
headdresses	P	P	A
"trophy skulls"	P	P	A
ocean shell	P	P	A
copper celts	P	R	A
Formal Mourning Gift			
stone celts	P	R	P
platform pipes	P	R	P
flint bifaces	P	R	P
obsidian bifaces	P	A	P
mica effigies	P	A	P
bone effigies	P	A	P
Laying-In Warrant			
cones	P	A	P
buttons: copper-covered wood, clay, or bone	P	A	P
Personalty AND 'Personalty'- Mourning Gift			
ear spools	P	P	P
shell beads	P	P	P
pearl beads	P	P	P
bear canines	P	P	P

Table 16.4 Summary—Non-Crypt Burials with at Least One Item of
Custodial Regalia

	Custodial Regalia	Other
Mound 2		
Burial 1	copper headdress plate	2 copper ear spools
Burial 3	ocean shell, two copper plates, copper axe	several thousand shell beads, 2 copper ear spools
Burial 4	a copper plate, ocean shell	2 copper ear spools, many shell beads, a few pearl beads
Burial 5	ocean shell container, a copper plate, "trophy skull," copper headdress	2 copper ear spools, many shell beads, a shell spoon, a small mica plate
Mound 4		
Burial 6	ocean shell	
Mound 26		
Burial 6	copper headdress, copper plate, ocean shell, "trophy skull"	large pearls, platform pipe, many shell beads and shell disks

Table 16.5 Mound 25 Crypt Burials with at Least One Item of Custodial Regalia

Mound 25 Floor	Custodial Regalia	Other
Burial 6	3 copper plates	100s pearl beads, 50+ buttons of stone, clay, wood copper-covered, 2 copper ear spools, 6 bear canines at neck, two copper "skewers," copper nosepiece
Burial 7	2 copper plates	50+ ear spools, 100s pearl beads, many copper-covered buttons encircle; 2 copper bracelets, 2 copper "skewers," and copper nosepiece
Burial 11	ocean shell, 2 copper plates, copper headdress	8 bone deer awls, 4 copper ear spools, 6 bear incisors, numerous pearl beads
Burial 12	2 copper plates	seed pearl beads at neck, copper pan flute
Burial 13	2 copper head plates	bead bracelets
Burial 22—double	No. 1—copper headdress, 2 copper plates	100s pearl and shell beads, 4 bear canines, large biface, 22 bear canines, 2 ear spools, cannel coal celt
	No. 2—none	No. 2—shell and pearl beads, 4 bear canines at neck, 2 ear spools, some pearl beads
Burial 23—double	South—crescent copper plate	
	North—none	bladelet
Burial 24	2 copper plates, ocean shell	2 copper ear spools, numerous pearl beads, mountain lion jaw, 4 bear canines
Burial 34	"trophy skull," copper plate	wild-cat jaw, many bear canines, 100s shell and pearl beads, human maxillary (worked), 4 large and 12 small bear canines, 4 copper ear spools, 3 mica effigy spear heads, headless female mica effigy
Burial 35	"trophy skull," 2 copper plates, copper headdress	wild-cat jaw, shell beads at wrist, 2 copper ear spools, bear canine necklace, cremation

Burial 41—triple	Skeleton 1—south: copper plate	limestone cone between knees, 14 bone imitation bear canines at hips, neck, 4 bear canines, pendant of barracuda jaw, shell and pearl beads, curved bone needle
	Skeleton 2—middle: "trophy skull"	human maxillary and mandible—worked, raccoon teeth, 25+ bear claws, shell and pearl beads, 4 bladelets, 3 bone awls
	Skeleton 3—north: none	large hollow antler tine, numerous shell beads, black steatite pulley-shaped ear ring
Burial 47—double	South—ocean shell, 2 copper axes	several bone needles, pearl and shell beads, mica effigy eagle's 3-toe foot-claw, large biface, mica effigy human hand, 2 shield-shaped mica items
	North—none	eagle foot-claw mica effigy with four toes, two mica circles, curved mica figure, two copper ear spools, number of shell beads
Burial 242	242—none	large pile pearl and shell beads, copper spoon ornament, 1 ear spool
Burial 243—double on single platform	243—copper plate	numerous beads, copper spoon ornament
Burial 248	3 copper plates, copper headdress	platform pipe, large biface, spool-shaped ornaments, copper-covered buttons, bears teeth, garment neck to knees with several 1000 pearl and shell beads
Burials 260 & 261 double—single platform	260—66 copper celts, 23 copper plates,	large jaw, many pearls, shell beads, engraved human
	261—copper head ornament, ocean shell	bones, worked copper, mica effigies, meteoric iron, etc.
Burial 270	copper axe, copper plate	200 shell and pearl beads, cut mica ornaments, bear canines
Burials 265 & 266— double	265—copper head-dress	
	266—none	
Burial 277	copper plate	
Burial 279	ocean shell	3 mica sheets, 3 lumps galena
Burial 281	3 copper plates, copper headdress	copper ear spools, copper-covered clay beads, copper beads, carved human femur, tortoise shell ornament, otter or beaver and bird effigies

Chapter 17

THE OFFERING ALTARS OF THE HOPEWELL SITE

The four major burnt deposits to be examined are the most thoroughly reported. They were discussed fully by Greber and Ruhl, whose description and comments are summarized below.[1] Particularly important is the cosmological perspective that Greber and Ruhl claim is supported by both the manner of formation of these deposits and the patterning and distribution of their artifactual contents. Two of the burnt deposits are found on the floor of Structure D/E, one in the clay basin of Sector D, the other in the clay basin of Sector E. The two other burnt deposits are found on the floor of Shetrone Mound 17. It is postulated that these clay basins are appropriately termed *burnt offering altars* and their contents are appropriately described as *sacrificial offerings*. It is further claimed that these two sets of burnt offering altars and their contents mediated two separate collective memorial/world renewal rites. Therefore, they were essential constituent elements of the mourning→spirit-release→world renewal ritual process at the Hopewell site.

It should be noted that there are no clear indications in the form of human cremations, body parts, and so on, that would unproblematically mark these deposits as the result of mortuary ritual. However, under the Mourning/World Renewal Mortuary Model it was argued that representations of the deceased can be treated as manifesting them, thereby transforming the behaviors these representations were used to mediate into mortuary rites. Confirming that these deposits can be interpreted in these terms can be effected by demonstrating that the artifacts making up their contents had been used in mediating mortuary/world renewal rites—renewal events in which human deceased figured. If this can be demonstrated, and since their terminal use in this way can be understood as manifesting some important aspect of the deceased, then the burning events in which they figured can be quite adequately characterized as memorial/world renewal rites. If many such symbolic warrants were utilized simultaneously, then this usage would constitute a collective memorial/world renewal rite.

Two points then must be demonstrated. First, it must be shown that the same artifact types making up these burnt deposits are also found undamaged by fire and in direct association with mortuary deposits. Second, it must be demonstrated that the patterning of the deposits expressed an immanentist cosmology of the type postulated under the World Renewal Model of the Ohio Hopewell. From these demonstrations it follows that these deposits could be interpreted without strain to be sacrificial burnt offerings mediating collective memorial/world renewal rites, thereby feeding back to enhance the validity of the Laying-In Crypt Model. In this regard, in fact, the first point was firmly established in the last chapter. The purpose of this chapter, then, is to demonstrate the second point.

A. THE SHETRONE MOUND 17 BURNT DEPOSITS

Shetrone Mound 17 was a rather small earthwork feature in the extreme northeast corner of the Hopewell site, just short of where the eastern embankment wall climbs to the upper terrace (fig. P.5). As fig. 17.1 indicates, there were three clay basins on the floor of this mound and burnt deposits were found associated with two of these. The central basin was badly damaged, a condition that Shetrone claims was the result of deliberate mutilation. To its south was another basin and piled beside it on a prepared base of "greenish-yellow clay" was Deposit 1, a large aggregation of burnt artifacts, both broken and unbroken, mixed with charred organic remains and earth. (81) To the immediate north of the damaged central basin was the third, also associated with burnt artifacts. However, rather than the aggregation making up Deposit 2 being beside the basin, it completely covered it. Although the artifacts in both deposits were badly fire-damaged, they had been burnt elsewhere, and then deposited in association with these two basins.

Just this part of the description suggests that these deposits and the three basins constitute the focal elements of a set-piece. This would make these basins the primary frame and the artifacts the primary focal components of an important ritual or complementary set of rituals. However, these will not be treated as custodial regalia since, of course, they were subjected to significant degrees of destruction. Although Greber does not recognize these deposits as mortuary in nature, she uses the differential patterning of the artifact categories that make them up to support a cosmological interpretation, which will be summarized and drawn on here to establish reasonable grounds that these deposits were a critical part of a collective memorial/world renewal rite.

Although both deposits displayed the characteristic range of Ohio Hopewell ceremonial artifacts, as Greber comments, "they are not identical. Rather, their contents reflect what I interpret to be a cultural complementarity" (81). At the base of the southern Deposit 1 there was a large stone plate. Over it were placed all the platform pipes (in fragments) found in Mound 17, as well as all of the many copper, meteoric iron, and silver artifacts (e.g., ear spools), but no copper plates or headdresses (table 17.1). The only obsidian biface found was in Deposit 1. It was a ceremonial point in the style that has been dubbed the Ross Barbed Point by James Griffin.[2] The deposit also contained the only shark tooth, as well as all the flint bifaces, all the engraved bone items, and almost all the stone celts (49), stone tablets, stone bird effigies, and ear ornaments.

Deposit 2 was placed directly on and spilling over its associated basin to form a circular pile 4½ feet in diameter. It covered a stone disk placed at the center of the altar. This may have been the comple-

ment of the large stone plate at the base of Deposit 1. It was also
similar to one in a major burnt deposit at Tremper. Deposit 2 contained
all the cut mica artifacts found in Shetrone Mound 17, as well as all
the bone tools (as contrasted to the engraved bone items of Deposit
1), all the Hopewellian bladelets (Flint Ridge), bear claws, stone bars,
plummets, and gorgets, many of which were made from chlorite
schist.[3] This contrast even extended to those categories that were
shared between the deposits. One deposit usually had only one, two,
or three items of a given category, while the other deposit had many
of the same category. For example, stone cones and fragments of cones
were plentiful in Deposit 2, thirty made from chlorite schist and fifty
from pyrite, while only three were found in Deposit 1. In reverse,
Deposit 1 had 49 stone celts while Deposit 2 had only 2 or 3.

Greber considers that the almost complete separation of some
categories of artifacts between these two deposits and apparently the
complete separation of copper and cut mica artifacts as marking a
complementary duality. By this she means that the patterning manifests
the core of the Hopewellian world belief, which was based on structuring the cosmos as a fundamental, multidimensional duality. "These
assemblages may be seen as representing two basic forces or qualities
which contribute to a whole life force."[4] To underscore this interpretation she points out that mica and copper artifacts were segregated
between these two deposits, and since these are the two most widespread of the Hopewellian material resources, being effectively ubiquitous in most regional Hopewellian sites, such a separation must have
been the result of a cognitive-normative distinction. It was argued earlier that prehistoric Native American peoples treated natural materials
as participating in the sacred powers of those areas and sectors of the
cosmos with which they were seen as "naturally" associated, shell and
copper probably being associated with Underwater and the Underworld, respectively, flint with the land (Middle World), and so on. Furthermore, the separation of mica and copper might implicate a belief
that mixing these two together in the context of these deposits would
be polluting to the cosmos since it would be mixing together the cos-

mic elements that were naturally segregated. To this observation can be added the fact that all the Hopewell bladelets were found in Deposit 2 while all the bifaces were found in Deposit 1. Earlier it was argued that the bladelet complex embodied symbolic pragmatic proscriptions by which sacred pollution caused by mixing the use of the same tools across ritual spheres could be avoided. It is indicative that all these bladelets are found in Deposit 2 and all are Flint Ridge (Van Port) types, while all the bifacial tools are found in Deposit 1. This separation reinforces the notion that the copper/mica separation may also indicate strong ritual proscription.

B. THE MOUND 25 BURNT DEPOSITS

The two burnt deposits on the floor of Mound 25 differ from the above in several ways. First, their context is different. While the former were found in a locale well separated from any (other) mortuary activity, these two were found on the common floor of Structure D/E and surrounded by the residue of mortuary activity—although no human mortuary remains, per se, were found directly with these deposits. Each deposit is found in a different sector, D and E, thereby being physically separated. Unlike Shetrone Mound 17, which seemed to be specific to the event or events mediated by the burnt deposits, Structure D/E mediated different types of events and it is possible that the two deposits, one in each sector, were not part of either a single integrated event or two related events. However, it is highly probable that they were complementary aspects of the same event, as will be argued shortly. Second, while there was a separation of different categories of artifacts, it was carried out differently. For example, mica and copper were not separated. Finally, in this case each deposit was directly over its associated basin and the burning was done directly in and over them.

While considerable wood ash and charcoal were mixed with these artifacts, Willoughby, as quoted in Greber and Ruhl (76), suggests that the fire of Deposit 1 (Sector E) was enhanced by the addition of some type of oil or oil-like material.

A highly inflammable substance, possibly pitch or animal fat, was also evidently used for fuel, and the confined, penetrating smoke permeated through and through the carvings of bone, antler and ivory, the teeth of animals, certain varieties of stone, and other objects, rendering them a deep, lustrous black throughout.[5]

Deposit 2 (Sector D) was also thoroughly burned, again with the addition of similar "fuel." It should be noted that this added material may be something more than simply a "fuel." The use of materials to produce smoke intentionally is a common method in many religious rituals of evoking the powers that are identified with the cosmos. Also found in these two very extensive deposits were mica blocks and artifacts, over 500 copper ear spools, some covered with meteoric iron and silver, nuggets of copper and silver, copper- and silver-covered balls, some potsherds, many carvings in wood, bone, antler, and shell, many bladelets, quartz crystals and fragments, literally thousands of pearls, a few platform pipes, many shell beads, bear claws, shell gorgets, over 100 obsidian bifaces, several thousand bird bone beads, bone needles, cut animal jaws, some textile residues, and a number of small mammalian foot bones. Many of these items were of the same type as found in the Shetrone Mound 17 deposits, and, interestingly, they were also similar to those found in the Turner deposits. As Greber comments, the sheer size of these two deposits was eclipsed only by the large deposit of Wyandotte chert of Mound 2.[6]

Again, although there is considerable overlap in the categories of artifacts in these two deposits, there are some important noncongruencies. This is particularly clear in the case of obsidian bifaces, platform pipes, and pearls. All examples of each were in only one or the other of the two deposits. All the thousands of pearls, for example, were in Deposit 1, while the five known platform pipes were all in Deposit 2. Although mica was found in both deposits, the worked mica items, in the form of abstract symbols and effigies, were all in Deposit 1, while the unworked mica blocks and mica debitage were in Deposit 2. Again, although some categories were represented in both deposits, there was a distinct bias in numbers. For example, Deposit 2 had only three bird bone beads and many of the bone awls and needles, while Deposit 1 had 325 bird bone beads and may have had no awls or needles. The foot bones of small mammals form another skewed distribution. Deposit 1 had 110 while Deposit 2 had 690.[7] Finally, copper ear spool distribution was distinctly asymmetrical. Deposit 1 had about 500, many strung into a bundle, while there may have been none in Deposit 2, although Greber leaves open the possibility that a small number were included.

In their conclusions, Greber and Ruhl maintain that the two paired deposits, the set under Shetrone Mound 17, and the above set, each constituted complementary sets and, although displaying different structuring, they were variations on the same cosmological themes. These slight differences are perfectly consistent with the symbolic pragmatic approach taken in this book. As argued earlier, the cosmology→ideology relation is contingent, so that while the cosmology can remain constant, the ideological expressivity can change over time. In this case, the burnt deposits are expressive representations and, therefore, manifest ideological postures.

When these complementary patternings are perceived in the context of the Hopewell embankment earthwork itself, this supports the view that these deposits were the media and outcome of sacrificial offering rites directed to the particular aspects of the cosmos manifested in these earthworks. Therefore, as postulated earlier, the deposit basins can be adequately characterized as burnt offering altars. When this conclusion is combined with the earlier demonstration that the artifactual contents had, indeed, been associated with the deceased as elements of laying-in shrouds and as mourning gifts, then it can be concluded that these burnt offering altars likely did mediate collective memorial/world renewal rites. Summarizing the interpretation of the Great Copper Deposit by Greber and Ruhl confirms this interpretation by showing that the construction of Mound 25 was probably the ritual abandonment of the complex of structures on its floor and that these counted as a World Renewal Lodge Complex. In terms of the notion of framing device discussed fully in chapter 14, this would mean that, since Structures C/D, E and A, B and F constituted a Renewal Lodge Complex, then the vaulted chambers as the primary frame of the mortuary deposits that they contained would be appropriately termed laying-in crypt altars.

The Great Copper Deposit

> [Four] feet from the base line, were found one hundred and twenty pieces of copper . . . all laid flat, and occupied a space 3 feet long by 2 feet wide, with layers of bark above and below. There were no skeletal remains connected with the deposit nor was any altar found near it. (Moorehead 1922, 109)

This is part of Moorehead's comments on the fifth major deposit which needs to be examined in order to support the cosmological framework implicated by the Laying-In Crypt Model. This is often referred to as the Great Copper Deposit and it nicely contrasts with the burnt offering altars and their associated sacrificial burnt offerings. It particularly contrasts in that the symbols, cut and shaped from copper sheets, were carefully placed and not damaged. "All of these objects were hammered from native copper and many of the designs were duplicated".[8] These bear a close relation to the types of copper artifacts found in association with the Burial Numbers 9 and 12 set-piece mortuary deposits at Mound City Proper. However, as Moorehead comments, this deposit was not found in association with human remains. Instead, it was "buried beneath sixteen feet of solid earth" (100) and embedded about the central point of Mound 25. This would put it at more than 1 m above the floor. The positioning of the deposit, the forms of the copper items, and the ordered placement echo the deposit of the two copper celts that "topped off" Mound 7 of Mound City Proper. This suggests that it was part of the ritual construction of the mound and, as such, would have played an important symbolic pragmatic role in constituting this construction as a major world renewal rite.

The large number of iconic effigies and abstract forms cut from copper sheets making up the Great Copper Deposit were carefully placed on a layer of bark sheets, much like the bed of bark covering the crypt floors, and further covered by bark sheets before the earth covering was placed (111). They included simple concave disks with central holes, bracelets, rings, pendants, serrated strips, animal effigies (fish, bear), effigies of bear claws, canines, and deer antlers, conventionalized human faces, eyes, and such abstract items as swastikas and quartered circular and rectangular objects, as well as one trefoil. Most of the objects had small edge holes drilled into them, probably for suspension purposes.

Greber and Ruhl comment that the relative position as well as the care taken with their placement suggests that these were a total set of regalia and that, as expressive symbols, the combination of the different items and their associations manifest a structured set of concepts about the cosmos as a coherent totality. They point out that certain animal motif forms, for example, the bear, deer, and, possibly, bird, are combined into a single complex motif item. This is illustrated by what is traditionally referred to in the literature as the serpent's head effigy.[9] However, they interpret the item quite differently. The overall form depicts a deer's ear flap and they support this claim by reference to an engraved human femur found in direct association with either Burials 260 and 261 or 278,[10] and on which is depicted what was probably intended to be a shaman (or a shamanic priest) wearing a mask that displays the antlers and ear flaps of a deer, the latter having precisely the same shape as the item in their fig. 4.33. They argue further that the distal end of the ear flap shape represents the muzzle of a bear. The deer ear flap/bear muzzle combination is quartered, expressing a direct relationship between a four-sectored cosmos, and two major animal species (113). They point out that the upper half is dominated by the eye motif, which they suggest might be the owl, and this contrasts with the distal or lower half dominated by the bear motif.

> The upper portion of the quartered circle [i.e., as depicted in the "ear flap"], in the positive aspect of the design, repeats a common Hopewell "comma" motif, with an added projection. Willoughby identified this mo-

tif with the owl, yet it is also reminiscent of the more realistic bird talons
cut from mica found in Mound 25. (Greber and Ruhl 1989, 281–82)

They also do not exclude the possibility that, because of its shape, this
item implicates the serpent. Greber and Ruhl further comment on the
dominance in the artifactual data of the Hopewell site of materials
related to deer, bear, and bird, either by being derived from the animals
involved, such as bear canines, claws, and the thousands of bird bone
beads found in Deposit 2 of Mound 25, or as effigy depictions (277–78).
Of course, other animal species, and therefore other aspects of the
world associated with them, are also manifested by one or the other
of these means at the Hopewell site (e.g., pearls and ocean shells).

Such a complex integration of different animal motifs with cosmological motifs can certainly be supported by what we know of historic Native American world beliefs as expressed in their myth and
ceremony. The quadripartite cosmos is often envisioned as a serpent
and the essential powers of the different elements of the world are
often thought to be manifested in different species, each associated with
a different aspect of the cosmos, the earth, the sky, the water, the
different quarters, and so on. The same type of expression of the sacred
structuring of the cosmos is found in the mortuary set-pieces at Mound
City. All this suggests that the ritual usage of this complex of copper
custodial regalia was taken to express and, thereby, elicit the essential
powers that animated the cosmos in all its structured complexity. The
symbolic pragmatic point of this deposit, then, would be to invoke,
enhance, and sustain the ongoing and intrinsic connection between the
cosmos and the Hopewell site, and its resident custodial cult, reconstituting the embankment earthwork as an axis mundi, and this would
ensure that the labor invested in constructing this mound, and all the
post-mortem sacrifices that it embodied in the form of burnt deposits,
crypt and non-crypt burials, cremations, and so on, would serve to
enhance the sacredness of the cosmos, thereby reversing the polluting
effects that the pursuit of organic reproduction and survival by humans
necessarily entailed, given the essential contradiction of human existence in an immanently sacred world.

Critical Hermeneutics

Significant qualitative and quantitative evidence has been presented that supports the Laying-In Crypt Model. In terms of the hermeneutic spiral, this is not sufficient by itself since it is necessary to show that this account is preferred to alternative accounts of the same data. While these latter accounts, funerary and modular in character, have not been summarized, the above presentations have drawn on the interpretations of these same data as given within the scope of the funerary view. It is notable in this regard that the total body of data that the Laying-In Crypt Model addresses is divided by the funerary/modular accounts into two separate categories, one category addressing the variation in the burials and the other addressing the content and variation of the burnt offering deposits and the Great Copper Deposit. For example, Greber and Ruhl's excellent interpretation of the burnt offering and the Great Copper deposits was presented by them quite separately from any discussion of the burials. Similarly, Lloyd has examined the burial data, but without relating these to the burnt offerings.[11] That is, accounts that have focused on the burial data from the perspective of the funerary paradigm have largely assumed that the "burnt deposits" and other "nonmortuary" data mark ritual that was unrelated or only peripherally related to the mortuary ritual.

Indeed, the treatment of the burnt deposits and the Great Copper Deposit independently of the burial data might be a deliberate choice on the part of Greber and Ruhl in reaction to the tendency of the funerary view to be extended to the total body of ritual data. Greber has made it quite clear that she considers that this mischaracterizes the social nature of the great embankment earthworks. Her point is well made, when the funerary paradigm is assumed. However, when the data are treated in terms of the mourning/world renewal perspective, dividing the data into mortuary and nonmortuary categories marking separate ritual spheres becomes, in turn, a mischaracterization of the social nature of these earthworks. Therefore, implicit in the presentation of the evidence and argumentation in support of the Laying-In

Crypt Model is a critique of the alternative accounts of these same data.

As specified earlier, there are several rational bases for choosing among alternative models. One of the most important is a straightforward comparison of the relative coherence of the accounts of alternative models. The model that can give not only the most coherent account of the same data but can also eliminate puzzles raised but not resolved by the other accounts is favored—until its own puzzles are resolved by another model. However, there is also the question of the scope of the data that the alternative models can cover. Accounts given in terms of the funerary paradigm fall short in both ways. First, the scope of the funerary paradigm is typically too narrow to account for the complexity of the total patterning and variation of the ritual data that make up Hopewell sites. Therefore, it encourages its proponents to postulate models that characterize the data in terms of multiple, largely mutually autonomous ritual spheres. This might be justifiable if the mortuary and nonmortuary data were balanced in quantitative and qualitative terms. However, when the mortuary dimension is as pervasive in its presence and overwhelming in its quantity as at the Hopewell site, excluding the few instances of "nonmortuary" data by claiming they manifest a different ritual sphere suggests that the funerary paradigm is inadequate in scope. In contrast, the same complex set of ritual data has been accommodated by the Laying-In Crypt Model, while largely avoiding internal contradictions.

Even within its own narrow terms, the funerary view does not generate models that can give an adequate account of the very data they are designed to explain, namely, the variation in burial treatment. According to the funerary paradigm as applied by Greber and discussed fully earlier, for example, the social standing of the deceased at the time of death is measured by the ranksum gained across the various dimensions of post-mortem treatments. Therefore, the burial treatments of deceased associated with the same range of custodial regalia should be largely equivalent. However, this was established not to be the case when crypt and non-crypt burials associated with the same

range of custodial regalia are compared. Most of the crypt burials associated with custodial regalia also had a greater variety and larger
quantity of noncustodial regalia than had the non-crypt burials with
equivalent custodial regalia. There was also a clear difference in distribution of decorated garments that were referred to earlier as laying-in
shrouds, suggesting that only crypt burials "deserved" these coverings,
while non-crypt burials were impoverished in this regard, even if they
were associated with the same range of custodial regalia, and such
formal artifacts as copper celts and ear spools.

In contrast, the Laying-In Crypt Model accounts for these anomalies without strain or contradiction by characterizing the mortuary
data in terms of the post-funerary aspect of the incremental
mourning→spirit-release→world renewal ritual process. Therefore, in
terms of scope and coherence, it can be concluded that the Laying-In
Crypt Model is the preferred account. In these terms, the artifactual
patterning across the Burnt Deposit, Crypt Burial, and Non-Crypt Burial Contexts is the outcome of a complex mourning/world renewal
ritual process that was rather abruptly truncated, at least twice, as a
result of two successive ritual abandonments of Renewal Lodge Complexes. These ritual abandonments served to constitute major world
renewal events, leading to the conclusion that the total Hopewell site,
along with its Renewal Lodges, constituted a major world renewal cult
locale.

Implications for the Ecclesiastic-Communal Cult Model

Treating the burials in the Renewal Lodge Complex under
Mound 25 as resulting from the truncation of a complex funerary
→mourning→spirit-release→world renewal ritual process resolves an
important problem for the Ecclesiastic-Communal Cult Model that is
raised by Greber's ranksum analysis of these same Hopewell site burial
data. As discussed earlier, she clearly sees meaningful parallels between
the R-R-C pattern of the Great Houses of Seip and Liberty Works and
the spatial patterning of Structures C and D/E. There is substantial
agreement with her view here. Where the two approaches differ, of

course, is that the Civic-Ceremonial Center Model treats the R-R-C pattern as manifesting a ranked clan system and the Ecclesiastic-Communal Cult Model treats the same pattern as manifesting an ecclesiastic-communal cult. Given the similarity of patternings, it could be expected that the overall ranksum and ranksum profiles of Seip and Hopewell would be similar. However, this is not the case. As Greber points out, at the Hopewell site

> individuals ranking high within the present framework are scattered among the groups and a range of ranksums is found within each group. The patterns of the distribution of artifact classes and the index sum itself are similar across the three major groupings. (Greber and Ruhl 1989, 56)

In short, there were no statistically significant differences in the ranksum profiles of the mortuary groups across these sectors. This is quite different from the patterning of the overall ranksum and ranksum profiles she found at Seip-Pricer and Seip-Conjoined. While not accepting the premises of the Civic-Ceremonial Center Model, nevertheless, the Ecclesiastic-Communal Cult Model used these profiles to empirically ground its claims that the Great House CBL was patterned by the conjunction of the senior/junior age-grade and the laity/clergy structures. Since this model also postulates that the Hopewell site was the locale of an autonomous ecclesiastic-communal cult, the lack of parallelism with the Seip and Liberty Works sites would appear to be a problem.

Certainly, Greber recognizes this lack of similarity as a problem. To accommodate it, she suggests "that each of these three burial groups [Sectors D and E of Structure D/E and Structure C] represents a social component of the whole society represented within the Central Mound".[12] Since each of the three components displays differential distribution of individual ranksums while displaying no statistically significant differences among their ranksum profiles, she concludes that they were a set of "differentiated but nonranked social components" (57). By "differentiated" it is assumed she means ritually specialized. She also treats the burials in the smaller features, A1, A2, F, and B, as individuals who "may have been specially chosen from the larger components as representing some needed or valued social purpose" (58). Before presenting this conclusion, she briefly explored the possibility that the equivalent ranksum profiles may have been the result of the burial groups being related diachronically. However, given the sharing of a common prepared floor under Mound 25, her preferred view is that these burial groups were largely contemporary. All this means that, under the Civic-Ceremonial Center Model, the social structures of the Hopewell site group and those responsible for the Seip and Liberty Works sites would be significantly different. This latter conclusion, however, would not be a problem for the Civic-Ceremonial Center Model since Greber has consistently argued that there was a great deal of social structural variation across Ohio Hopewell.[13] This claim, of course, arises from the basic assumption of the funerary view that mortuary variation directly maps funeral events and the latter would rather directly reflect social structure.

This approach to prehistoric mortuary practices in the Eastern Woodlands has been challenged by the Mourning/World Renewal Mortuary Model. For this very reason, however, the contrast between the ranksum profiles of the Hopewell site, on the one hand, and the Seip and Liberty Works sites, on the other, does pose a problem for the Ecclesiastic-Communal Cult Model. This model characterizes all major Ohio Hopewell sites as cult locales based on the same fundamental social structures. Fortunately, the Laying-In Crypt Model can, in fact, be used to preserve the Seip and Liberty Works mortuary data in support of the Ecclesiastic-Communal Cult Model, while adding the Hopewell site data to this support. In its terms, it becomes quite clear that all or at least most of the different stages of the same type of mourning/world renewal mortuary process are clearly indicated at the Hopewell site, while at the Seip and Liberty Works sites, the mortuary data consist primarily of terminal burials following cremation—postmortem sacrificial offerings—and the latter would have occurred after the laying-in period.

Here the differences in the timber constructions between Hope-well, Seip, and Liberty Works become important. As pointed out ear-lier, at the Hopewell site, apparently all mortuary activity was per-formed at ground level. However, at Seip and Liberty Works, the same range of activities would have been divided between the upper and lower levels of the Great House Renewal Lodges. It is perfectly rea-sonable to assume, therefore, that at these latter locales much of the laying-in period ritual would have occurred by cycling the deceased from the upper level to the lower level and back up again, with the terminal rites occurring in the log crib crypts on the ground-level floor. According to the Laying-In Crypt Model, the crypts of the Renewal Lodges on the floors of Mound 25 and Mound 23 of the Hopewell site would have been used for all the laying-in period rituals and these would have occurred prior to the series of rites leading to the terminal non-crypt burial rites, many of the latter being full cremations. Since it is postulated that many deceased would have been at different stages in their individual laying-in periods when the ruptures in the process occurred, these would have been "frozen" into the complex variation that was noted in detail earlier. Thus, it could be expected that the ranksum profiles of Mound 25 at the Hopewell site would be different from those of Liberty Works and Seip, even though, as postulated under the Ecclesiastic-Communal Cult Model, all three were the locales of equivalent cults.

The Laying-In Crypt Model supports the Ecclesiastic-Communal Cult Model in another way, in this case by reinforcing support for the Split Age-Grade Mortuary Model. On the floor of Mound 25 there were 101 or 102 burials, in total. At least 75 of the deceased were extended crypt burials. The other 25–26 were non-crypt cremation burials. If the crypt burials were the result of the sudden truncation of the normal laying-in period, then the above figures suggest that only about one-quarter of the deceased cycled through the laying-in crypts would have become cremated floor burials—in the normal course of events. If this is the case, then this would suggest a 1:4 retention rate. That is, it could be expected that on the average out of every four deceased processed through the laying-in crypts only one would have been processed to the terminal post-mortem sacrificial cremation stage and deposited on the floor. Those who were not retained, of course, would have been buried in clan CBLs, as will be discussed later.

This is simply a ballpark and conservative figure for the Hopewell site since (1) it does not include the non-crypt extended and cremated burials under the lesser mounds; (2) it does not include the burials on the floor of Mound 23, which had 49 extended burials, of which only one was a cremation; and (3) it collapses the temporal structuring of the total site. However, it does support the temporal structuring that was argued earlier with regard to the Seip-Pricer and Seip-Conjoined Great House Renewal Lodges. At that time, differential retention rates of 1:6 and 1:4 were estimated for the senior and junior sectors, respectively, of an autonomous ecclesiastic-communal world renewal cult. The Laying-In Crypt Model gives some support for the notion of a retention rate, while displaying possibly a higher estimated mean level.

This may be as much as can be done to ground the Laying-In Crypt Model by using the mortuary data of only the Hopewell site. However, in terms of the hermeneutic spiral, one way of enhancing the relevance of a model is to take a body of data that was not part of its original construction and apply the basic premises. The Turner site serves this end. It is a major Hopewell embankment earthwork on the Little Miami River in Hamilton County, Ohio (fig. 2.11). This site is a good candidate for applying the insights gained from analyzing the Hopewell site because of several similarities: (1) the majority of the burials at both sites consisted of extended inhumations; and (2) both sites display similar major burnt offering deposits.

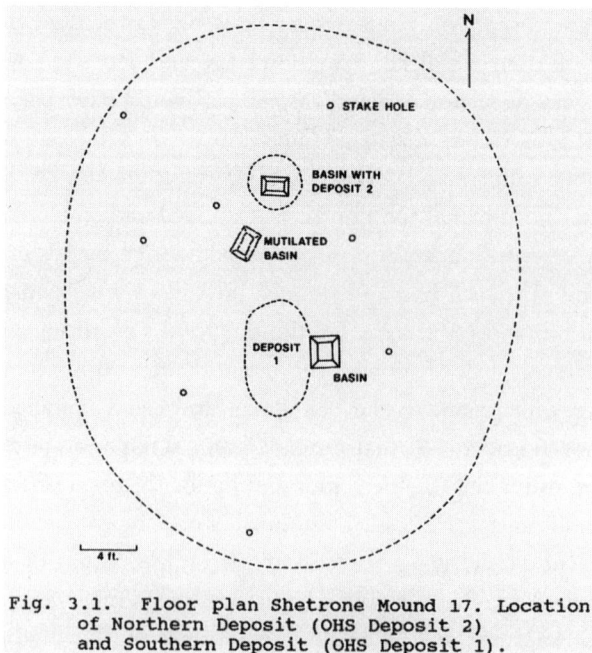

Fig. 3.1.   Floor plan Shetrone Mound 17. Location
of Northern Deposit (OHS Deposit 2)
and Southern Deposit (OHS Deposit 1).

FIG. 17.1. Shetrone Mound 17 Floor (Greber and Ruhl 1989, 82, figure 3.1)

Table 17.1 Summary Overview of Deposits 1 and 2,
Shetrone Mound 17

Deposit 1	Deposit 2
Male	Female
Copper, Silver, and Meteoric Iron	Mica
Platform Pipes	Bladelets
Bifaces	Cones
Engraved Bones	Bone Tools
Stone Celts	Bear Claws
Bird Effigies	Stone Bars
Stone Tablets	Plummets and Gorgets

Chapter 18

THE LAYING-IN CRYPT AND
BURIAL ALTARS OF TURNER

The Laying-In Crypt Model of the Hopewell site addressed an issue that was pinpointed by Shetrone, namely, "Cremation of the dead preponderated at Mound City, Tremper, Harness and Seip, while the reverse is true at Hopewell and Turner".[1] This model has given an initial account of this inversion. Given these mortuary parallels between Hopewell and Turner, applying the Laying-In Crypt Model to the mortuary data of the Turner site seems reasonable. However, this immediately highlights a radical contrast between the mortuary programs of these two sites, this being the absence at the Turner site of the vaulted chamber. Since both sites are characterized by the preponderance of extended inhumations and the Laying-In Crypt Model accounts for the complex mortuary data of the Hopewell site in terms of the funerary→mourning→spirit-release→world renewal process, it is possible that a symbolic pragmatic equivalent of the Hopewell vaulted chamber type of laying-in crypt existed at Turner (fig. 2.11). It is postulated here that the feature referred to in the literature as the stone-lined pit "grave" served the same purpose as the vaulted chambers at the Hopewell site, and may even have served additional purposes.

Willoughby's cross-sectional sample of these mortuary features is illustrated in fig. 18.1. He gives a generalized description of them.

The interments of the uncremated bodies were usually in comparatively shallow graves, many of which were carefully constructed and outlined with flat limestones set upright. In some instances, a wall was laid at the head and foot. . . . The bottoms of a few of the more carefully made graves were paved with flat stones, and upon this paving the body was extended, usually upon its back. Comparatively few artifacts were found with the skeletons. . . . The graves were usually filled with earth, and covered with one or more layers of limestones, which in some instances extended beyond the limits of the grave. Other graves were filled with stones. Still others were without the side and end stones. These, however, were usually, not always, partially or wholly covered with a layer of stones near the surface. (Willoughby and Hooton 1922, 14–15)

By drawing on the Mourning/World Renewal Mortuary Model, I postulate that, rather than being graves mediating terminal funerary rites, these were mortuary crypts that did double service. Initially, each may have been used as a laying-in crypt, probably cycling several deceased through their laying-in periods. Then, the usage of the stone pit feature would have been changed to serve as a mortuary depository facility mediating terminal world renewal post-mortem sacrificial rites, as indicated by either an extended inhumation and/or one or more cremations. In effect, it was used both as a laying-in crypt altar and as a world renewal altar. I will call this the Laying-In/World Renewal Crypt Model of the Turner site. It follows from this that another of the major differences between the Turner site and the Hopewell site, as well as the other sites of the Chillicothe region, is that while in the latter most (not all) of the mortuary remains were found on the mound floors, at Turner most (not all) were found in these pit features, and these were physically separate from the known mounds. These features made up the major, although not the exclusive, collective burial locales (CBL) of the site, often referred to as "cemeteries" or, less misleadingly, burial places.

The Laying-In/World Renewal Crypt Model postulates a variation in the material warranting of the same basic mortuary process that was examined at the Hopewell site under the Laying-In Crypt

Model. Demonstrating this claim is going to require considerable argumentation and empirical grounding since it is clear from Willoughby's description, and from more recent work, that the funerary view figures centrally in interpretations of this site.[2] However, Willoughby recognized that more than funerary practices were probably involved, since he pointed out that a number of the pit features contained no burials. Nevertheless, he continued to use the classification system initiated by Putnam, who referred to almost all of the features that he and his team excavated as graves.[3]

The labeling system that Putnam developed combined numbers and letters according to the sequence in which the pit features were excavated. This resulted in four separate recorded sets (fig. 18.2), making up basically two nonmound burial places. Putnam and Metz excavated the largest number of these features and labeled them numerically as Grave 1 through Grave 32 with no letter added. Some "cache pits" are also included in this group. The second set was referred to as the "a" sector, excavated by M. H. Saville, and these are labeled Grave 1-a through Grave 9-a, also with several "cache pits" being included. The third is the "b" set, labeled from Grave 5-b through Grave 12-b. These three sets make up the major part of the "Turner Burial Place within the Great Enclosure" (fig. 18.2), and form the Western CBL "inside" the main embankment earthwork, just south of the northwestern sacred circle. The second main extra-mound CBL is immediately northeast of Mound 14. It was excavated by Ernest Volk and forms the "c" sector, labeled Grave 1-c through Grave 4-c. These features had one extended burial, several cremations, and various human parts, including an indeterminate but not large number of fragmented infants' and children's bones, making a total difficult to estimate.

Of the estimated 101 mortuary deposits of this site most were found in the "cemeteries" or Burial Place areas while the minority were found in the mounds. Greber estimates that the first three sets contained 55 individual burials, 37 extended and 18 cremations.[4] The other mortuary deposits were found in the different mounds, the majority being extended burials. Together, Mound 1 and Mound 12 had about

17 deposits, and Mounds 2 and 11 contained 4. Mound 1, near the
northeast entrance to the Great Enclosure component (fig. 18.2), had
6 mortuary deposits involving 9 or 10 deceased, and Mound 12, the
larger of the two mounds in the upper circular portion of the earthwork, had 9 mortuary deposits, some extended and some flexed burials,
as well as several human skeletal parts separately distributed on its
periphery (e.g., pelvises and femurs).[5]

Of the forty-eight "graves" making up the primary part of the
"Turner Burial Place within the Great Enclosure," 9 had either poor
indication, or, in one or two cases, no indication of human remains.
For example, Grave 26 (fig. 18.2) had no human remains. It was in
every other respect similar to a number of those that did contain human remains.

> An irregular area covered with flat stones, 15½ feet in length, and 5 feet
> in width at its widest point. Beneath the stones a pit 9 feet long and 4
> feet wide had been dug to a depth of 3 feet. This contained black soil,
> sand, and gravel. On the bottom near its center lay a piece of mica; near
> its northern end a post-hole, 10 inches in diameter and 19 inches deep,
> had been sunk beneath the level of the floor. No human remains were
> found. (Willoughby and Hooton 1922, 20)

Also, in some cases, the unlabeled "cache pits" noted above displayed
the same attributes as a number of the features referred to as "graves,"
the only difference being that most of the latter had human remains
associated with them and the former did not. To confuse matters more,
a few feet southeast of Grave 2-a there is a feature labeled as a "cache
pit" that apparently did contain human remains. Its stone topping pavement, about 6 inches below the earth surface, was 4 feet long and 2
feet wide. At its western end, "10 inches from the surface, was a mass
of burnt bone, presumably human, in small fragments." It is puzzling
why this was not termed a "grave" instead of a "cache pit," particularly
since Willoughby's general description of a cache pit as being "usually
4 to 6 feet deep, and 30 to 40 inches in diameter . . . [containing] charcoal, ashes, burnt clay, potsherds, and animal bones" (23), would have
excluded this feature.

The weight of the funerary view is not eliminated simply by acknowledging that some of the pit features labeled "graves" did not contain burials. It continues to weigh heavily today, although Greber has reiterated her position that there "is no logical necessity for assigning all evidence of rituals with these [classic Ohio Hopewell] mounds to burial rites. I would extend this observation to include those mounds which do have burials."[6] Of course, she is not implying here that the Turner "grave" burials, per se, are anything other than funerary in nature. Rather, as pointed out earlier, she is suggesting that those features and artifacts having no direct association with human remains but found within the site could be the result of rites of a nonmortuary nature.

> There are many individual life crises or social functions which are associated with ritual. The societies responsible for the many mounds and earthworks leave material remains which give indications of a society rich in rituals. The view of assigning all such rituals to burial rites (since burials are frequently associated with ritual areas) may be the result of the particular culture of the viewer rather than a dichotomy recognized by the living societies. (Greber 1976, 141)

Hence, the funerary view has been effective in entrenching the assumption that the pit features having mortuary remains were, for the most part, single-use "graves," and, as an extension, that all or many of the artifacts found in association with human remains were the result of deliberate funerary burials.[7] However, when the examination of these pit features and their contents is done within the theoretical framework of the Mourning/World Renewal Mortuary Model, it becomes clear that the total mortuary patterning of the Turner site is so varied that skepticism is appropriate toward the view that the pit features containing human remains are adequately characterized as graves. This variation will be examined in some detail below under the Laying-In/World Renewal Crypt Model.

DEMONSTRATING THE MODEL

The method and expectations of the Laying-In Crypt Model of the Hopewell site can be used to initiate the demonstration of the

Laying-In/World Renewal Crypt Model. The former hinged on comparing the three artifact deposit contexts, the Crypt Burial Context, the Non-Crypt Burial Context, and the Burnt Deposit Context. A similar or equivalent patterning should be discernible at Turner. In addition, however, there is the critical claim of the Laying-In/World Renewal Crypt Model that the pit features were used for two separate parts of the total funerary→mourning→spirit-release→world renewal mortuary process: the laying-in period, or parts of it; and the terminal burial rites, either extended or cremated. This suggests that there was an in-between part of the process that may have been performed elsewhere. In this regard, it can be expected that activities on the floors of the major mounds would have played an intrinsic part in this process. Therefore, the floors must be examined for evidence to link the pit features and their contents to the floor activities. Critical evidence would be burnt deposits similar to those at the Hopewell site, indicating that a similar process of collective memorial/world renewal ritual occurred at Turner.

Furthermore, it can be expected that, although rare, evidence should be forthcoming of misplaced laying-in period materials in the pit features as a result of the mortuary cycling process, much in the manner discussed earlier for the Hopewell laying-in crypt. The best such evidence would be finding artifacts of the laying-in warrant or formal mourning gift type and demonstrating that these were misplaced, suggesting that the deceased were typically removed at the end of the laying-in period and then returned as part of the terminal postmortem sacrificial world renewal rite. Other indicators of a cycling process would be signs of cumulated burials in the same pit feature, and additional construction or modification of pit features. Finally, since these features are postulated as altars, the Laying-In/World Renewal Crypt Model of the Turner site predicts that the pit features should display properties attributable to iconic cosmic congruency. These are the primary expectations that this model can generate.

To prevent confusion, in the following summary description and analysis of the stone-lined pit features and their contents, the terms given in the original report for the pit features, Grave 1, Grave 2-b, and so on, will continue to be used, keeping in mind, however, that they enumerate features that, as postulated here, would not be appropriately characterized as tombs or graves.

1. Cosmological Congruency

It was suggested above that the pit features could be expected to display patterning that can be attributed to iconic cosmological congruency rules. In fact, there are two lines of evidence that support this. First, Greber has argued that there are two clusters of pit features in the Western CBL based on contrasting North-South and East-West axial orientations.[8] These were not perfectly oriented to the cardinal directions. Rather, for the East-West orientation, she allowed a range falling within the horizon turning points of the summer and winter solstices and she allowed for a lesser range with regard to the North-South orientation. With one exception, all the rectangular pit features in the Western CBL fall within one of these two orientations. She uses this duality to suggest that there were two social components manifested at Turner. She applied her ranksum methodology to these two but was not able to discern any statistically significant differences between them. However, in regard to the alignment data, her claim supports the view that iconic congruency rules are being manifested. Furthermore, as she mentioned, a dual component structure may be indicated here. Certainly, this could be used to support the Ecclesiastic-Communal Cult Model, for instance, one set of crypts being particular to the senior age-grade sector and the other to the junior age-grade sector.

Second, the manner of constructing and filling these pits is further evidence for iconic congruency. The use of large flat stones to line the walls, and, in a number of cases, to lay down the floors and to cover these pits is particularly interesting since, in fact, the use of this stone seems superfluous to funerary burial purposes. These pits were dug into naturally packed gravel. It seems implausible to claim that flat stones were carefully placed edge up for some "practical" purpose, such as to prevent erosion, even if they were used as crypts for extended public ritual. There are also a number of pit features, both burial and "cache pits," without the stone lining, suggesting that, in fact, the use of stone was of symbolic pragmatic relevance. This suggestion is reinforced when the use of stone at this site is more fully described. The structures on the floors of the Turner mounds were not particularly impressive in terms of overall size, nor were the mounds themselves, for that matter. The careful and prolific use of the same flat limestones, however, all of which were procured with considerable labor from the Little Miami River, several thousand feet away, suggests that iconic congruency rules are being manifested in both cases.

2. The Contents and Patterning of the Pit Features

One of the above expectations was that there should be some evidence of reuse of the pit features, resulting from a complex history of usage, first as laying-in crypts and then as terminal world renewal crypt altars. Supporting evidence would be (1) "extraneous" or "stray" artifacts, suggesting that they had been lost or dropped with the removal and/or displacement of the deceased during the mortuary cycle; (2) cumulated but separate burials; and (3) evidence of structural modification to the pits, allowing for changes during their usage. This also can be extended to the "cache pits."

One of the most illuminating of the pit features in support of reuse is Grave 5 (fig. 18.2). Greber treats the extended adult female burial as the second-highest-ranking deceased in the Western CBL, basing her estimation on the relative number and arrangement of the esteem-endowing artifacts found within the pit.

> There were numerous artifacts . . . deposited in a rather elaborate fashion in the grave, at the corners and under the four large flat stones which nearly covered the bottom of the grave. Some of the items included were copper ear spools, a copper celt, a Busycon shell, and "four bear's teeth with pearls." (Greber 1976, 125)[9]

Willoughby's summary description can be used to clarify the nature of the artifact distribution that Greber characterized as being "deposited in a rather elaborate fashion." Because of the importance of Grave 5, his description is quoted almost in full.

> This was 6½ feet long, 33 inches wide, and 20 inches deep, and was outlined with the usual upright stones. Four large flat stones, upon which the body had been placed, nearly covered the bottom of the grave. The body was extended, head to the east. With the bones of each hand lay a spool-shaped ear-ornament of copper. At the side of the grave near the left shoulder were two pairs of similar ornaments, and a bone point. At its foot near the southern corner lay a conical stone object, bear teeth ornaments, and a piece of galena. Two of the teeth were inlaid with pearls. Under the large floor-stone near the foot of the grave were several flaked flint knives, and a small copper blade ... Beneath the second floor-stone lay the two discs of a spool-shaped ear-ornament, and a portion of a copper-covered bead. Under the third floor-stone were a copper bead, and a band of thin copper. (Willoughby and Hooten 1922, 16–17)[10]

This description can be used to reconstruct the steps that most probably occurred to bring about the arrangement of the contents. First, the materials found under the four large floor-stones would have been deposited before these stones were laid. Rather than characterizing the distribution of these as the result of being "deposited in a rather elaborate fashion," in Greber's terms, they can be more adequately described as being scattered or even randomly dispersed. Such a pattern would be more likely the result of being misplaced or unwittingly dropped rather than being deliberately placed. Furthermore, apparently three of the items were broken, a copper-covered bead and two separate copper ear spool discs. None of these objects is unique, as such, and, in fact, one (unbroken) ear spool was placed in or beside each hand of the buried female, and there were two unbroken pairs placed near her left shoulder. Similar positioning of ear spools was noted at the Hopewell site. Thus, finding ear spools in the hands of deceased or by their heads is not unusual. Finding them broken under floor-stones is unusual. Also, while finding bladelets in these burials is usual, finding copper beads is unusual.

All this constitutes the need to establish criteria by which to assess the distribution of these and other artifactual materials with the Turner burials. This can be done by summarizing their placement in other burials. Willoughby describes one bladelet in Grave 15 (fig. 18.2) as being beside the left shoulder of the deceased and, without giving a particular provenance, simply says that several others were found.[11] This suggests that they may have been caught up in the burial fill or possibly dropped when the tasks they were used for were finished. He also comments that this Grave 15 burial was associated with a copper band, not under the body but on its breastbone. Willoughby later comments that this was similar to the copper band found under "the third floor-stone" of Grave 5.

Grave 24 also had bladelets (fig. 18.1, feature f). Again, however, Willoughby does not specify exactly where they were found in this feature. He commented that the Grave 5-b burial had one near its right foot. Along with the reported finding of "several flaked flint knives" under the floor-stone near the foot of Grave 5, the placement of bladelets suggests a rather casual pattern, as would be expected under the Dual Ritual Spheres Model that claims the Hopewell bladelets would typically be tossed or dropped once they had served their immediate ritual purpose. However, there were several cases where a more deliberate placement is suggested. For example, the extended burial in Grave 1-a had several bladelets near its left shoulder and a copper band (3 inches by 2 inches) also on its upper breast "similar to the bent copper bands from graves 5 and 15, although somewhat larger" (21). The deceased in Grave 25 had several bladelets that "rested on the right scapula, which was much decayed" (19). The clearest expression of deliberate placement of bladelets is in Grave 1-c in the Eastern CBL. There were 15 bladelets in a small cluster near the left shoulder of the skeleton. Willoughby suggests that these had been deposited in a bag. With regard to bladelets, then, the evidence suggests that, while there were a few cases of apparently deliberate association, the typical method was to be dropped, tossed to one side, or casually left near the body once they had been used.

The distribution of galena has a similarly mixed patterning of apparently deliberate and unwitting placement. Grave 5, of course, had a single galena piece in the southwest corner. This would place it near the feet of the burial. There were galena pieces found at the feet of the two extended burials in Grave 5-a. However, this was a complex burial involving several deceased. The two "main" burials were extended side by side and "between the lower limbs" there was "a considerable quantity of human remains, two pieces of galena, and a thick piece of mica, which had also been subjected to the action of fire" (22). The wording is ambiguous but it will be assumed that all these items, and possibly the "human remains," were subject "to the action of fire." Between the two extended burials "were the remains of a child, a shell vessel, and several pieces of copper" (22). Also, Willoughby states that a Busycon shell was placed at the feet of each extended burial, implying that these may have been deliberately placed. His description gives the distinct impression that the two extended burials were focal participants and the rest of the materials, both "human remains" and artifacts, were accompanying them. The alternative possible interpretation is that the "considerable quantity of human remains," as well as the two pieces of galena and the mica, were either in place when the extended deceased were added, or were added after and independently of these deceased. In contrast to this apparent nonintentional or, at most, secondarily intentional form of galena placement, Grave 6-b burial had a galena piece in its right hand and two small galena pieces in its left hand. This grave had another pit under it, again suggesting that a series of events had occurred prior to its deposition.

In sum, the normal or deliberate ritual placement of galena, copper strips, bladelets, and copper ear spools would be in the hands (galena, ear spools), on the breast (copper strips), or near the shoulders or head (ear spools, bladelets). In contrast, while the extended deceased in Grave 5 had ear spools in her hands and clustered near her head (and a bone point), the conical stone, bear teeth inlaid with pearls, and galena were near her feet in the corner of the pit, which would not appear to be the typical ritual placement, suggesting that these were not deliberately placed. In the same terms, the partial copper bead, the copper band, bladelets, copper celt, and two discs of one or two broken ear spools under the floor-stones indicate that these were not the result of having been placed in order to be in association with the deceased lying on top of these floor-stones. Rather, these were probably unwittingly dropped or displaced during an earlier, pre-floor-stone ritual period. This is particularly indicated by the copper band and copper blade. These would appear to be items of custodial regalia. The probability of their being misplaced when the deceased was being radically maneuvered (e.g., being placed into or removed from the pits), would be very low, unless the circumstances, such as poor lighting conditions (e.g., night-time) made it more likely. Under such circumstances, and others could be imagined, small items could easily go astray in placing or removing the deceased. All this suggests the following five-stage history of Grave 5 pit feature usage.

Stage 1. The pit was constructed.

Stage 2. At least one but more likely several sequential laying-in periods occurred with the individual deceased being wrapped in laying-in shrouds and placed directly on the pit floor along with associated mortuary artifacts.

Stage 3. After an indeterminate period of usage, the floor-stones were put in place.

Stage 4. Again, at least one but more likely several sequential laying-in periods occurred with the individual deceased being wrapped in laying-in shrouds and placed on the floor-stones installed during Stage 3, along with associated mortuary artifacts.

Stage 5. The final use of the feature was the Grave 5 female burial. The pit was filled with earth and its stone capping placed.

In support of Stage 2, the copper band, which normally would have been on the breast of the deceased (e.g., Grave 15, Grave 1-a), was found "under the third floor-stone" (17), probably being displaced in the process of removing the deceased associated with it. Also under the floor-stones were found two discs of one or two broken copper ear spools, a copper bead, and a broken copper-covered button. These latter two items would probably have been part of one or more laying-in shrouds similar to those postulated for the Hopewell site crypt burials. In removing the body (or bodies), these various items were unwittingly misplaced and left behind.

In support of Stage 4, the conical stone, the bear's teeth, and the galena piece found in the south corners were, again, probably the result of the removal of one or more deceased at the end of the laying-in period(s). The conical stone is consistent with a laying-in garment, and the bear's teeth are consistent with either a garment (Pattern B) or a necklace (Pattern A). The galena piece would normally be in the hand or beside the hand of the associated deceased. This accumulation of disparate elements suggests a considerable period of usage before the placement of the terminal burial. If this is accepted, then Stage 5, the terminal burial, marks the transformation of the pit from being a laying-in crypt altar having a fairly complex history to being a terminal world renewal crypt altar. This means that the placement of the extended female deceased would have been the final step of the cycle starting with her earlier laying-in period, not necessarily in the same feature, her post-laying-in period treatment elsewhere, and then her final placement in Grave 5, constituting her as a post-mortem sacrificial offering of a world renewal rite that was terminated by the ritual filling of the pit.[12]

If this interpretation is close to being correct, then there should be other examples of multiple usages of these pit features and extensive post-mortem manipulation of the deceased. Indeed, there is considerable evidence of both. Grave 6-b displays stratification similar to that of Grave 5 pit feature. This was a typical rectangular, stone-lined pit containing an extended burial. Below the "foot-stone" of the extended burial another major pit was found, about 1 m in diameter and about 3 m deep. It had been carefully filled with burnt clay and charcoal with animal bones and flaked stones, as well as small pockets of stones, about 5–6 in each. It is clear that Grave 6-b was first built, probably used for ritual, then the second pit was dug into the floor and used, and then filled, returning Grave 6-b to its "original" usage as a pit feature; a stone floor was then laid; and the pit with this added stone floor was finally used for terminal burial ritual.

Grave 15, with two extended burials, suggests at least two separate, largely independent usages. It is possible that the "considerable quantity of human remains," all apparently subjected to prior burning, were either the residue of ritual prior to the two extended burials being placed or else additions made after their placement. In either case, a sequence of rituals was carried out in this feature prior to its being finally closed. Another example is Grave 26. This had a flat-stone paving over 15½ feet long by 5 feet at its widest. While it was irregular in shape, it covered a regular pit that was 9 feet long by 4 feet wide and 3 feet deep, filled with dark earth, sand, and gravel, probably stratified. A piece of mica was on the base, about center, and there was a post hole at the northern end, 10 inches in diameter by 19 inches deep. "No human remains were found."[13] This suggests that the pit was modified after initial construction by the addition of a standing post and then it was reused.

Grave 27 (fig. 18.1, feature h) was also a large pit with a regular rectangular flat-stone paved top, about 12 feet long by 6 feet wide, and it contained two large pits side by side. Pit 1 was 4 feet wide by 4 feet deep, and filled with dark earth having a few animal bones and stones (presumably) mixed in. Conjoined Pit 2 was filled with flat stones laid horizontally. This pit had two "cists" formed as the stones were laid.

> A little north of the centre . . . was a cist (a'), 18 inches long and 12 inches high, containing fragments of human bone; and against a portion of the temporal bone rested one-half of a copper ear-ornament. Shell beads were also found with the bones. At the southern end of the pit, and near its bottom, a second cist (b') was discovered, containing cremated human bones and the remains of a pair of copper ear-ornaments. There were no indications that the burning of the bodies had been carried on in the pit. The cremating had evidently taken place elsewhere, and the ashes and partially burnt bones placed in the tomb made to receive them. (Willoughby and Hooton 1922, 20)

Willoughby's claim that the "burnt bones [were] placed in the tomb made to receive them" can be questioned. Pit 2 was between 6 and 8 feet long, at least twice the size of Pit 1. If only cremations or bundled bone burials were intended for these pits, then they could have been considerably smaller. In fact, almost all of the cremated burials were placed in small stone-lined features. Therefore, Grave 27 displays a size and complexity that was surplus to what was normally associated with a cremated or bundle burial, suggesting that when it was initially constructed, it was not planned specifically to contain the two cists. For example, while Grave 28 was only slightly smaller than Grave 27, the former contained an extended adult burial. Thus the double-pit feature making up Grave 27 was probably designed for laying-in purposes and terminal extended burial. The two cist burials, then, probably demarcate the final phases of a complex history of ritual usage. This final phase was itself complex, involving the transformation of the large pit from being a major laying-in crypt to being a major world renewal crypt altar mediating at least two sequential events of post-mortem sacrificial ritual involving the deposition of deceased who had been subjected to radically prescriptive post-mortem manipulation.

The Laying-In/World Renewal Crypt Model would anticipate burials indicating extensive bodily exposure in a locale other than the pit in which the burials were found. In fact, this is indicated in several examples. Grave 27, above, of course, is one such example. Grave 1-a also displays the accumulation of burials in different states. As Willoughby (21) describes it, while the "principal burial" was an extended one, there were two other deposits of human remains. One was a disarticulated skeleton scattered across the lower part of the "principal burial." Its pelvis was broken into three pieces and the skull was lying near its feet with the jaw displaced 12 inches from the skull. The third burial was a cremation. Before the cremated remains were deposited, about a foot of earth was placed over the first two burials. Signs of burning on this earth just above the head of the "principal burial" were associated with the cremation. The latter was covered with flat stones, and, presumably, the rest of the pit was then filled with earth. Each of the three deceased was treated differently and, apparently, sequentially. The extended burial seems to have received the least degree of manipulation, and was also associated with the major artifacts. It was also the "base" on which the second skeleton was deposited and the state of deterioration of this deceased was sufficiently advanced to involve the breaking of the pelvis and the scattering of the skull and lower jaw. The disarray of this deposit might very well have been the result of these bones being tossed in as a loose bundle. In any case, it is clear that it had been subjected to incremental stages of manipulation outside this pit, this being a good indicator of post-mortem sacrifice. The deposition of the cremated remains as part of the topping up simply confirms that this pit feature had a complex ritual history that can probably be more adequately characterized in mourning/world renewal ritual terms than in funerary terms.

Along the same lines, the single deceased in Grave 12-b was treated to a staged series of manipulations, although it cannot be deduced from the data if some of these occurred elsewhere. The upper body was extended and associated with a *Busycon* ocean shell container near its skull. However, its skull was not in a natural position. Instead, it was twisted to the right and lying on the upper right humerus. The lower portions of the legs appear to have been deliberately broken off below the knees with the right tibia and foot attached, placed almost at right angles across the right femur while the broken lower bones of the left leg, with foot attached, "extended diagonally downward from near the centre of the left femur" (25). Hall has reported that deliberate bone modification, e.g., drilling, is known to be a form of spirit-release rite and that this rite is often associated with both reincarnation of the released spirit and world renewal.[14]

Grave 29 (fig. 18.1, feature a) was another case of a deceased subjected to incremental post-mortem manipulation prior to final deposition. In this case, the deceased was placed in a prone position, with badly shattered legs that were flexed. Willoughby reports that this breakage was caused by four large, flat stones that were placed immediately on the body. While the breakage can be attributable to these stones, it is also the case that the bones of the deceased must have been in a somewhat fragile state when these stones were placed on top. Combined with the prone position of the body, the broken bones suggest that it had been subjected to extensive exposure elsewhere prior to final deposition. It should be noted, in this regard, that the broken legs of the deceased in Grave 12-b, as described above, cannot be attributed to heavy flat-stones since these were on top of the pit, separated from the deceased by three feet of earth. Clearly, the breakage of the lower legs was a deliberate part of the deposition process.

Reuse of these pits for mediating post-mortem sacrificial rites is also clearly indicated by the mortuary patterning of the Eastern CBL. Grave 2-c was demarcated by a circular stone pavement about 5 feet in diameter. Below the stones, scattered skeletal and skull fragments of both infants and children were noted, none of which were cremated. Under the western sector was a shallow pit that was filled with cremated adult bones, a platform pipe, and a shell bead. As Volk commented, while adults were cremated, "None of the infant bones from this grave were burned" (27). Grave 3-c contained a few bones that may have represented at least two and possibly more individuals. Also a large standing stone dividing this "grave" covered fragments of infant and adult bones.

As remarked in chapter 14, lethal human sacrifice, although probably rare, when it did occur in a polyistic locale-centric social system, would probably entail consensus on the part of those closely related to the victims. For this reason, infants offered by parents or senior adult relatives might be the most likely candidates as lethal sacrificial offerings. The bones of infants and children might be considered in these terms. In support of this speculation: at Liberty Works, an extended six- to nine-month-old infant burial was found deposited in a rectangular pit (Feature 84, fig. 2.6, p. 20) dug into the Edwin Harness floor at the junction between the circular Section 3 and the rectangular Section 2.[15] It seems too convenient to find an infant deposit at such a critical location marking the "liminal entry" between the R'-C junction of the Great House. Therefore, lethal sacrifice may have been called for in order to constitute a foundation construction ritual. However, it is also possible, of course, that these infants and children had died naturally and, therefore, were "conveniently" seconded as post-mortem sacrificial offerings.

SUMMARY

The indicators for the patterning of the stone-lined pit features of the Western and Eastern CBLs in support of the Laying-In/World Renewal Crypt Model have been given. Probable iconic cosmological congruency has been identified in the form of the selected axial orientation of the pit features, and the use of flat stones derived from the river to build and cover them. Even more compelling was the finding of indicators of reuse of these pit features in the form of (1) "displaced" valued raw materials and artifacts; (2) structural modifications of and additions to the pit features, suggesting a change during their history of ritual usage; (3) numerous indications of reuse of the pit features; and (4) multiple indications of incremental post-mortem manipulation, such as the cumulative deposition of deceased in different states of deterioration, including extended burial with deliberate bone breakage, and the sequential addition of extended and cremated burials. These are most reasonably accounted for in terms postulated by the model, namely, that the pit features did double duty as crypt altars serving laying-in period mourning/world renewal ritual, and as world renewal crypt altars mediating post-laying-in period terminal post-mortem sacrificial ritual. The next step in grounding this model, then, is to establish that the mortuary patterning of the pit features and the materials and features found on mound floors are systematically linked as the material media and outcome of the funerary→mourning→spirit-release→world renewal ritual process of this site.

PEABODY MUSEUM PAPERS   VOL. VIII, No. 3, PLATE 4

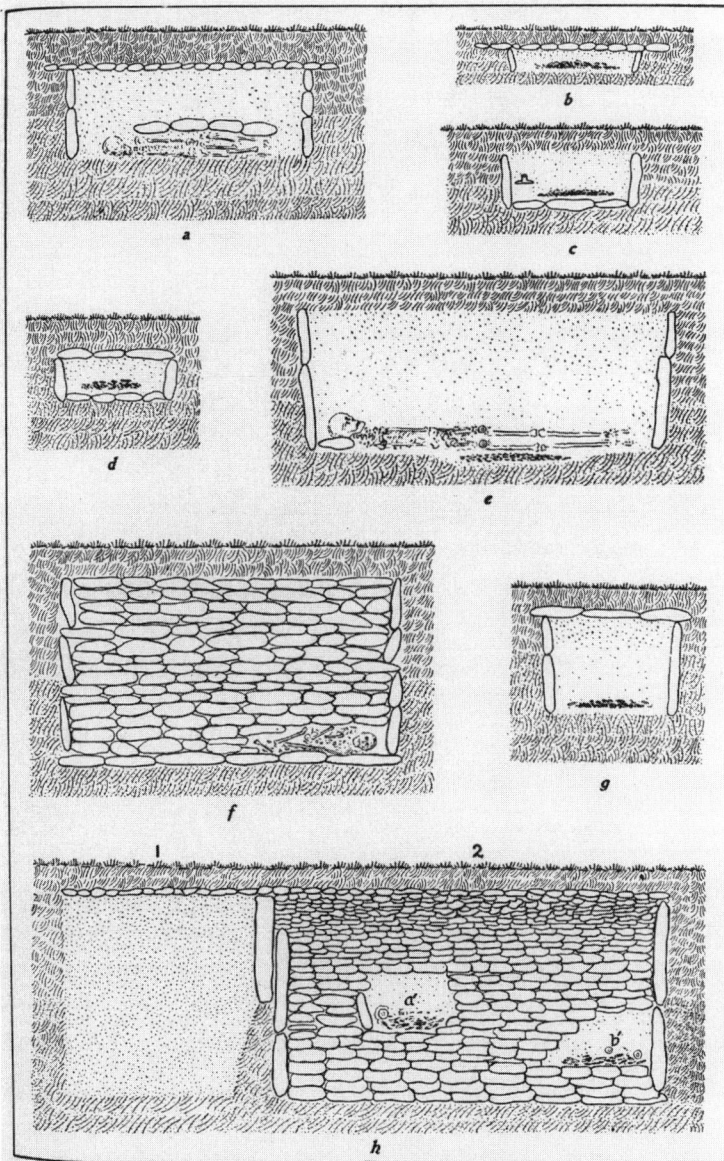

Cross-sections of graves containing skeletons or cremated human remains, Burial Place in the Great Enclosure: a, Grave 29; b, Grave 32; c, Grave 7; d, Grave 2; e, Grave 1; f, Grave 24; g, Grave 21; h, Grave 27.

FIG. 18.1. The Turner Stone-Lined Pits (Willoughby and Hooton 1922, plate 4)

PEABODY MUSEUM PAPERS
VOL. VIII, No. 3, PLATE 3

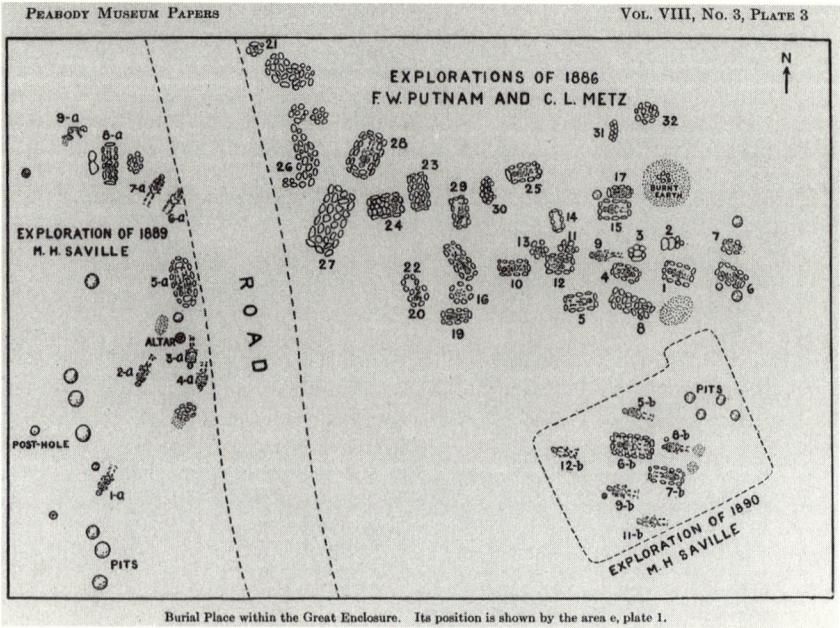

EXPLORATIONS OF 1886
F.W. PUTNAM AND C. L. METZ

EXPLORATION OF 1889
M. H. SAVILLE

ROAD

ALTAR

POST-HOLE

PITS

EXPLORATION OF 1890
M.H SAVILLE

PITS

Burial Place within the Great Enclosure.   Its position is shown by the area e, plate 1.

FIG. 18.2. The Turner Burial Place in the Great Enclosure (Willoughby and Hooton 1922, plate 3)

Chapter 19

THE CONTROLLED FIRE REDUCTION FEATURES (CFRS) OF TURNER

The second line of evidence in support of the Laying-In/World Renewal Crypt Model of the Turner site is based on the contents of the major mounds of this site, sometimes referred to as altar mounds. Mound 1 and the Conjoined Mound will be the focus. What is named here as the Conjoined Mound was numbered by Willoughby and Hooton as several individual mounds, Nos. 3, 4, 5, 6, 7, 9, and 14 (fig. 19.1). Treating them as a conjoined set of mounds might be more appropriate since, as will be argued, this earthwork was probably the result of the simultaneous abandonment of the conjoined floors with associated structures found at its base. In keeping with the earlier terminology, the total set of structures represented by the floors of this conjoined mound will be termed the Renewal Lodge Complex. Of these seven mound components, the focus will be on Mound 3 and Mound 4 and the major features on their floors (figs. 19.2 and 19.3), in particular, first, the set of compound pits that is unique to Turner (fig. 19.4), and then the two burnt offering altar deposits. To support the Laying-In/ World Renewal Crypt Model, it is necessary to link the Western and Eastern CBLs to the compound pits and the burnt offering deposits. The burnt offering deposits should fit what would be expected of the residue of collective memorial/world renewal rites of the type already demonstrated for the Hopewell site. The contents then would be primarily the remains of laying-in warrants and mourning gifts, both personalty-mourning gift and formal mourning gift types. Completing the grounding of the model will prepare the basis for arguing the nature of the relationship that linked the Turner and Hopewell sites.

THE CONJOINED MOUND

The Conjoined Mound is found in the southeast sector of the Great Enclosure (fig. 2.11). Its depiction in figure 19.1 is somewhat idealized. According to Greber, the relative positioning of Mounds 3 and 4 is slightly skewed and not 90 degrees as shown here.[1] However, the axis is oriented in a slightly east of north direction. Mound 3 is the southernmost and Mound 14 is the northernmost component. According to the plan design, if a line were drawn through the approximate centers of Mound 3 and Mound 14, it would roughly divide this complex into mirror-image halves. The component floors covered by these mounds are defined by low, flat, conjoined stone walls, reinforcing the notion that these floors were part of a single, complex structure. Each floor had one or more defined burning areas: stone hearths, burnt clay areas, altars showing signs of burning, and/or linked pits, tunnels, and flues displaying marks of sometimes intensive fire.

The floor of Mound 3 (fig. 19.2) is about 34 m in diameter (100 feet). It is one of the two largest of these conjoined floors and it displays the most complex set of ritual burning features. The oblong floor of Mound 4 (fig. 19.3), about 36 m by 21 m (108 feet by 60 feet), is second only to the floor of Mound 3 in complexity of activity areas. Before focusing on these two floors, the features on the other floors will be briefly summarized in order to demonstrate that the core symbolic pragmatic meanings of all these floors were probably very similar. The floor of Mound 5 (fig. 19.1) had four rectangular altars arranged in a vertical sequence, indicating that the floor was added to during its use-history. The floor of Mound 6 had a circular altar and three rectangular baked clay hearths, again at different levels. The floor of Mound 7 might be considered the "keystone" in that it was the central link of the floors covered by Mounds 3 and 4 on its southerly and northerly sides, respectively, and by Mounds 5 and 6 on its southeasterly and southwesterly sides, respectively. A circular depression bearing signs of burning was found in the center of the floor of Mound 7. West of its center there was a circular basin and east of it, a rectangular altar. This was built on top of another, again indicating a history of ritual additions.

The floor of Mound 9, immediately joined to the northern side of the floor of Mound 4, had a large burnt area in its center covered by a layer of black ashes. There was a western altar, and east of the center, at a lower level, there was a circular basin displaying signs of heavy burning.[2] The floor of Mound 14 was approximately 30 feet in diameter, about equal to the floor of Mound 7. Near its center was a small rectangular altar. These basins, hearths, and altars were devoid of any artifacts. However, the complementary circular and rectangular altars usually had their basins deliberately filled with sand, ashes, and/ or earth, and then were covered by the initial layer of the stratified materials making up the low mound that covered each floor. This initial layer, about 2 to 4 inches thick, was often a mixture of sand and course gravel that was "cemented into a mass. . . . The principal cementing material is probably lime, produced by the accidental reduction of limestones in the great fire. The lime thus produced, mixed with the wood ashes, probably formed the cementing ingredient."[3] This mixture of sand and gravel seems to be the most common form of initiating the closing of a floor and the construction of the mounding.

The mounds at Turner were typically built in a series of contrasting layers: earths of different types, clay, mixed clay and earth, sand, ashes, gravels, and so on. Sometimes a total layer was made up of different components. For example, a layer of clay would cover a large fraction of the mound surface, and then another material would be spread over the rest of the surface, as well as being partly spread on top of the clay, and so on. Typically the termination of the mound construction was marked by either a clay layer or a carefully laid stratum of flat stones or pebbles locally available.

THE CONTROLLED FIRE REDUCTION (CFR) FEATURE

The floors of Mound 3 and Mound 4 stand as complementary with the Mound 3 floor being a circle and the Mound 4 floor being an oblong. The main axis of the floor of Mound 4 is at 90 degrees to the main east-of-north axis of the Conjoined Mound. In terms of specialized areas for practices mediated by fire reduction processes, the floor of Mound 3 is the most complicated known locale in Ohio Hopewell. Its complexity relates to both the number and the variety of fire reduction features. There are at least four formal feature types: altar, pit, burnt clay "hearth," and a set of connected pit-tunnel-flue features. These latter will be referred to as Controlled Fire Reduction features (CFRs).

The CFRs are almost exclusive to the floor of Mound 3 (indeed, Turner is the only known Ohio Hopewell site that has CFRs of this type). They appear to be designed and built in order to control the conditions of firing, both in terms of intensity of heat and the control of the flow of oxygen and combustion gases (fig. 19.4). Willoughby indicates that there were 30 in total (31 if a single pit without a tunnel flue but with vertical chimney flues is included; fig. 19.2).[4] As a unit the CFR is a complex made up of covered pits connected by a long connecting tunnel flue and one or more chimney pit flues. It required both digging into the natural matrix of the floor for the pits, and adding onto it for the tunnel flues and chimney flues, using tree logs and branches as molds.

CFRs come in two basic types, the simple CFR (N=25) and the complex CFR (N=5). The simple CFR (fig. 19.4, feature b) was made up of a single circular pit located next to the stone wall encircling the floor, about 2½–3 feet deep and 1–1½ feet in diameter. It had an attached tunnel flue, about 1 foot in diameter and 7–8+ feet long, starting from or near the top of the pit and angled downward and toward the center region of the floor, terminating in a small, subsurface pit with a basin-like bottom. From the top of this small pit there was usually one or two short vertical chimney flues, about 2 feet long and 2–4 inches in diameter, that opened at the surface of the mound floor. In some cases, a small clay dish, flat or conical, was found placed upside down over the chimney flue exit and sometimes a clay dome or dish was also placed over the top of the large pit. It is speculated here that these were the means of producing the heat and they worked by controlling the flow of gases during firing.

In contrast to the simple CFR feature type, the complex CFR feature had two large pits, and one had three pits. In the former case, one pit was on the side closest to the perimeter wall and one at the inner end of the tunnel flue leading toward the center of the floor. As in the simple CFR case, the tunnel flue was also angled downward to the second pit. This second pit was the larger of the two, 4–4½ feet deep and 1–1½ feet in diameter (fig. 19.4). It was not uncommon to find clay cone dishes or plates covering the openings of the large pits at the end of the tunnel furthest from the perimeter wall encircling the floor. The three-pit complex CFR, referred to here as CFR 6 (fig. 19.4, feature c), had its second large pit, in this case, the first inner pit, 6 feet deep and 14 inches in diameter, and its third pit, closest to the floor center, 9 feet deep and 4 feet in diameter at its widest.

This set of simple and complex CFRs was distributed across the floor in a large subrectangular layout. Willoughby, quoting from the notes of Dr. Metz, Putnam's field archaeologist, indicates that "The space between the encircling stone wall and the edge of the hearth was filled with black ashes and loose soil, to a depth of 2 feet."[5] However, as can be seen in figure 19.2, there is no indication of this loose fill and the pits appear to have been dug into the natural base of the floor level. This consisted of a thick layer of dark clay under the surface stratum and a bed of gravel under the clay stratum. This is confirmed by the claim that some of the pits were sufficiently deep to cut completely through the upper clay stratum and into the under stratum of gravel (43). However, a loose gravel fill is indicated between the natural dark clay stratum of the floor into which the pits were dug and the encircling stone wall, this being apparently part of the construction of the wall. Clarification of precisely what kind of matrix surrounded the perimeter pits, "black ashes and loose soil" or the natural dark clay and gravel substratum, is not resolved by the profile cross-section of Mound 3 (fig. 19.5). Obviously more research will be required. This slight ambiguity does not affect the rest of this description and analysis.

The total set of thirty CFRs can be divided into two almost-equal subsets by a wide layer of black ashes that was covered by a large semicircle of heavily burnt earth about 3 to 4 inches deep, referred to by Willoughby as the "Hearth of burnt clay" (fig. 19.2). The ashes and the superimposed "Hearth of burnt clay" covered CFR 1 through CFR 16, constituting these as a set demarcating a discrete zone. Their linear distribution approximated a J-Form. Therefore, this set will be referred to as the Southern CFR J-Form. The rest of the CFR features, which are effectively numbered from CFR 17 through CFR 30, were not covered by the "Hearth of burnt clay." Their linear distribution also approximates a J-Form, forming a second and complementary set that will be termed the Northern CFR J-Form.

The relative spatial and linear distribution of simple and complex CFRs in each J-Form was different. CFRs 1 to 6 of the Southern CFR J-Form were aligned roughly NW–SE, making up its "base," and CFRs 7 to 16, making up its "back" of this J-Form, were aligned roughly SW–NE. CFR 6, the most prominent and complex of the CFR features, is the "hinge" linking the base and back of this Southern CFR J-Form. CFRs 6 to 16, the "back" of this J-Form, can be divided into two units of five each, demarcated by complex CFR 11 and complex CFR 16. This distribution suggests that there might be a real structural difference internal to the Southern CFR Sector. This may be the case for the Northern CFR J-Form also, although the spacing and relative distribution of its constituent CFRs would suggest a different and possibly simpler structuring. CFRs 22 to 30, making up the "back" of the Northern CFR J-Form, were all simple types and did not display an internal spatial division similar to the "back" of the Southern CFR J-Form. The short arm or "base" of the Northern CFR J-Form, demarcated by CFRs 17 to 21, had only one complex type, CFR 18. With these differences noted, it is possibly more important to recognize the probable complementary nature of these two J-Forms, a claim that is reinforced by the "Hearth of burnt clay" that, as described above, covers only the Southern and not the Northern CFR J-Form and also stops at the gap separating CFRs 16 and 17.

The pits, tunnel flues, inner basins, and chimney flues were subjected to different degrees of heat and their contents are different. The pits displayed significant fire hardening of the bases and walls and were found almost filled with a mixture of ashes and coarse charcoal. Metz's description of CFR 1, as reported by Willoughby (fig. 19.4, similar to feature b), can be used as an example. He states that this 30-inch-deep pit contained "ashes and coarse charcoal, its bottom showing marked evidence of the action of fire, being burned hard and red" (37). While the eight-foot-seven-inch-long tunnel flue made of "yellow clay" showed no signs of fire hardening, its concave base was "covered with pure white ashes, 2 to 3 inches in depth, containing small fragments of bone" (38). The smaller terminal pit with its basin-like bottom had no contents. However, the two vertical chimney flues had an ash coating and indications that "heat and smoke probably at one time passed through them" (38).

Metz comments on the contents and condition of several other pits, tunnel flues, and chimney flues. For example, the main pits of CFRs 2 and 3 were the same as that of CFR 1 and the pits of CFRs 4, 5, and 6 "were filled with loose black earth," while a small quantity of white ashes was found on the base of the associated tunnel flues. CFRs 5 and 6 each had "a hood of clay showing evidences of having come in contact with fire at their tops" (39). CFRs 5 and 6 had pits located at the inner end of the connecting tunnels. The inner pit of CFR 5, 4½ feet deep, was filled 21 inches deep with "ashes, gravel, several hard lumps of earth and ashes, and a few flakes of mica, the remaining portion being empty" (40). The first inner pit of CFR 6, 6 feet deep and 14 inches in diameter, "contained ashes, sand, black friable soil, and a fragment of pottery," and its immediate neighbor, 9 feet deep and, at its widest, 4 feet in diameter, "was nearly filled with ashes, sand, charcoal, and several pieces of hard burnt clay" (40).

Willoughby gives Metz's general assessment of the CFR features and their state and contents.

> The clay composing the inner surface of the sides and roof of the tunnels is very dry and friable, and does not show any sign whatever of having come directly in contact with fire. It seems the heat had been conducted through the tunnel to the flues at the back of each of them. In the flues no traces of fire can be found, except a very little white ashes adhering on the sides. In the pits directly in front of the tunnels, coarse charcoal and ashes are found in abundance, the bottom of the pits being burned red and hard, also the sides part way up. (Willoughby and Hooton 1922, 41)

Finally, he mentions Putnam's written comments that at the inner end of some of the tunnel flues "the walls were covered with a thin glossy incrustation, evidently formed by the condensation of vapours . . . [and] that in some instances there was a hard lime-like lining at the bottom and sides of the tunnels," and that about "a half bushel of 'ashes' from the different tunnels" was analyzed and it contained "many irregular porous masses of various sizes up to two inches or more in length, which are fragments of the deposit from the floor or side . . . [and proved] to be composed principally of carbonate and phosphate of lime" (40).

Discussion

This complex of CFRs has been a puzzle to Ohio Hopewell archaeology. An interpretation can be attempted in the light of the symbolic pragmatic perspective and the reader can determine whether this resolves the puzzle, or at least, opens a pathway to exploring a reasonable resolution. First, as alluded to above, the patterning of the complex would be amenable to a social structural interpretation under the Ecclesiastic-Communal Cult Model. As such, this complex patterning of the Southern and Northern CFR J-Forms would be compatible

with a dual age-grade structure.[6] The fact that the "concrete" characteristic of the initial construction stratum of the covering mound is
equally distributed across both CFR J-Form Sectors further reinforces
the conclusion that all of these CFR features were used and abandoned
at the same time. This means that similar tasks, those in which these
complex features played a significant symbolic pragmatic role, were
carried out by both postulated groupings during the same time period.

Assuming that these sectors index a dual senior/junior age-grade
laity structure, it is suggested that, because the Southern CFR J-Form
displays greater internal structuring, it represents the junior age-grade
sector. This is based on the Ecclesiastic-Communal Cult Model which
claims that, while both age-grades would be internally structured by
age-cohort hierarchies, this would be more distinct in the junior than
in the senior age-grade sector. Extending the logic of the model to this
set of CFR features, the three sets making up the Southern CFR J-
Form, CFRs 1-5, CFRs 7–11, and CFRs 12–16, with CFR 6 as the "pivotal
hinge," could be reasonably interpreted as manifesting the age-cohort
hierarchy of the junior age-grade sector. A similar but simpler hierarchy
is manifested in the Northern CFR J-Form, with the most complex set
being CFRs 17–21, directly opposite CFRs 1–5(6) of the Southern CFR
J-Form Sector. Thus CFRs 17–21 might represent the ranking age-cohort
of the senior age-grade, while the homogeneously organized set of
simple types, CFRs 22–30, diametrically opposite to CFRs 7–11 and CFRs
12–16 of the Southern CFR Sector J-Form, may represent the rest of
the cohorts of this postulated senior age-grade for whom age cohort
hierarchy is relatively unimportant.

This social structural interpretation is not simply speculative. It
is a logical application of the Ecclesiastic-Communal Cult Model to the
CFR pattern as empirical data. Of course, the interpretive conclusion
is not definitive. The best that can be said at the moment is that this
interpretation is not inconsistent with the Ecclesiastic-Communal Cult
Model. Clearly, this is not sufficient since it could be argued that the
dual complementary CFR Sector J-Form pattern is also not inconsistent
with some other plausible dualistic social organization, such as a

moiety-clan. Therefore, as part of the hermeneutic spiral, to adjudicate between these alternatives, a further line of evidence will be examined: the complex patterning of burnt deposits on the floors of Mound 3 and Mound 4. When this and the CFR patterning are seen together, it becomes clear that they are more supportive of the Ecclesiastic-Communal Cult Model than of the alternative moiety-clan social organization view.

Despite the practical design of these features, the regularity of their form and layout reinforces the view that they manifested strong symbolic pragmatics by which the material transformations that their fire-reduction capacity was used to effect were transformed into ritual of the intended type. The sheer quantity of burning that they appeared to have been used for would also suggest that this ritual had the practical purpose of producing a material product that was critical for assuring the felicity of further ritual. This product would have been bone and wood ash, "pure white ashes," ashes of other colors, and lime concretions. It is postulated that these CFR features might have been a type of "industrial-strength" ritual production of a range of ash materials valued for their symbolic pragmatic capacity to mediate further types of rituals. This means that the "mass" production process was itself a ritual activity, since only in being ritualized could the productive process ensure that the desired output of ashes would have symbolic pragmatic capacity to mediate further rituals. An alternative way of putting this is to treat the "raw materials" that fed the CFRs, for example, animal bones, as the outcome of the ritualized reduction of animals, and the further reduction of these materials for subsequent rituals as itself part of the ritual sphere regularly enacted at the Turner site.

The evidence for this claim is found in the ubiquitous use of black, white, and gray ashes in other components of the site. For example, ash was liberally used in the production of the mound strata. Mixed clay, ash and sand, or ash and clay, as well as alternate ash and black earth strata were used in different mounds (e.g., Mound 1, Mound 4). The lowest stratum of Mound 9 was made of black ash. Of course, it has already been pointed out that a 3-inch-thick black ash stratum was laid over the Southern CFR Sector J-Form, with the "Hearth of burnt clay" placed over it. Ash strata were also often used as bedding on which altars were built. "Cache-pits" characteristically had a great deal of ash placed in them, usually in stratified form. Grave 3, one of the pit features without any visible human remains, was filled with black soil, charcoal, and ashes. Although Willoughby did not specifically state that it was stratified, he commented that "it had much the appearance of a cache-pit" (16). Grave 30 was covered by a crescent-shaped set of flat stones about 5 feet long covering a pit that was 2 feet deep and 8 inches wide. It was filled "with black soil and ashes, a fragment of unio shell, and a few animal bones. No human remains, recognized as such, were found" (21). The bottom of Grave 31 was covered by a stratum of "white ashes containing a few fragments of animal bones; above the ashes was a stratum of gravel, the remainder of the pit being filled with black soil" (21). About six "cache-pits" southwest of sector "a" of the Western CBL, usually 4–6 feet deep and 30–40 inches in diameter, contained "charcoal, ashes, burnt clay, potsherds, and animal bones" (23). Finally, abandoned basin altars were carefully filled with different-colored ashes. All these associations suggest that ash was not simply a by-product of ritual but an essential part of it.

Indeed, there are many data in support of the view that the burning of animal parts played an important role in the ritual practices of this site. As will be detailed later, hundreds of canine teeth and foot bones of both small and large mammals (from foxes and lynxes to bears), as well as hundreds of deer and elk astragali were found in the major burnt deposits on the floors of Mounds 3 and 4. Given the subsistence midwifery view, the inedible parts of these animals must not be wasted, and therefore the rest of the skeletal material of these animals may very well have been transformed through intensive ritual burning. Just north of the main altar in the center of the floor of Mound 3 was found a "refuse pile of ashes mixed with animal bones, potsherds, and other waste material" (45), that, based on its nearness to this altar, may have been the accumulated residue of earlier ritual practices. Thus, it is possible that, in the normal course of ritual events, the organic elements of this material would have been reprocessed in one of the nearby CFRs and transformed into the "pure white ash" or the blue ash, or the black ash that is found making up the stratified fill of "cache-pits," the fill of the burnt basin altars; and the strata of the mounds.

In this regard, Grave 1, a major burial feature on the floor of Mound 1, was a double burial in a rectangular pit dug into the floor. The deceased were covered by about 12 inches of sand and gravel and immediately over this was placed a stratum of blue ashes, 1½ inches thick. According to Metz, "In these blue ashes were two lines of white ashes, 2 inches wide, and extending the full length of the grave. In these white ashes, *small hard concretions, similar to the white substance taken from the flues under mound 3, were found*" (30, emphases added). The use of this ash indicates a direct connection between ash materials produced under CFR conditions and mortuary practices. Of course, it cannot be definitively proven that the operations on the floor of Mound 3 were active during the time that this burial was carried out. More importantly, whether the ashes used in this burial were produced in the CFRs on the floor of Mound 3 or in others that were not found, they indicate a direct relation between mortuary practices and the use of ash produced by such methods. Furthermore, as Willoughby notes, the particular source of the material for these "two lines of white ashes" would be from the tunnel flues, and these were not directly accessible from the connecting pits (fig. 19.4). Collecting these ashes would probably have required a partial dismantling of one or more of the CFR units, further suggesting that the ash played a symbolic pragmatic role that was superfluous to funerary purposes, as will be discussed in more detail shortly.

Putnam particularly comments on the use of ash, suggesting that its relation with both the "sacrificial altars" and burials marks it as serving a particularly important ritual purpose.

> [Under] one of the altar mounds a large ashpit, six feet deep, . . . was discovered, and under another altar mound were several pits of smaller size but of similar character. . . . Beneath a small mound containing skel-

etons, was an excavation six feet wide and twenty-seven inches deep, filled with ashes mixed with animal bones, potsherds, and other objects. (Putnam 1973, 206)

It seems that the CFRs, then, were specialized ritual facilities in their own right, and the variety of ashes that they probably produced made them additionally valuable. Above all, they are evidence of the fact that the activities occurring on the floor of Mound 3 were systematically related to different components in the rest of the site. Finally, experiment may prove useful here. Reproducing these features and using them to reprocess partially burnt materials or to transform animal bone into "fine" ashes of different colors, as well as other ways, could serve to enlighten us on their possible uses and meanings.

THE CONJOINED MOUND AND THE BURNT DEPOSITS

The Turner burnt deposits are key integrating data by which to strengthen the confirmation of the Laying-In/World Renewal Crypt Model of the Turner site. This is because their contents significantly overlap and replicate the range of items found in the burnt offerings at the Hopewell site. These latter were key to the confirmation of the patterning predicted by the Laying-In Crypt Model of the Hopewell site. Therefore, as in the latter case, the Laying-In/World Renewal Crypt Model of the Turner site can be used to predict that there should be a range of artifacts that was ritually deposited independently of the pit features found in the Western and Eastern CBLs. These artifacts are postulated to be the curated laying-in warrants and mourning gifts, both personalty-mourning and formal mourning types, derived from the laying-in period of the deceased. The central altars on the floors of both Mound 3 and Mound 4 contain just such deposits. It is postulated that these burnt deposits can be most coherently treated as mediating complementary collective memorial/world renewal rites, and that, therefore, they were synchronic parts of the same complex ritual. This implies an alternative postulate, of course, this being that the events producing these two burnt deposits were separate rites that occurred independently, one much earlier than the other. These two postulates can be called the Synchronic Postulate and the Sequential Postulate of the two burnt deposits, respectively.

If the Sequential Postulate is upheld by the data, it would suggest that these seven conjoined floors and their associated features were constructed, used, and abandoned more or less in a sequential manner. If this were the case, then each floor space could be treated as equivalent, used by the same type of simply structured group as it was reproduced over time. The content of each floor and the type of stratification of each mound should be largely similar. Certainly, given the earlier description of the conjoined floors, this is not an implausible thesis. Each of these floors had a similar but individually different mix of features, hearths, altars, etc., and in several cases, some of the features were rebuilt and/or superimposed, indicating a history of reuse. Also, each mound covering was stratified, although slightly differently. If, however, the Synchronic Postulate is upheld by the analysis of the burnt deposit data, there would be grounds to conclude that this total set of ritual spaces marked a complexly structured social group, such as a complex cult, that used and then abandoned the total set simultaneously.

In short, the Synchronic and Sequential Postulates implicate different types of Renewal Lodges: a complexly structured social order or a simply structured social order, respectively. The former would support the Ecclesiastic-Communal Cult Model, and the latter would not. Of course, even should the evidence support the Synchronic Postulate, it does not follow that it would translate directly into support for the Ecclesiastic-Communal Cult Model over the Civic-Ceremonial Center Model. This is because, as suggested with regard to the complementary Northern and Southern CFR J-Forms, both models would be consistent with internal structuring of this nature. Therefore, it is the particular manner in which the relevant data support the Synchronic Postulate over the Sequential Postulate that is important. Unexpectedly, the patterning of the two burnt deposits, when combined with the complementary dual CFR J-Forms, makes it clear that the social nature of the groups that they mark would be consistent with an ecclesiastic-communal cult, but inconsistent with a dual clan-moiety structured group.

The material contents of each deposit are summarized below in table 19.1. Many of the types that were found were the same as those making up the Hopewell burnt deposits. This is precisely what could be anticipated under the premises of the Laying-In/World Renewal Crypt Model. Most of the items listed would easily qualify under the two categories of laying-in warrants and mourning gifts, both personalty-mourning and formal mourning types, as defined at the Hopewell site. At the same time, if the deceased in the stone-lined pits mark terminal world renewal rites, then they should have only a few items of personalty and mourning gifts, the other items having been removed and curated following the laying-in period. What is particularly indicative of this is the variety and quantity in the burnt deposits of copper-covered wood, clay, and stone buttons, copper cones, and antler tip cones, along with flint and obsidian bifaces, mica ornaments and effigies, and bone effigies and engravings, on the one hand, and the effective absence of these types of items from the "graves," on the other.[7]

Importantly, effectively all of the above categories were also found in both the Crypt Burial Context and the Burnt Deposit Context at the Hopewell site. This strongly argues that these played the equivalent symbolic pragmatic role at the Turner site, being part of what was entailed in the laying-in period. Reinforcing this conclusion was the finding that a number of the *misplaced* items in several of the "graves" of the Western CBL would fall under the category of laying-in warrants and formal mourning gifts. The misplaced items found in Grave 5 are particularly relevant in this regard. However, the same pattern was replicated in a number of other pit features, indicating the use of these initially as laying-in crypts and then as terminal world renewal crypt altars.

However, what is characteristic of the Hopewell site and the sites of the Chillicothe Tradition and is not characteristic of the Turner site is the quantity and distribution of the copper plates and the copper celts. At Turner only a few copper celts were found, one of them probably being mislaid (Grave 5, the "copper blade"). The large copper plates common to the Chillicothe region are not found at Turner (although one small example was). The closest equivalent at Turner might be the equally rare, small, thin copper plate (band) normally found on the breast of the deceased. Nevertheless, no small copper plates were found in the burnt deposits. Their being absent is what could be expected of custodial regalia.

The Section 3 burnt deposit had copper-covered wood, stone, clay, and bone buttons, literally thousands of shell beads and pearl beads, and an estimated 17,000 shell beads specific for "embroidery," according to Willoughby. The thousands of such beads found in the Section 3 Altar suggest a process of systematic, massive use of these items for the laying-in period and their equally systematic removal and curation.[8] Copper cones, copper beads, and cones made of the tips of deer antler tines, all of which would probably be attachments to various garments, were also found only in the burnt deposits and, typically, not in the "grave" burials, except for those that were apparently misplaced.

However, found in the Turner site burnt deposits and absent from the Hopewell site burnt deposits are unique clay human figurines, cannel coal, and unusually shaped natural stones and concretions. Such differences can be anticipated, given the contingent nature of artifact presence or absence across sites. This variation and lack of replication in these important material artifact categories (copper plates, celts, copper headdresses), rather than discouraging the extension of the findings of the Hopewell site to the Turner mortuary program, actually reinforce the significance of the replications and similarities that do occur, suggesting that the Laying-In Crypt Model, with the addition of the world renewal crypt meaning, can be generalized to Turner. That is, although a laying-in crypt burial population is apparently lacking at

Turner, nevertheless, given the earlier demonstration that the pit features had complex ritual histories, the patterning of both the artifacts of the stone-lined pit features and the deposit altars fit the patterning that could be expected under the Laying-In/World Renewal Crypt Model.

Interestingly, as a reinforcement of this conclusion, there is the distribution of galena and mica. Galena is not found in the Turner burnt deposits but is sometimes found with the Turner burials. Mica is found in both, and it is found with and without human burials in the "grave" pit and "cache-pit" features of the burial places. As pointed out earlier, with respect to galena, an indication that it was serving as a warranting aspect of the event would be to find chunks placed in the hands of the deceased. It was also pointed out that there were several cases in which galena chunks were found in the "grave" but not associated with the deceased in this way. Grave 5 had a galena piece and a mica block in the southern corner. Grave 5-a had two galena pieces and a thick piece of mica associated with "a considerable quantity of human remains" (22) between the legs of the two extended burials. The absence of galena and the presence of mica in the burnt deposits suggests that while both mica and galena could be part of the laying-in period, only mica, not galena, was appropriate as a mourning gift or a laying-in warrant. If galena were appropriate as either of the latter, it would have been removed and curated at the end of the laying-in period, and subsequently found in the burnt deposits. Since only a few deceased were found directly associated with it in their hands, it would appear to be a rather rare form of personalty or custodial regalia, and, typically, was placed with the deceased during the laying-in period, and kept with her/him to the final burial. This reinforces the earlier conclusion that the few instances of finding "stray" galena suggests that it was the result of being misplaced during the removal of the deceased at the end of the laying-in period. In the same terms, finding "stray" mica in the pit features, as well as its being an important component of the burnt deposits, clearly links these two pit and burnt deposit contexts, and, as a linking element, it is most reasonable to conclude that it was part of what was entailed in constituting the laying-in warrant and/or mourning gift aspect of the laying-in period.

Evidence that the burnt deposit practice was not a "one-shot" event but a tradition at Turner, confirms this model. A series of excavations by Putnam and the Peabody Museum removed extensive southern and northwestern sections of the western part of the Great Enclosure embankment and revealed evidence of very intense ritual activity: large burnt hearths, crematory altar, pits, major ash beds, post holes, and so on. In fact, a small burnt offering altar was found just northwest of the Graded Way in the southwestern sector of the Great Enclosure (fig. 2.11). Associated with this offering altar was a large hearth and a number of small "cache-pits." The burnt deposit in the altar contained engraved human ulnas, among other items.[9] Willoughby describes this offering altar as 36 by 25 inches, in the typical form of those found on the floors under the Conjoined Mound, and summarizes this deposit in the following terms.

> Mixed with the ashes in the basin were the perforated canine teeth of small mammals, beads of bone and copper, copper-covered buttons, fragments of copper ear-ornaments, perforated fossil teeth of shark, a flint knife blade, pieces of mica, fragments of carved bone, and various other articles, all more or less injured by contact with fire. (Willoughby and Hooton 1922, 8)

The contents of this deposit largely replicate the burnt deposits on the floor of Mounds 3 and 4. All this marks an active period prior to the construction of this major embankment.

COMPARATIVE ANALYSIS

Were these burnt deposits synchronic or sequential? Under the Synchronic Postulate they would have been complementary, each mediating a ritual that was required to complete the other, or, alternatively, each was a component of a larger complex ritual. Under the Sequential Postulate, the two rituals were performed at different times and, therefore, were two separate rites of the same type. In this case, the Section 3 burnt deposit may have been generated first, and then, at some indefinite time later, the Section 4 deposit may have been produced, or vice versa. The Sequential Postulate is favored if the contents of the two deposits display strong redundancies. This is because, as independent instances of the same type of ritual, it could be expected that the same basic range of material categories and styles would be required. The Synchronic Postulate is favored if the opposite is the case, namely, a certain degree of mutual exclusivity of artifact types such that the two patternings can be treated as complementary. Of course, since the premise is that both deposits mediated similar ritual, judgments about the significance of either redundancies or non-redundancies have to be made. Under the Sequential Postulate, a temporal gap is premised, allowing time for innovation in the symbolic pragmatics of the same ritual, and this could account for *some* divergence in styles and types. Under the Synchronic Postulate, since the two deposits mark complementary rituals, difference is expected. However, too much would begin to question that these two deposits mark complementary rituals.

Assuming the Sequential Postulate, what would the divergences and lack of overlap between these deposits have meant in ritual terms? The absence of copper ear spools from the Section 4 burnt deposit (table 19.1) suggests that during the period when this section was used, these items were not in ritual use. In reverse, the absence of all fossils and unique stones and concretions from the Section 3 burnt deposit suggests that during the period when this section was used, these items were not in ritual use. Equally, since no phalanges of pawed animals were found in the Section 4 burnt deposit and no astragali of hoofed animals were found in the Section 3 deposit, then there would be sequential periods when only one and not the other category of animal species was used as part of the ritual practices. Furthermore, under the Sequential Postulate, the Section 4 period would have seen such typical Ohio Hopewell items as shell and pearl beads being almost absent, since there were few of these reported in this burnt deposit. In contrast, during the Section 3 period there would be huge quantities of these items used. Also, the Section 4 period would have been characterized by the group's having copper and iron nuggets but not using them to produce iron beads, copper icons, and so on.

This analysis under the Sequential Postulate would suggest that there were radical swings in the material media of ritual practices occurring from one period to the next. By radical here I mean that ritual would be performed without the use of some of the most widespread Ohio Hopewell artifactual categories: copper ear spools, pearl and shell beads, and so on. This is not plausible. The Synchronic Postulate, however, can easily accommodate these differences. In fact, as suggested above, some such non-redundancy would be anticipated. For example, copper-covered (some iron-covered) wood, clay, and stone buttons, copper ear spools, and copper effigies are reported as being found exclusively in the Section 3 burnt deposit. The buttons would be an important part of the laying-in shrouds or garments, and, therefore, they should end up in the burnt offering deposits. Particularly interesting is the fact that all the paw foot bones are also in the Section 3 deposit. If these were also the remains of small-animal skins used as laying-in warrants, this, combined with the exclusive deposition of copper-covered wood, clay, and stone buttons, suggests that laying-in garments and shrouds may have been burnt exclusively in this deposit. Copper ear spools would be important personalty-mourning gifts. The absence of the copper cutout symbolic effigies and icons from the Section 4 deposit and their reported exclusive presence in the Section 3 deposit clearly support the specialized complementary ritual view (the Synchronic Postulate), suggesting that the two sections were ritually interdependent, and therefore, contemporaneous.[10]

Thus, these two burnt offering deposits demarcate a distribution of critical raw and finished materials that marks two complementary groups, each specialized in performing an aspect of a total ritual regime. This supports the conclusion that these two burnt deposits were more or less simultaneously produced as part of a collective memorial/world renewal rite. All this favors interpreting the conjoined floors as marking a single, contemporaneously used Renewal Lodge Complex under the responsibility of a complexly structured group. Earlier, however, it was pointed out that this conclusion does not by itself support the claim that Turner was the locale of an ecclesiastic-communal cult since a dual structuring would not be inconsistent with the moiety-clan organization. In the earlier CFR feature analysis, it was pointed out that the Southern CFR J-Form displayed an internal patterning that was not replicated in the Northern CFR J-Form. The former had its complex CFRs distributed so as to sector and anchor the simple CFRs in a manner that would not be inconsistent with the age-cohort hierarchy of a junior age-grade sector of an ecclesiastic-communal cult. The Northern CFR J-Form had its complex CFRs making up the major part of the base of the J-Form (fig. 19.2). It was argued much earlier that, since age differences among the members of a senior generation tend to be overridden and suppressed by their lifetime of shared ritual experiences, the age-cohort hierarchy of the senior sector would have a much reduced organizational impact. It could be expected, then, that the ritual nature of the CFR-mediated behavior of this group would tend to reflect specialization and not age-differentiation. The clustering of the complex CFRs to make up the northeastern base of the Northern CFR J-Form would be consistent with this expectation.

In contrast, because the two sectors of a moiety-clan duality would have equivalent age structuring and the same range of ritual duties would be equally distributed, it is suggested that equivalent patterning would be displayed across the two CFR J-Forms. Of course, it is possible that the internal clan order of the moieties could be ranked. Allowing for this possibility would partly account for the greater complexity of the Southern CFR J-Form patterning. However, when the degree of specialization implicated in the discontinuities of the contents of the two burnt deposits is added as a second line of evidence, these together reinforce the conclusion that the two sectors stand in complementary and not rank order to each other. This translates into the conclusion that the differential patterning of the Northern and Southern CFR J-Forms and the burnt deposits manifest that type of ritual specialization between nominally equal sectors of a cult manifesting a junior/senior age-grade structure.

SUMMARY

The presence of the copper cutout effigies and symbolic icons in the Section 3 deposit presents a puzzle. While similar items were used to mediate important ritual at the Hopewell and Mound City sites, their ritual deposition did not involve their destruction. Because they were not destroyed, it was concluded that they were custodial regalia. Similarly, at these sites ocean shell containers were used to mediate mortuary ritual and were not destroyed. However, since both copper effigies and fragments of several ocean shell containers were found in the Section 3 burnt deposit, it is clear that Turner had its own protocols in this regard. That is, these mark a difference in the ritual symbolic pragmatics between these two traditions. This conclusion will be reinforced in the next chapter when a particularly important feature is examined, the Turner deposit of "trophy skulls." Like those at the Hopewell site, these were not subjected to ritual destruction. However, they were treated in a quite distinctly different manner, reinforcing the view that there are some real divergences in ritual protocol distinguishing Turner from the Hopewell site and its neighbors. Despite the differences in treatment, it is suggested that the destroyed items of copper cutout effigies and symbolic icons had an equivalent importance as custodial regalia for the Turner site group as they had for those at Mound City, Hopewell, Seip, and Liberty Works, except that they may have been recognized as *group* rather than the *usufruct* custodial regalia. It was earlier argued (chapter 14) that, like stewardship artifacts, group custodial regalia would normally be withheld from terminal ritual that would separate the group from these artifacts. However, as with any symbolic pragmatic rule, this general rule can always be overridden by pressing circumstances, and, as will be discussed in more detail in the following chapter, the above may indicate just such circumstances.

The analysis of the contents of the Conjoined Mound has empir-

ically anchored the claim that there was a structural connection between the activities on the floors of this mound and those that generated the pit features and their contents making up the Western and
Eastern CBLs, as postulated by the Laying-In/World Renewal Crypt
Model. The validity of a model can be further demonstrated by applying it to new data. To this end, two more analyses of the Turner data
are presented in the next chapter, followed by a comparative analysis
of the Hopewell and Turner data. This analysis and its conclusions
serve to initiate the interpretation of the inter- and intra-cult relations
of the Ohio Hopewell, which is the focus of part 4.

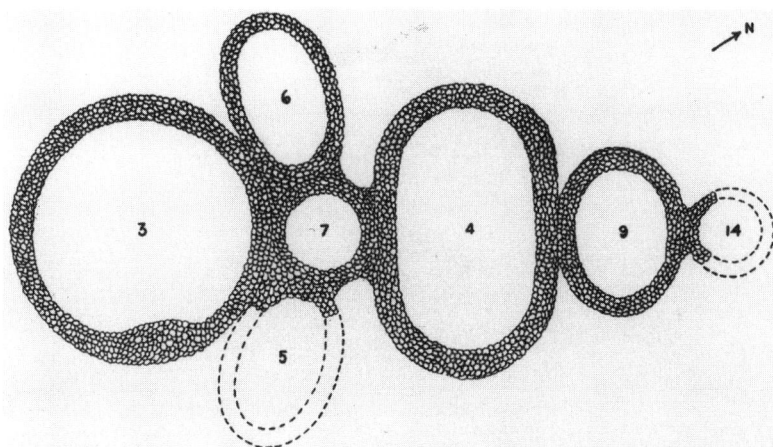

FIGURE 15

Plan of the seven connected altar mounds within the Great Enclosure, showing wall
outlining base of each.

FIG. 19.1. Floor of the Turner Conjoined Mound (Willoughby and Hooton 1922,
33, figure 15)

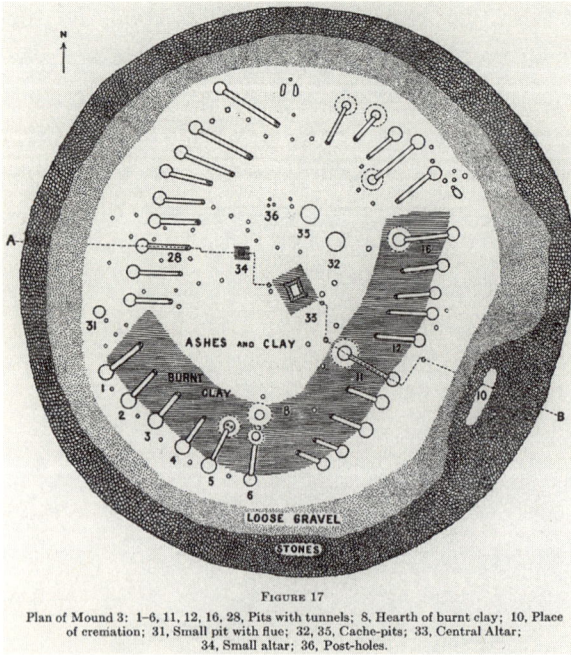

FIGURE 17

Plan of Mound 3: 1–6, 11, 12, 16, 28, Pits with tunnels; 8, Hearth of burnt clay; 10, Place of cremation; 31, Small pit with flue; 32, 35, Cache-pits; 33, Central Altar; 34, Small altar; 36, Post-holes.

FIG. 19.2. Floor of Mound 3 (Willoughby and Hooton 1922, 36, figure 17)

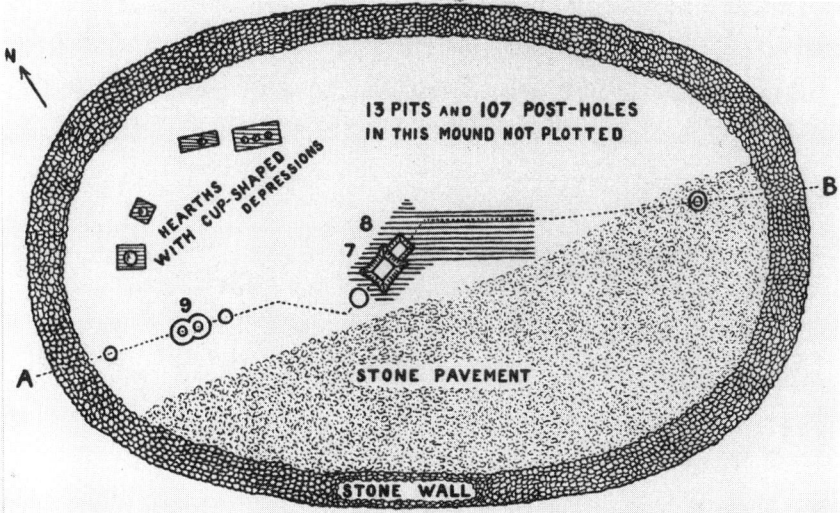

FIGURE 28

Plan of Mound 4: 7, Altar 1, from which many artifacts were taken; 8, Altar 2; 9, Adjoining pits with clay cone covers.

FIG. 19.3. Floor of Mound 4 (Willoughby and Hooton 1922, 64, figure 28)

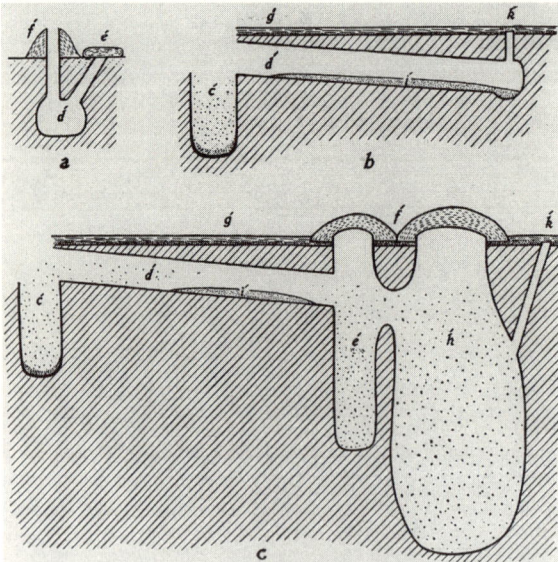

FIGURE 18

Mound 3. Cross-sections of pits and tunnels: a, Small pit (31) and flues; é, Stone covering flue; f′, Clay cone; b, Pit and Tunnel 28; c, Pit 6, and connecting tunnel; g′, Layer of burnt clay covering thin stratum of black ashes; f′, Clay cones covering top of pits; i′, Layer of ashes mixed with carbonate and phosphate of lime.

FIG. 19.4. Controlled Fire Reduction Features (Willoughby and Hooton 1922, 38, figure 18)

FIGURE 16. Cross-section of Mound 3: 1, Intrusive burial; 2, Intrusive pit
of charcoal and ashes; 5, Clay; 6, Clay and gravel, mixed; 7, Yellow clay with
with tunnel and outer pit; 12, Layer of limestones; 28, Pit with tunnel and flues;

FIG. 19.5. Cross-Section of Mound 3, no. 2, the "Trophy Skull" Pit (Willoughby
and Hooton 1922, 34–35, figure 16)

containing two skeletons and sixteen skulls; 3, Top soil; 4, Clay with specks
thin covering of sand; 8, Concrete; 9, Post-holes; 10, Place of cremation; 11, Pit
33, Central altar; 34, Small altar.

Table 19.1 Summary Content of Section 3 and Section 4 Burnt Deposits

Categories	Section 3 Altar	Section 4 Altar
Raw Materials		
Copper Nuggets	35—small	7, 1 large (56 oz.)
Meteoric Iron	several small	1—27 oz.
Gold	15 sheets	absent
Mica	3 large sheets	absent
Cannel Coal	absent	26 pieces
Artifacts		
Copper Ear Spools		
Some Silver, Iron Onlay	50	absent
Copper Adze/Celt	1	absent
Copper Bracelet	3	1
Copper Beads	700	present
Copper Cones	present	present
Copper Symbols	28	absent
Iron—Pan Flute	1	Specifically reported as no iron artifacts in deposit
-Head Piece	1	
-Beads	several	
-Ear Spools with Iron On-lay	several	
Flint Blades	16 bifaces	many fragments
Obsidian	11 bifaces	absent
Goldstone Bifaces	11	absent
Human Bones	2 engraved parietal	absent
Figurines	absent	6+
Mica Effigies and Symbols	about 50	many (Horned Serpent)
Copper-Covered Buttons	several	absent
Beads	700	
Shell Beads	22,000	fragments
Pearl Beads	36,000	absent
Pearls	12,000	200
Small Animal Teeth	2,000	fragments—many
Paw Foot Bones	about 600	absent
Hoofed Foot Bones	absent	284—deer, elk
Concretions		many noted
Bear Canine	36+	absent
Alligator Teeth	12	absent

(continued)

Table 19.1 (*Continued*)

Categories	Section 3 Altar	Section 4 Altar
Antler Cones	34	44
Fossils	absent	many noted
Hollow Effigies	absent	two
Animal Claws	absent	fragments
Ocean Shell Vessels	fragments—several	absent
Clay Ear Rings	3	absent
Tortoise Shell	spatula-like	absent

Chapter 20

THE TURNER-HOPEWELL
IDEOLOGICAL AXIS

Mound 1 (fig. 20.1) was located "inside" the Great Enclosure at the northeast entry (fig. 2.11). It was a low mound (ca. 1.75 m) and its perimeter covered a low circular wall of flat stones, 0.35 m high and about 0.7 m thick, with a diameter of ca. 19.6 m, enclosing a floor covered with a layer of "concrete" composite of ashes, gravel, and coarse sand. A bit east of the center of the floor there was a 2.5 m circular fireplace containing four pottery vessels and ashes with mixed charcoal, flint flakes, and, interestingly, "charred corn and corn-cobs."[1] There were six mortuary deposits, five immediately under and one above the floor (fig. 20.1). Four major mortuary pit features circumscribed the central fireplace. Three were located to its northeast, Grave 1, Grave 12, and Grave 13, and one immediately to its southwest, Grave 16. All were aligned in the same SE–NW direction. These mortuary deposits serve as a good test case because most involved extended burials, suggesting a minimum of post-mortem manipulation. For this reason they challenge the claim of the Laying-In/World Renewal Crypt Model.

Graves 12 and 13, each containing one adult extended inhumation, were aligned on the same SE–NW axis, separated by about 1.4 m. The deceased in Grave 12 was placed with the head to the northwest and the deceased in Grave 13 had the head oriented to the southeast, suggesting that these two pit features were deliberately aligned along the same axis, and that the two deceased were placed so as to be complementary opposites in terms of head-feet direction. No artifacts were associated with either. While they display no post-mortem manipulation, supporting the view that they were focal participants rather than symbolic warrants of the mortuary practices their burials mediated, several other aspects speak against this conclusion. For example, the positioning of the two pits on the same axis and the reverse head-feet orientation of the deceased suggest that these were part of a single, complex, and coordinated event that manifested ritual intentions surplus to funerary concerns. Seeing these mortuary deposits within the total context, as further described below, supports the conclusion that these mediated mortuary rites that had little funerary relevance.

Grave 16 was a large pit feature, about 2.8 m long and 1.6 m wide, containing three adult extended burials, all three placed with the heads northwest and level with one another. This is interesting since, as seen from figure 20.1, the individuals were apparently of different heights. The heads were about one foot apart. Given that the pit was only about 1.6 m wide, this suggests a rather tight fit. The two "outer" deceased had associated artifacts. One had about two hundred pearls by the right tibia and a platform pipe with two bowls at its right knee. The other "outer" deceased had an ocean shell container, several shell ornaments and large pearls at its feet, a shell spoon item at its left knee, and in its left hand a cut shell container. The central inhumation had no associated items (31).

With the exception of the crowded nature of this compound mortuary deposit, all this would suggest a compound funerary event. However, since a number of the items listed with the "outer" inhumations were on the sides of the bodies furthest from the wall, these could easily be associated with the central inhumation. Alternatively, given that all but one of these items were found by the feet and legs, it is possible that both the deceased and the material artifacts were intended either as offerings or else, as was the case for a number of the mortuary deposits in the Western CBL, that the artifacts were inadvertently misplaced from prior mortuary events. This would be more consistent with the view that Grave 16 was the outcome of a post-mortem sacrifice. Before completing the assessment of this feature, however, several other mortuary deposits will be described, in particular, Grave 1.

Grave 1 was briefly described earlier. Its axis was parallel with the axes of the other three mortuary pits already described (fig. 20.1). It contained two extended deceased, an adult female and a child, both with heads oriented to the southeast. The pit was about 2.9 m long and 1.4 m wide, only slightly narrower than Grave 16 with its three adult inhumations. Grave 1 is described as having one standing stone at the southeast end and five at the northwest end, replicating the patterning found in a number of the stone-lined pit features in the Western and Eastern CBLs. The female adult had a unio shell spoon at her knee level. The child was immediately to her right and near its head "lay three bone awls, a bundle of bone needles, and several flaked knives of flint. At the left foot was a unio shell filled with a red substance. Several flaked knives, some broken, were also found near the shell" (30).

The layer of sand and fine gravel immediately over the bodies must have been about one foot deep since it covered the single standing southeastern stone, which is described as about one foot high. The blue ash layer immediately over the fine gravel and sand layer was about 1½ inches thick. Running from one end to the other were two lines of white ash, each two inches in width, containing concretions similar to those found in the white ash material deposited in the tunnel flues of the CFRs on the floor of Mound 3. Over this bed of blue and white ashes were 18 inches of gravel. This suggests that the total depth of this mortuary deposit pit would be about 32 inches.

From the cross-section diagram of this mound, these pits appear to have been dug from the surface level straight down and were about the same depth (fig. 20.2). The one-foot-high stone wall that encircled the floor is indicated as also being sunk into this surface. This wall and the mortuary pits are all shown covered by the hard concrete-like floor, which was apparently built over an earlier occupation level, as indicated by four post holes in the northwest sector that were also covered by the floor. Just north of the large central hearth described above, and on the same axis but northwest of Grave 12, there was a small altar built on the floor (fig. 20.1). While the altars on the base of the Conjoined Mound were usually dug into the floor and then lined with clay, this one was built up with clay 12 inches high and sloping inwards as it rose so that the base was 1.8 m in diagonal and the top was 1 m.[2] It was heavily burnt. Over the top surface of this altar was a dome of clay, about 15 cm thick, covering a mortuary deposit of a flexed burial in a prone position. The cross-section diagram indicates that a thick layer of white ash surrounded this crematory altar as far as the central hearth and that over this ash layer and covering the domed burial were alternate layers of ashes and black earth. The excavators also report that below the floor in the southwestern sector and toward the periphery there was a possible human cremation covered by a set of slightly overlapping flat stones.

DISCUSSION

Taking these extended inhumations individually, an argument could be made that the treatment of each marks the deceased as a focal participant in the burial event that each mediated and, therefore, that these were funerary burials. When placed in the context of the mound floor as a set-piece within the larger frame of the Turner site, a stronger argument can be made that Graves 1, 12, 13, and 16 were mortuary deposits mediating nonfunerary rites, possibly mortuary rites initiating the final phase in the use of this locale for world renewal ritual. This would constitute the pits as laying-in/world renewal crypts, as indicated by the careful alignment of these mortuary features along the same SE–NW axis, the placement of the central hearth on the floor so that it was largely bracketed by these sub-floor deposits, the termination of the mortuary rites with the building of the concrete-like floor, and, in particular, the special treatment accorded to Grave 1, namely, its strata of gravel and bluish ashes with the white ash strips, its base of sand and fine gravel, and the presence of standing stones. With the exception of the two white ash strips, these echo the fill content of a typical "cache-pit" and the standing stone lining of the pit features in the Western and Eastern CBLs. The view that the terminal use of the latter was as world renewal altar crypts has already been empirically grounded, and these four burial pits display similar characteristics.

Indeed, there are data indicating that Grave 1 may have had dual use as both a laying-in crypt and a terminal world renewal altar. As described above, at the foot of the child there is an assortment of materials, a unio shell filled with a red substance, some bladelets, and some broken items, suggesting material that was overlooked when a previous usage period was terminated. Furthermore, as pointed out in the previous chapter, it is likely that one or more CFRs had to be dismantled in order to procure the white ashes from the tunnel flues, implying that these burials served ritual needs that were surplus to funerary purposes. Finally, if these had been funerary burials, then it could be anticipated that those inhumations having ocean shell containers, which are assumed to be the primary type of custodial regalia characteristic of Turner, would also have other regalia, such as, possibly, copper bands. In any case, at least significant formal personalty could be anticipated. Other than the few pearls, the platform pipe, some awls, and a few scattered bladelets, the range of items identified is not that different from that found with a number of burials in the Western and Eastern CBLs. All this suggests that, in fact, the rich range of artifacts that make up the contents of the burnt deposits examined earlier were removed from or never placed with these mortuary deposits, suggesting that they mediated the terminal world renewal rites following the laying-in period.

Furthermore, while Grave 1 and Grave 16 are almost the same size, Grave 1 had two extended deceased, one child and one adult female, and Grave 16 had three adults. It is suggested that the pit of Grave 1 would be the "standard" crypt for two deceased. But Grave 16, with almost the same dimensions, contained three adult deceased. Therefore, when it came time to complete the ceremony that these deceased mediated, it would appear that an unplanned extra was added, possibly the middle one, thereby accounting for its apparent lack of any artifacts. This suggests that the compound mortuary deposit of three deceased marks a complex ritual event that, combined with the other pit features, constituted sacrificial world renewal rites. Finally, as pointed out above, most of the materials in Grave 16 may have been only incidentally associated with the burials. This follows from the fact that their distribution might be quite adequately characterized as scattered, overlooked from previous mortuary rituals when these earlier deceased had been removed.

The clearest example of nonfunerary mortuary ritual is found in the altar of Mound 1, which had a flexed body in a prone position placed on it and was then covered with clay to form a dome. When used for mortuary burning, it has been argued that the basin altar is the primary symbolic pragmatic device by which to constitute cremation as a post-mortem sacrifice. The fact that this deceased was given a flexed burial in what appears to be basically a crematory altar implies, then, that it was the altar that served as the primary focal symbolic warrant here while the deceased was a noncremated post-mortem sacrificial offering marking its terminal use. The fact that the altar is being used in a novel manner suggests that this activity was special.

THE "TROPHY SKULL" AS GROUP CUSTODIAL REGALIA

The largest known aggregate deposit of Hopewell "trophy skulls," sixteen, was placed in a large pit that was carefully dug into the eastern sector of Mound 3, after completion of the latter (fig. 19.5). They were placed in a circle surrounding two extended inhumations lying in the center of the base of the pit. Most show marks of having been worked, decorated, and curated. This merits a full quotation by Willoughby.

> This [intrusive pit] was on the eastern side of the mound [Mound 3] . . . , and had been dug to the depth of 7 feet. The clay filling was like the upper strata of the mound, but the layers of stone had been removed. A coating of ash-like substance covered the bottom of the pit, and extended 3 feet up its sides. The extended skeletons of a man and a woman lay upon the bottom near its centre. Covering their tibiae were ashes

containing bits of burnt bone, both human and animal. Near the skeletons was a flat stone upon which rested a rounded sheet of mica about
12 inches across. About the two skeletons were sixteen crania of men
arranged at regular intervals.[3] Two or three other fragments were also
recovered, one of which was part of an infant's skull. . . .

The sixteen skulls were unaccompanied by other bones. They were
probably family relics, connected with or belonging to the man whose
skeleton occupied the center of the grave. Thirteen of them have superficial scratches or cuts on their surface, apparently made with flint
knives in the process of removing the flesh. Some of the skulls had been
painted red, and red ocher still adheres to the surface of six . . . Five of
the skulls have one or more perforations, about ⅛ inch in diameter, in
the vault of the cranium. The sixth example has eleven perforations,
and another apparently started. . . . The position of the holes seems to
indicate that at least a part of them were intended for the passage of a
suspending cord. Others may have been used for the insertion of feathers
and other decorations. (Willoughby and Hooton 1922, 60–61)[4]

With the exception of the curated skulls, the rest of this mortuary
feature fits the overall Turner pattern, including the ash layer, the
addition of ashes containing burnt animal remains and burnt human
bones, and the fragments of infant bones, all of which suggest post-
mortem sacrificial ritual. Furthermore, the flat stone, rounded sheet of
mica, use of the clay, and, of course, worked and painted skulls are
characteristic elements of Ohio Hopewell. Willoughby's characterization of these "trophy skulls" as "probably family relics" is typical of the
funerary view, of course.

This deposit raises two intriguing problems. First, given the Sacred Earth Principle, dismantling embankment earthworks was
avoided. It is logical to extend this proscription to include the mortuary
mounds as among the cosmological icons of their associated embankment earthworks. Thus, a major intrusive disturbance of this nature is
not what would be expected. Second, since at the Hopewell site "trophy skulls" were identified as items of custodial regalia, there is good
reason to assume that they were also recognized as custodial regalia at
Turner. However, at the Hopewell site these items were in individual

burials, suggesting that they were treated as usufruct custodial regalia. At Turner, where "trophy skulls" were conspicuously absent from individual burials, this collective deposition suggests that "trophy skulls" were recognized and used as group custodial regalia. Group custodial regalia is the type of potent artifact that defined the nature of the group that used it and, therefore, was normally withheld from terminal mortuary deposition. If so, this might go some distance to account for the overall paucity of custodial regalia in the Western and Eastern CBLs. Imagine these sixteen "trophy skulls" distributed across these mortuary deposits, and a pattern of custodial regalia distribution more similar to the Hopewell site would emerge.

This large burial deposit associated with two extended deceased would appear to be the symbolic pragmatic equivalent of the "treasury" of 66 copper celts found with the dual crypt containing Burials 260 and 261 on the floor of Mound 25 of the Hopewell site. A similar "treasury" of copper plates was found at Seip. It was argued earlier that artifacts falling under the category of group custodial regalia would be rare in the archaeological record and would occur only at major ruptures, either institutionally called for by the nature of these ruptures, or else being the only reasonable alternative given the unexpected and disruptive nature of the rupture. The Seip and Mound 25 examples would suggest institutional ruptures; the Turner case, a system disrupture. In any case, this analysis suggests that certain artifact types, such as copper celts, plates, and "trophy skulls," could be emically structured into both types of custodial regalia, group and usufruct.

Since normally items of group custodial regalia are withheld from terminal deposition by being passed on from one generation of custodial keepers to the next, in this case, it would appear that they were accompanied by their last custodian (or custodians).[5] For this reason, this pit feature and its contents may implicate a very serious state of affairs, one that warranted two major deviations from the norm: (1) overriding the proscription against dismantling earthworks; and (2) suspending the rule that, as group custodial regalia at Turner, the "trophy skulls" be passed on to the next generation of custodians. Such a radical

social discontinuity, occurring some time after the completion of the
Conjoined Mound, might have been equivalent to the rupture that
has already been suggested at Seip. In short, for Ohio Hopewell, this
unique, even anomalous, mortuary deposit marks a local instance of
what will be postulated as a widespread institutional collapse. This was
earlier alluded to as the Middle Woodland–Late Woodland Transition
and is discussed in considerable detail in chapter 23.

THE TURNER-HOPEWELL IDEOLOGICAL AXIS
EMBANKMENT EARTHWORK FORMS AND
INTER-CULT RELATIONS

Despite the many similarities between the Hopewell and Turner
mortuary programs, there is a major aspect of these two sites that, on
first appearances, separates them quite radically, and this is the difference in the embankment earthwork forms. The Turner site embankments seem unique (fig. 2.11). They appear to display neither the C-
Form nor the T-Form attributes that are characteristic of the Miami
Fort Tradition of this region, nor do they display the circle and square
components that are found attached to the main C-Form of the Hopewell site, representing that site's use of the C-R Configuration attributes
of the Chillicothe Tradition.

Setting aside the plan forms for the moment, the Turner and
Hopewell earthwork sites do share some interesting parallels. The
Turner site displays both the SR-Profile embankment attribute and the
K-Profile attribute (fig. 2.11). It also displays two major embankment
components, one being the elongated subrectilinear, and the other the
Complex G-Form. The latter does not display the C-R configuration
typical of the Chillicothe Tradition, but it does display a Circle-Oblong
(C-O) pattern (also replicated in the Mound 3 and Mound 4 floors).
This will be termed the C-O Configuration. It has a dual-level arrangement with the smaller subcircular embankment earthwork (the C component) on the remnants of an upper terrace and the larger oblong
earthwork, the Great Enclosure, on the lower terrace. The two components were connected by an earthen ramp bordered by two K-Profile

embankments, that is, a variant of the aggregation neck component (the Graded Way). It apparently was built out of the earth that was removed to make the two deep and broad trenches that parallel the perimeter of the upper subcircular C component. The SR-Profile attribute is displayed by the two sacred circles "inside" the Great Enclosure. According to the temporal priority of the SR-Profile over the K-Profile, the two sacred circles would have preexisted the Great Enclosure and the subcircular upper component. Based on earlier discussions of integrating older embankments with more recent construction, whatever the symbolic pragmatic meanings of these two circles, they would not have been incompatible with the symbolic pragmatic meanings of the Great Enclosure.

It has often been noted that the Hopewell site was built on two levels with the back of the C-Form along the edge of the upper terrace and the two arms of the C-Form extended down to the edge of the lower terrace. I have argued elsewhere that this may be the result of a prescription, namely, that the C-Form type must incorporate within its embankment the highest local land form.[6] Therefore, if a terrace or ridge is next to and overlooking the lower terrace on which the embankment was to be built, then a dual-leveled embankment earthwork would be required. It may be, of course, that this is not so much a prescriptive requirement as it is a preference. That is, there may be only a few cases in which the positioning of a lower terrace and a higher terrace is such as to enable the construction of a dual-level system. When all these conditions were available, they would be treated as the opportunity to effect a preferential dual-level type of arrangement that we see at the Hopewell site.

This dual-leveled construction attribute is clearly evident at the Turner site. This arrangement apparently was deliberate since, as described above, in constructing the subcircular component, two deep trenches were cut across the ridge and parallel to each side of the subcircular embankment, effectively isolating the latter from the rest of the ridge (fig. 2.11). These trenches are interpreted here as the result of combining borrow pitting to procure earth for the Graded Way and producing a berm in order to constitute the upper earthwork as a modified G-Form/K-Profile embankment. Thus, a deliberate combination of the "traditional" and the "new" was carried out: the "traditional" Miami Fort rule or preference, and the "new" Complex G-Form. Of course, this "new" Circle-Oblong (C-O) contrast was not quite the same as the Circle-Rectilinear (C-R) contrast characteristic of the Chillicothe Tradition. However, it is still considered here to be a modified version of the latter, probably innovated so as to be acceptable within the Miami Fort Tradition. Because of the combined use of the aggregation neck, the K-Profile attribute, and the C-O Configuration, this construction can be dated as contemporary with the period of the addition of the C-R components at the Hopewell site. These are considered to postdate High Bank and Seip, the latter demarcating the initial Late Ohio Hopewell phase. This means that the Turner C-O embankment earthworks date to the middle Late Ohio Hopewell phase. It is reasonable to conclude that the total set of embankment earthwork components of the Turner site cover a chronological span from the later Early Woodland (the sacred circles of the Mt. Horeb Tradition) to at least the middle Late Ohio Hopewell phase.

In an important sense, then, even though the embankment earthworks of the Hopewell and Turner sites are formally different, the main embankment components share several attributes, implicating a deeper historical-structural linkage: (1) they manifest the same concern for dual-leveled construction; (2) they share the notion of balanced symmetry; (3) they share the use of both the S-Profile and the K-Profile embankment attributes; and (4) they reveal parallel developmental sequences. Although the Hopewell site does not display the aggregate neck attribute and the Turner site does (the Graded Way), the latter shares this component with other Chillicothe Tradition sites, such as High Bank and Newark. Furthermore, the C-O Configuration displays the skewed/off-centered attribute that is characteristic of the Paint Creek C-R Motif. If the axis of symmetry of the upper embankment component is drawn through the aggregation neck, it would cross the axis of symmetry of the Great Enclosure at an acute angle.

This reinforces the above conclusion that, while the horizontal plan forms are different, the contrasts that are manifested in the Turner and Hopewell sites are structurally equivalent. Since it has already been argued that the C-R Configuration components of the Hopewell site constitute this as a variant of the Paint Creek C-R Motif, it follows that the Turner site is probably also just such a variant. This further reinforces the postulated dating of the construction of this large embankment earthwork as the middle Late Ohio Hopewell phase, since the latter would follow the earliest balanced symmetrical expression of this configuration at Seip and/or Liberty Works. It also suggests the possibility that the Great Enclosure was built so as to circumscribe the sacred circles, respecting the proscription against disturbing preexisting embankments, while incorporating the relevant symbolic pragmatic meaning into the total construction.

Finally, it must be recalled that a considerable history of ritual activity occurred at the Turner site prior to the building of this Complex C-O embankment earthwork. As pointed out earlier, a dense patterning of hearths, post holes, pits, and so on, was found under the southwestern and northern embankment sectors of the Great Enclosure, along with the third burnt offering deposit containing materials that were similar to the burnt offering deposits on the floors of Mounds 3 and 4. Since this activity would have occurred prior to the construction of the Great Enclosure, it probably was carried out in conjunction with related activities in the sacred circles. The building of the Complex C-O embankment earthwork was, therefore, the continuity of these past practices in a new formal context. The newness of this context was simply an innovated expressive form that manifested the same, traditional symbolic pragmatic meanings.

There may have been one modification of an expressive nature that was new, however, and again, this is evident at both the Hopewell and the Turner sites. Prior to the construction of this Complex C-O embankment earthwork at the Turner site, the ritual that took place outside the sacred circles would have overlooked the river valley. This may have been a critically important aspect of the ritual since, at the

Hopewell and other C-Form embankment earthwork sites, the open side overlooks the valley bottom. However, with the construction of the C-O Configuration, there seems to have been a "turning away" from this perspective. The entryways of both the upper level and the lower level components are oriented away from the river valley. A similar reorientation is displayed at the Hopewell site, this being the "closing" of the open side of the C-Form by building the low stone wall along the terrace edge.

The Turner site is not alone in southwestern Ohio in combining the Paint Creek C-R Motif with older attributes. Willoughby pointed out that Turner shares some attributes with two other nearby sites, Milford Works (fig. 2.9) and Camden Works (fig. 2.10). Both are north of Turner, the former about 5 km north on the East Fork of the Little Miami and the latter about 7 km. While noting that on surface appearances "there is little resemblance between the Turner, Milford, and Camden Works" (1), Willoughby goes on to make several points, noting that all three have a small circular component attached by a long aggregate neck to a larger component and that in two instances (Turner and Milford), the small circular component is on a raised terrace. Milford also displays the basic attributes of the Paint Creek C-R Configuration while Camden lacks the infix, somewhat similar to the Seal site south of Chillicothe. However, all display the skewed/off-centered conjunction. It is because of these elements that Willoughby suggests that they are only "superficially" different from Turner. The number of parallels among these three suggests that they share the same symbolic pragmatic meaning. As variant Paint Creek C-R Configuration sites, they would embody the solar motif meaning. The link with the Hopewell site, with its original C-Form, also suggests that they are all material variations of the same configuration, and because they display a closer morphological similarity to an actual Paint Creek C-R Configuration than does the Turner site, it is suggested that the latter was built first and the other two came later, possibly near the end of the Late Ohio Hopewell phase.

In sum, the Turner and the Hopewell sites are complementary

versions of a similar transregional history of embankment construction. Each embodies elements of two separate but historically related ritual embankment earthwork traditions, the Miami Fort Tradition of southwest Ohio and the Chillicothe Tradition of south-central Ohio. Each also appears to be the sacred locale that connects these two regions: Turner rooted in the southwestern Ohio Miami Fort Tradition; and Hopewell on the western periphery of the central Scioto/Paint Creek drainage, the core region of the Chillicothe Tradition. The long-term historical relations between these two sites may have constituted the axis that linked together the cults that made up these two embankment earthwork traditions.

As mentioned earlier, Greber has suggested that the Hopewell site is the locale of a set of allied communities, particularly with reference to the structures found on the floor of Mound 25. I can fully support the core of her claim, namely, that the Hopewell site was at the center of a rather extensive alliance network. However, in terms of the Ecclesiastic-Communal Cult Model, as enriched by the above analysis, some modifications need to be made to her view. First, of course, there is the nature of the social organization of those groups involved. Where the Civic-Ceremonial Center Model recognizes monistic communities meeting and interacting as modular peer polity allies being responsible for the construction, use, and abandonment of the Hopewell site as a "primate center," the Ecclesiastic-Communal Cult Model would recognize the Hopewell site as the locale of an autonomous cult that was a key player in a network of similar autonomous cults. Its geographical position on the western periphery of the central Scioto drainage placed it centrally on pathways linking the central Scioto Valley to the middle and lower Great and Little Miami drainages. As indicated by the above analysis, its relation to Turner, and through it to Fort Ancient and other Miami Fort Tradition sites in this region, may have been as critical to the development of the Ohio Hopewell as was its relation to the Chillicothe group of cultic Renewal Lodges, including Mound City, Seip, and Liberty Works, and to the more distant Newark site in the upper Muskingum drainage of central

eastern Ohio. If the former can be termed the Turner-Hopewell ideological axis, then the latter can be termed the Chillicothe-Newark
axis.[7] The history of the relations between the Hopewell site and the
C-R Configuration sites in the Central Scioto/Paint Creek drainage
region (Mound City, Liberty Works, Seip, Baum, Works East) would
apparently constitute the interface of these two Ohio Hopewell axes.

This means that neither the Turner site nor the Hopewell site
was a locale of an emergent "super-cult," any more than the postulated
Chillicothe-Newark ideological axis constituted an autonomous "super-
cult." Rather, these were the earthwork locales of autonomous resident
cults that were in close interaction with each other, and, in many cases,
probably cooperated through shared labor, resources, and ritual, and
also competed for reputation. A number of the mortuary and related
data of the Turner and the Hopewell sites would suggest that these
autonomous resident cults tended to be very open, promoting widespread cooperative interaction: for example, embankment patternings
that mark mutual but arm's-length influence; a set of mortuary forms
that, as morphological variants of each other, mediated the same range
of mourning/world renewal practices; extensive and overlapping geographical contacts, as marked by the Hopewell artifactual stylistics, and
so on.

This claim of possible cooperative and competitive interaction
mediated through the North Fork/Paint Creek interface of the Central
Scioto region is supported by the fact that Turner apparently was the
channel through which modified versions of the Paint Creek C-R Motif
came to be present in the Little Miami River (e.g., Camden and Milford), and, similarly, the Hopewell site may have been the channel
through which both C-Form types (e.g., Cedar Bank) and versions of
the ridge-top types (e.g., Fort Hill, Spruce Hill), came to be present
within the Chillicothe region (as well as in the more distant Upper
Muskingum region where the Newark site is located).

Despite this extensive and historical interaction between southwestern and central Ohio as mediated through the Turner-Hopewell/
Chillicothe-Newark axes, the two resident cults of Hopewell and

Turner shared many of the same symbolic pragmatics of mourning/ world renewal rites and these set them somewhat apart from the cults responsible for Mound City/Hopeton, Seip, and Liberty Works. This suggests that the prevalence of extended over cremated burials at Turner and Hopewell is not fully accounted for by the Laying-In Crypt and the Laying-In/World Renewal Crypt Models. These two cults probably shared other social and cultural factors that set them apart from the Chillicothe Tradition cults (e.g. Mound City Proper, Seip, Liberty Works, and Tremper). What these might be is explored in part 4.

CRITIQUE

The analyses of the mortuary patternings of the Hopewell and Turner sites have enhanced the validity of the Mourning/World Renewal Mortuary Model by empirically grounding the premise that much of the mortuary patterning of these sites was the consequence of their being the monumental contexts of incremental funerary→mourning→spirit-release→world renewal ritual processes mediated by facilities that are adequately characterized as specialized crypt altars. At the Hopewell site, these were postulated to be single-duty laying-in crypt altars, and at the Turner site the equivalent mortuary facilities, the stone-lined pits, were postulated to have served double duty as both laying-in crypt altars and world-renewal crypt altars.

In terms of the hermeneutic spiral, this double duty claim implicates a critique of the original Laying-In Crypt Model of the Hopewell site. If the stone-lined pits of the Western and Eastern CBLs of the Turner site served double duty, it is desirable that the Hopewell "vaulted chamber" crypts and their contents should be reexamined with this possibility in mind. Space, however, prevents pursuing this interesting challenge. It is interesting because it is suspected that some anomalies in the Hopewell site mortuary data implicated by the Laying-In Crypt Model might be worth exploring in terms of the Laying-In/World Renewal Crypt Model. One anomaly, for example, is the variable condition of the extended burials in the Hopewell crypts. Many of the extended skeletons were in fairly good and equivalent condition. In several cases, however, the skeletons were seriously deteriorated. If all these deceased were partway through the laying-in period when abandoned simultaneously, as postulated under the Laying-In Crypt Model, their state of deterioration (or preservation) should be approximately the same. This raises the possibility that the model is not fully adequate, and that some of these crypts may also have served double duty similar to the claim for the pit features at Turner. While this possibility does not detract from the basic adequacy of the Laying-In Crypt Model of the Hopewell site and the dual truncation of the Renewal Lodges it postulates, the empirical grounds for expanding this model along the double-duty dimensions will have to be pursued in a subsequent work.

These analyses have given reasons for the similarities linking these two sites and how they contrast with the sites making up the Chillicothe Tradition. In short, these two analyses have sufficiently confirmed the Mourning/World Renewal Mortuary Model and the Ecclesiastic-Communal Cult Model so as to allow exploring the social organizational dynamics of Ohio Hopewell along the *Congregationalist↔Presbyterian↔Episcopalian continuum (C↔P↔E continuum)* as articulated earlier. Since this continuum implicates competition and negotiation for power arising from both internal and external structural contradictions, it seems appropriate to explore the possibility that, given the contingent nature of the cosmology-ideology relation, factions sharing the same basic cosmology but differing in ideological commitments were intrinsic to Ohio Hopewell cults. This is the core premise of part 4.

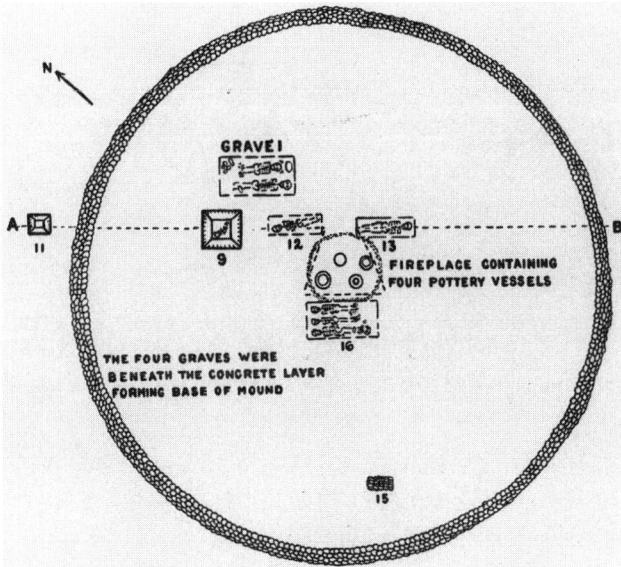

FIGURE 13

Plan of Mound 1: 9, Altar with enclosed skeleton; 11, Small altar; 1, 12, 13, 16, Graves beneath concrete; 15, Stone-covered space.

FIG. 20.1. Floor of Mound 1 (Willoughby and Hooton 1922, 31, figure 13)

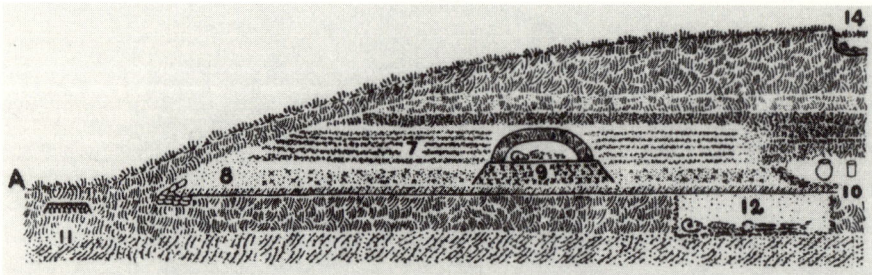

FIGURE 12. — Cross-section of Mound 1: 1, Clay; 2, Yellow clay; 3, Mottled clay, sand ashes and black earth; 8, White ashes; 9, Altar with enclosed skeleton; 10, Fireplace with

FIG. 20.2. Cross-Section of Mound 1 (Willoughby and Hooton 1922, 28–29, figure 12)

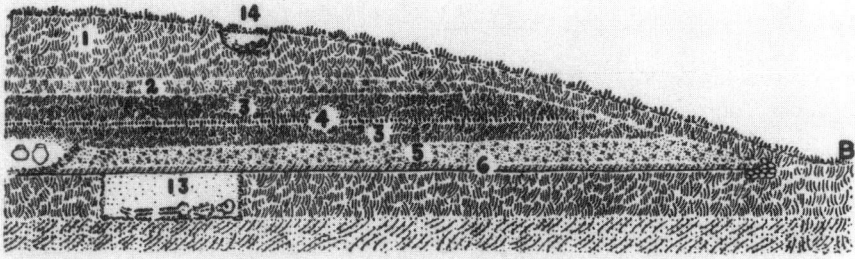

and ashes; 4, Dark earth; 5, Brown burnt clay; 6, Concrete; 7, Alternate layers of pottery vessels; 11, Small altar; 12, 13, Graves beneath concrete layer; 14, Intrusive burials.

Part 4

FACTIONAL COMPETITION, CONFLICT, AND RUPTURE

Chapter 21

THE IDEOLOGICAL
IMPERATIVE

Elizabeth Brumfiel has theorized factions as vertically organized and structurally equivalent groupings within and/or cutting across the formal components of organizations.[1] She argues that they are based on face-to-face, personal, patron-client relations, largely motivated by self-interest. A successful faction is marked by its leaders' occupying the focal positions of the organization, and, as patrons, they reward their client followers by appointing them to lesser positions. In her view, the pursuit of self-interest and personal reward guarantees that factions are necessarily competitive. However, despite competition being divisive, Brumfiel argues that it has a strong conservative effect since (1) factions share the same beliefs about the nature of the world, and (2) factions are structurally equivalent.

> [While] factions compete for resources, their structural similarity insures that they will hold similar ideas about what the world is like and what it should be like. Factional competition tends to be non-revolutionary in interest. The objective of factional competition is to achieve a favourable allocation of existing benefits; each faction hopes to gain more while its competitors gain less. Participants conceptualize factional competition as a zero-sum game in which one party's gain is another's loss. Thus, in factional competition debate generally centers upon the relative legitimacy of each faction's claim rather than the merits of substantively different social programs. (Brumfiel 1994, 5)

She points out, however, that factional competition may be unwittingly socially transformative or even "revolutionary."

> [Factional] competition has been regarded as non-revolutionary . . . , a temporary response to changing conditions . . . or as an impediment to constructive, meaningful social action. However, given the proper environmental and social context, factional competition expands in scale and intensity until it transforms the conditions of its own existence. It can then be a major force in social transformation. (Brumfiel 1989, 127–28)

While Brumfiel's view that factions are structurally equivalent is indisputable, the argument that their "structural similarity insures that they will hold similar ideas about what the world is like and what it should be like" collapses or fuses what are autonomous intentional states (e.g., beliefs, wants, hopes, and intentions) into a single, monolithic intentionality that easily slips into belief→action determinism. Indeed, this determinism is implicit in her view that factions will be conservative and nonprogressive in their overall orientation. This strongly contrasts with the position taken in this book, namely, that because of the contingent nature of the cosmology-ideology relation, different cult organizations, and even different factions within the same cult, can and most likely do share the same cosmological perspective while, simultaneously, staking out contrasting ideological positions. Hence, alternative and often *contradictory* ideological positions can logically and probably will coexist in the same cult. Support for these alternatives will be driven and sustained not by differences in cosmology but by differences in ideology. These differences will be fed and reproduced by the conflicting interests that the contradictory tendencies of the social structuring of the cult generates. In short, factions and factional competition can be quite revolutionary.

Ideologies are complex structures of constitutive and symbolic pragmatic rules, protocols, and rationales that stipulate the range of forms that the collective behaviors of the cult and its material media must display in order to constitute the ritual by which it can discharge its duties and fulfill its raison d'être. Ideologies are historically negotiated and constructed by the factions that make up the cult. Therefore, the total ideology of a cult will embody a spectrum of constitutive rules and protocols and these can range along a continuum that has mutually opposed extremes. It follows that the ideological position of any given faction will often be only incompletely manifested in the material projects that it supports. It will actively pursue promoting its ideological position through negotiation with other factions and these sometimes bitter negotiations will focus on the forms, make-up, and magnitude of the material projects that a cult ought to or must undertake. Typically, although not always, the negotiations are conducted in practical terms so that, given the circumstances, a working compromise that all factions can live with is achieved. This compromise will be called the cult's ideological stance. Since this is a working compromise, the stance is constantly subject to modification. The ideological stance of a cult at any given time must be distinguished from its ideological positions, the latter being the total range of ideological positions desired and promoted by its various factions. Therefore, at any given moment significant components of the total cult ideology will exist in suppressed or materially unexpressed states. This means that there are always some ideological positions that are either not manifested at all, or they are manifested in only a partial or compromised manner.

THE IDEOLOGICAL CULT FACTION MODEL

The Ideological Cult Faction Model will be treated as an auxiliary of the Proscriptive/Prescriptive Ecological Strategy Model, particularly designed to address the factional dynamics postulated to be characteristic of the ecclesiastic-communal cults. As suggested above, cult factions are treated here as arising out of and reproducing the contradictory tendencies generated by the conjunction of the major social structures constituting the organization. Its factions are identified with and promote particular ideological rules and protocols that favor the interests of their participants As an auxiliary model, the deontic proscriptive/prescriptive contrast is useful to characterize the relative ideological positions of factions as proscriptive or prescriptive. In an immanently sacred world, the dynamics promoting alternative proscriptive to prescriptive ideological positions would arise from the ambiguity over the polluting costs of ritual in comparison to its claimed sanctifying gains. Polluting costs would necessarily arise from earthwork and timber construction, the building of mortuary facilities, the procuring of raw materials (both exotic and local) for the production of ritual artifacts and facilities, and so on. These costs could not be completely avoided since the effectiveness of the sanctifying moment of ritual would largely hinge on producing material media as felicitous iconic warrants of ritual. In effect the essential contradiction of human existence that is intrinsic to an immanently sacred world applies as much to ritual as to ecological practices, and, of course, they are intrinsically related.

In these terms, it is logical for the Ideological Cult Faction Model to postulate that a faction inclined toward a prescriptive ideological position would typically argue for a ritual strategy that would maximize the material moment of ceremonial interventions on the grounds that the greater sanctifying gains achieved would compensate for and even overshoot the increased polluting costs that such escalating construction would incur. A faction inclined toward a proscriptive ideological position would promote the opposite strategy by arguing to minimize the material moment of ceremonialism, thereby minimizing the polluting output while still, in its view, achieving the collective sacred goal of world enhancement.

A proscriptive cult faction would usually not oppose earthwork construction or radical post-mortem mortuary manipulation, as such, but it would encourage restraint within the current context. Of course, under the appropriate conditions, a proscriptively inclined faction might reject the current situation and retrogress by abandoning older constructions and opting for lesser or none at all. This historical process of proscriptive and prescriptive negotiations among factions would be demarcated by an oscillating and/or spiraling of material outcomes. Because the factions cannot deny the prime raison d'être of the cult without denying the intelligibility of their ideological positions and competition, this will tend to ensure, but not guarantee, a general consensus, or, possibly better stated, a negotiated contingent consensus concerning what precise modalities should be taken to satisfy the major material needs demanded for the cult's ritual to be effective. As pointed out above, these needs can be broadly treated as earthwork and timber construction, mortuary facilities, and material resource procurement. However, the potential is always there for each faction to perceive the other as promoting inadequate or even damaging activity, either more polluting than sanctifying, or vice versa, thereby reducing the cult's capacity to effectively discharge its collective commitment, its raison d'être. Thus competition could become quite fierce and, under intransigent conditions, it can lead to major organizational ruptures, that is, disruptures. This is precisely what the Ideological Cult Faction Model will postulate led to the "collapse" of the Ohio Hopewell.

THE CONGREGATIONALIST↔PRESBYTERIAN↔ EPISCOPALIAN (C↔P↔E) CONTINUUM

In terms of the Ideological Cult Faction Model, the source of factional dispute is the contradictory tendencies generated by social structural conjunction. In elucidating the Ecclesiastic-Communal Cult Model (chapter 12), it was postulated that the emergence of the ecclesiastic-communal cult resulted from the conjunction of clergy/laity and senior/junior age-grade social structures and that this combination aggravated the preexisting contradictions embedded in the cult-clan relations. It is these structural contradictions that are postulated to be the source of proscriptive/prescriptive factional crystallization. The integration of senior and junior age-grades into a single organization exacerbated intergenerational tensions, of course. The clergy/laity structure promoted competition between shamanic priest and laity leadership for control of the cult's affairs. The contradictions in the above led to aggravating the cult-clan negotiations over the mortuary treatment of shared deceased. Because senior laity leaders also would tend to have important positions in their clans, they would promote proscriptive or moderate post-mortem manipulation of the deceased. In contrast, having less direct involvement in their clans, the junior laity leadership, in alliance with the clergy, would promote radical or prescriptive post-mortem manipulation of the deceased. Directly correlated with these proscriptive/prescriptive post-mortem mortuary regimes would be corresponding proscriptive/prescriptive positions across the material features, facilities, and artifacts that made up the custodial iconic warrants of mortuary/world renewal ritual: embankment earthworks, Renewal Lodges, custodial regalia, as well as the labor and long-distance procurement and exchange practices that these materials required.

These proscriptive tendencies of the senior age-grade laity leadership and prescriptive tendencies of the junior age-grade and clergy leadership formed the major proscriptive/prescriptive factional divisions and the correlation of these divisions with the material media of ritual as the expression of the ideological stance of a cult allows estimating the position of the cult on the C$\leftrightarrow$P$\leftrightarrow$E continuum. As discussed earlier, the presbyterian position is a "middle-of-the-road" stance, characterized as a fluctuating balance of power between the proscriptive and prescriptive factions. Therefore, the senior age-grade leadership, generally leading the proscriptive faction, would favor a congregationalist stance and, should it prevail in the cult affairs, there would be a left-of-center ideological stance manifested in the material media of world renewal ritual. An episcopalian stance is favored by the prescriptive faction. This faction, as pointed out above, would tend to be supported by the leadership of the junior age-grade in alliance with the clergy. Therefore, should the prescriptive faction prevail, the cult would move right-of-center toward the episcopalian stance.

For the Middle Woodland of the Central Ohio Valley, then, a dynamic and potentially dangerous balancing process between congregationalist and episcopalian bifurcation was at play. Factional negotiation would not guarantee that the "middle way" would prevail. If either extreme emerged, the cult would destabilize. An extreme congregationalist stance would mean that the integration of senior and junior age-grades as "equals" had broken down. The cult would either split or revert to being a simple unitary communal cult under the dominance of the senior sector. Similarly, if trends toward the episcopalian extreme developed strongly, this would generate such conflict both within the cult and between cult and clan that it would probably lead to an abrupt rupture, also probably at the senior/junior age-grade link. Indeed, it is postulated that, as a result of polarization of factions, these two tendencies did occur, simultaneously, and caused a rupture that initiated a cascading "collapse" of the system of Ohio Hopewell cults across the Central Ohio Valley.

The next logical step would be to use the premises of the Ideological Cult Factional Model to postulate a set of material indicators by which to map the fluctuation of the ideological stances of the major ecclesiastic-communal cult locales along the C↔P↔E continuum. However, before moving to this stage, a major question that has been postponed until now must be addressed, this being the organization of the clan and its ceremonial aspect. This is an important dimension by which to round out the total model and clear the way for a focus on the variable stances of the ecclesiastic-communal cults themselves.

THE CLAN CBL MOUNDS

In his apparently exhaustive compilation of the literature on the known Adena and Ohio Hopewell CBLs, Fischer computed that there are a total of 1,559 Adena and 1,474 Hopewell known burials.[2] Of the latter, he claimed only 78 were nonmound burials. Of the 50 "completely" excavated Hopewell mounds, he calculated an average of 22.62 burials per mound. Assuming that these numbers represent the demographics of the Central Ohio Valley, and assuming at least seven hundred years of history, 300 B.C.–A.D. 200 for Adena and 100 B.C.–A.D. 400 for Ohio Hopewell, the overall population would be too small to be able to provide the labor that these earthworks implicate, not to mention the labor for timber construction, long-distance expeditions, specialized artisan production, and so on. To account for the disparity between the labor involved and the relatively small number of burials, it has been argued that intercommunity alliances would allow for pooling the labor of several separate communities.[3]

Certainly, cooperative pooling of labor was postulated under the Autonomous Cult Model as part of the dynamics that led to the emergence of Ohio Hopewell. However, this still does not account for the small burial numbers, since the labor required to produce the earthworks, *in toto*, still radically exceeds the total known burial numbers and the demographics that can be extrapolated from them, given the funerary and modular views. Therefore, this suggests that a significant proportion of the regional populations was systematically not included in the terminal burials of these Adena and Ohio Hopewell CBLs. The argument that can follow is that this exclusion registers social segregation as being intrinsic to the Ohio Hopewell communities. This means that those buried in the major Ohio Hopewell CBLs were privileged and all the others were excluded inferiors. Prufer makes this quite explicit, axiomatically declaring that "all those buried in the great mounds are those of the privileged, for only they and their bereaved survivors could have marshalled the amount of communal labor outlay and wealth necessary for the construction and provision of such tombs." Hence, in his view there is a pressing need "to find out where the common people were buried".[4]

There is no question that Prufer is correct in saying that only a small proportion of the overall Ohio Middle Woodland population is represented in the Ohio Hopewell CBLs and that there is a pressing need to find out where the majority of the deceased were buried. However, this problem has been addressed only from the perspective of the Dispersed Hamlet / Vacant Center Model. Since Prufer first proposed his Vacant Center Thesis, subsequent development of this model has made the elite/commoner argument even less credible than it was then. Under the Dual Clan-Cult Model and its ancillary Dual Ritual Spheres Model, the ecclesiastic-communal cults and the clans constituted autonomous ritual spheres that would have negotiated the appropriate mortuary treatment of the deceased to whom both had legitimate responsibility. As discussed under the Split Age-Grade Mortuary Model, the result of this negotiation would be a rather low overall mortuary retention rate for the cults. The logical answer to the above question, then, is to say that the vast majority of the Middle Woodland deceased would be in Clan CBL locales. The question, then, is not where are the "commoners" buried but where are the clan CBLs located?

This raises the important point: What would be the material correlates of clan CBLs? In terms of the Mourning/World Renewal Mortuary Model, mortuary rites vary along the focal participant↔ symbolic warranting continuum with the former pole marking funerary and closely related rites (e.g., kinship mourning, naming, adoption rites), and the latter pole marking world renewal rites. This latter pole, of course, is particularly indicated by regimes of prescriptive post-mortem manipulation and the former, by regimes of proscriptive post-mortem manipulation. Clan-based CBLs, then, would tend to display the indicators of a proscriptive or minimizing post-mortem manipulation strategy. They would be relatively small burial locales, probably mounds, distributed in accordance with the dispersed domestic settlement pattern. The age and gender make-up should reflect a domestic demographic structure and the total number in any given mound would be commensurate with the dispersed, relatively small extended families postulated for the Middle Woodland.[5] However, as part of custodial usufruct and inclusive tenurial relations, a mound might include the deceased of different extended families that were related by marriage or descent. Thus, small to relatively large is possible, ranging from five to possibly fifteen or twenty burials.

While these should emphasize the funerary orientation with extended inhumation dominating, a number of, although probably not many, terminal burials could have occurred following an extensive laying-in period held at the locale of the cult to which the deceased belonged. Thus, the removal of the deceased from the cult crypt followed the termination of the cult's laying-in period, and its disposition in the clan CBL would result in some disorder of the skeletal bones.

Rather minor personalty would accompany the burials, demarcating the social standing of the individual *within the clan* rather than within the cult. Since these are social contexts in which leveling mechanisms would be at work, some artifacts such as necklaces and bracelets of shell, copper, and bone beads would be expected, but not the type of formal items that are found in the burnt deposits at the Hopewell site or in the crypt burials and, certainly, no custodial regalia. Burial features would be minor and/or standard for most deceased, possibly limited to log cribbing with collective burials.

According to the most current data, since the Middle Woodland domestic locales were dispersed primarily across the valleys, higher terraces, and slopes, and not across the more distant uplands, the distribution of the postulated small clan CBL mounds would have largely correlated with the distribution of early lowland Euro-American farming settlements. They would also have been small enough for a private, American family farm operation to plough across or remove if they impeded planting and harvesting. Of course, their relatively small size would also make them interesting targets for amateur diggers. The result would be progressive destruction. Indeed, in this regard Moorehead comments on the number of small mounds in the zone around Chillicothe and in the Miami drainages during the late nineteenth century, saying that the mounds were "almost as numerous as farmsteads now," and pointing out that scores and even hundreds had been opened and destroyed by his time.[6] Interestingly, there is more than a quantitative comparison implied in his comment. These numerous American farmsteads would have been dispersed and the implication here is that the many mounds were also dispersed, precisely along the lines suggested above.

This widespread destruction of small, dispersed mound earthworks encouraged him to proceed to do the same thing by excavating a number of the few that had been left undamaged—although in this case he did carry out some minimal amount of recording. The reported results, for the most part, conform to what was postulated above. The

Redman-James Group, Moorehead's Mounds 34, 35, and 36, as well as the nearby Mounds 37 and 39, were five lesser mounds located about 5 km east of Chillicothe and spaced between 150 to 300 m apart, more or less conforming to the average distance between dispersed "hamlets" as postulated by Pacheco.[7] Mound 34 was the smallest, with only one extended burial, about one foot below the surface. Mound 35, about 170 m northeast of the latter, was about 12 m in diameter and 3 m high. It contained five extended burials dispersed across the floor. The floor had a large ash pit. Apparently, during the construction of the mound, a necklace made of 42 copper beads, 1 mussel shell, and 5 snail shells was dropped or placed on the surface. It was found 1 m above the floor level. Mound 36 contained fifteen extended burials, fourteen adults and one child, on a prepared and red-burnt floor. The child had two copper rings and three shell beads. Most of the adults had no surviving artifacts associated with them. Those who did had flint knives (i.e., bladelets [?] and bifaces), and a stone celt and gorget. The bodies were covered with a thick layer of charcoal, suggesting some form of fire-related ritual as part of the funerary ceremonies, since this fire ritual did not involve human cremation. The association of these burials with the use of copper and major burning rites supports the interpretation that these were Middle Woodland and not Late Woodland burials.

Mound 37 was a fairly large oval mound, 15 m by 30 m and 4 m high, with an east-west axis. On the prepared floor there was an 8-cm-thick mixed stratum of burnt bones and ashes, about 6.3 m by 8.75 m in extension (no suggestion that the bones were human). On the center of the floor was a child, about 10 years, with a necklace of 119 marine shell beads. About 2 m above this burial was an extended adult burial, which Moorehead specifically described as not being intrusive since about .6 m above and slightly south of this burial a wolf skull had been deposited. The extended burial, as he comments, had some bones missing (e.g., some of the cervical and lumbar vertebrae), and some improperly placed. All this is clear evidence that this was a burial that occurred after an extensive period followed by the transportation of the body, possibly only as bones, and its reconstructed placement in this burial. The individual was buried with a bracelet of 40 shell beads at his wrist.

About 420 m northeast of Mound 37 was Mound 39. It contained two extended burials, one being accompanied by a slate gorget on its head. The head of one was placed over a pit in which Moorehead found some mica flakes. A copper bracelet and a chalcedony "spear head" were found, the latter near the edge of the floor and the former on the floor, about 6 m from the edge. Apparently these were not associated with burials. A pit, 1 m deep by .6 m wide, was found with pieces of human bones, probably the remains of either a flexed or bundle burial. Potsherds were found. Unfortunately, no illustration or any comments sufficient to identify it by type were given.

The use of mounds with prepared floors and features, burnt deposits, and thick layers of mixed animal bones and ashes; the presence of necklaces and bracelets of shell beads and some copper beads; the presence of mica, the use of the wolf skull; and the bifaces; evidence of ritual involving the construction of the mounds, and so on, all echo but are distinctly different from Ohio Hopewell mortuary practices. Therefore, while these data suggest Middle Woodland as the probable temporal period, they also suggest that the groups responsible sustained arm's-length relations with those groups responsible for mortuary events in the nearby Ohio Hopewell earthworks. The almost exclusive dominance of extended inhumations, the evidence of some manipulation that, nevertheless, was "corrected," indicating care to emphasize the person as participant rather than as a symbolic warranting medium, manifest a symbolic pragmatics oriented to a funerary rather than a post-mortem sacrificial mortuary program. The only exception is, of course, the possible flexed or bundle burial—a rare event that may have been expressive of a type of midwifery ritual. When all these attributes are correlated with the spatial distribution of these mounds, it definitely points to a dispersed domestic clan-based CBL pattern. The overlaps and divergences from the Ohio Hopewell mortuary program are just about what could be expected if cult and clan systems operated independently but in parallel with each other, as postulated by the Dual Clan-Cult and Dual Ritual Spheres Models.

The dispersed cluster of lesser, probably Middle Woodland mounds described above is not unique. Moorehead reports a similar pattern about 6.5 km west of Chillicothe, a set of three conical mounds referred to as the Slate Mills Mound Group.[8] Mound 45 was the largest and contained five extended inhumations, four adults, one child. They were contained in a large timber cribbing with bark sheets over and under them. Associated artifacts were shell bead necklaces, a hematite cone, cut mica crescents and a copper bracelet and beads. Another cluster of five mounds was reported by Moorehead, his Mound 17 being the largest of the five. These were relatively isolated on table land about 6 km southwest of the nearest major Hopewellian embankment, the Frankfort earthworks.[9] Apparently Moorehead focused on this set of mounds shortly following his excavation of the Porter Mound 15, found in the infix of the Frankfort embankment earthworks. Up to that time, this latter mound may have been the "richest" Ohio Hopewell mound that he had excavated. Although the artifactual data were not quantitatively as dense as he was to find a few years later at the Hopewell site (published in 1922, although he first excavated there in 1891), the full range of Ohio Hopewell custodial regalia was represented. Therefore, his expectations were probably high when he tackled these nearby lesser mounds of the Coiner Farm group (Mound 17). Instead, what he found was more akin to what was reported above, only with fewer indications of Ohio Hopewell affiliations. In Mound 17 he found five extended burials, one with a broken projectile point, a diamond-shaped stone gorget, and 32 shell beads. At the center of the mound another skeleton was revealed. Before he had advanced very far, however, the walls collapsed on him and, apparently, Moorehead abandoned any further digging there.

It might be argued that much of the content of these lesser mounds is more appropriately labeled Adena or Early Woodland. Indeed most of these isolated mounds are classified by Fischer as Adena. A number of flint bifaces associated with one of the fifteen extended

burials of Mound 36 could comfortably fit within the Late Adena Robbins and related types. Also associated with the extended burials of this
mound were two Adena style tubular pipes, as well as stone celts and
tablets that would also fit under the Late Adena label. The use of mica
crescents, copper bracelets (Slate Mills Mound 45), modified "bow-tie"
slate gorget (Mound 39), and so on, are also noted. The Coiner Farm
Mound 17 is placed by Fischer in his undetermined Adena period category, meaning that it could be either Early or Late Adena, in his dating
system. Mound 45 of the Slates Mill site is classified by him as Late
Adena. He does not list the Redman and James Group, Moorehead's
Mounds 34, 35, 36, 37, and 39.

On the other hand, there are a number of mortuary attributes
associated with these lesser mounds that do not at all fit within the
"classic" Adena burial mounding. These extended burials were all laid
on a prepared burnt floor and covered by a thick charcoal layer—
distinctly not the type of mortuary attributes associated with the "classic" Adena CBL mounding and its accumulated "stacking" of prepared
tombs, crypts, and platforms. Rather, it is much closer to Hopewell.
Also there were indications that the final placement of the bodies followed extensive periods of exposure. For example, an extended inhumation in Mound 37 had a number of bones missing and/or placed
improperly, suggesting that the bones had been gathered, cleaned, bundled, and then brought to the burial locale where they were reassem-
bled. While the final burial being the end result of significant post-
mortem manipulation is expected for both Adena and Ohio Hopewell
CBLs, such displacement and careful reassembling would be more characteristic of the clan CBL pattern, as postulated above. That is, following the public laying-in period rites at the Renewal Lodge of a local
cult, the deceased would be removed from the cult locale while still
in "participant" mode, and a terminal funerary burial would be carried
out by the clan's CBL, reconstituting the deceased in her/his participant
role by reassembling the bones.

In any case, Fischer's identification of these groups of lesser
mounds as Adena does not present a particular problem for him, tem-

poral or cultural, since he identified Late Adena as being largely contemporary with Ohio Hopewell. Using the radiocarbon dates available at that time, he dated Late Adena and Early Hopewell as initiated at the same time, ca. 200 B.C., with Late Adena terminating ca. A.D. 500, and Late Ohio Hopewell terminating ca. A.D. 550. Although these latter dates are not accepted today, if Clay's temporal model for Adena is accepted, then Middle and Late Adena, encompassing the period 150 B.C. to A.D. 250, still places much of the Adena and Hopewell archaeological records as overlapping.[10] Furthermore, even if these were "Adena," they are less similar to classic Adena and more similar to classic Ohio Hopewell.

There is another point that must be made here. Fischer's study is firmly anchored within the monistic modular view. For him both the Adena and Hopewell records mark autonomous communities maintaining distinct, boundaried territories, and marking their separateness through opposing stylistics. This paints the interesting sociopolitical picture of possibly mutually hostile modular Late Adena and Ohio Hopewell communities—based on the principle of exclusive territorialism—practicing different ideologies and living beside each other for many generations. Given the intercommunity aggressiveness that is assumed to be intrinsic under this monistic modular view, all this does stretch credibility a bit. It is clear that the structuring of the archaeological record that Fischer has summarized is much more complex than his modular approach can handle. Put plainly, the Adena and Hopewell records are most fully realized in the major Adena and Ohio Hopewell CBLs, not in these lesser mound sites. The former are quite distinctive. These major CBLs are complemented by the dispersed domestic habitations, which (it has been claimed here) mark autonomous clan-based practices. All this suggests that we should not expect that the material correlates of the clan-based mortuary ritual replicate the material correlates of either the Adena or the Hopewell cult types. There will be some overlap because of the shared use of the deceased in their differentiated mortuary ritual. The data presented above indicate these overlaps and differences. In fact, the use of Adena-like artifactual material in these postulated clan-based mounds existing only a "stone's throw" from major Ohio Hopewell sites can easily be explained on the grounds that clans would wish to sustain an arm's-length stance from the Ohio Hopewellian world renewal cults. One means of doing so would be to draw on preexisting Adena-like stylistics, thereby clearly constituting these mortuary acts as distinct from the type of mortuary acts promoted at the Ohio Hopewell CBLs.[11]

This discussion was initiated by pointing out that the major Ohio Hopewell CBL mortuary data are inadequate to account for the amount of labor required to build the monumental earthworks and related constructions of this region and period. A solution, admittedly insufficiently grounded empirically for the moment, has been sketched out by arguing from the perspective of a polyistic locale-centric social system as postulated by the Dual Clan-Cult Model. Of course, the clan CBL aspect must still be treated as tentative and more research will be required.

This analysis can be applied retroactively to the earlier critique of Mills's funerary view. Much of his criticism of Squier and Davis's sacrificial model was grounded on his claim that cremation was simply the normal funerary mode of this region, and he reinforced this by making reference to the mortuary data of Tremper, which were also primarily cremations. "The sites of the Mound City group were found to be similar in every way to that of the Tremper mound, on the lower Scioto, where the sacred structure, with its crematories and depositories was used *solely for the cremation and burial of the dead, and for the attendant funereal ceremonies.*"[12] It was pointed out earlier (chapter 14) that Mills ignored what would appear to be a major difference between the two mortuary practices, this being that the cremated mortuary deposits were primarily individual at Mound City but collective at Tremper, aggregated in four major altars. However, in expanding the Ohio Middle Woodland mortuary sphere to include the multiple lesser mounds as described above, it becomes clear that Mills's invoking the Tremper mortuary data to assert that cremation was simply the way funerals were conducted in that region at that time is not tenable. In quantitative terms, it would appear that, regionally, cremation was the exception. It became the rule only in the context of the major Ohio Hopewell earthworks, as argued under the Mourning/World Renewal Mortuary Model. With this clan-based CBL perspective outlined, the next chapter can refocus on the material correlates of the Ideological Cult Faction Model of the Ohio Hopewell world renewal cult system.

Chapter 22

TIME AND THE MATERIAL CORRELATES OF IDEOLOGICAL FACTIONALISM: A HISTORICAL PERSPECTIVE

Since any interpretive assessment of the data in terms of the Ideological Cult Faction Model hinges critically on the diachronic dimension, a summary and refinement of the earlier chronological framework proceeds by postulating and applying the material correlates of proscriptive/prescriptive ideological postures. I have commented earlier that in her various publications Greber has raised the importance and difficulty of controlling the diachronic dimension in Ohio Hopewell—in particular, the sequences related to the development of individual locales and how the latter can be used to map the historical relations among these locales. Many different aspects of the archaeological data have been used as chronological markers. In terms of the Ideological Cult Faction Model, the embankment earthworks would have figured so prominently as monumental iconic warrants in the production of critically important ritual that their forms, material make-up, spatial and directional orientation, and magnitude—as well as absence—would have been major matters at the core of ideological negotiation. This would make their production and abandonment major occasions involving collective decisions mediated by intense factional discussion and, possibly, even confrontations.

This argument reinforces the rationale for the initial chronological seriation of the sites based on embankment attributes of plan and profile forms.[1] At that time the two S-Profile versions, the SR-Profile and the SL-Profile, were postulated as equivalent variants that largely overlapped in time. It may be that the SR-Profile, best represented in the Adena Sacred Circle, preexisted the SL-Profile. But Clay's analyses of Peter Village, which displays the SL-Profile attribute, suggests the opposite.[2] That is, the SL-Profile attribute may have preexisted the SR-Profile attribute. In any case, there would be a significant period of overlap with the Middle Adena, circa 150 B.C.–A.D. 1, and Early Ohio Hopewell, ca. 100 B.C.–A.D. 1.[3] In this system, then, the Mt. Horeb embankment earthwork of the Middle Adena would be best represented by the sacred circle (Simple G-Form/SR-Profile), and representing Early Ohio Hopewell would be the C-Form/SL-Profile configuration. Since Cedar Bank Works (fig. 1.5) on the east bank of the Scioto River, just 2 km north of Mound City, displays all the requisite C-Form/SL-Profile attributes, then—despite the fact that it has a platform mound—it would be contemporary with the early period of the Hopewell site, marking the Early Ohio Hopewell phase.[4] Using this initial seriation, the chronological scheme of the Chillicothe region can now be refined, starting with the Mound City/Hopeton Complex. Given that the SL-Profile attribute demarcates the Early Ohio Hopewell phase, this would place Component C (the Shriver Works) of the Mound City Cluster as Early Ohio Hopewell, along with Cedar Bank Works to the northeast of Mound City.[5] Confirming this date is the burnt deposit in a large, 2-m-diameter crematory altar found in the mound at the center of Component C.[6] There were actually several superimposed altars. In the large basin of the last used altar there was the residue of human cremations, the only recognizable bones being two humeri with five copper bracelets each. While Greber notes that such artifacts are recognized as Adena, she points out that finding ten in one deposit is more akin to Hopewell than Adena. Finally, she quite rightly stresses that the other mortuary features, the cremation, the basin, the mica, the sand stratification, and so on, are Hopewellian traits.

> This context indicates to me behavior patterns which are closer to the remains of actions seen in several of the mounds at the Hopewell site (and at Mound City) than in the Adena Mound. The fine workmanship and the total number of bracelets deposited also appear closer to a more common Hopewellian behavior than a Scioto Adena one. (Greber 1991, 17)

Given this set of features and artifacts that it shares with the contents of the mounds of Mound City Proper, and its propinquity to the latter, the cultural and organizational affiliation of these two seems unquestioned. Most importantly, the temporal affiliation is also indicated. Assuming that the cremation and its associated materials were contemporary with Component C, it would have predated the Component A (K-Profile) embankment that encircles Mound City Proper. Component C would also be largely contemporary with Cedar Bank Works, and partially contemporary with but extending later than Component B of the Mound City Cluster, the Adena sacred circle that is immediately northwest of Mound City Proper. The Component A embankment, displaying the subsymmetrical K-Profile attribute (KPp-Profile), can be dated to the Middle Ohio Hopewell phase.

This does not mean that the mounds of Mound City Proper, those encircled by Component A, are all Middle Ohio Hopewell. Baby and Langlois have addressed this question.[7] Working with the results of a series of excavations during the 1960s and 1970s, they have recovered the ground plans of all the known structures under the mounds of Mound City Proper (fig. 22.1, but Mounds 7, 3, 18, and 2 were not reexcavated). The majority, they noted, were rectilinear—square, rectangular, parallelogram, trapezoidal—while one was oval. They also pointed out that Mills's mapping of the post holes may be inaccurate. He tended to indicate that they formed circular or oval plans when, in fact, they are rectilinear. From this Baby and Langlois identified two major patternings that may have chronological implications: the wall construction method and the axial orientations. Of the two types of wall construction methods, one using single-post walls on all four sides and one using the "classic" double-post side walls and single-post end walls, they suggested that the former was earlier and the latter later.

With one exception, each Renewal Lodge had a central entry at each end and a screening arrangement "blocking" the entrance. By drawing the axes through the double doorway system of the majority of the buildings, they noted that three axial orientations prevail: one set on the gateways of Component A, one N–S set, and one set pointing towards the complex made up of Mounds 7, 3, and 18, situated slightly southwest of the center of the space encircled by the Component A embankment. They suggested that these sets "provide a partial chronology, since it must be assumed that the target of orientation predated the oriented structure" (29). From this they divided the mounds into a pre–Component A and a post–Component A clustering. The pre–Component A set of buildings were (1) the N–S oriented structures, (2) the buildings under Mounds 7, 3, and 18, and (3) the buildings oriented toward them, these being under Mounds 12, 13, 15, 16, 17, and 23. The post–Component A buildings had their axes oriented toward the gateways, and these were under Mounds 5, 8, 9, 19, 21, and 22. Since the buildings on the floors of Mounds 7, 3, and 18 were the targeted mounds, they conclude that these may be the oldest and, of course, preexisted the Component A embankment. They also comment that a set of post molds extending west from Mound 10 may demarcate a timber wall that preexisted the construction of Component A.

Based on the presence of midden under the southeast embankment wall, Brown and Baby come to the same conclusion. Mound City Proper was initially occupied and used prior to the building of the K-Profile Component A embankment.[8] They also noted, following Mills, that the structures under Mound 18, Mound 13, and Mound 7 (and possibly a fourth, Mound 12) marked two periods of construction, with the second activity floor being built on top of a lower activity floor. Based on ceramic evidence they divided Mound City into early and late phases and these correlated with the pre–Component A embankment and the post–Component A embankment periods. Based on limited radiocarbon dates, they suggested the pre–Component A early phase dates up to ca. A.D. 1, and the late post–Component A phase dates up to ca. A.D. 150.

It seems reasonable to claim then that the Early Ohio Hopewell phase at the Mound City Cluster is marked by the S-Profile embankment earthworks and dates between ca. 100 B.C. to ca. A.D. 1. Although the Cedar Bank earthwork is about 2 km upriver, its open side is oriented toward Mound City Proper. This suggests that Cedar Bank and Component C were important material constituents of the ritual that occurred at Mound City Proper during its Early Ohio Hopewell phase. This means that the later Early Ohio Hopewell mortuary ritual, illustrated by the patterning on the second-phase floor of Mound 7 as discussed earlier, should be treated as integrated with the activity mediated by the crematory altars found in the center of Component C. If the C-Form/SL-Profile works are correlated with solar-related rites and the sacred circles generally mediated lunar-related renewal rites, then the solar/lunar duality that was to be expressed in the later Sacred Dual C-R Motif (e.g., the Newark site) would have been displayed for the first time in the above combination of the monumental embankment earthworks of the Mound City Cluster, marking the Early Ohio Hopewell phase. Therefore, from the beginning of Ohio Hopewell in this immediate region, the complex of embankment earthworks making up the Mound City Cluster manifested an integration of lunar-related and solar-related symbolic pragmatic meanings. This is precisely the condition postulated by the Autonomous Cult Model as generating Ohio Hopewell, namely, the integration of mutually autonomous Adena-type senior and junior communal cults.

As argued earlier, the Middle Ohio Hopewell phase is marked here by the Mound City/Hopeton Complex, demarcated by the Component A encircling Mound City Proper along with the K-Profile components of Hopeton. Indeed, the Mound City/Hopeton Complex may actually be the first explicit monumental expression of the Sacred Dual C-R Motif. This would have developed in stages. The Hopeton site sector is made up of two preexisting sacred circles, thereby manifesting lunar-related symbolic pragmatic meanings. The construction of the Hopeton Circle would have initially supplemented the lunar-related meanings of these sacred circles, possibly representing the Heavens, under the lunar aspect, to complement the sacred circle as the Underworld, also under the lunar aspect. The Hopeton Rectilinear would have been added to integrate the latter two circles as representing the Underworld in a rectilinear form, and, therefore, it would have constituted the first, albeit subsymmetrical, expression of the C-R Motif Configuration, and possibly at the same time, combined both solar and lunar aspects, since there are solar alignments embedded in the Hopeton Square.

The final stage was probably the construction of both the Component A embankment and the two parallel embankments or, as it has been termed, the aggregation component, leading from the Hopeton C-R juncture to the riverside and oriented directly at the ramp that leads to the eastern gateway of Component A, thereby integrating Mound City Proper and Hopeton into a single Complex G-Form. As argued earlier, this aggregation element may have been intended to have Component A "infixed" between the Hopeton Circle and the Rectilinear. This would make up the first (subsymmetrical) expression of the Paint Creek C-R Configuration. Thus, the Mound City/Hopeton Complex would have embodied a dual symbolic pragmatic meaning, becoming the first manifestation of the Sacred Dual C-R Motif in the C-R Configuration form. This "original" Sacred Dual C-R Motif locale clearly demarcates the Middle Ohio Hopewell phase, ca. A.D. 1–150.

Recent work at the Hopeton Rectilinear has revealed supporting chronological evidence. Excavations have exposed burnt residue on top and under the embankment. The radiocarbon dating of this and of charcoal samples taken from "sub-mound and sub-earthwork features at the Mound City Group" closely match. The former uncalibrated date is 1930 ± 60 B.P. (Beta-96598). Bret Ruby, the excavator, comments that "if this chronology proves accurate, it suggests that these two mound and earthwork complexes may have functioned together *as a single closely related unity.*"[9] There seems to be a convergence of separate lines of evidence, morphological, ceramic, and radiocarbon, that supports referring to this set of sites as the Mound City/Hopeton Complex. This sequence also nicely fits the Ecclesiastic-Communal Cult Model, as is discussed in the next chapter.

The emergence of Late Ohio Hopewell phase in this immediate region can be seen as initiated by the construction of High Bank, probably the first symmetrical version of the C-R Configuration. This configuration would have been fully entrenched with the subsequent construction of the first Paint Creek C-R Configuration site. This could explain the construction anomalies at Seip and Liberty Works that were pointed out in chapter 4. It was argued then that whereas at Seip (fig. P.4), the small K-Profile circle found in the asymmetrical part of the infix was completed before the construction was started on the Paint Creek C-R Configuration, in the case of Liberty Works (fig. P.3), the equivalent small K-Profile circle (found within the lobed circle) was still not completed when the decision was made to suspend work on it and build the large Paint Creek C-R Configuration that we now see. In chronological terms, this suggests a joint scenario in which the two cults at Seip and Liberty Works monitored each others' activities and changes in the construction practices in one promoted changes of plan in the other. In this case, then, it would not be unreasonable to argue that Seip had already completed its small K-Profile circle when Liberty Works, emulating its near neighbor Seip, started building its own. Before it had completed the small circle, however, the Seip group initiated the building of the large Paint Creek C-R Configuration earthwork. In this scenario, Liberty Works would have abandoned the building of this small circle and turned to building their own version of the Paint Creek C-R Configuration.

This suggestion of a rapid shift in construction strategy gains some empirical support when it is noted that the construction work at

Liberty Works was "sloppy." Squier and Davis described this as the builders' "carelessly scooping up the earth . . . [and] leaving irregular pits," and so on.[10] That is, the "carelessness" is consistent with the view that the construction of the earthwork was part of a world renewal rite that was itself under time constraints. Of course, if this is the case, then it could be asked why there was no correction of this "careless" work once the ritual construction was completed. Such corrective work would have been physically simple. It was earlier suggested that this was not a "failure." As part of the total construction project, once formed the "inner" borrow pits could not be eradicated any more than an embankment normally could be dismantled.

It could be reasonable to conclude, then, that the construction of High Bank was the first expression of the symmetrical C-R Motif Configuration and that following this construction, Seip was the first expression of the symmetrically balanced Paint Creek C-R Configuration. Liberty Works emulated Seip, and therefore, it was the second formal expression of the Paint Creek C-R variant. This also suggests that these two contemporary groups may have been caught up in a form of rivalry in the pursuit of cult reputation, partly the result of their being the outcome of the postulated amicable split of the original Mound City/Hopeton cult. Each cult would challenge the other through earthwork construction, instigating a further escalation, and so on.

Since the Hopewell site later emulated the construction of Seip and Liberty Works by adding the key C-R components, this may also have correlated with the abandonment of the Renewal Lodge complex and the building of Mound 25 inside the "half-moon" SL-Profile feature. Indeed, this may also be why the "half-moon" SL-Profile feature was constructed—without any gateways. It was a symbolic closing of the C-Form/SL-Profile period of the Hopewell site. This suggests that the Hopewell site would have continued in its C-Form/SL-Profile guise when the K-Profile era at Mound City/Hopeton emerged, demarcating the Middle Hopewell phase at that site. If this is the case, then Mound 25 and temporally related mounds at the Hopewell site (whichever these were, e.g., Mound 2, Mound 11, among others) incorporate both

Early and Middle Ohio Hopewell phases. If, as stated above, the add-on C-R components can be used to imply that the Hopewell cult was emulating the Paint Creek C-R Configuration, then the latter already must have been in place at Seip and Liberty Works. This does not mean that the Hopewell site was abandoned with the construction of Mound 25, marking the end of the Early and Middle Ohio Hopewell phases at this site, only to be subsequently reoccupied in the middle Late Ohio Hopewell, as marked by the C-R Configuration, although this is a possibility. It is more likely that cult occupation following the construction of Mound 25 and the add-on C-R components was continuous until the site was abandoned in the later Late Ohio Hopewell phase of this site, probably corresponding with the construction of Mound 23.

To summarize this discussion (see table 22.1 below), the first era of Mound City was marked by the Mound City Cluster of Components B and C and Cedar Bank. The Middle Ohio Hopewell phase was marked by the K-Profile Component A and the emergence of the Mound City/Hopeton Complex. These two construction styles and phases, together, correspond to the C-Form/SL-Profile era of the Hopewell site. That is, while the Mound City/Hopeton Complex prevailed in the central Scioto, marking the Middle Ohio Hopewell phase, the Hopewell site retained its "pure" C-Form/SL-Profile earthwork orientation. High Bank, Seip, and Liberty Works, and similar C-R Configuration sites, overall, delineate the Late Ohio Hopewell phase, both the early and the middle of the Late phase, in the Central Scioto zone. The add-on C-R components at the Hopewell site are postulated as marking the equivalent early Late Ohio Hopewell phase at this locale and the building of Mound 23 as marking the termination of the middle Late Ohio Hopewell phase at this site, probably when it was abandoned. Baum and Works East for the Central Scioto, and Frankfort for the North Fork, demarcate the later or terminal years of the Late Ohio Hopewell phase, and, of course, the last of the major embankment earthwork constructions of the central Scioto region.

It was earlier postulated that the construction of the Complex

C-O embankment earthwork of the Turner site would probably have occurred just prior to or during the early Late Ohio Hopewell phase, about the same time as the addition of the C-R components at the Hopewell site. Thus the Circle-Oblong era of Turner and, probably, the Conjoined Renewal Lodge Complex correlate with the terminal years of use of the Mound 25 Renewal Lodge Complex and the construction of the Renewal Lodge on the floor of Mound 23 at the Hopewell site.

This hardly exhausts the range of embankment earthwork data that can be used to entrench this chronological model. For example, it has already been suggested that the Newark Circle-Octagon, the Wright Square, the large Cherry Valley C-Form / K-Profile embankment earthwork with its cluster of mounds, and the three aggregation elements collectively constitute a complex expression of the Sacred Dual C-R Motif. This complex was probably produced partly as a result of interaction of the cults in the Chillicothe and Newark regions, constituting what was earlier referred to as the Chillicothe-Newark axis. It was argued earlier that the Newark Circle-Octagon emulated the High Bank Circle-Octagon and that Seip started constructing its Paint Creek C-R earthwork shortly after Newark started its Observatory Circle. This activity at Seip stimulated the change of plan at Newark that brought about the Feature A / Observatory Mound Complex, suggesting that the first construction phase of the C-R components of the Newark site were initiated after the initial phase of High Bank was built and finished about the same time as Liberty Works.

Based on a number of radiocarbon dates, the ridge-top T-Form sites seem to cross-cut this scheme, with a number of them possibly bracketing the whole time period of the Ohio Hopewell. For example, it is clear that Fort Ancient was constructed and used over a number of periods. Construction attributes that support this include the southern, middle, and northern units. These can also be correlated with complementary patterns at Chillicothe. The balanced symmetry of the gates at Fort Ancient compared to mirror-image symmetry of High Bank would count as one complementary pair. The addition of the

K-Profile aggregation to the northern unit of Fort Ancient, suggesting that this addition marks the Late Ohio Hopewell phase at this site, would complement the use of similar constructions at Newark and Mound City/Hopeton.

It is clearly shown in table 22.2 that the original embankment earthworks defining the Ohio Hopewell would be those primarily identified with the Miami Fort Tradition, namely, the C-Form and T-Form embankment earthworks. Embankment sites of the former tradition are found in both southwest and central Ohio, e.g., the Hopewell and Cedar Bank sites. Near Newark there is the Granville site. Also in the Chillicothe region there appear to be two instances of the T-Form type embankment work, Fort Hill and Spruce Hill.[11] It is out of the interaction between these two SR-Profile and SL-Profile traditions that the dual C-R Configuration of the Chillicothe Tradition first emerges as manifested at the Mound City/Hopeton Complex.

MATERIAL CORRELATES OF IDEOLOGICAL POSTURE

With the elucidation of the Ideological Cult Faction Model in the last chapter, and the refinement of the chronological scheme completed above, it is now time to postulate a set of material indicators by which to interpret the variation of the archaeological record in terms of the historical shifts of the ideological stance of the cults responsible for these sites. The two major areas that would be strongly subject to factional negotiation would be construction and mortuary practices, along with supporting practices, such as long-distance procuring of exotic resources. With respect to mortuary practices, one important measure of the ideological stance of cults would be the spectrum of mortuary treatments of the deceased. The proportion of cremations to noncremations and the variation in this proportion among contemporary cults and over time would be a rather sensitive measure of relative dominance of proscriptive to prescriptive ideological factions. An increase of extended inhumations over cremations, for example, would mark a proscriptive post-mortem regime and indicate the rising power of the proscriptive faction. The reverse, marking a prescriptive post-mortem manipulation regime, would indicate the rising dominance of the prescriptive faction.

The context nature of the locale in which the mortuary event occurs is also a critical constituent element of the mortuary event and, therefore, differences in this area can nuance the symbolic pragmatic meaning of the post-mortem manipulation criterion. For example, while extended burials in clan CBLs would constitute the mortuary event as strongly funerary in nature, in the cult CBL context, that is, in a major embankment earthwork, it would indicate a compromise between proscriptive and prescriptive factions. In this case, the prevalence of inhumations that display signs of extensive exposure while largely sustaining the skeletal integrity, as marked, for example, at the Turner and the Hopewell sites, might be best treated as registering a relatively middle-of-the road presbyterian stance, a balancing of the funerary focus of the clans and the post-mortem sacrificial focus of the cults.

Related to this "balanced" proscriptive/prescriptive or presbyterian stance would be the possible development of an indirect mortuary form of sacrifice, such as the performance of mortuary rites marked by burnt offering deposits of valued artifacts. This compromise would see the kin and companions of the deceased simultaneously discharging clan and cult duties in proportion to the magnitude of the burnt offerings. Therefore, an indicator that the views of the proscriptive faction were strong would be when significant numbers of deceased associated with major custodial regalia were given terminal burial rites involving extended inhumation, and correlated with major burnt deposits not associated directly with human deceased, while a smaller proportion of deceased were given cremation burials. The reverse would be to find a high proportion of cremated burials and lesser burnt offering deposits. Instead, there would be a tendency for such items as formal personalty and mourning gifts, relatively few in number, to be cremated with the deceased. A strong exception to this would be custodial regalia, which would be buried largely unscathed and in association with the cremated deceased.[12]

The development of construction programs, both in terms of morphological patterning and overall magnitude of embankment earthworks, would also figure centrally in ideological negotiation among competing factions. Radical change in form in the same site, such as juxtaposing the C-Form/SL-Profile and C-R Configuration attributes, or the SR-Profile and the K-Profile embankment attributes, would indicate historical shifts in what would count as appropriate monumental warranting devices. However, since these could indicate either proscriptivist or prescriptivist shifts, the direction of the shift would be indicated by the relative magnitude of construction that this morphological transformation marked. The assumption here is that a prescriptive shift in balance of power would be indicated if the formal change correlates with an increase in the magnitude of the overall construction program. A proscriptive shift would be suggested if such expressive formal changes were realized by a non-growth or a relative reduction in the magnitude of the additional embankment construction compared to preexisting forms at the same site. However, caution has to be exercised here since the program could include mound earthwork construction. Therefore, while the embankment additions might remain the same magnitude as the older embankments, or even mark a relative reduction in magnitude, the mounds might significantly expand in size compared to earlier mounds, marking countervailing tendencies, the latter being prescriptive and the former being proscriptive.

There is another aspect of construction that must not be overlooked. This is the use of timber. It is clear that the Seip-Pricer Great House Renewal Lodge took considerable timber to build and there is also evidence that trees were treated as important symbolic pragmatic elements of these structures. Greber has detailed the range of data derived from the re-excavation of the Edwin Harness site that supports this view. The soil stains found around specifically located structural and nonstructural post holes indicate selective painting of posts, not only demarcating the entrances and exits between rooms and sections, but also specific corners and undoubtedly other specially potent areas. Furthermore, in terms of the symbolic pragmatic approach, not only the use of mineral coloring, but also the use of specific species of trees would be part of the constitution of that potency. The variation of tree species also supports the view that there was selective choice involved.[13]

Wymer discusses the impact that dispersed gardening would have had on local environments, transforming mature white oak forests into secondary mixed forests that may have enhanced overall reproductivity of the multiple food and utility species that they exploited. She goes on to point out that while gardening would actually allow for a recycling of land, this would not be the case for monumental earthworks, which would have necessitated clearing the mature forest from the terraces on which they were built.

> What [the Middle Woodland Ohio peoples] did on a small scale at their habitation sites, they also did on a grand scale at their ceremonial centers. The Newark Earthworks are located on the same glacial outwash terraces as are many of the Hopewellian habitation sites. Available data indicate that if left undisturbed by human populations, a rich and dense white oak forest would have once covered this terrace environment. . . . [It] seems likely that a large number of trees must have been removed prior to the architectural planning and construction of the immense embankments and mounds. (Wymer 1997, 159–60)

An alternative view holds that major embankment earthwork sites were located on prairie pockets that had been sustained in selected areas from the early Archaic period by regular burning practices of the inhabitants, particularly in the bottom lands where major streams converged.[14] This would ensure the maintenance of open land for grazing and browsing species and, of course, with the emergence of the Mt. Horeb and, in particular, the Ohio Hopewell, these would come to be ideal for embankment earthwork construction. Such historically sustained intervention would not contradict a deontic ecological regime arising from within an immanentist cosmology since these forest and brush burning practices could be central to world renewal as a form of ritual as well as practical cycling of both the grazing and browsing animals and the resources they consumed. The Proscriptive/Prescrip-

tive Ecological Strategy Model could accommodate either the maintenance of pocket prairies as the outcome of long-term cycling or Wymer's view that primary forest had first to be cleared from the areas
occupied by the embankment earthworks. If future research favors the
former, then, in general, a proscriptive orientation would be indicated,
and if the latter is favored, then it would suggest a more prescriptivist
orientation and might account for the Ohio Hopewell as a prescriptive
surge resulting from intensified pursuit of cult reputation.[15]

Both timber and earthwork construction would require major
forms of intervention in the sacred natural order. Clearly, some forms
of construction required the use of timber, such as building the Renewal Lodges, while the use of earth, clay, stone, and sand was largely
limited to embankment and mound construction. However, it is now
becoming clear that timber was also used in building screens and even
free-standing "stockades," sometimes in combination with earthen embankments and sometimes alone. Through the use of resistivity analysis, Greber has recently established what appears to be the residue of
a "stockade" buried under the aggregation neck and the associated part
of the High Bank Circle-Octagon.[16] Since this analysis did not do a full
circuit of the Circle and Octagon, she admits to not knowing if this
feature extends completely beneath the High Bank earthwork embankments. However, she points out that similar buried timber features
have been found at other embankment earthworks. Riordan, for example, has reported excavating the burnt residue of a timber wall feature at the Pollock site in southwestern Ohio. He argues that fairly
mature trees were used to build a heavy stockade-like feature on top
of a preexisting earthen embankment and that it was extended as a
lighter, screen-like construction beyond the original embankment and
along the edge of the cliff overlooking the creek.[17] Subsequently, the
standing timbers were burned and another stratum of earth was used
to cover them.

While maintaining (correctly, in my view) that Pollock was a
ceremonial locale, he suggests that this free-standing timber construction may mark a short period of social instability during which the

locale was converted into defensive use, followed fairly shortly by a more normal period that permitted its removal. The Ideological Cult Faction Model would give a different account. In general, it postulates that when a cult was planning projects that could use either timber or earth, as in the above case, it would be highly probable that manipulating the relative reliance on these resources would be used as part of the negotiating between proscriptive and prescriptive factions. In general, exploiting the forests would be favored by the proscriptive faction as the less extreme form of intervention in that forests would regrow and earth could not, while exploiting elements of the earth (surface/ deep stratum, sand, clay, and so on) would be favored by the prescriptive faction for the flexibility in symbolic iconic content that it enabled.

Thus, in the case of Pollock, the original use of earth and then the addition of timber and the switch back to earth suggests a history of shifting prescriptive to proscriptive and back to prescriptive postures. The same applies to the use of timber in the initial construction of Miami Fort (248). Further evidence of the possible sequential use of timber and earthwork construction in Ohio Hopewell has been presented by the recent excavation of an apparent "Woodhenge" at the Stubbs Earthwork site that has clearly demonstrated that cycles of earthwork and timber construction occurred.[18] In terms of this model, both timber and earthwork construction practices must have figured in the negotiations of cult factions.

The need to carry out an initial clearance of forest for construction may be one factor that could override the prescriptive preference since it would be difficult to rationalize the "wastage" of trees that would have been cut. Of course, one way of rationalizing this would be to declare an equally important sacred need, for example, expanding the building construction program. This latter tactic could promote the construction of a major set of buildings making up a Renewal Lodge Complex, such as the Great House Renewal Lodges at Seip and Liberty Works. Furthermore, although Seip-Pricer and Seip-Conjoined Great House Renewal Lodges were the focal timber buildings at this site, there is clear evidence that support buildings were associated with such major timber constructions, and that, although smaller, there may have been more of them than the current data indicate.[19] Clearly more research is required in this area. In any case, timber construction cannot be left out of the ideological negotiations between proscriptive and prescriptive factions.

As stated above, the second aspect of the archaeological record that would be sensitive to the interplay of factionalism would be the artifacts that mediated the ritual program. Both factions would be united in agreeing that these artifacts were necessary for a full ritual program since they were symbolic warrants and without them ritual practices would not be fully felicitous. Still, proscriptive factional strategies would be informed by a minimalist sacred pollution calculation and the prescriptive by a maximalist sanctification calculation. In an immanently sacred world, human travel would have grave import for the sanctity of both the individual and the cosmos since it would require crossing sacred natural boundaries. Such long-distance travel would be strongly ritualized in order to minimize the sacred transgressive aspect of travel. Hence, an intrinsic part of this long-distance procurement process would be to treat it as a sacred quest and this would entail the wearing and displaying of recognizable Hopewellian items as sacred "passports and visas" (e.g., copper ear spools, copper plates, copper celts, and so on), thereby minimizing pollution and ensuring that those passing from one region to another would be recognized by local custodial groups as legitimate questers with whom they would cooperate.

In this regard, Seeman's comments about the function of Hopewellian regalia and formal personalty as symbols facilitating communication that reaches beyond the capacity of words is an excellent insight.[20] He is thinking primarily in terms of the problems of communicating across languages. Certainly, this would be among some of the difficulties that long-distance travel would have to solve. However, his views can be supplemented by emphasizing that this perspective is, at base, a treatment of Hopewell stylistics in symbolic pragmatic terms. The Hopewellian material culture served as a complex set of pragmatic devices whereby the individuals' using of sacred paths was warranted, thereby ensuring that their crossing natural boundaries would not count as transgressions of the sacred order. As pointed out above, they would do this by constituting the wearers as occupying the widely recognized social position of quester of sacred resources, thereby endowing them with a range of action powers as representatives of a recognized although distant world renewal cult. All this would be in addition to, but also partly to resolve, the problems involved in the crossing of language boundaries as the travels progressed.

Physically procuring resources from the locales where they were naturally found, such as copper from pits, flint from quarries, shells from the sea, and so on, also had grave import for the sanctity of the cosmos since it entailed deliberate intervention in the sacred natural order. Such procurement would require considerable warranting ritual regalia as part of the procuring process itself. For example, Clay points out that the Peter Village site was directly associated with galena/barite mining.[21] It would be perfectly consistent with the symbolic pragmatics informed by an immanentist cosmology for responsible groups to build a major sacred locale in association with this material resource, since it could be argued that warranting such intervention would require presencing the sacred powers that would authorize it, and at the same time, would constitute the procuring behavior as ritual that would at least partially rectify the pollution that such activity would necessarily generate. Based on the pottery evidence, Clay points out that the Peter Village site attracted groups from considerable distances.

In equivalent terms, Lepper et al. have recently argued that it is highly likely that the Flint Ridge quarries, the source of the most valued chert used for ritual bladelets, was probably viewed as having special potency. The flint beds are very extensive, estimated to cover about 1,000 ha. Notably, lesser earthworks and mounds are distributed across the zone, possibly marking sacred locales associated with the ritual required to rectify the type of sacred pollution that such intensive physical interventions as quarrying entailed. Based on their careful assessment of the quarry lithics, it is suggested by them that, following the Middle Woodland period, this region may have been avoided by later generations.

> It is possible that after the Hopewell decline, symbols so powerfully associated with the Hopewell culture were abandoned and Flint Ridge flint was no longer useful in the new rituals of the agricultural societies. Perhaps Flint Ridge flint was deliberately deemphasized because of its association with Hopewell rituals and ideology. (Lepper et al. 2001, 71)

Clearly, a complex calculation of the polluting costs and the sanctifying benefits that both direct exploitation of raw materials and long-distance travel generated would be a continual aspect of cult life, and this calculation would inform factional negotiation. In general, the prescriptive faction would favor long-distance travel and direct procurement since it would enhance large-scale procurement of exotic and powerful raw materials, such as copper, obsidian, ocean shell, galena, mica, selected cherts required for bladelets, and so on, as well as being a major mode of pursuing cult reputation and personal renown through transforming these expeditions into sacred ordeal quests. This would be a powerful reason for the junior age-grade sector to tend toward prescriptive procurement and travel strategies. It would call for the specialized intervention of the shamanic priests in ritually preparing the youth for these expeditions and organizing their ritual purification on their return by presiding over and even conducting any rites of sacrifice that were required. The proscriptive orientation would be favored by the senior age-grade sector. This does not mean that the senior age-grade leadership would deny the necessity of this exotic material. Rather, they would prefer minimizing the polluting costs involved in procurement by minimizing the actual quantity and, possibly, the variety of resources and by minimizing the traveling distance required. Both of the latter would call for "down-the-line" exchange mechanisms involving gift exchange among closely spaced world renewal cult CBLs.[22]

In sum, there are a number of material correlates that can be used to map the postulated ideological negotiation for power between competitive factions. Of course, taken individually, these criteria are fairly straightforward, as indicated in table 22.2. These variables, however, cannot be treated independently, summed up for a site, and read off as the ideological stance of the cult. Although presented as polar opposites, each pairing should be treated as simply marking the extremes of a continuum with most sites fluctuating along the continua in a complex process of negotiation. This is because of the various factors already outlined. First, the factions are united in terms of their basic cosmological beliefs about the nature of the world, humanity's place in it, and what the raison d'être of the cult was. They would also have a shared understanding of the alternative ideological symbolic pragmatics that each faction preferred. Indeed, the idealized material expressions of these alternative ideological positions would be at the core of the factional negotiations. Therefore, while agreeing on the relevant aspects of the cosmos that must be made manifest and present in the ritual locale, they would dispute over how this should be done: whether the locale should be built of timber or earth or a combination, and how large it should be, and so on. Continual negotiations would be required and these different material correlates would be differentially weighted, with prescriptivists willing to acquiesce to the minimizing demands of the proscriptivists in one or more areas in return for the latter's acquiescence to the prescriptivist demands in other areas.

In short, as with any hermeneutic process, these material correlates can be used only as heuristic guidelines. Just as the actual factions involved could reassess and reorder their priorities, there can also be serious disagreement among the archaeologists interpreting the record, one giving greater weight to embankment earthworks, another to mortuary practices, and so on. Interpreting the data is never an easy task. However, the criteria and their rationale have been made clear so that the interpretations to be given in the following chapter can be critically examined by others who can also give alternative logical readings supported by their theoretical rationales, thereby cooperatively participating in a hermeneutic spiral of expanding knowledge of the Ohio Hopewell.

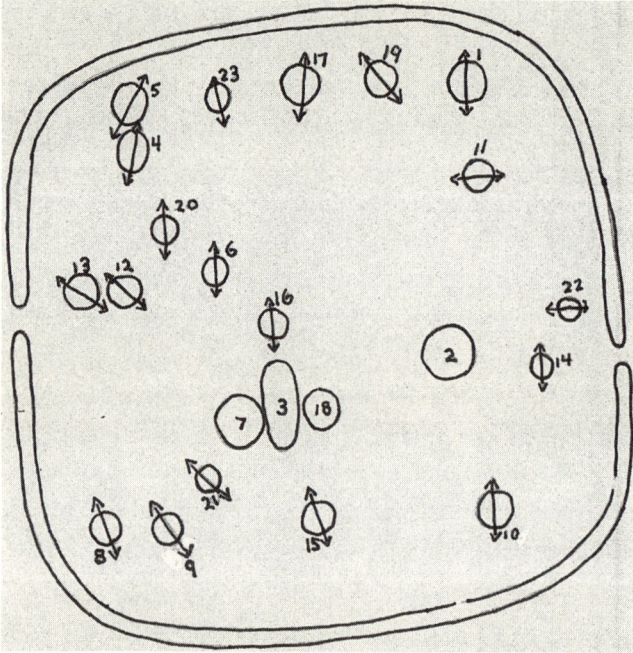

FIG. 22.1. Summary of Orientation of Structures on Floors of Mound City Proper (Baby and Langlois 1977, 39, figure 28)

Table 22.1 Schematic Chronology of Major Ohio Hopewell Sites

Ohio Hopewell Phases	Central Scioto/ Paint Creek	Hopewell/North Fork	Turner and Related Sites	Newark
ca. A.D. 400	**Complex C-R Era (symmetrical)**	**Complex C-R Era (symmetrical)**	**Complex C-R Era**	**Complex C-R Era (symmetrical)**
Terminal			Camden, Milford	
	Works East and Baum	Frankfort		
		Mound 23	Conjoined Renewal Lodge Complex	
		Renewal Lodge under Mound 23 and Add-on C-R components		
			Great Enclosure	Wright Square Cherry Valley Infix Circle-Octagon
Late	Seip and Liberty Works			
	High Bank	**SL-Profile Era** Mound 25 and SL-Profile "Half-Moon" embankment earthwork	**SR-Profile Era** Sacred Circles	**SR-Profile Era**
				Fairground Circle (Great Sacred Circle)
ca. A.D. 150	**Complex G-Form Era (subsymmetrical)** Mound City/ Hopeton Complex			Lesser Sacred Circles
Middle	Hopeton			

ca. A.D. 1	**SL-Profile Era**			
Early	Mound City Proper/Compo-nents B and C Cedar Bank	Original C-Form/ SL-Profile and Renewal Lodge Complex (Struc-tures D/E and C)	Ritual Complex	Lesser Sacred Circles
ca. 100 B.C.	Adena Mound and related earthworks			

Table 22.2 Material Cultural Correlates of Proscriptive and Prescriptive Ideological Factions

		Prescriptivist Posture	Proscriptivist Posture
Mortuary Form		cremations: partial or full	extended inhumations
Earthwork	Mound	maximize size and internal structure	minimize size and internal structure
	Embankment	maximize size and internal structure	minimize size and internal structure
Timber Construction		maximize, promote abandonment, burning, and reconstruction	minimize, promote re-use
Artifacts	Variety	maximize elaborate and exotic varieties	keep it simple and minimize exotic varieties
	Procurement	direct	indirect (exchange)
	Quantity	maximize	minimize
	Distribution	individualized	collective

Chapter 23

THE SHIFTING
IDEOLOGICAL POSTURES
OF OHIO HOPEWELL

The chronological framework and the material criteria of factional competition can now be used to interpret the historical and regional variation in the Ohio Hopewell in terms of ongoing ideological negotiations between competing factions of given cults as postulated under the Ideological Cult Faction Model. First the Mound City/Hopeton Complex and Tremper, and then the Hopewell, Seip, and Liberty Works sites will be examined in these terms. The interpretive application of the model will be completed with an account of the collapse of the Ohio Hopewell and the emergence of the Late Woodland period in this region. Confirmation of this interpretation by means of the hermeneutic spiral requires presenting and critiquing an alternative temporal and social organizational model of these same data. This will be done in the following chapter.

I. THE MOUND CITY/HOPETON COMPLEX
AND TREMPER

Greber has argued that much of the stylistic innovation identified as Hopewellian may have emerged through the work of the populations responsible for the Adena earthworks in the zone immediately north of Chillicothe.[1] These make up a cluster that she has referred to as the Chillicothe Northwest Group. Important earthworks in this group are the Adena Mound site, the Worthington Mound group, and the Carriage Factory/Miller Mound to the immediate southwest of Mound City Proper. Her insight can be supplemented with the claim that this "explosion" of cultural expression would not have occurred in a social vacuum. While the creative imagination manifested in the more fascinating aspects of Ohio Hopewell may have been more localized here, it is likely that it was stimulated by the interaction between this Central Scioto region and the southwestern Ohio region, possibly even mediated through the Turner-Hopewell axis. The primary point of this interaction, of course, would be innovation in world renewal ritual, and the dynamics of this relation promoted a competitive pursuit of cultic reputation, as postulated under the Proscriptive/Prescriptive Ecological Strategy Model.

More specifically, the auxiliary Autonomous Cult Model argues that if a system of complex, ecclesiastic-communal cults were to have emerged (i.e., the Ohio Hopewell), it would have required the preexistence of a regional system of autonomous simple communal cults (e.g., the Adena). As the motivating condition would be an escalating pursuit of cultic reputation, it could be anticipated that one obvious way of enhancing reputation would be cross-generational sharing in the construction of a monumental project that, at the same time, could be claimed to advance the world renewal goals that both the senior and junior age-grade cults shared, while—initially—respecting their traditional arm's-length relations of mutual autonomy.

It is postulated that the Mound City Cluster of SR-Profile and SL-Profile embankments would be among the earliest known expressions of a successful integration of the construction programs of at least two autonomous senior and junior age-grade communal cults, and that this cluster constitutes one of the earliest ecclesiastic-communal cult locales and demarcates the Early Ohio Hopewell phase in the central Scioto region. Under the Mourning/World Renewal Mortuary Model a complex set of rites is postulated as occurring, mediated by possibly joint complementary solar and lunar renewal ceremonies. The former would be performed at both Cedar Bank Works and Component C, and the latter at Component B and, possibly, the two sacred circles on the east bank of the Scioto. Indeed, that these four sets of earthworks straddle the Scioto River may have figured in the symbolic pragmatics of the cycle of solar/lunar world renewal rituals that they jointly performed.

Associated with these works and probably immediately preceding them would have been the constructions covered by the Worthington Mound group and, possibly a little later, the Carriage Factory/Miller Mound, although the latter may actually be contemporary with the earliest renewal lodges associated with Mound City Proper. Unfortunately, the evidence required to confirm this possibility is inadequate, but it is certainly a reasonable possibility (18, 22). Therefore, because of full-scale cremation at Mound City Proper, because of the major labor that was required to build Cedar Bank Works and the Component C embankment earthwork, and because of the possibility that there were "off-site" laying-in crypt renewal lodges (e.g., the Carriage Factory/Miller Mound), all this implicating an escalation of both construction and full post-mortem sacrificial rites, the claim can be made that the Early Ohio Hopewell phase in the Mound City Cluster was more prescriptive than proscriptive in orientation.

Thus, it would appear that during this Early Ohio Hopewell phase in the central Scioto region, the cultic agenda was driven more by the ideological views of the prescriptive faction than by those of the proscriptive faction. The former would be under the joint leadership of the shamanic clergy and junior age-grade sector. This prescriptive orientation was enhanced by the construction of the Hopeton earthworks, constituting the emergence of the Middle Ohio Hopewell phase in this region. The latter earthworks were subsequently followed by the integration of Mound City Proper and Hopeton to constitute the Mound City/Hopeton Complex. It is this period that may have witnessed the rebuilding and expansion of the major Renewal Lodges in Mound City Proper and the performance of world renewal rites that generated the complex set-pieces, some of which were described earlier. Not only do these set-pieces display elaborate custodial regalia, but the latter were all accompanied by cremations. According to material criteria stipulated earlier, the treatment of the artifactual material as focal custodial regalia and the treatment of deceased more as supporting symbolic warrants than participants suggest a strong prescriptive post-mortem manipulation regime, even to the point that the deceased who were subjected to this extreme reductive transformation may have been shamanic priests and/or laity custodians, constituting these rites as primary post-mortem sacrifices.

This reinforces the view that the Middle Ohio Hopewell phase in the central Scioto, as marked by the construction and use of the Mound City/Hopeton Complex, may have not only been largely dominated by the prescriptive faction, but may have been in almost a fanatic frenzy, combining expansionary construction with extreme post-mortem sacrifice. This claim is reinforced by the qualitative and quantitative nature of the changes in the earthwork construction program. In qualitative terms, the innovation of the Complex G-Form and its symbolic pragmatics of complementary circle/rectilinear contrast, celestial alignments, selection of types of earths, and so on, highlights a simultaneous radical break with previous earthwork traditions, both the SR-Profile and SL-Profile attributes. While this alone is not sufficient to indicate the ideological direction of this shift, the quantitative changes marked by the large Hopeton Circle matched by the Hopeton Rectilinear and the building of Component A suggest that this was generated by the prescriptive faction in almost a fanatic frenzy.

Further reinforcing the interpretation that the Middle Ohio Hopewell phase at the Mound City/Hopeton Complex marks the dominance of the prescriptive faction are the range and quantity of artifactual materials. Not only did these elaborate set-pieces implicate major procurement expeditions for copper, obsidian, and mica, in all probability direct expeditions carried out partly as ordeal quests by junior laity cult members, but also the production suggests skilled shamanic clergy control of the sacred work of the artisans, who would themselves be classed among the shamanic clergy. It was mentioned earlier that elaborate productions of this sort cannot be characterized simply in technical terms; they also required elaborate ritual in order to ensure that every stage of production was properly enlivened by sacred powers.

As postulated above, the emergence of Ohio Hopewell has to be seen in structural terms, as the emergence of new expressive ideologies in competitive pursuit of cultic reputation but within a largely unchanging region-wide cosmology. The escalating construction and postmortem sacrificial practices of these early and middle phases cannot be explained exclusively in terms of internal dynamics, even when these are articulated in ideological factional terms. Part of the "fanatic frenzy" would have been the result of a focused awareness of the activities at neighboring sites. This may account for the parallels between Tremper and Mound City.

Tremper is on the west bank of the lower Scioto River, only a few kilometers north of Portsmouth on the Ohio River. As mentioned earlier, this site displays attributes that indicate a close relation with Mound City Proper at the Component A time, and possibly earlier. They share the same type of simple subsymmetrical K-Profile embankment earthwork. Both their mortuary programs display strong prescriptive treatment of the deceased in the form of full body cremation. In fact, Tremper might be considered to set the standard in this regard since, as pointed out earlier, all the cremations were deposited in four major platform altars constituting an aggregated post-mortem sacrificial offering.[2] In contrast, the cremation deposits of the Mound City Proper were largely individual, although there may have been a few instances of compound cremation deposits.[3] Another direct indication of the closeness of relationship between Tremper and Mound City is that the platform pipe figured as a major artifact category. Although there were both plain and effigy pipes, the latter were in the majority in both cases. They shared the same range of effigy styles and used the same Ohio pipestone quarries, found a short distance from Tremper on the east bank of the Scioto River. The parallels that clearly link Tremper and Mound City/Hopeton and place them in the same Early and Middle Ohio Hopewell phase suggest not only that they were in close harmonious ideological interaction but that the interaction was partly facilitated by both cults being right-of-center, having the prescriptive faction dominant in cult affairs.

However, there are divergent attributes that suggest the prescriptive faction at Tremper was even more radical than at Mound City. While the total number of cremations at Tremper cannot be calculated because of their collective deposition, it is clear from the aggregate that there may have been between three hundred and four hundred.

> The communal depositories, peculiar to the Tremper mound, were four in number, consisting of a main depository, located in the east end of the structure, and three smaller ones, in the western end. . . . The contents of the [main] depository no doubt represented the remains of hundreds of cremated bodies, indicating the use of the grave for a long period of time. (Mills 1916, 277)

The number of cremations at Mound City is much smaller, possibly no more than ninety.[4] If both sites are considered to have covered roughly the same time span of the Early and Middle Ohio Hopewell (ca. 100 B.C. to ca. A.D. 150), the difference in numbers must be due to a real difference in the mortuary programs. Despite the fact that full cremation was effected at Mound City Proper, there must have been many deceased who did not end as cremations on the floors of the Renewal Lodges. Therefore, as suggested earlier, the mortuary process for many may have terminated with extended burial, possibly in clan CBLs or in structures under major mounds of the Chillicothe Northwest Group, such as the conjoined Worthington Mounds or the Carriage Factory/Miller Mound.[5]

This means that despite the strong presence of prescriptive elements in the Mound City/Hopeton mortuary program, this complex site also manifests a less radical prescriptive posture than does Tremper, when measured in terms of the mortuary regime. There may have been an ongoing accommodation at Mound City between cult and clans in which the latter would cooperate in order to have their deceased given public honor, thereby enhancing the position of clans in their alliances, while the cult would selectively limit full cremation, focusing on major notable deceased, those having custodial obligations with respect to major ritual regalia, as well as deceased members who held lesser standing in both the cult and the clans. As argued earlier, if standing in a cult is largely time sensitive, then, everything else being equal, there would be fewer junior age-grade members having accumulated major standing than senior members, as marked in terms of custodial regalia. The mortuary residue on the floors of the known Mound City Proper mounds has this interesting mixture of a minority of major set-pieces (i.e., elaborately deposited regalia with one or two cremations) and a majority of "ordinary" cremation deposits accompanied by few if any custodial regalia artifacts on spaces that showed minor preparatory modification.

Thus, if Tremper is used as the extreme example of a cult in which the prescriptive faction dominated, as indicated by full cremation and collective deposition, then the Mound City/Hopeton Complex can be assessed as not quite so extreme. The dynamics that motivated this latter cult to carry out the major innovation and expansion demarcating the Middle Ohio Hopewell phase at this site and that drove it on to the next innovative step, the full C-R Motif Configuration, might be the result of equal external and internal forces. The external source would not have been the clans of its members, since this relationship would appear to have been relatively balanced, if the mortuary assessment is treated as a good measure of this. Rather, it must have been the presence of its nearest cult neighbor, the resident cult of the Hopewell site.

2. THE HOPEWELL SITE

It is postulated that the autonomous cult at the Hopewell site had the same basic structure as the Mound City cult. Fortunately, as discussed earlier in great detail under the Laying-In Crypt Model, the mortuary stages that the record displays are more complete than those displayed at the Mound City/Hopeton Complex. The implications of the Laying-In/World Renewal Crypt Model of the Turner site can also be tentatively extended to the Hopewell site. Still, compared to those at Mound City Proper, the nature of set-piece burials at the Hopewell site suggests a significant difference in the balance of factional power. If the majority of the highly elaborate crypt burials at the Hopewell site were in the laying-in stage, then, as argued earlier, these should not be directly compared to such Mound City Proper set-pieces as those on the floor of Mound 7. Rather, the closest equivalent of the latter set-pieces at the Hopewell site would appear to be non-crypt extended burials, such as those on the floor of Mound 2, where the large deposit of Wyandotte chert preforms was also found. All five burials were extended and accompanied by major custodial regalia, the equivalent of the regalia associated, for example, with the cremated Burial Number 13 mortuary deposit of Mound 7. At the Hopewell site, then, the custodial regalia burials that are comparable to those at Mound City were associated with major cache deposits, such as the Wyandotte chert deposit in Mound 2, or the obsidian deposit of Mound 11, although the burials at the latter were cremations. In short, the mortuary set-pieces at the Hopewell and Mound City Proper sites seem to be quite different. While at the latter the focal component seems to be a major set of custodial regalia accompanied by one or more cremations, at the former, it is a major deposit of highly valued raw materials accompanied by extended burials displaying significant custodial regalia.

Since almost all of the non-crypt custodial regalia burials at the Hopewell site are extended inhumations, relative to Mound City Proper, this suggests a proscriptive post-mortem mortuary orientation. The set-piece deposits of raw materials of the Hopewell site, however, including copper and silver nuggets in the burnt offering deposits and the large deposit of the obsidian debitage under Mound 11, suggest major long-distance interaction, indexing a prescriptive orientation. Reinforcing this view is the large quantity of finished copper items in the form of axes and adzes, copper plates, copper headdress bases, and ear spools, indicating a strong influence of the clergy through the control of artisan ritual production. However, Mound City Proper also contained major quantities of exotic items, in particular the mica sheets used as a major component of a number of the set-pieces, as described earlier. The overall impression at Mound City Proper is that it procured some of its raw materials through long-distance interaction (e.g., mica and other materials, possibly in both raw and finished form) and through inter-cult exchange (e.g., in finished form, as illustrated by the deposit of two hundred plus platform pipes in Mound 8).

In terms of procurement of raw materials, then, both the Hopewell and the Mound City Proper sites display similar prescriptive tendencies during the Early and Middle Ohio Hopewell phases. However, if the burnt deposits of the Hopewell site are added to the above, this suggests a more collectivist than individualist distribution of warranting artifacts. The latter marks a prescriptive orientation and the former, a proscriptive orientation. The burnt deposit containing the platform pipes and other materials of Mound 8, Mound City Proper, would be only an approximate equivalent of the Hopewell site burnt deposits, since the latter had a broader range and greater quantity of artifact types which, according to the Laying-In Crypt Model, mediated *collective* memorial/world renewal rites. All this serves as important evidence that there was a proscriptive tendency at this site that was not as evident at Mound City Proper.

Turning to the construction programs, both sites would also appear to be about equal during the Early Ohio Hopewell phase, even if the Carriage Factory/Miller Mound construction is included. It was postulated that the two SL-Profile embankments, Component C of the Mound City Cluster and Cedar Bank Works, demarcate the Early Ohio Hopewell phase at Mound City Proper. When combined, these would probably be roughly equivalent in magnitude to the Hopewell C-Form embankment. However, at Mound City, the Middle Ohio Hopewell phase is marked by the escalating construction program that brought about the Mound City/Hopeton Complex. There was no equivalent construction program at the Hopewell site until much later with the construction of Mound 25, which is considered to correlate with the add-on C-R Configuration components. Thus, while the Mound City/Hopeton Complex was the result of a major construction expansion demarcating the Middle Ohio Hopewell phase, the Hopewell site remained rather conservative in this regard during this same Middle Ohio Hopewell phase.

Overall then, the Mound City and Hopewell site cults can be assessed as about equally positioned on the C↔P↔E continuum during the Early Ohio Hopewell phase, both right-of-center, with Mound City being more zealously prescriptive in its mortuary program, thereby tending a bit more toward the episcopalian orientation, and Hopewell hovering a bit closer to the presbyterian center. Both promoted prescriptive construction strategies, suggesting a strong clergy influence along with a prevalence of the interests of the junior age-grade laity sector. Despite this, the proscriptive ideological orientation and the senior age-grades in both cults maintained a strong presence and tethered some of the prescriptive tendencies. However, the Middle Ohio Hopewell phase saw a clear divergence developing between these two sites, as manifested by the two construction programs: the Mound City/Hopeton Complex construction program marked a major escalation in factional power in favor of the prescriptive orientation, which was not emulated at the Hopewell site. Therefore, during the Middle Ohio Hopewell phase the Hopewell site registered a more central presbyterian stance, given the low incidence of cremation and reduced earthwork construction during this phase. An alternative way of putting this would be to say that it remained stable, defending the received "tradition," while Mound City/Hopeton was forging the way to the emergence of High Bank, Seip, and Liberty Works.

The addition of the C-R components to the C-Form embankment marks a major attempt on the part of the Hopewell site cult to reorient itself within the ambient of its prescriptive neighbors at Seip and Liberty Works. As argued earlier, these additions probably occurred as part of the abandoning of the Renewal Lodge Complex on the floor of Mound 25, and the construction of the Renewal Lodge on the floor of the future Mound 23. However, it should be noted that the Paint Creek Square is less than half the size of the standard type. This could not be the result of the lack of space, since the area east of the C-Form embankment appears to be adequate for a full-sized Paint Creek Square.

This also applies to the small circle. The addition of the C-R components and the building of a new Renewal Lodge following the local Hopewell site tradition, then, are interpreted as largely a strategic compliance on the part of the dominant proscriptive faction with the demands of the prescriptive faction to emulate their more prescriptivist neighbors at Seip and Liberty Works.

Not only are the add-on C-R components much like a token expression of the "classic" Paint Creek C-R Configuration, but the mortuary program on the floor of Mound 23 during the Late Ohio Hopewell phase continues to be dominated by extended inhumations and, in fact, almost completely excludes cremation, suggesting a marked shift toward a more proscriptive mortuary regime. Moorehead refers to a number of extended burials as "charred," these being deceased who were laid on platform altars that had been fired and were still very hot at the time of the placement, thereby "cooking" the posterior parts of the bodies. Furthermore, it should be recalled that while the Renewal Lodge on the floor of Mound 23 formed a pattern equivalent to the pattern of Structure D/E under Mound 25, there were no equivalents to Structures A, B, F, and C, the latter being the possible equivalent of the clergy sections of the Seip and Liberty Works Great House Renewal Lodges. This suggests that the dual patterning of the structure on the floor of Mound 23 probably manifests only the senior/junior age-grade structure. That is, the laity/clergy structure was not being manifested here. This is not to suggest that the laity/clergy structure was dissolved but that the balance of power between clergy and laity leaders was such that the former, as a group, could not declare their own space. Their position would be closer to that of the shaman, a religious specialist contracted by the laity leadership. The reduced role of the shamanic priest would ensure that he was important in ritual but not in managing cult affairs. This means, of course, that the floor patterning under Mound 23 manifests a cult that was moving toward the congregationalist position and, corresponding to this, manifesting a strong proscriptive factional dominance under the leadership of the senior age-grade sector.

3. SEIP AND LIBERTY WORKS

The cults of Seip and Liberty Works are postulated as being directly "descended" from the Mound City/Hopeton Complex cult, and this descent corresponds with the emergence of the Sacred Dual C-R Motif Configuration demarcating the Late Ohio Hopewell phase. As argued earlier, the original Great House Renewal Lodges of these two sites, Seip-Pricer and Edwin Harness, respectively, probably preexisted for a brief period the associated Paint Creek C-R Configuration earthworks. This suggests that, while the small circles that are incorporated within the infixes may have served some lesser ritual purposes, major rites for both of them may have been cooperatively performed at Mound City/Hopeton and High Bank. It was only when the Paint Creek Configurations were constructed that the Mound City/Hopeton Complex may have been completely abandoned while, as it was suggested earlier, it is likely that High Bank continued to be cooperatively shared and expanded by these two cults during the Late Ohio Hopewell phase.

It was suggested that the layout of the R-R-C type of Great House Renewal Lodge combined the total funerary→mourning→spirit-release→world renewal process in a single dual-leveled timber construction, the upper level being used for the more extended laying-in period, and the ground floor sections being used for, possibly, both the post-mortem sacrificial cremation rite and the terminal burial rites of these cremations. In short, the dual-leveled Great House Renewal Lodge would have been a ritual as well as an organizational innovation that, when combined with the newly innovated Paint Creek C-R Configuration, speaks loudly for the dominance of the prescriptive faction. Effectively, almost all mortuary deposits on the floors at all three Great House Renewal Lodges of Seip and Liberty Works were partial or full cremations. This echoes the earlier Mound City Proper mortuary program with its focus on cremation. Therefore, in the terms used earlier, the log-crib features that contain these cremated mortuary deposits would have been primarily not laying-in but world renewal crypt altars.

It was argued that the C component of the R-R-C pattern of the Great House clearly marked a "professional" clergy. This is not to say that clergy were absent in those locales that lacked such Great House Renewal Lodges. Rather, it simply says that the R-R-C pattern most clearly articulates the laity/clergy and senior/junior age-grade structures and that this articulation probably manifests the dominance of the prescriptive faction. Although neither of these sites displays major depositions of raw materials, such as the obsidian and Wyandotte chert deposits at the Hopewell site, they have a respectable quantity and variety of Hopewell goods, displaying all the typical materials. Furthermore, a number of Section 1 burials on the floors of both the Seip-Pricer and Seip-Conjoined Great House Renewal Lodges were associated with multiple copper plates and celts, suggesting individual custodians of multiple usufruct custodial regalia. There is also the large deposit of 12 copper plates with one large 28-lb copper celt centered on the floor of Section 1 of Seip-Pricer that is not accompanied by mortuary remains. This deposit is placed in a log crib that is the same as the features containing the mortuary deposits that surround it and, as suggested earlier, may be an example of a deposit of group custodial regalia, rarely found in the archaeological record.

It is notable that the same type of log-crib facility was used for both this deposit of custodial regalia and mortuary deposits. This suggests these facilities are appropriately characterized as altars rather than as burial platforms. Seen in these terms, the fact that at Seip and Liberty Works, as well as at Mound City Proper, they typically contain human mortuary remains simply reinforces the earlier claim that these are adequately characterized as material warrants of post-mortem sacrificial rites. Also, the particular context of this large copper plate+celt deposit, centered on the floor of Section 1 and surrounded by many mortuary/custodial regalia artifact deposits, suggests a parallel with the great Wyandotte chert deposit of Mound 2, the Hopewell site. This was also accompanied by the five major mortuary/custodial regalia artifact deposits, only in this case, the mortuary aspect was made up of extended inhumations. In the Seip-Pricer case, it might be that the

burials of Section 1 were treated as the collective senior age-grade custodians of this "treasury" of postulated group custodial regalia.

In terms of exotic raw material procurement, the Seip and Liberty
Works cults are treated as largely equivalent to their predecessor at the
Mound City/Hopeton Complex, that is, more than moderately inclin-
ing to the prescriptive pole of the continuum. Add to this the fact that
cremation, major timber construction, and maximum earthwork construction figured importantly at these sites and this clearly locates them
as strongly prescriptive. However, this assessment applies to these sites
only during the occupation of the Seip-Pricer and the earlier occupation
of the Edwin Harness Great House Renewal Lodges. It will be argued
below that most of the occupation period of Seip-Conjoined, as well
as for the latter part of Edwin Harness, was characterized by a strong
simultaneous left-right ideological polarization. The left orientation is
indicated by the absence of mortuary deposits in Section 3 at Seip-
Conjoined (the clergy section), and the right orientation is indicated in
Section 2 by the shift from the traditional split-log terminal burial rite
to the fire-clay terminal burial rite, as discussed fully in chapter 13.
Another indicator of this polarization may be suggested by the change
in entryways. Whereas the floor-plan diagrams of both the Edwin Harness and Seip-Pricer Great House Renewal Lodges (figs. 12.1 and 12.2,
respectively) clearly show that the entryway in each was through
Room 3, the postulated room of the shamanic-priests, the floor plan
diagram of Seip-Conjoined (fig. 12.3) shows an entryway through Room
1, the postulated senior sector room. This suggests that during the later
occupation period of Seip, and possibly also of Liberty Works, the
senior laity leadership had gained a control over the use of the Great
House that previously had been the privilege of the clergy.

With the above qualifications, the dominance of the prescriptive
factions at both Seip and Liberty Works, at least for the major part of
their occupation during the Late Ohio Hopewell phase, is supported
by the geographical scope of their influence. This is particularly well
attested to by the distribution of the C-R Configuration. Based on the

similarities in C-R morphology, layout, directional orientation, the Great Hopewell Road, and so on, Newark and Chillicothe were postulated as constituting a primary axis of alliance.[6] If this thesis turns out to be empirically supported, then both the construction of the Newark Circle-Octagon and the Great Hopewell Road clearly mark this alliance as under the influence of a strong prescriptive faction. Portsmouth (fig. 1.12), which may be the continuation into the Late Ohio Hopewell phase of the resident cult situated at Tremper during the Early and Middle Ohio Hopewell phases, would also have been influenced by this alliance, although the local cult maintained its autonomy. It has been argued elsewhere that the Portsmouth complex of rectilinear and circular constructions is a deliberate and ingenious variation of the C-R Configuration, clearly setting this locale apart from those at Chillicothe while suggesting similar episcopalian, right-of-center leanings.[7]

THE MIDDLE WOODLAND-LATE WOODLAND TRANSITION

The Rupture

Here it will be postulated that the shift from Seip-Pricer to Seip-Conjoined occurred during an initial polarization of the proscriptive and prescriptive factions along the senior/junior age-grade and laity/clergy structural axes and moving in opposed ideological directions simultaneously. The development of this simultaneous and contradictory ideological shift characterized the Seip-Conjoined occupancy. This strained the integration of the cult and then, as earlier postulated, a rather abrupt rupture of the cult occurred. However, at the time of the abandoning of the Seip-Conjoined Great House, it is postulated that the Seip cult was still nominally integrated. This claim is based on the assumption that the ritual closing and abandoning of one locale would occur only when the new locale was ready for occupation, thereby ensuring continuity of the organization and its ritual practices. Therefore, it is probable that the actual rupture of the Seip cult occurred only after the completion of the initial construction phase of the Baum embankment earthworks (fig. 2.2) and before the terminal mantle of the Seip-Conjoined mound could be completed.

Although the rupture may be most clearly expressed at Seip-Conjoined, it is postulated that the same polarizing tendencies were at work across the Ohio Hopewell cults, all of which would be experiencing similar structural stresses. Given the occurrence of a widespread polarization of proscriptive and prescriptive factions across the ecclesiastic-communal cult system, all that was needed was for an important cult to rupture and this could initiate a cascading series of ruptures along the senior/junior and clergy/laity structural fault lines. Therefore, the postulated Seip rupture might have initiated equivalent ruptures at other major sites, such as Liberty Works (or Works East), Newark, Frankfort, and so on. At Turner a rupture is implicated by the large mortuary deposit of "trophy skulls" accompanied by two deceased, either or both of whom probably had served as the final custodians of these group custodial regalia items. The process may have been manifested in different forms in different locales. In some cases, the proscriptive faction, dominated by senior laity, may have physically withdrawn and focused their attention on clan and domestic affairs. In other cases, the prescriptive faction, dominated by junior laity in association with clergy, may have simply physically withdrawn. Alternatively, a given embankment earthwork locale may have been abandoned by both factions simultaneously.

For example, this rupture may have been manifested at Frankfort by the abandonment of this locale and the construction and occupation of the structure under the nearby Raymond Alter Mound. Greber has argued that the Raymond Ater patterning is distinctly different from Seip, Hopewell, and Liberty Works. She treats this site as indicating a single component, kinship-based community. Because it displays an absence of infant or child burial as well as a high 2:1 proportion of male over female burial, she characterizes it as having "a relatively strong age hierarchy and sex differentiation."[8] She also interprets the data as indicating that, while the community was a mono-component, it was internally ranked, not in terms of subcomponents as she claims was the case for Seip and Liberty Works, but in terms of individuals.[9] In contrast, under the Ideological Cult Faction Model, it can be postulated that the patterning of Raymond Ater manifests a world renewal cult that has shifted into a strong congregationalist stance. This is indicated by its two conjoined rooms, one elongated and much larger than the other, clearly a break from all the known Great House Renewal Lodges. Furthermore, the major custodial regalia items (copper plates and marine shell), as well as copper ear spools, are more prevalent in the smaller of the two sections, marking this as probably the senior age-grade sector—despite its burial population being much smaller than that of the large room.[10] Finally, the lack of any embankment earthwork reinforces this proscriptive assessment. However, this congregationalist stance would be unstable since, in effect, it constitutes a straightforward communal cult, on the one hand, and, on the other, it contradicts the companionship principle. Raymond Ater, and other similar locales, might indicate the last of the autonomous Ohio Hopewell cults in which the senior laity had become dominant, almost inevitably leading to their assimilation into the emerging nucleated communities demarcating the early Late Woodland period.

The conclusion, which may be better characterized as a proposal for future research, is that the senior laity leadership of the ecclesiastic-communal cults found themselves in a chronic conflict of interest arising from being leaders in both clan and cult affairs. This encouraged a strong tendency toward innovating and promoting a comprehensive alternative strategy, namely, the integration of cult and clan ritual practices. To effect this, senior clan and cult leadership, often the same people, would probably have promoted nucleated settlements that would take on the collective task of world renewal, possibly integrating the latter with traditional clan ritual. A dual and synergistic process would then have been set into motion. The integration of clan and cult practices in the same locales promoted dissolving the bifurcated ceremonial earthwork/domestic habitation settlement pattern that had prevailed during the Early and Middle Woodland periods, and possibly from the Middle Archaic period in this region, and the emergence of an integrated ceremonial/domestic settlement pattern, realized as the formation of nucleated villages. The creation of compact villages generated by the integration process brought about a partial deontic ecological inversion. The "implosion" of dispersed domestic habitations into nucleated units that simultaneously piggybacked the material facilities of ritual while maintaining essentially the same regime of dispersed gardening and foraging transformed the traditional proscriptive subsistence-settlement/prescriptive ceremonial posture into a prescriptive settlement-proscriptive subsistence/ceremonial posture.

This integrated ceremonial/domestic settlement mode unwittingly dissolved the condition that had favored the continuity of the spatially isolated autonomous cult pattern. Under the bifurcated ceremonial/domestic settlement pattern, youths from different valleys regularly aggregated in the uplands where they would have cooperated in collective tasks, thereby forming lifelong bonds of companionship that would lead them to join preexisting cults. With the shift to the integrated settlement pattern, the autonomous nucleated communities of a local region would easily have had enough youth to mobilize for these special dispersed logistical tasks. Combined with this greater localization of cooperative foraging, the integrating of the world renewal practices within the overall ritual of the local nucleated communities would have further discouraged and largely eliminated the need for long-distance procurement, intensive construction, and production of elaborate artifacts, spelling the rather rapid disappearance of the Hopewellian stylistics and practices from the archaeological record.

Importantly, the integrated ceremonial/domestic pattern would not change the basic polyistic *locale-centric* nature of the social system. Rather, it would promote new forms by which senior and junior age-grades, male and female sodalities, lineage and clan communal cults, and so on, could maintain the traditional arm's-length relations. For example, both clan and nonclan communal cults might continue in the same village, each having different and specialized sacred lodges, one for ancestral ritual and the other for renewal ritual. Or the junior age-grade groups of local villages might share certain non-kin ritual practices while maintaining separate autonomous kin ritual practices. In short, it is postulated here that the Middle Woodland-Late Woodland transition, characterized as the transformation from the bifurcated to the integrated settlement system, would leave the historically rooted, polyistic locale-centric social system in place. But now it would be constituted as a complex and ramifying regional network of autonomous villages, each village manifesting a complex interlacing of kinship and nonkinship social structures and relating to its near neighbors in equally complex ways. Each village would also have custodial usufruct rights and obligations over its surrounding patchwork of gardens and each would also maintain custodial responsibilities over the pathways that connected it to the others in the region. This custodianship would be shared by kinship and sodality leaders.

In effect, given the nucleation of settlement, a form of practical unitary territorialism based on custodial usufruct would emerge. Each Late Woodland nucleated settlement would have practical access to and, for this reason, de facto control of its immediate surroundings with its dispersed gardens. But I want to stress that this practical unitary territorialism would simply mimic and not be the type of exclusive territorialism typical of a monistic modular social system. Clearly, much more could be said about the nature of this postulated integrated ceremonial/domestic settlement system. But this would require exploring the archaeological records of the Late Woodland and Late Prehistoric periods of this region. Therefore, it is well beyond the scope of this book. I will simply speculate that it is likely that a cycling of integrated and bifurcated settlement patterns may characterize these latter periods, possibly with strong deontic forces in the form of ritualized subsistence innovations (e.g., emergence of the bow and arrow technology and of maize agriculture as a staple field crop) favoring the entrenching of the integrated mode.[11] This would account for what may have been predominantly an integrated polyistic locale-centric settlement system being in place when the societies of this region were first confronted with European adventurers and, later, permanent settlers. To complete the hermeneutic spiral, the following chapter will critically examine models of the same Late Ohio Hopewell phase sites representing the alternative Exclusive Territorial/Proprietorial Domain Paradigm.

Chapter 24

A CRITIQUE OF THE CIVIC-
CEREMONIAL CENTER VIEW
OF OHIO HOPEWELL

N'omi Greber has presented a model postulating the chronological and developmental relations linking Seip, Liberty Works, Works East, Baum, and Frankfort Works. The model sets A.D. 100 as the initial construction date of ceremonial facilities at both Seip and Liberty Works, although it claims that, prior to this period of ceremonial usage, domestic occupation occurred in the immediate areas around these sites. Following an extensive period of occupation, these two sites were abandoned and Works East, Baum, and Frankfort were constructed and occupied, possibly all contemporaneously. The chronology is based on a suite of important radiocarbon dates derived from Liberty Works and Seip, with weighted mean dates being ca. A.D. 300.[1] This median date is claimed to be bracketed by a temporal spread of ±200 years (A.D. 100–500) that would allow for developmental stages (table 24.1).

The chronological development of Seip is postulated by this model as an incremental series of "major construction events," which are abbreviated as MCEs. For convenience, this will be referred to here as the Major Construction Events or MCE Model (fig. 24.1). Greber suggests that each developmental stage is roughly correlated with two generations of the occupying kinship-based community group, with a total of 12 to 13 generations, about 30–35 years per generation being assumed. This would encompass a four-hundred-year occupation span divided into seven periods. The model claims that two extended families may have initiated occupation, each estimated as having four nuclear family units and each of these assigned an average of five individuals. This means that each extended family would average about twenty individuals, making a founding population of about forty.

The model postulates that rather than any major construction, the founding period produced only dispersed domestic debris and minor ritual locales, such as those associated with the residue of the buildings found in the northwest sector of the infix, including House 7, Feature 5, having the early date of A.D. 90 ± 85 (DAL-280) (table 24.1). Generations 1 and 2 spanned this founding occupation period and by its termination, the two original extended families had expanded to ten. Therefore, only starting with Generation 3 was each occupation period characterized by a major construction event (MCE). These are labeled as MCE I through MCE VI (fig. 24.1). MCE I of Generations 3 and 4 witnessed the first major construction, this being the Seip-Pricer Great House. The total number of burials found on and off the floor of the Seip-Pricer Mound is used to estimate the MCE I population at ten extended families, or about two hundred. Generation 5 (MCE II) abandoned the Seip-Pricer Great House and built the initial mounding by using "culturally sterile soils and a gravel cap."[2] Generations 6, 7, and possibly 8 were responsible for MCE III, which was marked by building and occupying the Seip-Conjoined Great House. This period also witnessed the completion of the Seip-Pricer Mound, indicated by the series of strata of earth matrix mixed with cultural debris and the final gravel mantle. The model also postulates that MCE III witnessed a population reduction to possibly eight extended families. This postulated reduction accounts for the smaller number of burials associated with the Seip-Conjoined Great House compared to the Seip-Pricer Great House. The abandoning of Seip-Conjoined and the building of its first-stage mantle initiated and characterized MCE IV. The model "fuses" MCE IV and MCE V by suggesting that the initial construction of the embankment earthwork started with the large and small circles, possibly making up MCE V, while the building of the square is attributed to MCE VI, thereby completing the embankment earthworks, as shown in figure 1.4. "I suggest that planning and building the square-shaped enclosure was the final major construction event. . . . Unlike the circular walls, these relatively low walls contained no midden but were made of carefully chosen soils" (215).

Greber relates the Seip and Baum embankment earthworks chronologically and genealogically, pointing out that their forms are so similar that they "suggest . . . that these two sites represent a single local society that used the land around and between the two enclosures and considered the entire region to be their social space" (217). There is also an important claim made here that a shift in focus from Great House construction to embankment earthwork construction occurred. This has chronological implications since Baum has no apparent Great House, or at least there is no known mound located in the infix that would mark the abandonment of a Great House (Renewal Lodge). Therefore, if the shift in construction was from Great House to embankment, then Seip would have to have been occupied first and Baum second.

Using this claim, the model treats the morphological parallels of the embankments as marking important ties among three modular kinship-based communities, each with its own "social space." Liberty Works and Works East embody the historical development of one community, Seip and Baum the second community, and the third community was Frankfort, which was not historically "twinned" with another and earlier embankment earthwork site, although it is located on the North Fork of Paint Creek just a few kilometers upstream from the large Hopewell site.

> I suggest that the existence of the five remarkably similar complex geometric enclosures record one such set of contemporary interregional relationships. Cultural and social ties among three corporate groups were strong enough to allow or to cause the use of some overarching design principles in the construction of this series of enclosures. (Greber 1997, 220)

This "triple alliance" postulate will be treated as an ancillary of the MCE Model, and for convenience sake, it will be termed the Triple Alliance Model.

I. THE TRIPLE ALLIANCE MODEL CRITIQUE

The Triple Alliance Model will be critiqued first, and then critical attention will be focused on the MCE Model. It is notable that while the Hopewell site is recognized by Greber as a sixth example of a site in this region having the "square-shaped enclosure," she does not consider it to be among these "remarkably similar complex geometric enclosures." As she puts it: "A square-shaped enclosure was added to the eastern side of the great enclosure wall at the Hopewell site, about 9 km downstream from Frankfort. However, this square differs in size and iconographic details from the five I consider to be a regional unit" (220). She does not specify how the Hopewell square differs in "iconographic details" from the five at Frankfort, Seip, Baum, Works East, and Liberty Works. Possibly what is implied is that one of its four walls differs. Of course, it was pointed out earlier that this is probably the result of the builders' appropriating the east wall of the Hopewell C-Form/SL-Profile embankment and using it as the west wall of the square, thereby incorporating a "ditch" in this wall. It was also earlier noted that the Hopewell Paint Creek Square is smaller than the norm. However, these differences do not appear to constitute compelling reasons for ignoring the possible relations between the Hopewell earthwork and those forming the postulated "triple alliance." Indeed, since the spatial positioning of Frankfort and Hopewell is equivalent to that of Seip and Baum and of Liberty Works and Works East, and given the morphological parallels between the C-R components of the Hopewell site and Frankfort, it would seem perfectly reasonable to match these two sites.

Setting this point aside for the moment, it is clear that two important aspects of the Triple Alliance Model and MCE Model are in close harmony with the argument that has been developed in this book, namely, that these Complex G-Form embankment earthworks had temporal significance, and that, as symbolic structures, they de-marcated important social and cultural relations and ritual postures. As Greber puts it, these "five remarkably similar complex geometric en-closures" demarcate "[c]ultural and social ties among three corporate groups [that] were strong enough to allow or to cause the use of some overarching design principles in the construction of this series of enclosures" (220). However, this quotation suggests an under-characterization of the symbolic nature of these material features. Here, the ties that bound these groups together are being treated as the independent variables that "allow" or "cause" the material features to be built, thereby expressing the "overarching design principles" that they shared. This implies that the symbolic dimension is simply an "add-on" factor, an embellishment of the preexisting social relations that they represented. This threatens to reduce the role of symbolic construction to serving the somewhat anemic function of being the medium for conveying information about a social set-up that would already have been widely known and recognized.

The symbolic pragmatic approach would claim that such con-struction was a critical part of the constitution of that very social set-up, not because it displayed a sharing of some abstract symbolic cal-culus that "conveyed information" about an alliance, but because it manifested the realization in corporate collective activities of common ideological commitments, that is, enhancing the sacred powers of the cosmos. If this is not accepted, and, instead, this "information convey-ance" role is all we can claim for monumental construction, then such claims are subject to serious criticism on the grounds that there would be no point in such massive construction endeavors. If earthworks serve merely to publicize social ties that preexist the features, then what purpose could building them have served other than decorative public relations? The rationality of those responsible could be questioned. In effect, this "information conveyance" notion empties earthwork con-struction of any symbolic pragmatic relevance and, therefore, reduces it to being a mere public relations show. By characterizing the role served by the earthworks in symbolic pragmatic terms, however, replicating "an overarching design" does not simply reflect a preexisting set of alliance relations. It is a part of the social strategy that constituted these relations, and the rationale for these relations is firmly grounded in the ideological task of world renewal that these groups took themselves to be promoting and realizing through their cooperative construction efforts.

This also means that the "iconographic details" do take on an importance, just as Greber suggested. However, rather than these being grounds for ignoring the Hopewell site, they become grounds for a more subtle interpretation of the relationships that the morphology partly constituted. It is claimed here that while there are attributes that the Hopewell and Frankfort sites share with the other four sites of Seip, Baum, Liberty Works, and Works East, there are attributes that Hopewell and Frankfort share between them that distinctly set them apart from the other four sites. In this regard, Romain has pointed out that the large Frankfort circle, the infix, and its connected Paint Creek Square display an *ad quadratum* geometrical relation. This means that the Square fits relatively neatly inside the infix—the diameter of the infix equals the diagonal of the Paint Creek Square. However, as he argues, in contrast to the *ad quadratum* infix-square relationship, the infix-square relations of Seip, Baum, Works East, and Liberty Works are based on *ad triangulum* geometry. For example, an equilateral triangle with sides equal to the length of the side of its Paint Creek Square fits within the perimeter of the Baum infix.[3] This *ad quadratum/ad triangulum* contrast reinforces the earlier claim that the Hopewell site on the North Fork and the Mound City/Hopeton Complex on the central Scioto represent competing ecclesiastic-communal cults. In historical terms, then, the Hopewell and Frankfort sites would manifest a "descent" relation that was partly constituted and reproduced via shared morphological detailing of the embankments, and it was equivalent to but separate from the "descent" relations that Greber claims historically linked Seip and Baum, and, similarly, Liberty Works and Works East. The shared differences between these two units, on the one hand, Hopewell-Frankfort on the North Fork and, on the other,

Seip-Baum and Liberty Works-Works East on the Paint Creek-Scioto, were part of what was required for these two to maintain a mutual arm's-length relation.

2. THE MCE MODEL CRITIQUE

In the MCE Model, the Great House Renewal Lodges and the embankment earthworks are treated as the results of independent and alternative construction programs bearing only a historically contingent relation: first the Renewal Lodges and their moundings and then the embankment earthworks. By contrast, the symbolic pragmatic approach would argue that the construction of embankment earthworks, Renewal Lodges, and mortuary moundings formed a single coherent historical tradition of mortuary/world renewal practices. The occupation of the Seip site, then, would have been carried out by a group already steeped in this complex tradition such that no proper ritual life could be complete without the appropriate complex of material cultural warranting facilities. This suggests that Seip would have had an embankment earthwork from the beginning of its active use as a ritual locale rather than having one that was added only toward the end of its occupying the site. This is not to claim that the Paint Creek C-R Configuration embankment earthwork was built as a founding exercise. The claim of the MCE Model is accepted in this regard, namely, that the construction of the Seip-Pricer Great House Renewal Lodge probably occurred prior to the construction of the Paint Creek C-R Configuration embankment earthwork, or, at least, that the construction of the Renewal Lodge was initiated prior to the construction of these major embankment earthworks. However, it was also argued that the small K-Profile circle served this initial ritual stage while, as suggested in the previous chapter, the cult continued to participate with Liberty Works, its neighboring allied cult, in carrying out major rituals at both the Mound City/Hopeton Complex and the High Bank Works.

As postulated by the MCE Model, at least four generations were required to build the embankment earthworks (fig. 24.1). During this span of 120+ years, or at least during most of it, there would have been no access to a Great House in this locale, since the MCE Model claims both Seip-Pricer and Seip-Conjoined as having been abandoned and mounded over. Furthermore, since these two sets of components, circles and square, were constructed by different generations, they would have been built as two independent units. This implies that each would have had independent symbolic meaning. Certainly, this would seem to contradict the earlier claim that these embankments expressed some "overarching design principles in the construction of this series of enclosures." In support of this set of individuated constructions, Greber points out that the circular elements were built of fill having cultural materials mixed within it while the walls of the square "contained no midden but were made of carefully chosen soils."[4] Presumably, this difference marks a period of everyday occupation during the construction of the circle embankment components, while the period of constructing the square was one of infrequent occupation. The latter relates to the claim of the MCE Model that, while the community was building the square element, it was actually using Baum, and, therefore, there would be no domestic debris being accumulated at Seip during this construction.

In terms articulated earlier, the Paint Creek C-R Configuration is a holistic, monumental solar-related icon of the cosmos with the perimeter circle (the C component) representing the Heavens and the infix, the large "inner" circle, the Middle World. Therefore, the "overarching design principles" that Greber refers to would correspond to this cosmological view. In these terms, the differential earth fill patterning that she identifies would manifest emic notions about different categories of earth fill appropriate for different components, as argued earlier with regard to the Newark site. The use of surface-stratum fill (rich with "midden") would be consistent with this view, since the C component and the large infix component would represent and participate in the powers of the Heavens and the Middle World, respectively. Similarly, the use of "carefully chosen soils" or deep-stratum earth, for the square would be consistent since this latter element, as the R component, would represent and manifest the powers of the Underworld.

THE SYMBOLIC PRAGMATIC ALTERNATIVE ACCOUNT

In short, the Seip construction program was an intricate and expanding program in which embankment earthworks and Great House Renewal Lodges worked as a whole and not as disparate and unrelated parts. This requires a significant reconfiguring of the construction stages postulated by the MCE Model, both by shortening the chronological span it postulates for the occupation of the Seip site and by moving the initial C-R embankment construction to a time shortly after the initiation and/or completion of the Seip-Pricer Great House Renewal Lodge. In terms of the Split Age-Grade Mortuary Model, a single-cycle Great House CBL would embody a 25–30-year time span. Each recycling of the same Great House CBL would add another 25–30 years per age-grade promotion. Since it was concluded under the Split Age-Grade Mortuary Model that Seip-Conjoined was a dual-cycle CBL, it was suggested that Seip-Pricer, with its larger burial population, probably incorporated more than two cycles. This suggests treating the four linear and one circular burial patternings on the floor of Seip-Pricer, Section 2, as demarcating the retained deceased of four or five sequential junior age-grade sectors. This means that the space had been utilized for at least five generations.

The Seip-Pricer Mound contained 123 individuals, including both floor and above-floor burials (table 24.2). There were 48 floor and above-floor burials in association with the Seip-Conjoined Mound. To make a reasonable comparison of these two burial populations, only the floor burials should be included, and, since there were no burials in Section 3 of Seip-Conjoined, the individual burials in Section 3 of Seip-Pricer should also be excluded. This means that the total number of mortuary floor deposits in Sections 1 and 2 (Seip-Pricer) would be 91 [(Section 1) 53 + (Section 2) 38 = 91], and for Sections 1 and 2 (Seip-Conjoined) 43 [(Section 1) 24 + (Section 2) 19 = 43]. Based on the split-log/fire-clay rite distinction, the 43 individual burials of Seip-Conjoined mark a dual-cycle split age-grade CBL incorporating a two-generational spread of 50–60 years. The total number of individual burials in Sections

1 and 2 (Seip-Pricer) would index slightly more than double this (91/43=2.11), or more than twice the period of occupation compared to Seip-Conjoined, making it a four+ generation CBL. Rounded up to whole numbers, Seip-Pricer is estimated to be a five-cycle CBL. This is within the estimate of four or five clusters making up the Seip-Pricer Section 2 burials. As demonstrated earlier, these two CBLs were sequentially used. Therefore, they incorporate an estimated spread of seven generations (5 + 2 = 7 generations), or between 175 and 210 years (7 × 25 years = 175 or 7 × 30 years = 210). This will be treated as a 200-year spread. If the generational span of 30–35 years is used, as postulated by the MCE Model, this would expand the period of occupation to about 210–245 years.

Based on the radiocarbon dates for Seip-Pricer of 330 ± 40 (table 24.1), and allowing a minimum of 50 years (2 × 25 years) for the occupation of Seip-Conjoined following the abandonment of Seip-Pricer, this would suggest a five-generational occupation of Seip-Pricer from ca. A.D. 200 to 340–350, and a total occupation of Seip-Conjoined from ca. A.D. 340–350 to 400. This also neatly brackets Greber's weighted average for Seip and Liberty Works of A.D. 300, but with the lesser scope of ± 100 years, or A.D. 200 to 400 (compared to the scope of ± 200 years of the MCE Model). Prior to A.D. 200 and following A.D. 400, the Seip site would be characterized by low-level ritual usage. Earlier it was argued that the border of the Middle/Late Ohio Hopewell phase was ca. A.D. 150, with the construction and use of High Bank marking the emergence of the Late Ohio Hopewell phase. As pointed out above, Seip and Liberty Works would have been built later, or ca. A.D. 200 (± 15), which would nicely fit the above estimates.

The difference between the proportion of individual burials in Sections 1 and 2 of Seip-Conjoined compared to Sections 1 and 2 of Seip-Pricer may be historically significant. The former is 1.26:1 (24/19), while the latter is 1.4:1 (53/38). With five generations postulated for Seip-Pricer, the 1.4 figure would appear to be a more adequate estimate of the normal proportion of Section 1 to Section 2 individual burials than the 1.26 proportion of Seip-Conjoined. This would suggest that Seip-Conjoined was abandoned prior to the normal period, this being when the final senior age-grade retired and made way for the promotion of the junior age-grade. This possible anomaly will be added to the earlier anomalies that were listed for Seip-Conjoined. These were (1) the absence of burials in Section 3, (2) the lack of a Section 4, (3) the dual split-log/fire-clay rites, (4) the apparent absence of an entryway indicated through Room 3, and (5) the unfinished state of the mound. Anomaly 6, then, is the possible abandonment of Seip-Conjoined prior to the age-grade promotion rituals. All this reinforces the view that, as postulated earlier, a serious rupture in the normal cyclic process of cult reproduction occurred.

THE MARRIAGE EXCHANGE MODEL

DeBoer has constructed a complex social and chronological model of the embankment earthworks using a seriation technique based on morphological attributes of the embankment earthworks combined with radiocarbon dating.[5] As his analogical basis he uses his study of the ceremonial centers of the Chachi, a forest-dwelling riverine-oriented horticultural people of modern Ecuador, claiming that the Middle Woodland peoples were organized into a similar type of social system. The Chachi currently live in small dispersed gardening households along the Ecuadorean rivers and within territories that they collectively use. These individual households regularly gather at a permanent but only periodically used civic-ceremonial center, occupying residences that are spatially related so as to replicate in "miniature" the relative positioning of the riverine territory that each regularly uses during the annual cycle. The purpose of their gathering is to conduct ritual, such as funerals, other rites of passage, and presumably renewal rites, since DeBoer also characterizes these centers as microcosms of the cosmos. When the groups disperse, the ritual center is once more "vacant." He suggests that a similar set of domestic, cosmological, and alliance meanings may be manifest in the great Ohio Hopewell embankment earthworks, referring specifically to the Dispersed Hamlet/ Vacant Center Model of the Ohio Hopewell as first postulated by Prufer and revised by Pacheco and Dancey and Pacheco.[6]

DeBoer's analysis of the Ohio earthworks in these terms is two-dimensional. First, he gives a basis for constructing the earthworks by interpreting them, in general, as displaying cosmological congruencies. He expands his analogical base by pointing out that most Native North American peoples identify their gathering places as "Great Houses," synthesizing their domestic arrangements with the cosmological structures through modes of congruency. In this regard, he points out that the rectilinear layout of the buildings and plazas of the southeastern Creek and Muskogean were conceptualized by them as manifesting elements of both residential structures and the cosmos, for example, by noting that the perimeter structures delineating the plaza were referred to as "beds" by some southeastern peoples. He then shifts to the Ohio Hopewell earthworks (fig. 24.2). The particular patterning that he focuses on is the C-R Configuration. He also recognizes variations on this theme, while interpreting the significance of these quite differently from the Sacred Dual C-R Motif approach.

To link the historical analogy to the Ohio Hopewell earthworks, he draws on Faulkner's observation that Native North American cultures of the southeast often had both winter and summer domestic dwelling places, often side by side, with the winter house being circular and the summer house being square.[7] He suggests that the circle-square (C-R) pattern of the Chillicothe Tradition may simply be an alternative metaphorical expression of the same conceptualization, linking ordinary kinship winter and summer residences to the cosmos via the embankment earthworks.[8]

He also brings in other elements of the archaeological record to show that a dual patterning is quite pervasive. As was done earlier in this book, he also draws on Greber and Ruhl's structural analysis of the two deposits on the floor of Shetrone Mound 17.[9] He makes a persuasive argument that the animal effigies that make up many of the forms of the two well-known platform pipe deposits at Mound City and Tremper, as discussed earlier, can be interpreted as entailing a vision of the world as a tripartite structure of heavens, waters, and land.[10] This could also be easily reconstrued into the terminology suggested for interpreting the C-R Configuration as the cosmos structured into the Heavens and the Underworld mediated by the Middle World in both its land and water elements.

Having argued that the contrasting circle and square pattern effectively grounds a complementary opposition with a manifold meaning of both cosmological and everyday social life (e.g., winter house/summer house, male/female, consanguine/affine, friend/enemy, heavens/underworld, and so on), he then suggests that the morphological variation might plot significant shifts in alliances mediated through marriage practices. This is the basis of his Marriage Exchange Model in accounting for the earthworks. He suggests that the circle-square duality represents not only the cosmos as a "Great House" but also the structures immanent in the domestic unit, with the circle representing the "home" group based on patrilineal descent, and the square representing the in-marrying wives from distant patrilineal kin groups, whom he characterizes as the foreigners or "enemy." He sees the alliance system as going through three stages, and these form the sequence that underwrites and warrants his seriation so that the relative size and orientation, including mode of attachment, of the embankment elements, and the transformations of the alliance relations are expressed in and cause the morphological transformations of the embankment earthworks.

It is important to note that this seriation largely focuses on the C-R Configuration. It does not address the C-Form, T-Form, or the distribution in time and space of the S-Profile and its SL-Profile and SR-Profile versions, and so on. The Mt. Horeb sacred circle, however, is critical and it is used by him to anchor the seriation. He argues that, because this earthwork preceded the C-R Configuration, the square or rectilinear element added onto the "traditional" sacred circle counts as the addition of a "foreign" element. He then suggests that the known radiocarbon dates seem to support this order. He recognizes that the

Complex G-Form has variations. However, rather than noting that these might manifest a complementary sacred duality, he treats them as variations of the same cosmological meaning of Great House/Cosmos (fig. 24.2).

He then uses kinship alliance theory as the theoretical grounds for interpreting and temporally ordering the earthworks according to changing morphology. He uses the contrast between generalized or "indirect" spouse exchange and reciprocal or "restricted" spouse exchange, arguing that the latter is most effective in small, face-to-face populations and the generalized is most effective in societies having large populations. He then suggests that a generalized system of spousal exchange emerged in Ohio Hopewell out of a prior reciprocal system, not because there was an expansion of population but because it was highly dispersed, and, as a result of greater reliance on gardening, the value of female labor had increased. Therefore, he postulates that greater reliance on gardening combined with dispersed hamlets were conditions that generated a shift from direct, reciprocal spousal exchange, this being marked by "symmetrical" circle-square patterns, to indirect, generalized exchange, this being marked first by distortions in size and orientation of the C-R Configuration, and then by the emergence of what has been termed here the Paint Creek C-R Configuration. He argues that the infix is the expression of the reassertion of the home group over the foreign group.

It is the postulated shift in marriage exchange system, then, that is mapped in the formal seriation of the earthworks. The early phase is the circle-square pattern, and it includes Hopeton, Circleville, High Bank, and the Wright Square of Newark, simply because, in each case, the circle and square are "equivalent" in size and symmetrically related. It should be noted, however, that, in fact, the Wright Square does not have an "equivalent" circle in the immediate manner that Hopeton, High Bank, and Circleville do. The scheme puts Liberty Works, Baum, and Works East as marking the late period, primarily because of the asymmetries already noted, these marking the "generalized exchange" system, and in between is postulated the "indecisive period," based on those embankment earthwork sites that display anomalies in relative size, skewing, missing features, and so on. These include the Newark Circle-Octagon, which is the near twin of High Bank, along with Seal and Marietta, Seip and Frankfort.

Critique

It is clear that DeBoer has treated the morphology of the earthworks as both socially and chronologically relevant. He is also right to emphasize the cosmological sense of the earthworks. However, there are several problems with his overall account. First, it is explicitly premised on the Dispersed Hamlet/Vacant Center Model. Because of this, it is also committed to the monistic and modular view, and, therefore, to the assumption that the primary metaphor on which such a society will materially construct its cosmological expression is necessarily the kinship-based domestic residence. Hence, the domestic "family" locale becomes embodied in the "Great House" of the embankment earthworks, and these latter become the material means of synthesizing and sanctifying the domestic domain by embedding the sacred structures of the cosmos in them.

While the perspective that this book has developed certainly recognizes the earthworks as manifesting and participating in the sacred powers of the cosmos, it does not accept the basic premise on which DeBoer grounds his model, namely, the identification of the earthworks with the domestic sphere. Certainly, it is not denied here that those responsible for the domestic sphere were very concerned about the manner in which their material relations with the local environment ramified through the sacred structuring of world that humans occupied. This concern is central to the Proscriptive/Prescriptive Ecological Strategy Model and the essential contradiction of human existence on which it is grounded. However, under the latter conceptualizations, rather than this concern serving to synthesize the domestic sphere and the cosmos via the symbolic meaning of the earthworks, the domestic arrangements would be seen as the primary source of sacred pollution rather than sanctification. Therefore, while it is certainly the case that ritual was performed in the context of domestic settlements (the Dual Ritual Spheres Model clearly articulates this), the major world renewal ritual would be carried out in an area that would not be polluted by regular human domestic occupation. In this regard, the earthworks and their associated Renewal Lodges represent the antithesis of the domestic dwelling. Whereas the latter was ephemeral, largely in order to minimize the degree of pollution brought about by human occupation, the former, the "classic" Hopewell structures, were large, permanent, and the prescriptive medium for world renewal rites by which the cosmos was sanctified and resanctified. Therefore, it is asserted here that the domestic dwelling would *not* figure as a model for public ritual centers. Instead, it would be the Renewal Lodge/Embankment Earthwork Complex, embodying the social structures of the cult and integrating these with the cosmos, that would mediate the synthesis of cult and cosmos.

This is not to claim that affinal alliance construction and systems of spousal exchange did not play an important part in the social world of the individual members of the cults responsible for the embankment earthworks and the Renewal Lodges. As argued earlier, the age-cohort companions making up a cult would normally come from different and widely dispersed domestic locales, and, therefore, many would also become brothers-in-law to each other so that the cults, largely unwittingly, would have been the medium for reciprocal marriage alliance characteristic of small, face-to-face populations, even though the populations were dispersed over large regions. From this, however, it does not follow that the morphological variation in the embankment earthworks maps changing marriage systems. Instead, as argued earlier, this variability is simply the historical development of the Sacred Dual C-R Motif, most fully expressed by the High Bank and the Paint Creek C-R Motif Configurations. Therefore, rather than the High Bank C-R Configuration mapping a state of reciprocal marriage alliances, and the Paint Creek C-R Configuration marking an evolving state of affinal alliances, as DeBoer claims, these are here reiterated as the complementary constituents of the Sacred Dual C-R Motif. As argued earlier, the ecclesiastic-communal cult required both types of locales, or their equivalents, in order to discharge its ongoing sacred duties to the cosmos.

In the chronological scheme promoted by this book, the emergence of the Paint Creek C-R Configuration slightly later than the "symmetrically aligned and equally sized circles and squares," the High Bank C-R Configuration, is simply the development of the ideological protocols manifested in the morphology of the Dual Sacred C-R Motif.[11] As argued earlier, the original C-R Configuration may have been the less than symmetrical Hopeton Circle-Rectilinear. Therefore, contra DeBoer's claim that Hopeton and High Bank are chronologically equal, the former marks the emergence of the Middle Ohio Hopewell phase, ca. A.D. 1, and High Bank marks the beginning of the Late Ohio Hopewell phase, ca. A.D. 150. This significantly differs from DeBoer's chronology.

What marks High Bank as unique is that it may have been the first (nearly) symmetrical C-R Configuration. It would have been modeled on the subsymmetrical Hopeton Circle-Rectilinear. As discussed earlier, while the High Bank and Paint Creek C-R Configuration, together, constitute the Dual Sacred C-R Motif, this dual motif was prefigured in the Mound City Cluster and more fully expressed in the formation of the Mound City/Hopeton Complex. Furthermore, rather than Seal, Marietta, and the Newark Circle-Octagon demarcating the "indecisive" middle period, as the seriation that DeBoer offers indicates (fig. 24.2), the Newark Circle-Octagon would be effectively contemporary with High Bank, based on the fact that these are near mirror-image twins. Marietta would be slightly later, and certainly Seal, as a variant Paint Creek C-R Configuration site, would have been built after Seip and Liberty Works and, indeed, probably it was roughly contemporary with the add-on C-R components of the Hopewell site. This critical summary of alternative chronological and social organizational accounts of Ohio Hopewell will be completed in the epilogue by subjecting the Proscriptive/Prescriptive Ecological Strategy Model of Ohio Hopewell to an internal critique. This is in complete accordance with the hermeneutic spiral.

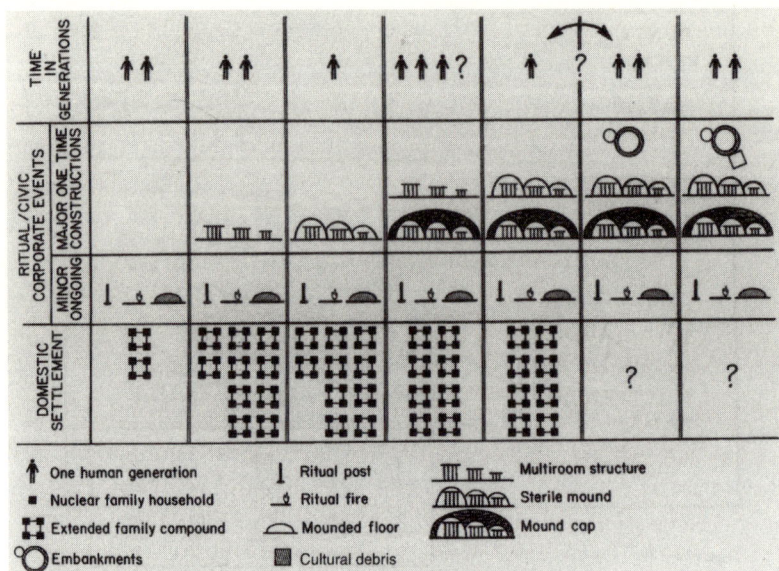

FIG. 24.1. The Postulated MCE Chronology of Seip (Greber 1997, 225, figure 8.3)

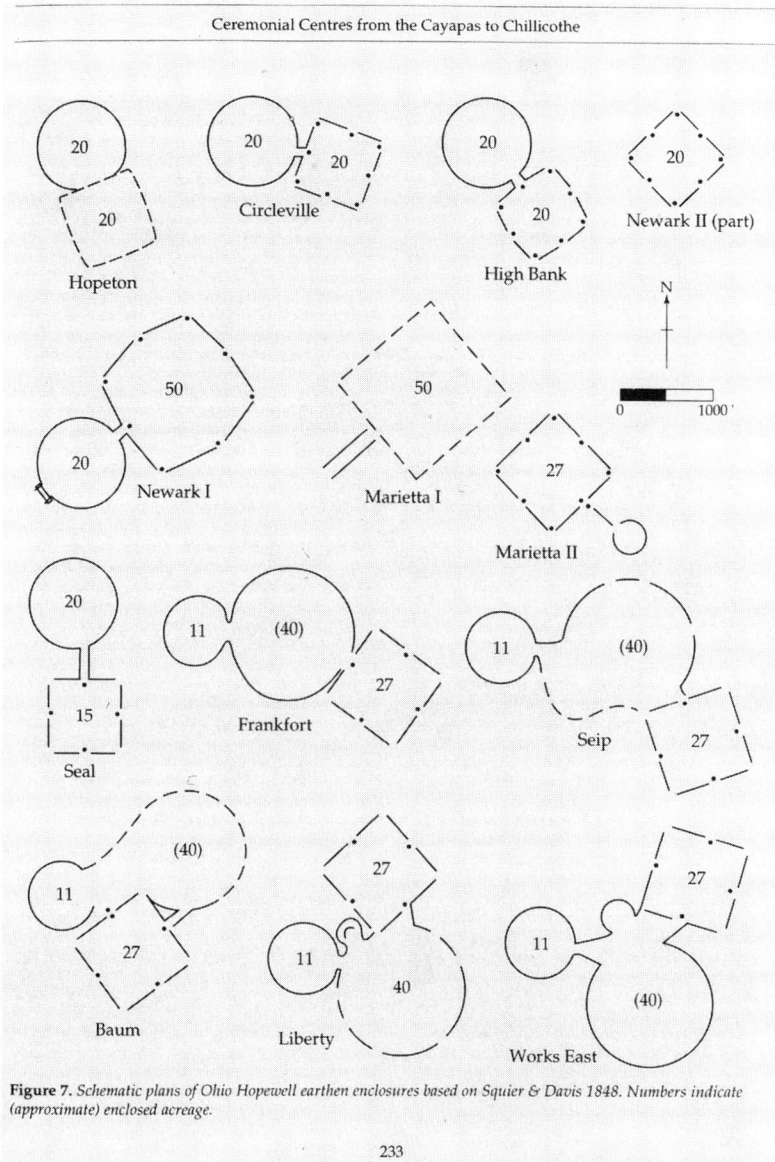

Ceremonial Centres from the Cayapas to Chillicothe

Figure 7. Schematic plans of Ohio Hopewell earthen enclosures based on Squier & Davis 1848. Numbers indicate (approximate) enclosed acreage.

233

FIG. 24.2. Postulated Chronology of the C-R Motif Configuration (Deboer 1997, 233, figure 7)

Table 24.1 Suite of Seip Radiocarbon Dates (Derived from Greber 1983, 91, and 1997)

Suite of Radiocarbon Dates		
Seip-Pricer Mound	A.D. 55 ± 100	UCLA-292
House 7, Feature 5	A.D. 90 ± 85	DAL-280
Houses 1, 2, and 3 Unit D Midden Layer	A.D. 230 ± 80	DAL-116
Burial 32, Seip-Pricer Great House	A.D. 330 ± 40	DIC-2473
Locality 20, Feature 2	A.D. 350 ± 45	DIC-1725
Locality 20, Feature 4	A.D. 470 ± 55	DIC-1724
House 6, Feature 13	A.D. 1055 ± 110	DAL-282

Table 24.2 Summary of Individual Burials (Based on Greber 1979a, Table 6.3:34)

	Section 1	Section 2	Section 3	Off Floor
Seip-Pricer	53	38	19	13
Seip-Conjoined	24	19	0	5

EPILOGUE

The Autonomous Cult Model argued that the Adena Mt. Horeb and Ohio Hopewell Traditions differed in the ideological aspect of the cosmology that they shared—different symbolic pragmatics—because they differed in social structural terms, these social differences constituting the difference between communal and ecclesiastic-communal type cults. In these terms, it was argued that the Adena system of simple communal cults was constituted as a set of mutually autonomous senior age-grade and junior age-grade world renewal cults forming an open network spanning much of the Central Ohio Valley during the Early Woodland. It was argued that by respecting the same-generation companionship principle, these senior and junior age-grade cults maintained a well-defined arm's-length relation toward each other. As *communal* cults, that is, cults without the clergy/laity structure, they generally would have drawn on the specialized ritual know-how of local shamans when required. It was further argued that this system of cults was the social condition that made possible the emergence of the Ohio Hopewell system of complex cults, the latter termed in this book ecclesiastic-communal cults because of the clergy/laity structure. This system was also constituted as an open network of autonomous cults, in this case, spanning parts of the Central Ohio Valley, particularly defined in two general regions of Ohio, the southwestern region and the south-central and central-east region, thereby constituting the Miami Fort Tradition and the Chillicothe Tradition of Ohio Hopewell embankment earthworks, respectively.

The Autonomous Cult Model postulated a social innovation that brought about this transition. Under the pressure of competitive pursuit of cultic reputation, in some regions local autonomous senior and junior age-grade cults started to negotiate the "temporary" pooling of labor and skills in order to enlarge the earthworks that served as monumental icons mediating world renewal ritual, thereby enhancing their individual cultic reputations. Successful ad hoc arrangements served as precedents to extend the cooperative interaction, leading to linking the senior and junior age-grades into a single integrated cult. This integration, however, unwittingly transformed the shamanic religious specialists who traditionally would have been "contracted" into being more or less permanently associated with particular cooperating cults, thereby being transformed into shamanic priests constituting what became recognized as an official categorical position of clergy. This recognition would simultaneously transform the ordinary participants into laity, and thus the clergy/laity structure emerged, constituting the cult as a permanent but periodically aggregating congregation characterized by the conjunction of the dual senior/junior age-grade and laity/clergy relational structures. The differences between these two material cultural records, Adena Mt. Horeb and Ohio Hopewell, therefore, hinge on relatively minor but fundamental social structural differences that were generated by and escalated the dynamics that promoted the Ohio Hopewellian innovative stylistics.

The Ideological Cult Faction Model does not claim that further social structural changes caused the formal material variation within Ohio Hopewell. Rather, it claims that this variation was caused by the working out of the structural contradictions that characterized the newly emergent ecclesiastic-communal cults, this working out being realized as the escalation of competition among ideological factions of a cult in the pursuit of reputation. Because the standing of factions hinges on promoting alternate and sometimes contrasting symbolic pragmatics, their realization in collective behavior can produce diametrically opposite forms, manifesting contrasting ideological positions, while leaving the basic social structure and the basic, shared cosmology largely intact. However, just as the Adena Mt. Horeb/Ohio Hopewell dichotomy is claimed to be based on social structural transformations, it might well be asked if equivalent social structural differences could account for some of the variation within Ohio Hopewell, particularly the formal earthwork, mortuary, and artifactual differences marking the Miami Fort Tradition and the Chillicothe Tradition. Or are these differences simply the historical outcome of factional negotiations caught up in variable clan and cult relations? If the latter is the case, then the variation marking the Fort Miami Tradition and the Chillicothe Tradition would be simply the result of alternative ideological protocols, each subsystem of cults treating those protocols that they favor as correct and those of their more distant neighbors as misdirected. Or is it possible that both factors were at work: factional competition producing alternative ideological protocols *and* interregional social structural differences?

The discussion of Turner assumed that this site and the other regional sites of Fort Ancient, Miami Fort, Stubbs Works, Colerain, Pollock, and so on, were the sacred locales of world renewal cults having the same social structural make-up as those demarcating the Chillicothe Tradition, such as Seip, Liberty Works, Works East, Mound City/Hopeton Complex, and so on, and only the symbolic pragmatic protocols varied as a result of local factional and interregional competition. For example, Fort Ancient may simply be the unique southwestern Ohio expression of the Lunar Motif that is similarly manifested in the High Bank C-R configuration in south-central Ohio. Turner, with its rich mortuary record, may simply be the complementary solar-related ritual locale, so that Turner and Fort Ancient made up the dual sacred locales of a single ecclesiastic-communal cult of this region. This view is supported by the earlier argument that the mortuary program of Turner was similar to that at the Hopewell site, both being dominated by extended inhumations that were exposed for considerable periods. The burnt offering deposits of Turner are also very similar to those of the Hopewell site. Altogether, this suggests that both Hopewell and Turner may simply mark moderate left-of-center stances on the C↔P↔E continuum, implying that senior age-grade leadership was more dominant overall than junior age-grade and clergy leadership. Thus, it may be sufficient to conclude that the Turner and Fort Ancient sites manifest an integrated cult system that is differentiated from the Chillicothe Tradition only in ideological terms that sustained different and somewhat contrasting symbolic pragmatic expressive forms.

However, the radical morphological differences between the Miami Fort and Chillicothe Traditions may not be accounted for in such straightforward ideological factional terms. It may require postulating social structural differences. For example, while the T-Form and C-Form embankment earthworks (or their equivalents) may together embody a duality of the same order as that displayed in the High Bank/Paint Creek C-R duality, this might also be manifesting an arm's-length social relation equivalent to the solar circle/sacred circle duality postulated for the Mt. Horeb Tradition, that is, a system of mutually autonomous senior and junior age-grade cults. Still, given the morphological differences of the sacred locales between the Miami Fort and Chillicothe Traditions, on the one hand, and the Mt. Horeb Tradition, on the other, this possibility could be challenged. As a speculative suggestion, it is possible the Miami Fort Tradition of southwestern Ohio combined social structural elements of both the Adena Mt. Horeb simple and the Ohio Hopewell complex cult systems and it is this combination that is manifested in the T-Form/C-Form duality.

In these terms, it is possible that the Turner site and its equivalents, such as Colerain, Camden, and Milford, may have sustained a system of complex or ecclesiastic-communal cults while Fort Ancient and its equivalents (e.g., Miami Fort) sustained a simple cult system of communal cults. In this scenario, the Turner site would have a laity made up of senior and junior age-grade sectors with an attached set of shamanic priests. It would have responsibility for solar-related world renewal rites. At the same time, however, the junior age-grade cohorts may have traditionally held dual cult membership. As junior age-grade laity members in the Turner-type cults, they would also be members of autonomous world renewal cults responsible for the performance of lunar-related world renewal rituals, these being located in the T-Form ridge top embankment earthwork sites. Thus, the total system of cults would be constituted by mutually autonomous communal and ecclesiastic-communal cults.

In this more complex cult system, the left-of-center orientation of the Turner cult would be accounted for not only as a result of competition among ideological factions but also as a matter of recognized and accepted dominance by the senior age-grade sector in the affairs of the cult. The junior age-grade sector would play a relatively minor role in the cult's affairs, largely serving as the labor source. This might account for the relative paucity of mortuary artifacts at Turner, marking a reduced competition for esteem between senior and junior age-grade leadership and a reduced role for the clergy. Correlated with this would be the unique form of "trophy skull" custodianship, one which may have emphasized the special role of senior leaders as the chosen custodians of such potent group custodial regalia. The subordinate position of the junior age-grade in the Turner-type cult would have its compensation since, in this speculative suggestion, Fort Ancient and Miami Fort would have been the locales of the autonomous communal cults under the exclusive control of the junior age-grade sodalities. Indeed, if this were the case, then the junior age-grade cohorts of southwestern Ohio Hopewell may have maintained a much greater overall degree of autonomy with respect to the senior age-grade sodalities than did the junior age-grade cohorts in central and south-central Ohio.

These speculations raise plausible extrapolations generated from within the framework of models that has been discussed and elucidated in this book, while addressing some of the empirical problems that variations in the archaeological records of the Miami Fort and Chillicothe Traditions raise. All this would require a great deal more empirical research, theoretical construction, and test implication development before a plausible set of models might be generated and demonstrated. Indeed, it might finally be that the Miami Fort Tradition is most co-

herently accounted for in precisely the same terms as the Chillicothe
Tradition, that is, the symbolic pragmatic grammars of the two varied
because of different factional histories. On the other hand, there may
be significant differences demonstrated that, in fact, require a reassess-
ment of the possible cult structures of the Chillicothe Tradition.

Along these lines, the same process might be usefully applied to
the Hopewellian archaeological record *outside* the Central Ohio Valley.
Implicated in this suggestion, of course, is the central premise of this
book, that the major social and cultural framework outlined under the
Inclusive Territorial/Custodial Domain Paradigm can be generalized
across the prehistoric Eastern Woodlands: for example, an immanentist
cosmology, the essential contradiction of the human pursuit of survival,
inclusive territorialism, custodial usufruct, dispersed domestic settlement, polyistic locale-centric social networks, mourning/world renewal
post-mortem sacrifice, and so on. In this regard, the monistic modular
view of the Hopewellian archaeological record of the lower and central
Illinois Valley, well reported in the archaeological literature, may be
seen more coherently as delineating a simple-complex cult system along
the lines suggested above for southwestern Ohio. This type of modeling of variations may also be relevant to understanding the Pinson
Mounds site within its regional and even pan-regional context. Such
modeling, however, would sustain the general Inclusive Territorial/
Custodial Domain Paradigm. Therefore, this critique opens up rather
than shuts down the possibilities of this paradigm.

GLOSSARY

Behavior/Action Duality. In general, the action nature of a behavioral process is determined not by the objective changes it brings about but by the meaning that these changes have for the people of a given society. Thus, murder and execution are both forms of human killing but they are different social actions. The difference is constitutive and this is publicly constructed partly by the symbolic attributes of the material make-up of the killing process (e.g., warrants of execution, authorized killing mechanisms, and so on).

Belief-Action Contingency. The view that actions and beliefs are related contingently such that two people can have the same beliefs and act quite differently in respect to those beliefs. When cosmology is treated as a complex of world beliefs shared by a people, this becomes cosmology-action contingency. This means that to know the world in terms of a particular cosmology does not entail only a narrow range of possible ways to act that go along with that cosmology. Rather, an indefinite range of possible behaviors can count as equivalent ways of acting appropriately within the scope of the same cosmology.

Belief-Action Determinism. The view that actions are tightly tied to and determined by beliefs such that to have a particular belief or belief complex (e.g., a cosmology) also entails having the intention to perform certain types of actions that realize these beliefs. Therefore, this determinism assumes that to learn a belief or a system of related beliefs is to learn the actions that go along with it. When cosmology is treated as a complex of world beliefs shared by a people, this becomes cosmology-action determinism. In these terms, knowing the world in terms of a particular cosmology entails acting in ways that are part of that cosmology.

Context/Locale Duality. Similar to the Behavior-Action duality. A context is a symbolically constructed aspect of a locale or place. Therefore, to speak of the context of a place is to speak of its symbolic nature and this becomes part of defining and constituting the action nature of the behaviors regularly performed within that locale.

Cosmology-Ideology Contingency. This recognizes that cosmology and ideology are autonomous spheres with the former made up of collective beliefs of the world and the latter made up of constitutive rules, protocols, and rationales that stipulate the behavioral forms that will count as ritual actions. Just as beliefs and actions are contingently related, so cosmology and ideology are contingently related. This means that two or more peoples can share the same cosmology while differing in ideology, thereby differing in terms of the range of practices that each performs and that each views as appropriate modes of conduct in the world as believed.

Cosmology, Immanentist. This is any cosmology that envisions the world as immanently sacred. In this regard, the tangible order of nature is simultaneously the direct manifestation of the sacred order. The sacred natural world and human society are reciprocally transitive with human activity impinging on the sacred order, either by polluting and diminishing it or by sanctifying and enhancing it. Of course, the sacred state of the natural world similarly impinges back on human society so that good fortune manifests a well-balanced cosmos-human relation.

Cosmology, Transcendental. This is any cosmology that envisions a separation of the material and sacred worlds such that the latter transcends the former. In this case, the world can be taken to be a complex sign of the sacred order but is not itself essentially sacred. Normally, the sacred/natural world relation is asymmetrically transitive. The actions of the sacred world, the gods, can impinge on the human world, but human activity does not impinge on or modify the state of the sacred. There can be many permutations of this relation, however.

Deontics. Pertaining to the normative, that is, the ethical, moral, and constitutive aspects of social life. The deontics of social life are the rights, duties, obligations, and normative principles that are recognized and realized in the practices of a given society. All deontic phenomena, including most social relations, social positions, and social actions, rely on the expressive moment and, therefore, they are also symbolic in nature. Deontics can only be separated analytically from symbolic pragmatics.

Deontic Ecology. This is the view that deontic norms are an irreducible part of the ecological practices that characterize a social system. The behavior/action duality is implicit in this. For example, hunting is not a basic subsistence action. Rather, predatory killing is constituted as the basic subsistence activity of hunting. Hunting is subjects performing their predatory behaviors according to normative rules, thereby fulfilling the rights and duties by which a killing behavior is constituted as a hunting activity, rather than, say, a poaching activity, even though the two types of actions are objectively the same behaviors. What makes the difference that counts is the deontic moment as realized in expressive symbol, e.g., using hunting gear that displays the appropriate stylistics. In many cases, ecological deontics will be expressed as strong normative proscriptions and prescriptions governing the forms of conduct and, of course, the forms of the tools themselves. A basic deontic ecological dimension is rights and duties to land and resources, constituting relations of tenure.

Essential Contradiction. This treats the ecological pursuit of human survival in an immanently sacred natural world as profoundly contradictory. Since pursuing survival by means of foraging or cutting the forest necessarily disorders nature, it is always a sacred as well as dangerous mundane pursuit. Ecological practices come to have an irreducibly ritual moment undergirded by an immanentist deontics of proscriptions and prescriptions that are realized as critical midwifery ritual, often referred to in the anthropological literature as foraging taboos and imperatives. Furthermore, religious practices are sensitive to these practices since they often are recruited to reverse the sacred pollution that these mundane subsistence and settlement practices cause.

Human Sacrifice, Lethal. A form of mortuary practice in which the intended killing of a human is a constituent part of the social act itself. Typically, this is a religious act entailing world renewal.

Human Sacrifice, Post-mortem. A form of mortuary practice in which the death of the deceased does not figure as part of the sacrifice. Rather, the mode of post-mortem treatment becomes a critical part of the sacrificial ritual and, of course, the ideological rationale is usually some sort of renewal, the rebirth of the name of the person (name-reincarnation), of the family or descent group, of the society, of the world.

Ideology. This is treated as an autonomous body of rules, protocols, rationales, and commitments governing both the forms of behaviors and their material outcomes so that they count as among those actions that fulfill the collective commitments of those holding the ideology. An ideology presupposes a particular type of cosmology, but the cosmology does not determine the ideology (see Cosmology-Ideology Contingency).

Intentionality. The mental property of directedness at, toward, about, or of objects, states of affairs, events, and processes in the world. There are different types of intentional states: beliefs, perceptions, wants, wishes, fears, and so on. The mental state termed "an intention" is only one type of intentional state. Intentions, per se, are particularly important for archaeology since actions are their conditions of satisfaction. Therefore, to claim that certain prehistoric artifacts are "butchering tools" is to claim that they were used for butchering. In terms of intentionality, this means that their use partly fulfilled the subsistence intentions of the prehistoric subject. In short, any characterization of material artifacts in action terms is a commitment to the exercise of the corresponding intentions. Since archaeologists chronically label artifacts functionally, they cannot escape engaging with claims about prehistoric psychological states, e.g. intentions, beliefs, and so on.

Midwifery Ritual. In social systems characterized by immanentist cosmology, subsistence practices typically have a dual destructive/reproductive moment. Thus, it is common that the killing of the animal or the harvesting of the crop is simultaneously perceived as the reproduction of the species being exploited. The aspect of the subsistence practice that ensures the reproductive moment is termed midwifery ritual.

Mortuary/Renewal Ritual. In an immanently sacred world, the synthesis of mortuary and renewal ritual minimizes the polluting moment that would be required if these two rituals were done separately and maximizes the reproductive moment of both humanity and the cosmos.

Social System—Locale-Centric. A social system based on locales linked by criss-crossing paths into an open social network. No territory is recognized as exclusive. This creates a social system with no defined boundaries. Typically, locale-centric social systems are associated with polyistic structures.

Social System—Modular. A social system based on two-dimensional space having defined boundaries forming exclusive territories. Modular social systems are often monistic in nature.

Social System—Monistic. A social system in which a complex of social structural axes is fused into a single dominant social organization. Kinship as realized in tribal societies is the classic example.

Social System—Polyistic. A social system in which there is a disembedding of social structural axes. While kinship based on gender, age, and descent can still be important, autonomous social groupings can form based on one or more disembedded social structural axes, e.g., male gender and junior age-grade structures constituting autonomous sodalities specialized in ritual, military activity, and so on, and independent of kinship organizations, e.g., lineages and clans.

Socio-Intentional Perspective. All intentional states are realized within a network of social relations. Therefore, action intentions always require a social context in order to be constituted and exercised.

Speech Act Theory. The view that human communication is a form of action mediated by vocal or inscribed symbols, i.e., speech or writing. In many western societies, in uttering the terms, "I declare you husband and wife," if the speaker occupies the appropriate social position (priest or minister), she/he is not merely describing the relation of the couple. Rather, as a priest she/he is producing that relation by means of her/his declaration. The utterance of "mere" words, done properly and according to the pragmatic rules, constitutes a speech action that makes a real difference in that social world.

Symbolic Pragmatics—General. The claim that all social actions or practices are symbolically constituted. The difference between hunting and poaching is marked and thereby constituted by means of displaying the appropriate symbols of hunting or failing to display them.

Symbolic Pragmatics—Specific. The claim that material cultural style is the conventional or symbolic mode by which the action nature of material behavior is constituted. In this sense, the style of a spear is treated as the "warranting moment" by which the killing of the animal is constituted as the legitimate appropriation of its life, i.e., hunting.

NOTES

PROLOGUE

1. Ray Hively and Robert Horn, "Geometry and Astronomy in Prehistoric Ohio," *Journal for the History of Astronomy, Archaeoastronomy Supplement* 13 (1982): S1–S20.

2. Bradley T. Lepper, "An Historical Review of Archaeological Research at the Newark Earthworks," *Journal of the Steward Anthropological Society* 18 (1988): 118–40; Dee Anne Wymer, Bradley T. Lepper, and William H. Pickard, "Recent Excavations at the Great Circle, Newark, Ohio: Hopewell Ritual in Context," paper presented at the Midwest Archaeological Conference, Grand Rapids, Mich., 1992.

3. E. G. Squier and E. H. Davis, *Ancient Monuments of the Mississippi Valley: Comprising the Results of Extensive Original Surveys and Explorations* (1848; reprint, New York: AMS Press for Peabody Museum of Archaeology and Ethnology, Harvard University, 1973), 1–103.

4. Mark F. Seeman, *The Hopewell Interaction Sphere: The Evidence for Interregional Trade and Structural Complexity.* Prehistoric Research Series vol. 5, no. 2 (Indianapolis: Indiana Historical Society, 1979).

5. Robert C. Mainfort Jr., *Pinson Mounds: A Middle Woodland Ceremonial Center.* Division of Archaeology, Tennessee Department of Conservation Research Series no. 7 (Tennessee Department of Conservation, 1986); "Pinson Mounds: Internal Chronology and External Relationships," in *Middle Woodland Settlement and Ceremonialism in the Mid-South and Lower Mississippi Valley,* edited by Robert C. Mainfort Jr., Archaeological Report no. 22 (Jackson, Miss.: Department of Archives and History, 1988), 133–46; "Pinson Mounds and the Middle Woodland Period in the Midsouth and Lower Mississippi Valley," in *A View from the Core: A Synthesis of Ohio Hopewell Archaeology,* ed-

ited by Paul J. Pacheco (Columbus: Ohio Archaeological Council, 1996), 370–91; Robert L. Thunen, "Geometric Enclosures in the Mid-South: An Archaeological Analysis of Enclosure Form," in *Middle Woodland Settlement and Ceremonialism in the Mid-South and Lower Mississippi,* edited by Robert C. Mainfort Jr., Archaeological Reports no. 22 (Jackson: Mississippi Department of Archives and History, 1988), 99–115; "Defining Space: An Overview of the Pinson Mounds Enclosure," in *Ancient Earthen Enclosures of the Eastern Woodlands,* edited by Robert C. Mainfort Jr. and Lynne P. Sullivan (Gainesville: University Press of Florida, 1998), 57–67.

6. However, I discuss shortly the work of a number of archaeologists and other scholars who have profitably focused considerable attention and effort on the embankment earthworks.

7. Squier and Davis, *Ancient Monuments of the Mississippi Valley,* 1–103; also see Caleb Atwater, *Description of Antiquities Discovered in the State of Ohio and Other Western States,* (1820; reprint, New York: AMS Press for Peabody Museum of Archaeology and Ethnology, Harvard University, 1973), 136–95.

8. Warren K. Moorehead, *Primitive Man in Ohio* (New York: G. P. Putnam's Sons, 1892); *The Hopewell Mound Group of Ohio* (1922; reprint, New York: AMS Press for Field Museum of Natural History, Publication 211, Anthropology Series, vol. 6, 1980).

9. Olaf H. Prufer, "The Hopewell Complex of Ohio," in *Hopewellian Studies,* edited by Joseph R. Caldwell and Robert L. Hall, Scientific Papers vol. 12 (Springfield: Illinois State Museum, 1964), 35–83.

10. Tim Ingold, *The Appropriation of Nature: Essays on Human Ecology and Social Relations* (Iowa City: University of Iowa Press, 1987), 101–29, 130–64.

CHAPTER I

1. In support of the ceremonial view see Robert Patrick Connolly, "Middle Woodland Hilltop Enclosures: The Built Environment, Construction and Function," Ph.D. diss., University of Illinois at Urbana-Champaign (Ann Arbor, Mich.: University Microfilm International, 1996), 341–44; "Architectural Grammar Rules at the Fort Ancient Hilltop Enclosure," in *Ancient Earthen Enclosures of the Eastern Woodlands,* edited by Robert C. Mainfort Jr. and Lynne P. Sullivan (Gainesville: University Press of Florida, 1998), 112–13; also see Patricia Essenpreiss and M. E. Moseley, "Fort Ancient: Citadel or Coliseum?" *Field Museum of Natural History Bulletin,* (June 1984): 5–26; for the

"fortification" view see Prufer, "The Hopewell Complex of Ohio," 1964, 66–69; *The McGraw Site: A Study in Hopewellian Dynamics* (Cleveland: Museum of Natural History 1965), 135–36; "Fort Hill 1964: New Data and Reflections on Hopewell," in *Ohio Hopewell Community Organization*, edited by William S. Dancey and Paul J. Pacheco (Kent, Ohio: Kent State University Press, 1997b), 320; for an integrated "fortification-ceremonial" view, see Robert V. Riordan, "The Enclosed Hilltops of Southern Ohio," in *A View from the Core: A Synthesis of Ohio Hopewell Archaeology*, edited by Paul J. Pacheco (Columbus: Ohio Archaeological Council, 1996), 250–53; "Boundaries, Resistance, and Control: Enclosing the Hilltops in Middle Woodland Ohio," in *Ancient Earthen Enclosures of the Eastern Woodlands*, edited by Robert C. Mainfort Jr. and Lynne P. Sullivan (Gainesville: University Press of Florida, 1998), 82–84.

2. A. Martin Byers, "The Earthwork Enclosures of the Central Ohio Valley: A Temporal and Structural Analysis of Woodland Society and Culture," Ph.D. diss., State University of New York at Albany (Ann Arbor, Mich.: University Microfilms International, 1987), 53–71.

3. Robert Patrick Connolly, "Prehistoric Land Modification at the Fort Ancient Hilltop Enclosure: A Model of Formal and Accretive Development," in *A View from the Core: A Synthesis of Ohio Hopewell Archaeology*, edited by Paul J. Pacheco (Columbus: Ohio Archaeological Council, 1996), 267; "Architectural Grammar Rules at the Fort Ancient Hilltop Enclosure," in *Ancient Earthen Enclosures of the Eastern Woodlands*, edited by Robert C. Mainfort Jr. and Lynne P. Sullivan (Gainesville: University Press of Florida, 1998), 94–95.

4. William S. Dancey, "Village Origins in Central Ohio: The Results and Implications of Recent Middle and Late Woodland Research," in *Cultural Variability in Context: Woodland Settlements in the Mid-Ohio Valley*, edited by Mark F. Seeman, Midcontinental Journal of Archaeology Special Paper no. 7 (Kent, Ohio: Kent State University Press, 1992), 24–25; Jeff Carskadden and James Morton, "The Middle Woodland-Late Woodland Transition in the Central Muskingum Valley of Eastern Ohio: A View from the Philo Archaeological District," in *A View from the Core: A Synthesis of Ohio Hopewell Archaeology*, edited by Paul J. Pacheco (Columbus: Ohio Archaeological Council, 1996), 324; Eli Lilly, *Prehistoric Antiquities of Indiana* (Indianapolis: Indiana Historical Society, 1937), 51–52; John P. MacLean, *The Mound Builders: The Archaeology of Butler County, Ohio* (Cincinnati, Ohio: Robert Clarke and Co., 1879); Stephen D. Peet, *The Mound Builders: Their Works and Relics*, vol. 1 (Chicago: The American Antiquarian Society, 1892); S. Frederick Starr, "The

Archaeology of Hamilton County, Ohio," *The Journal of the Cincinnati Museum of Natural History* 28:1 (1960).

5. Byers, "The Earthwork Enclosures of the Central Ohio Valley," 67–73.

6. Lilly, *Prehistoric Antiquities of Indiana*, 51–52; MacLean, *The Mound Builders: The Archaeology of Butler County, Ohio*; Peet, *The Mound Builders: Their Works and Relics*; Starr, "The Archaeology of Hamilton County, Ohio."

7. Elliot M. Abrams, "Archaeological Investigation of the Armitage Mound (33-At-434), The Plains, Ohio," *Midcontinental Journal of Archaeology* 17 (1992): 88; "Woodland Settlement Patterns in the Southern Hocking River Valley, Southeastern Ohio," in *Cultural Variability in Context: Woodland Settlements in the Mid-Ohio Valley*, edited by Mark F. Seeman, Midcontinental Journal of Archaeology Special Paper no. 7 (Kent, Ohio: Kent State University Press, 1992), 20; R. Berle Clay, "Adena Ritual Development: An Organizational Type in Temporal Perspective," in *The Human Landscape in Kentucky's Past: Site Structure and Settlement Patterns*, edited by Charles Stout and Christine K. Hensley (Lexington: Kentucky Heritage Council, 1991), 30–31.

8. R. Berle Clay, "The Ceramic Sequence at Peter Village and Its Significance," in *New Deal Archaeology and Current Research in Kentucky*, edited by Donald Pollack and Mary Lucas Powell (Frankfort: Kentucky Heritage Council, 1988), 110.

9. For Marksville, see Dennis Jones and Carl Kuttruff, "Prehistoric Enclosures in Louisiana and the Marksville Site," in *Ancient Earthen Enclosures of the Eastern Woodlands*, edited by Robert C. Mainfort Jr. and Lynne P. Sullivan (Gainesville: University Press of Florida, 1998), 52–53; for Spanish Fort and the Little Spanish Fort, see H. Edwin Jackson, "Little Spanish Fort: An Early Middle Woodland Enclosure in the Lower Yazoo Basin, Mississippi," *Midcontinental Journal of Archaeology* 23 (1998): 217; for the Old Stone Fort see Willard S. Bacon, "Factors in Siting a Middle Woodland Enclosure in Middle Tennessee," *Midcontinental Journal of Archaeology* 18 (1993): 269–70; also see Charles H. Faulkner, "The Old Stone Fort Revisited: New Clues to an Old Mystery," in *Mounds, Embankments, and Ceremonialism in the Midsouth*, edited by Robert C. Mainfort and Richard Walling, Arkansas Archaeological Survey Research Series no. 46 (Fayetteville: Arkansas Archaeological Survey, 1996), 7–11.

CHAPTER 2

1. Robert P. Connolly and L. E. Sieg, "Prehistoric Architecture and the Development of Public Space at the Fort Ancient Hilltop Enclosure," paper presented at the Fifth International and Interdisciplinary Forum on Built Form and Culture Research, Second CSPA Forum on Architectural Practice, Cincinnati, Ohio, 1993, 7–8, 13.

2. Byers, "The Earthwork Enclosures of the Central Ohio Valley," 73–75.

3. Ray Hively and Robert Horn, "Hopewellian Geometry and Astronomy at High Bank," *Journal for the History of Astronomy. Archaeoastronomy Supplement* 15 (1984): S85–S100; also see William F. Romain, *Mysteries of the Hopewell: Astronomers, Geometers, and Magicians of the Eastern Woodlands* (Akron, Ohio: University of Akron Press, 2000), 39–41.

4. Romain, *Mysteries of the Hopewell*, 54–64; "Hopewellian Geometry: Forms at the Interface of Time and Eternity," in *A View from the Core: A Synthesis of Ohio Hopewell Archaeology*, edited by Paul J. Pacheco (Columbus: Ohio Archaeological Council, 1996), 200–203.

5. Romain, "Hopewellian Geometry,"198; *Mysteries of the Hopewell*, 73–76.

6. For the tripartite pattern, see N'omi B. Greber, "Within Ohio Hopewell: Analyses of Burial Patterns from Several Classic Sites," Ph.D. diss., Case Western Reserve University (Ann Arbor, Mich.: University Microfilms International, 1976), 94–98; for the quadripartite pattern, see N'omi B. Greber, *Recent Excavations at the Edwin Harness Mound, Liberty Works, Ross County, Ohio*, Midcontinental Journal of Archaeology, Special Paper no. 5 (Kent, Ohio: Kent State University Press, 1983), 87–89.

7. William C. Mills, "The Exploration of the Tremper Mound," *Ohio Archaeological and Historical Quarterly* 25 (1916): 262–398; "Explorations of the Mound City Group," *Ohio Archaeological and Historical Quarterly* 31 (1922): 423–584; Henry C. Shetrone, "Explorations of the Hopewell Group of Prehistoric Earthworks," *Ohio Archaeological and Historical Quarterly* 35 (1926): 1–227; Henry C. Shetrone and Emerson F. Greenman, "Explorations of the Seip Group of Prehistoric Earthworks," *Ohio Archaeological and Historical Quarterly* 40 (1931): 343–509.

8. A fourth point could be raised as a question. If the C-Form/SL-Profile embankment was transformed into the equivalent of the standard in-

fix, then why was the Circle not built "outside" the C-Form, say, attached to the northwest corner so that it could be skewed in orientation with regard to the eastern Paint Creek Square? All questions cannot be answered at once. More empirical analysis to discern the possible rules and principles underwriting the earthwork construction is required. This will be done in subsequent chapters.

9. In subsequent excavations, Hanson (1966) has confirmed that Squier and Davis's (1973) positioning of the eastern gate is accurate. In fact, in a series of excavations of Component A of the Mound City Cluster and of elements of the mounds making up Mound City Proper following Brown and Baby's (1966) exploratory work, the overall accuracy of Squier and Davis's map of Mound City seems well confirmed (Baby at al. 1971, 1975; Baby and Langlois 1977; Drennan 1972; Hanson 1966).

CHAPTER 3

1. A. Martin Byers, "Intentionality, Symbolic Pragmatics and Material Culture: Revisiting Binford's View of the Old Copper Complex," *American Antiquity* 64 (1999): 265–87.

2. Polly Wiessner, "Style and Social Information in Kalahari San Projectile Points," *American Antiquity* 48 (1983): 253–76; "Style or Isochrestic Variation? A Reply to Sackett," *American Antiquity* 50 (1985): 160–66; "Style and the Changing Relations between the Individual and Society," in *The Meanings of Things*, edited by I. Hodder (London: Unwin Hyman, 1989), 56–63; "Is There a Unity to Style?" in *The Use of Style in Archaeology*, edited by M. W. Conkey and C. A. Hastorf (Cambridge: Cambridge University Press, 1990), 105–12.

3. Roy Bhaskar, *The Possibility of Naturalism* (Atlantic Highlands, N.J.: Humanities Press, 1979), 44–47, 52–56; *A Realist Theory of Science* (Brighton: Harvester Press, 1978), 195–97; Anthony Giddens, *New Rules of Sociological Method* (London: Hutchinson, 1976), 75–86, 124–28; *Central Problems in Social Theory* (London: Macmillan, 1979), 117–20; *A Contemporary Critique of Historical Materialism* (London: Macmillan, 1981), 26–29; *The Constitution of Society* (Berkeley: University of California Press, 1984), 15–19.

4. Manfred Bierwisch, "Semantic Structures and Illocutionary Force," in *Speech Act Theory and Pragmatics*, edited by John R. Searle, Ferenc Keifer, and Manfred Bierwisch (Dordrecht, Holland: D. Dreidel Publishing, 1980), 3, 6; A. Martin Byers, "Structure, Meaning, Action and Things: The Duality of

Material Cultural Meaning," *Journal for the Theory of Social Behaviour* 21 (1991): 12–15; "The Action-Constitutive Theory of Monuments: A Strong Pragmatist Version," *Journal for the Theory of Social Behaviour* 22 (1992): 409–20; "Social Structure and the Pragmatic Meaning of Material Culture: Ohio Hopewell as Ecclesiastic-Communal Cult," in *A View from the Core: A Synthesis of Ohio Hopewell Archaeology*, edited by Paul J. Pacheco (Columbus: Ohio Archaeological Council, 1996), 179–84; 1999, 270–77; Stephen C. Levinson, *Pragmatics* (Cambridge: Cambridge University Press, 1983); "Interactional Biases in Human Thinking," in *Social Intelligence and Interaction: Experience and Implications of Social Bias in Human Intelligence*, edited by Esther N. Goody (London: Cambridge University Press, 1995), 221–60; John R. Searle, *Intentionality* (Cambridge: Cambridge University Press, 1983), 4–13; *The Construction of Social Reality* (New York: The Free Press, 1995), 59–78; *Mind, Language and Society: Philosophy in the Real World* (New York: Basic Books, 1998), 154.

5. J. L. Austin, *How to Do Things with Words* (Cambridge, Mass.: Harvard University Press, 1962), 92–101; Pierre Bourdieu, *Language and Symbolic Power*, edited by John B. Thompson (Cambridge, Mass.: Harvard University Press, 1991), 70–76; Rom Harré, *Social Being: A Theory of Social Psychology* (Oxford: Basil Blackwell, 1979), 49; Levinson, *Pragmatics*; "Interactional Biases in Human Thinking"; Searle, *Intentionality*, 4–13; *The Construction of Social Reality*, 59–78.

6. Bronislaw Malinowski, "The Problem of Meaning in Primitive Languages," Supplement in *The Meaning of Meaning*, C. K. Ogden and I. A. Richards (New York: Harcourt, Brace and Co. 1923), 470–75.

7. Lewis R. Binford, "Archaeology as Anthropology," *American Antiquity* 28 (1962): 217.

8. Searle, *Intentionality*, 1–13; Byers, "Intentionality, Symbolic Pragmatics and Material Culture," 267–68.

CHAPTER 4

1. Bradley T. Lepper, "The Archaeology of the Newark Earthworks," in *Ancient Earthen Enclosures of the Eastern Woodlands*, edited by Robert C. Mainfort Jr. and Lynne P. Sullivan (Gainesville: University Press of Florida, 1998), 130; Hively and Horn, "Geometry and Astronomy in Prehistoric Ohio," S1–S20; "Hopewellian Geometry and Astronomy at High Bank," S85–S100.

2. The precise order of these final steps need not be definitively postulated. For example, Feature A could have been closed after Feature B was

completed and before the Octagon was built or it could have been covered after the Octagon was completed. Even before the building of Feature B, Feature A could have been closed. Knowing the precise order would give insight into the thinking of the planners. However, as discussed below, what is important in Construction Scenario 1 is the claim that the original plan was changed before the Circle was completed.

3. It is widely recognized that these embankment earthwork construction programs often went in cycles. One type of cycle was to "rebuild" the embankment by adding new strata, what is here called a replication phase. Another cycle might see an extension of the horizontal plan, as at Hopewell and Fort Ancient (Connolly 1996a, 1996b; Essenpreis and Mosely 1984; Greber 1996, 1997; Lepper 1996, 1998; Riordan 1995, 1996; Wymer et al. 1992).

4. Squier and Davis 1973, 69.

5. Bacon, "Factors in Siting a Middle Woodland Enclosure in Middle Tennessee," 254; Connolly, "Prehistoric Land Modification at the Fort Ancient Hilltop Enclosure," 267–68, "Middle Woodland Hilltop Enclosures," 254–56; "Architectural Grammar Rules at the Fort Ancient Hilltop Enclosure," 97–99; Riordan, "The Enclosed Hilltops of Southern Ohio," 251–53.

6. Hively and Horn, "Geometry and Astronomy in Prehistoric Ohio," S-16, table 2; N'omi B. Greber and Katharine C. Ruhl, *The Hopewell Site: A Contemporary Analysis Based on the Work of Charles C. Willoughby* (Boulder, Colo.: Westview Press, 1989); Romain, *Mysteries of the Hopewell*, 101–42; Christopher S. Turner, "An Analysis of the Calendrical Sightlines at the Liberty Earthworks, Ross County, Ohio," paper in possession of author, 2000, 9; "A Probabilistic Analysis of Calendrical Sightlines at the Hopeton Earthworks, Ross Co., Ohio," paper presented at the Midwest Archaeological Conference, Columbus, Ohio, 2002, 3.

7. James A. Marshall, "Towards a Definition of Hopewell Core and Periphery Utilizing the Geometric Earthworks," in *A View from the Core: A Synthesis of Ohio Hopewell Archaeology*, edited by Paul J. Pacheco (Columbus: Ohio Archaeological Council, 1996), 212–19; Romain, "Hopewellian Geometry: Forms at the Interface," 206–7; *Mysteries of the Hopewell*, 80–100.

8. This was also commented on by Squier and Davis, *Ancient Monuments of the Mississippi Valley*, 69.

9. Robert L. Hall, *An Archaeology of the Soul: North American Indian Belief and Ritual* (Urbana: University of Illinois Press, 1997), 84–85.

10. Romain, "Hopewellian Geometry: Forms at the Interface," 208;

Mysteries of the Hopewell, 170–72. However, he recognizes that some of the embankment circles may be expressing and manifesting the Heavens. But he argues that this is an incidental association. "[In] the case of the Newark Observatory Circle, we note that the circular earthwork is connected to the octagon" (171).

11. N'omi Greber, "Astronomy and the Patterns of Five Geometric Earthworks in Ross County, Ohio," Abstract in *World Archaeoastronomy: Selected Papers from the Second Oxford International Conference on Archaeoastronomy Held at Merida, Yucatan, 13-16 January 1986*, edited by A. D. Aveni (Cambridge: Cambridge University Press, 1989), 495; Hively and Horn 1982; Romain 1996, 2000; Turner 2000, 2001.

12. I prefer to give High Bank priority. It is suggested later that Newark was drawn into the ambit of the central Scioto as a result of social influence from Chillicothe (High Bank) and that the construction program on which Newark embarked was part of its growing participation in this network.

13. Lepper, "The Archaeology of the Newark Earthworks," 126, 2,110 ± 80 B.P.

14. Bradley T. Lepper, "The Newark Earthworks and the Geometric Enclosures of the Scioto Valley: Connections and Conjectures," in *A View from the Core: A Synthesis of Ohio Hopewell Archaeology*, edited by Paul J. Pacheco (Columbus: Ohio Archaeological Council, 1996), 236–37; "The Archaeology of the Newark Earthworks," 120–22.

15. Alternatively, of course, the Cherry Valley Infix and the Wright Square could be imagined as remaining fixed while the Newark Circle-Octagon was "dragged" eastward by the aggregation element's progressively shortening.

CHAPTER 5

1. Bhaskar, *The Possibility of Naturalism*, 11–28, 164–69; *A Realist Theory of Science*, chapter 3, 143–227. Bhaskar's RRRE methodology, which is the basis of the hermeneutic spiral and which he takes as necessary when dealing with the open nature of social systems, is summed up in the following four phases. "(1) *Resolution* of a complex event into its components (causal analysis); (2) *Redescription* of component causes; (3) *Retrodiction* to possible (antecedent) causes of components via independently validated normic statements; and (4) *Elimination* of alternative possible causes of components" (1979:165).

2. Romain, "Hopewellian Geometry: Forms at the Interface," 208; *Mysteries of the Hopewell,* 170–72.

3. Lepper, "The Newark Earthworks and the Geometric Enclosures," 230.

4. Romain does not use the C-R Configuration terminology. However, his findings can be easily assimilated to it.

5. Romain, "Hopewellian Geometry: Formas at the Interface," 208.

6. Ibid., these are restated in Romain, *Mysteries of the Hopewell.*

7. Lepper, "The Newark Earthworks and the Geometric Enclosures," 230.

8. Lepper, "The Archaeology of the Newark Earthworks," 126.

9. Romain, "Hopewellian Geometry: Forms at the Interface," 208; *Mysteries of the Hopewell,* 170–72.

10. A reviewer of this book has kindly pointed out that there is at least one residual "borrow depression" still visible south of the Circle, suggesting that these were not minor effects of earth procurement. While it is maintained here that these were, in fact, "borrow depressions" and differed in emic categorical terms from the "borrow pits" associated with the Octagon, further research, particularly on the embankment contents themselves, is required.

11. Lepper, "The Archaeology of the Newark Earthworks," 126; Greber and Ruhl, *The Hopewell Site: A Contemporary Analysis,* 284.

12. Lepper, "The Archaeology of the Newark Earthworks," 126; Connolly, "Prehistoric Land Modification at the Fort Ancient Hilltop Enclosure,"

13. Lepper, "The Archaeology of the Newark Earthworks," 126.

14. More is said on this matter in part 2.

CHAPTER 6

1. William S. Dancey, "The Community Plan of an Early Late Woodland Village in the Middle Scioto River Valley," *Midcontinental Journal of Archaeology* 13 (1988): 223–58; "A Middle Woodland Settlement in Central Ohio: A Preliminary Report on the Murphy Site (33LI212)," *Pennsylvania Archaeologist* 61 (1991): 37–72; "Village Origins in Central Ohio: The Results and Implications of Recent Middle and Late Woodland Research," in *Cultural Variability in Context: Woodland Settlements in the Mid-Ohio Valley,* edited by Mark F. Seeman, Midcontinental Journal of Archaeology Special Paper no. 7 (Kent, Ohio: Kent State University Press, 1992), 24–29; "Putting an End to Ohio

Hopewell," in *A View from the Core: A Synthesis of Ohio Hopewell Archaeology*, edited by Paul J. Pacheco (Columbus: Ohio Archaeological Council, 1996), 396–405; William S. Dancey and Paul J. Pacheco, "A Community Model of Ohio Hopewell Settlement," in *Ohio Hopewell Community Organization*, edited by William S. Dancey and Paul J. Pacheco (Kent, Ohio: Kent State University Press, 1997), 3–40; Paul J. Pacheco, "Ohio Middle Woodland Settlement Variability in the Upper Licking River Drainage," *Journal of the Steward Anthropological Society* 18 (1988): 87–117; "Ohio Hopewell Settlement Patterns: An Application of the Vacant Center Model to Middle Woodland Period Intracommunity Settlement," Ph.D. diss., Ohio State University (Ann Arbor, Mich.: University Microfilms International, 1993); "Ohio Hopewell Regional Settlement Patterns," in *A View from the Core: A Synthesis of Ohio Hopewell Archaeology*, edited by Paul J. Pacheco (Columbus: Ohio Archaeological Council, 1996), 16–35; "Ohio Middle Woodland Intracommunity Settlement Variability: A Case Study from the Licking Valley," in *Ohio Hopewell Community Organization*, edited by William S. Dancey and Paul J. Pacheco (Kent, Ohio: Kent State University Press, 1997), 41–84.

2. See confirming evidence in Dee Ann Wymer, "Trends and Disparities: The Woodland Paleoethnobotanical Record of the Mid-Ohio Valley," in *Cultural Variability in Context: Woodland Settlements in the Mid-Ohio Valley*, edited by Mark F. Seeman, Midcontinental Journal of Archaeology Special Paper no. 7 (Kent, Ohio: Kent State University Press, 1992) 66–67; "The Ohio Hopewell Econiche: Human-Land Interaction in the Core Area," in *A View from the Core: A Synthesis of Ohio Hopewell Archaeology*, edited by Paul J. Pacheco (Columbus: Ohio Archaeological Council, 1996), 48–49; "Paleoethnobotany in the Licking Valley, Ohio: Implications for Understanding Ohio Hopewell," in *Ohio Hopewell Community Organization*, edited by William S. Dancey and Paul J. Pacheco (Kent, Ohio: Kent State University Press, 1997), 159, 161.

3. Ingold, *The Appropriation of Nature*, 19–20, 23.

4. James A. Brown, "Long-Term Trends to Sedentism and the Emergence of Complexity in the American Midwest," in *Prehistoric Hunter-Gatherers*, edited by T. Douglas Price and James A. Brown (New York: Academic Press, 1985), 207–8; "Food for Thought: Where Has Subsistence Analysis Gotten Us?" in *Foraging, Collecting and Harvesting: Archaic Period Subsistence and Settlement in the Eastern Woodlands*, edited by Sarah W. Neusius, Occasional Paper no. 6 (Carbondale, Ill.: Center for Archaeological Investigations, 1986), 316–17; Michael S. O'Brien, "Sedentism, Population Growth, and

Resource Selection in the Woodland Midwest: A Review of Coevolutionary Developments." *Current Anthropology* 28 (1987): 177–78.

5. Rom Harré, *Social Being: A Theory of Social Psychology* (Oxford: Basil Blackwell, 1979), pursuit of reputation 3–4, respect/contempt hierarchies 22–26, self-presentation 226.

6. Ingold, *The Appropriation of Nature*, 243–71; also see footnote 5, 275, in which he points out that hunting among many Boreal forest peoples is experienced as a hunter impregnating the herd of the animal-master, thereby ensuring its reproduction.

7. Ibid., 271.

8. Brown, "Long-Term Trends to Sedentism," 219; Douglas K. Charles, "Corporate Symbols: An Interpretive Prehistory of Indian Burial Mounds in West-Central Illinois," Ph.D. diss., Northwestern University (Ann Arbor, Mich.: University Microfilms International, 1985), 223–36; "Diachronic Regional Social Dynamics: Mortuary Sites in the Illinois Valley/American Bottom Region," in *Regional Approaches to Mortuary Analysis*, edited by Lane Anderson Beck (New York: Plenum Press, 1995), 78–79, 91; Douglas K. Charles and Jane E. Buikstra, "Archaic Mortuary Sites in the Central Mississippi Drainage: Distribution, Structure and Behavioral Implications," in *Archaic Hunters and Gatherers in the American Midwest*, edited by James L. Phillips and James A. Brown (New York: Academic Press, 1983), 117–22; Douglas K. Charles, Jane E. Buikstra, and Lyle W. Konigsberg, "Behavioral Implications of Terminal Archaic and Early Woodland Mortuary Practices in the Lower Illinois Valley," in *Early Woodland Archaeology*, edited by Kenneth B. Farnsworth and Thomas E. Emerson, Kampsville Seminars vol. 2 (Kampsville, Ill.: Center for American Archeology, 1986), 458–59.

9. A. Martin Byers, "Symboling and the Middle-Upper Palaeolithic: A Theoretical and Methodological Critique," *Current Anthropology* 35 (1994): 371–73; "Intentionality, Symbolic Pragmatics and Material Culture," 277–79.

10. Brown, "Long-Term Trends to Sedentism," 219; "Food for Thought," 318; Charles and Buikstra, "Archaic Mortuary Sites in the Central Mississippi Drainage," 117–22; Dancey and Pacheco, "A Community Model of Ohio Hopewell Settlement," 8–10; O'Brien, "Sedentism, Population Growth, and Resource Selection in the Woodland Midwest," 177–78; as well as; David S. Brose, "A Speculative Model of the Role of Exchange in the Prehistory of the Eastern Woodlands," in *Hopewell Archaeology: The Chillicothe Conference*, edited by David S. Brose and N'omi Greber (Kent, Ohio: Kent State Univer-

sity Press, 1979), 7–8; Gary H. Dunham, "Marking Territory, Making Territory: Burial Mounds in Interior Virginia," in *Material Symbols: Culture and Economy in Prehistory*, edited by John C. Robb, Occasional Paper no. 26 (Carbondale, Ill.: Center for Archaeological Investigations, 1999), 119–20; Jimmy A. Railey, "Woodland Settlement Trends and Symbolic Architecture in the Kentucky Bluegrass," in *The Human Landscape in Kentucky's Past*, edited by C. Stout and C. K. Hensley (Frankfort: Kentucky Heritage Council, 1991), 57–58, among others.

11. Tim Ingold, *The Appropriation of Nature: Essays on Human Ecology and Social relations* (Iowa City: University of Iowa Press).

12. Wiessner, "Style and Social Information in Kalahari," 269.

13. Ibid.

14. Lee Irwin, *The Dream Seekers: Native American Visionary Traditions of the Great Plains* (Norman: University of Oklahoma Press, 1994), 59–60, 68–69.

15. Pacheco, "Ohio Hopewell Settlement Patterns," 20–21; "Ohio Hopewell Regional Settlement Patterns," 22–23.

16. David P. Braun, "Midwestern Hopewellian Exchange and Supralocal Interaction," in *Peer Polity Interaction and Socio-Political Change*, edited by Colin Renfrew and John F. Cherry (Cambridge: Cambridge University Press, 1986), 121; John F. Cherry, "Politics and Palaces: Some Problems in Minoan State Formation," in *Peer Polity Interaction and Socio-Political Change*, 19, 24, 38; Colin Renfrew, "Introduction: Peer Polity Interaction and Socio-Political Change," in *Peer Polity Interaction and Socio-Political Change*, 1–4, 16.

17. Dancey and Pacheco, "A Community Model of Ohio Hopewell Settlement," 8–9; Pacheco, "Ohio Hopewell Settlement Patterns," 37–39; "Ohio Middle Woodland Intracommunity Settlement Variability," 43; Prufer, "How to Construct a Model," 124; Railey, "Woodland Settlement Trends and Symbolic Architecture," 58; R. Berle Clay, "Chiefs, Big Men or What? Economy, Settlement Patterns, and Their Bearing on Adena Political Models," in *Cultural Variability in Context: Woodland Settlements in the Mid-Ohio Valley*, edited by Mark F. Seeman, Midcontinental Journal of Archaeology Special Paper no. 7 (Kent, Ohio: Kent State University Press, 1992), 80; "The Ceramic Sequence at Peter Village," 13–14; N'omi Greber, "Two Geometric Enclosures in the Paint Creek Valley: An Estimate of the Possible Changes in Community Patterns Through Time," in *Ohio Hopewell Community Organization*, edited by William S. Dancey and Paul J. Pacheco (Kent, Ohio: The Kent State University Press, 1997), 214–19; James P. Fenton, "Early Woodland Burial

Mounds: Symbolic Elements in a Cultural Landscape," paper presented at the Conference on Integrating Highlands Appalachian Archaeology, New York State Museum, Albany, New York, 1996, 4–6; Frank L. Cowan, "A Mobile Hopewell? Questioning Assumptions of Ohio Hopewell Sedentism," paper presented at the symposium on Perspectives on Middle Woodland at the Millennium, sponsored by the Center for American Archeology, Père Marquette State Park, Grafton, Ill., 2000, 12–14.

18. Ingold, *The Appropriation of Nature*, 147–50.

19. Hurt, *Indian Agriculture in America: Prehistory to Present*, 67.

20. Ibid.

CHAPTER 7

1. Barbara Bender, "Gatherer-Hunter to Farmer: A Social Perspective," *World Archaeology* 10 (1978): 204–22; "Prehistoric Developments in the American Midcontinent and in Brittany, Northeast France," in *Prehistoric Hunters-Gatherers: The Emergence of Cultural Complexity*, edited by T. Douglas Price and James A. Brown (New York: Academic Press, 1985), 21–57; Brown, "Long-Term Trends to Sedentism and the Emergence of Complexity," 201–31; "Food for Thought," 315–30; Brown and Vierra, "What Happened in the Middle Archaic?" 1983, 165–95; Buikstra et al., "Diet, Demography and the Development of Horticulture," 67–85; Charles, "Woodland Demographic and Social Dynamics in the American Midwest," 175–91; "Shading the Past: Models in Archaeology," 75–99; Charles and Buikstra, "Archaic Mortuary Sites in the Central Mississippi Drainage," 117–41; O'Brien, "Sedentism, Population Growth, and Resource Selection," 177–97; Bruce D. Smith, "The Archaeology of the Southeastern United States: From Dalton to de Soto, 10,500-500 B.P.," *Advances in World Archaeology* 5 (1986): 1–92; "The Independent Domestication of Indigenous Seed-Bearing Plants in Eastern North America," in *Emergent Horticultural Economies of the Eastern Woodlands*, edited by William F. Keegan, Occasional Paper no. 7 (Carbondale, Ill.: Center for Archaeological Investigations, 1987), 3–47; *Rivers of Change: Essays on Early Agriculturalists in Eastern North America* (Washington, D.C.: Smithsonian Institution Press, 1992), 281–300; "Prehistoric Plant Husbandry in Eastern North America," in *The Origins of Agriculture: An International Perspective*, edited by C. Wesley Cowan and Patty Jo Watson (Washington, D.C.: Smithsonian Institution Press, 1992), 101–19; "Reconciling the Gender-Credit Critique and the Floodplain Weed The-

ory of Plant Domestication," in *Archaeology of Eastern North America: Papers in Honor of Stephen Williams*, edited by James B. Stoltman, Archaeological Reports no. 25 (Jackson: Mississippi Department of Archives and History, 1993), 111–25; "Seed Plant Domestication in Eastern North America," in *The Last Hunters-First Farmers*, edited by T. Douglas Price and Anne Birgitte Gebauer (Santa Fe, N.M.: New School of American Research Press, 1995), 193–214.

2. Brown, "Long-Term Trends to Sedentism," 208; "Food for Thought," 318; Brown and Vierra, "What Happened in the Middle Archaic?" 168.

3. Richard W. Jefferies and B. Mark Lynch, "Dimensions of Middle Archaic Cultural Adaptation at the Black Earth Site, Salina County, Illinois," in *Archaic Hunters and Gatherers in the American Midwest*, edited by James L. Phillips and James A. Brown (New York: Academic Press, 1983), 300; Charles and Buikstra, "Archaic Mortuary Sites in the Central Mississippi Drainage," 120–124, 126.

4. Smith, "The Independent Domestication of Indigenous Seed-Bearing Plants," 37–38; *Rivers of Change*, 291; "Seed Plant Domestication in Eastern North America," 202–4.

5. Brown, "Long-Term Trends to Sedentism," 223–24; Charles and Buikstra, "Archaic Mortuary Sites in the Central Mississippi Drainage," 117–20; Charles et al., "Behavioral Implications of Terminal Archaic and Early Woodland Mortuary Practices," 458; Harold Hassen, "Part One: Preliminary Evaluation and Interpretation of the 1984 Investigation at the Bullseye Site, 11-Ge-127," in *The Bullseye Site, 11-Ge-127: A Floodplain Archaic Mortuary Site in the Lower Illinois River Valley*, edited by Harold Hassen and Kenneth B. Farnsworth, Reports of Investigations no. 42 (Springfield: Illinois State Museum, 1987).

6. For the risk management view, see Brown, "Long-Term Trends to Sedentism," 206–7, and for the coevolution of settlement and subsistence see Smith, "The Independent Domestication of Indigenous Seed-Bearing Plants," 41.

7. Smith, "The Independent Domestication of Indigenous Seed-Bearing Plants," 31–32.

8. Brown, "Long-Term Trends to Sedentism," "Food for Thought"; Brown and Vierra, "What Happened to Middle Archaic."

9. For example, Ingold, *The Appropriation of Nature*, 130–44.

10. John A. Walthall and Brad Koldehoff, "Hunter-Gatherer Interaction and Alliance Formation: Dalton and the Cult of the Long Blade," *Plains Anthropologist* 43 (1998): 258, 266, 268.

11. Dan F. Morse and Phyllis A. Morse, *Archaeology of the Central Mississippi Valley* (New York: Academic Press, 1982), 90; "Northeast Arkansas," in *Prehistory of the Central Mississippi Valley*, edited by Charles H. McNutt (Tuscaloosa: University of Alabama Press, 1996), 119–35.

12. Walthall and Koldehoff, "Hunter-Gatherer Interaction and Alliance Formation," 263. Morse and Morse (1982, 1996) are convinced that the Sloan site is one of the earliest known collective burial sites in this part of North America.

13. Smith, "The Independent Domestication of Indigenous Seed-Bearing Plants," 37.

14. For a feminist theoretical critique of Smith's position, see Patty Jo Watson and Mary C. Kennedy, "The Development of Horticulture in the Eastern Woodlands of North America: Women's Role," in *Engendering Archaeology: Women and Prehistory*, edited by Joan M. Gero and Margaret W. Conkey (Oxford: Basil Blackwell, 1991), 269.

15. Charles and Buikstra, "Archaic Mortuary Sites in the Central Mississippi Drainage."

16. For some of the key mortuary literature of 1970s see Lewis R. Binford, "Mortuary Practices: Their Study and Their Potential," and James A. Brown, "The Dimension of Status in the Burials at Spiro." Both were published in *Approaches to the Social Dimension of Mortuary Practices*, edited by James A. Brown, *Memoirs of the Society for American Archaeology* 25 (1971): 6–29 and 92–111, respectively. Also, see James A. Brown, "Charnel House and Mortuary Crypts: Disposal of the Dead in the Middle Woodland Period," in *Hopewell Archaeology: The Chillicothe Conference*, edited by David S. Brose and N'omi Greber (Kent, Ohio: Kent State University Press, 1979), 211–19; Lynne Goldstein, "One-Dimensional Archaeology and Multi-Dimensional People: Spatial Organisation and Mortuary Analysis," in *The Archaeology of Death*, edited by Robert Chapman, Ian Kinnes, and Klavs Randsborg (Cambridge: Cambridge University Press, 1981), 53–69; Joseph A. Tainter, "Woodland Social Changes in West-Central Illinois," *Midcontinental Journal of Archaeology* 2 (1977): 67–98; "Mortuary Practices and the Study of Social Systems," *Advances in Archaeological Method and Theory* 1 (1978): 105–41. For more recent applications of this approach see Donald G. Albertson and Douglas K. Charles, "Ar-

chaic Mortuary Component," in *The Archaic and Woodland Cemeteries of the Elizabeth Site in the Lower Illinois Valley*, edited by Douglas K. Charles, Steven R. Leigh, and Jane E. Buikstra, Research Series no. 7 (Kampsville, Illinois: Center for American Archeology, 1988), 29–40, among many others.

17. For example, Clay, "Adena Ritual Development," 30–39; "Chiefs, Big Men or What?" 70–80; "The Essential Features of Adena Ritual," 1–21; Railey, "Woodland Settlement Trends and Symbolic Architecture," 56–77, among many others.

18. Saburo Sugiyama, "Burials Dedicated to the Old Temple of Quetzalcoatl at Teotihuacan, Mexico," *American Antiquity* 54 (1989): 98. For other Valley of Mexico examples see, Richard F. Townsend, *The Aztecs* (New York: Thames and Hudson, 1992).

CHAPTER 8

1. Geoffrey W. Conrad and Arthur A. Demarest, *Religion and Empire: The Dynamics of Aztec and Inca Expansion* (Cambridge: Cambridge University Press, 1984), 113–16.

2. Robert L. Hall, "An Anthropocentric Perspective for Eastern United States Prehistory," *American Antiquity* 42 (1977): 514–15; "In Search of the Ideology of the Adena-Hopewell Climax," in *Hopewell Archaeology: The Chillicothe Conference*, edited by David S. Brose and N'omi Greber (Kent, Ohio: Kent State University Press, 1979), 259, 261–62; "The Cultural Background of Mississippian Symbolism," in *The Southwestern Ceremonial Complex: Artifacts and Analysis—The Cottonlandia Conference*, edited by Patricia Galloway (Lincoln: University of Nebraska Press, 1984), 261–62, 272; *Archaeology of the Soul*, 22–24, 39–40, 129, 155–57, 167.

3. Conrad and Demarest, *Religion and Empire*; Sugiyama, "Burials Dedicated to the Old Temple"; Richard F. Townsend, *State and Cosmos in the Art of Tenochtitlan*, Studies in Pre-Columbian Art and Archaeology no. 20 (Washington, D.C.: Dunbarton Oaks 1979); Rudolf Van Zantwijk, *The Aztec Arrangement: The Social History of Pre-Spanish Mexico* (Norman: University of Oklahoma Press, 1985).

4. Hall, "Sacrificed Foursomes and Green Corn Ceremonialism," 250.

5. Charles et al., "Behavioral Implications of Terminal Archaic and Early Woodland Mortuary Practices," 458–59; Albertson and Charles, "Archaic Mortuary Component," 29–40; Douglas K. Charles, Steven R. Leigh, and Donald G. Albertson, "Burial Descriptions," in *The Archaic and Woodland*

Cemeteries of the Elizabeth Site in the Lower Illinois Valley, edited by Douglas K. Charles, Steven R. Leigh, and Jane E. Buikstra, Research Series no. 7 (Kampsville, Illinois: Center for American Archeology, 1988), 248–74; George H. Odell, "Preliminary Analysis of Lithic and Other Nonceramic Assemblages," in *The Archaic and Woodland Cemeteries of the Elizabeth Site in the Lower Illinois Valley*, 155–90.

6. Charles et al., "Behavioral Implications," 459.

7. Hassen, "Part One: Preliminary Evaluation and Interpretation," 12; Kenneth B. Farnsworth, "Part Two: Preliminary Evaluation of Bannerstones and Other Ground-Stone Artifacts from the Bullseye Site, 11-Ge-127," in *The Bullseye Site, 11-Ge-127: A Floodplain Archaic Mortuary Site in the Lower Illinois River Valley*, edited by Harold Hassen and Kenneth B. Farnsworth, Reports of Investigations no. 42 (Springfield: Illinois State Museum, 1987), 17.

8. Douglas K. Charles and Jane E. Buikstra, "Siting, Sighting and Citing the Dead," paper presented at the 64th annual meeting of the Society for American Archaeology, Chicago, 1999, 3.

9. Hassen, "Part One: Preliminary Evaluation," 9.

10. Charles and Buikstra, "Siting, Sighting and Citing the Dead," 2.

11. Cheryl P. Claassen, "Gender, Shellfishing, and the Shell Mound Archaic," in *Engendering Archaeology: Women and Prehistory*, edited by Joan M. Gero and Margaret W. Conkey (Oxford: Basil Blackwell, 1991), 296; "A Consideration of the Social Organization of the Shell Mound Archaic," in *Archaeology of the Mid-Holocene Southeast*, edited by Kenneth E. Sassaman and David G. Anderson (Gainesville: University Press of Florida, 1996), 243, 250–51.

12. Paul A. Delcourt et al., "Holocene Ethnobotanical and Paleoecological Record of Human Impact on Vegetation in the Little Tennessee River Valley, Tennessee," *Quaternary Research* 25 (1986): 347–48; Stanley Z. Guffey, "A Review and Analysis of the Effects of Pre-Columbian Man on the Eastern North American Forests," *Tennessee Anthropologist* 2 (1977): 127–28.

13. Watson and Kennedy, "The Development of Horticulture in the Eastern Woodlands of North America," 268; Smith, "Reconciling the Gender-Credit Critique," 117–18; "Seed Plant Domestication in Eastern North America," 213–14.

14. Claassen, "Gender, Shellfishing, and the Shell Mound Archaic," 243; "A Consideration of the Social Organization of the Shell Mound Archaic," 250–51.

CHAPTER 9

1. Smith, *Rivers of Change*, 201–48; also see Smith, "Prehistoric Plant Husbandry in Eastern North America," 101–19.

2. Abrams, "Archaeological Investigation of the Armitage Mound," 92–93; "Woodland Settlement Patterns in the Southern Hocking River Valley," 21; Jeff Carskadden and James Morton, "Living on the Edge: A Comparison of Adena and Hopewell in the Central Muskingum Valley of Eastern Ohio," in *Ohio Hopewell Community Organization*, edited by William S. Dancey and Paul J. Pacheco (Kent, Ohio: Kent State University Press, 1997), 377, 382; Flora Church and Annette G. Ericksen, "Beyond the Scioto Valley: Middle Woodland Occupation in the Salt Creek Drainage," in *Ohio Hopewell Community Organization*, 347; Clay, "Chiefs, Big Men or What?" 80; Cowan, "A Mobile Hopewell?" 12–14; Dancey and Pacheco, "A Community Model of Ohio Hopewell Settlement," 11–13; Pacheco, "Ohio Hopewell Settlement Patterns," 78–81; 1996, 24–32; John Waldron and Elliot M. Abrams, "Adena Burial Mounds and Inter-Hamlet Visibility: A GIS Approach," *Midcontinental Journal of Archaeology* 24 (1999): 102–3; Wymer, "Cultural Change and Subsistence," 141; "The Ohio Hopewell Econiche," 47–49; "Paleoethnobotany in the Licking Valley, Ohio," 158–61; Richard W. Yerkes, "Using Microwear Analysis to Investigate Domestic Activities and Craft Specialization at the Murphy Site, a Small Hopewell Settlement in Licking County, Ohio," in *The Interpretive Possibilities of Microwear Studies*, edited by B. Graslund, H. Knutsson, and J. Taffinder (Uppsala, Sweden: Societas Archaeologica Upsaliensis, Aun 14., 1990), 173; among others.

3. Charles H. Faulkner, "The Winter House: An Early Southeast Tradition," *Midcontinental Journal of Archaeology* 2 (1977): 143–44.

4. This region has only one known major earth and stone embankment construction, the Old Stone Fort. Bacon (1993, 260, 264) argues that it was constructed where it was specifically because it was the point of "balance" between the Underworld and the Upper World. In terms used in this book, this would make it a world renewal locale, possibly with particular concern for the reproduction of the fish dependent on the river. Bacon points out that these would have gathered every spring at the base of the two waterfalls that "embrace" this site.

5. Except see Charles M. Niquette, "Woodland Settlement Patterns in the Kentucky/West Virginia Border Region," in *Cultural Variability in Context:*

Woodland Settlements in the Mid-Ohio Valley, edited by Mark F. Seeman, Mid-continental Journal of Archaeology Special Paper no. 7 (Kent, Ohio: Kent State University Press, 1992), 17.

6. For the debate over the occupational status of the earthworks, see Robert Patrick Connolly, "The Evidence for Habitation at the Fort Ancient Earthworks," in *Ohio Hopewell Community Organization*, edited by William S. Dancey and Paul J. Pacheco (Kent, Ohio: Kent State University Press, 1997), 258–60, 266, 269; Sean Coughlin and Mark F. Seeman, "Hopewell Settlements at the Liberty Earthworks, Ross County, Ohio," in *Ohio Hopewell Community Organization*, edited by William S. Dancey and Paul J. Pacheco (Kent, Ohio: Kent State University Press, 1997), 238–40; Cowan, "A Mobile Hopewell?" 11–14; Frank L. Cowan, Theodore S. Sunderhaus, and Robert A. Genheimer, "Wooden Architecture in Ohio Hopewell Sites: Structural and Spatial Patterns at the Stubbs Earthworks Site," paper presented at the 65th Annual Meeting of the Society for American Archaeology, Philadelphia, Penn., 8–9; Dancey and Pacheco, "A Community Model of Ohio Hopewell Settlement," 13–15; Robert A. Genheimer, "Stubbs Cluster: Hopewellian Site Dynamics at a Forgotten Little Miami River Valley Settlement," in *Ohio Hopewell Community Organization*, 294–95; Greber, "Two Geometric Enclosures in the Paint Creek Valley," 217–19; Bradley T. Lepper and Richard W. Yerkes, "Hopewellian Occupations at the Northern Periphery of the Newark Earthworks: The Newark Expressway Sites Revisited," in *Ohio Hopewell Community Organization*, 187–89; Pacheco, "Ohio Hopewell Regional Settlement Patterns," 22–24; "Ohio Middle Woodland Intracommunity Settlement Variability," 59–60. Most of the above also touches on the debate over the occupational status of the domestic habitational locales. But, addressing this debate in particular, see Church and Ericksen, "Beyond the Scioto Valley," 341–47; Cowan, "A Mobile Hopewell?" 12–14; Dancey and Pacheco, "A Community Model of Ohio Hopewell Settlement," 15–19; Pacheco, "Ohio Hopewell Regional Settlement Patterns," 21–23; Yerkes, "Using Microwear Analysis to Investigate Domestic Activities," 167–76.

7. Harré, *Social Being*, 3–4.

8. Pacheco, "Ohio Hopewell Settlement Patterns," 184, Table X.

9. Also see Dancey, "A Middle Woodland Settlement in Central Ohio," 65–69.

10. Pacheco, "Ohio Hopewell Settlement Patterns," 118.

11. Ibid., 113–18.

12. Dancey, "A Middle Woodland Settlement in Central Ohio," for radiocarbon dates, 49–51, for settlement layout, 65–66.

13. Prufer, *The McGraw Site*, 127.

14. Dancey and Pacheco, "A Community Model of Ohio Hopewell Settlement," 8–10; Pacheco, "Ohio Hopewell Settlement Patterns," 9–10; "Ohio Hopewell Regional Settlement Patterns," 22–23.

15. Dancey and Pacheco, "A Community Model of Ohio Hopewell Settlement," 15–18.

16. Pacheco, "Ohio Middle Woodland Intracommunity Settlement Variability," 57–58.

17. Greber, *Recent Excavations at the Edwin Harness Mound*, 92.

18. Donald R. Cochran, "Adena and Hopewell Cosmology: New Evidence from East Central Indiana," in *Native American Cultures in Indiana: Proceedings of the First Minnetrista Council for Great Lakes Native American Studies*, edited by Ronald Hicks (Muncie, Ind.: Minnetrista Cultural Center and Ball State University Press, 1992), 38; "The Adena/Hopewell Convergence in East Central Indiana," in *A View from the Core: A Synthesis of Ohio Hopewell Archaeology*, edited by Paul J. Pacheco (Columbus: Ohio Archaeological Council, 1996), 344, 346–49; Kent D. Vickery, " 'Reluctant' or 'Avant-Garde' Hopewell: Suggestions of Middle Woodland Culture Change in East-Central Indiana and South-Central Southwest Ohio," in *Hopewell Archaeology: The Chillicothe Conference*, edited by David S. Brose and N'omi Greber (Kent, Ohio: Kent State University Press, 1979), 59–63.

CHAPTER 10

1. Hall, "An Anthropocentric Perspective," 506–9.

2. Thomas E. Emerson, *Cahokia and the Archaeology of Power* (Tuscaloosa: University of Alabama Press, 1997), 13–14.

3. Pacheco, "Ohio Hopewell Settlement Patterns," 49–51.

4. Raymond S. Baby and Suzanne M. Langlois, "Seip Mound State Memorial: Nonmortuary Aspects of Hopewell," in *Hopewell Archaeology: The Chillicothe Conference*, edited by David S. Brose and N'omi Greber (Kent, Ohio: Kent State University Press, 1979), 16–18.

5. Andrew C. Fortier, "The Emergence and Demise of the Middle Woodland Small-Tool Tradition in the American Bottom," *Midcontinental Journal of Archaeology* 25 (2000): 191–94; Robert A. Genheimer, "Bladelets Are Tools Too: The Predominance of Bladelets among Formal Tools at Ohio Hopewell Sites,"

in *A View from the Core: A Synthesis of Ohio Hopewell Archaeology*, edited by Paul J. Pacheco (Columbus: Ohio Archaeological Council, 1996), 94; N'omi Greber, Richard S. Davis, and Ann S. DuFresne, "The Micro Component of the Ohio Hopewell Lithic Technology: Bladelets," *Annals of the New York Academy of Sciences* 376 (1981): 490; Carol A. Morrow, "Blades and Cobden Chert: A Technological Argument for Their Role as Markers of Regional Identification during the Hopewell Period in Illinois," in *The Organization of Core Technology*, edited by Jay K. Johnson and Carol A. Morrow (Boulder, Colo.: Westview Press, 1987), 146–47; George H. Odell, "The Role of Stone Bladelets in Middle Woodland Society," *American Antiquity* 59 (1994): 102–4; Yerkes 1990, 171; Connolly, "The Evidence for Habitation at the Fort Ancient Earthworks," 262; Dancey, "A Middle Woodland Settlement in Central Ohio," 55–60; Pacheco, "Ohio Middle Woodland Intracommunity Settlement Variability," 45–52.

6. This is a position that Fortier (2000, 204–6) has recently argued with respect to the use and distribution of Hopewell bladelets in the Central Mississippi Valley, with focus on the American Bottom and the lower Illinois River Valley.

7. Odell, "The Role of Stone Bladelets," 117.

8. Anta Montet-White, *The Lithic Industries of the Illinois Valley in the Early and Middle Woodland Period*, Museum of Anthropology Anthropological Papers no. 35 (Ann Arbor: University of Michigan Press, 1968), 28; Morrow "Blades and Cobden Chert," 144–45.

9. Levinson, *Pragmatics*; "Interactional Biases in Human Thinking"; Searle, *Intentionality*, 4–13; *The Construction of Social Reality*, 59–78.

10. For example, Genheimer (1996, 97) notes that, of the bladelets he found at the Stubbs Cluster site, the majority was made of Wyandotte and Flint Ridge (Van Port) cherts, while the minority was made of Knox chert from eastern Tennessee (about 40, 20, and 10 percent respectively in a sample size of 374). A notable exception to this chert duality might be Liberty Works. According to Greber et al. (1981: 494), "thousands of pieces of debitage (largely Flint Ridge materials) have come from the site, as have numerous cores." I take this to mean that essentially only Flint Ridge (Van Port) chert was used for bladelet production at this site, a point confirmed by Vickery's analysis (1983, 80). While the Flint Ridge (Van Port) quarries are more than one hundred miles northeast of Chillicothe, there was a great deal of bladelet-quality chert within the local region.

11. Carskadden and Morton, "Living on the Edge," 371.

CHAPTER 11

1. Emerson, *Cahokia and the Archaeology of Power*, 14

2. Hall, *The Archaeology of the Soul*, 102–3.

3. Ibid., 62–63.

4. Clay, "Adena Ritual Development," 30–33; "The Essential Features of Adena Ritual," 14, 16–17, 19.

5. William S. Webb, *The Wright Mounds, Sites 6 and 7, Montgomery County, Kentucky*, Reports in Anthropology and Archaeology, vol. 5, no 1 (Lexington: University Press of Kentucky, 1940); *Mt. Horeb Earthworks, Site 1*, Reports in Anthropology and Archaeology, vol. 5, no. 2 (Lexington: University Press of Kentucky, 1941); *The C. and O. Mounds at Paintsville, Sites Jo 2 and Jo 9, Johnson County, Kentucky*, Reports in Anthropology and Archaeology, vol. 5, no. 4 (Lexington: University Press of Kentucky, 1942); William S. Webb and John B. Elliott, *The Robbins Mound, Sites Be 3 and Be 14, Boone County, Kentucky*, Reports in Anthropology and Archaeology, vol. 5, no. 5 (Lexington: University Press of Kentucky, 1942); William S. Webb and W. D. Funkhauser, *Archaeological Survey of Kentucky*, Reports in Anthropology and Archaeology, vol. 2 (Lexington: University Press of Kentucky, 1932); William S. Webb and Charles E. Snow, *The Adena People*. Reports in Anthropology and Archaeology, vol. 6 (Lexington: University Press of Kentucky, 1945).

6. Berle Clay, "Adena Ritual Spaces." Also see Mark F. Seeman, "Adena 'Houses' and the Implications for Early Woodland Settlement Models in the Ohio Valley," in *Early Woodland Archaeology*, edited by Kenneth B. Farnsworth and Thomas E. Emerson (Kampsville, Ill.: Center for American Archaeology, 1986), 576.

7. Railey, "Woodland Settlement Trends and Symbolic Architecture," 58, 61–64.

8. Clay, "The Essential Features of Adena Ritual," 16.

9. Brose, "A Speculative Model of the Role of Exchange," 7–9.

10. Clay, "Essential Features of Adena Ritual," 14; "Chiefs, Big Men or What?" 80.

11. Clay, "Adena Ritual Development," 33, 36.

12. Clay, "Adena Ritual Spaces," 589.

13. Clay, "The Essential Features of Adena Ritual," 9.

14. Gary Wright's (1990) analysis of the sacred circle effectively makes the claim that it embodies the cosmos in its initial creation. Like Hall (1997;

also see Romain 2000), he directly draws on the Earth Diver myth to argue that the embankment would embody the powers of the world rim, the "interior" ditch would represent the primal sea, and the flat interior zone, the middle world. The entrance ramps, which he suggests would represent the Milky Way, the road of the dead souls, were aligned in different directions. It would be interesting, therefore, to find if there was a tendency for these entrances to be aligned on one of the eight rising and setting turning points of the lunar cycle or, alternatively, the cyclic swing of the Milky Way.

15. Robert Mainfort Jr., "Adena Chiefdoms? Evidence from the Wright Mound," *Midcontinental Journal of Archaeology* 4 (1989): 171; Clay, "Adena Ritual Development," 34.

16. Ann C. Cramer, "The Dominion Land Company Site: An Early Adena Mortuary Manifestation in Franklin County, Ohio," M.A. thesis, Kent State University, 1989, 66.

17. Clay, "Adena Ritual Development," 34–35; N'omi B. Greber, "A Study of Continuity and Contrast between Central Scioto Adena and Hopewell Sites," *West Virginia Archaeologist* 43 (1991): 3.

18. E. Thomas Hemmings, "Investigations at Grave Creek Mound 1975–76: A Sequence for Mound and Moat Construction," *West Virginia Archaeologist* 36 (1984): 3–49.

19. It is possible, of course, that in some regions each of these autonomous cults may have maintained its own CBL. Clay (1991, 33–34) has pointed out, for example, that in eastern Kentucky and West Virginia, burial moundings have been excavated that show no residue of paired-pole circles on their floors. Therefore, the earlier scenario that described the abandoning of the Solar Circle and its appropriation as a CBL site was not played out in all areas that, nevertheless, practiced a version of the Adena ritual organizational type.

20. This figure is derived from the work of George Milner and Richard Jeffries who, using a computer-generated model, reconstructed the stages of the Robbins Mound. See George Milner and Richard Jeffries, "A Reevaluation of the WPA Excavation of the Robbins Mound in Boone County, Kentucky," in *Current Archaeological Research in Kentucky: Volume One,* edited by David Pollack (Frankfort: Kentucky Heritage Council, 1987), 33–42.

21. Although this patterning nicely fits the expectations, this same logic applied to the Ohio Hopewell mortuary data gives rise to complica-

tions that may have a significant recursive impact on this Adena analysis. This is not to claim that a splitting of age-grades does not account for the Adena data. Rather, it acknowledges that the data are insufficient at this point. Since this analysis is simply to introduce the notion of split age-grade mortuary populations, it can be treated as the first step in the hermeneutic spiral. The second step will be the application of the notion as a model to the Ohio Hopewell data, and what is learned from this step would logically lead to a recursive analysis of the Adena data. Obviously this latter step goes beyond the scope of this book and will have to be left for future work.

22. Milner and Jeffries, "A Reevaluation of the WPA Excavation of the Robbins Mound," 42. However, James Fenton (1996, 6–8) emphasizes the noncemetery nature of these same mounds, pointing out that they do not display a normal community patterning.

23. R. Berle Clay, "Pottery and Graveside Ritual in Kentucky Adena," *Midcontinental Journal of Archeology* 8 (1983): 110–11.

CHAPTER 12

1. Abrams, "Archaeological Investigation of the Armitage Mound," 92–93; "Woodland Settlement Patterns in the Southern Hocking River Valley," 21; Carskadden and Morton, "The Middle Woodland-Late Woodland Transition in the Central Muskingum Valley of Eastern Ohio," 320, table 19.1; Clay, "The Ceramic Sequence at Peter Village," 110; Waldron and Abrams, "Adena Burial Mounds and Inter-Hamlet Visibility," 101–2.

2. Cochran, "The Adena/Hopewell Convergence in East Central Indiana," 342, 346.

3. A. F. C. Wallace, *Religion: An Anthropological View* (New York: Random House, 1966), 86–87.

4. Irwin, *The Dream Seekers*, 72, 97–98; Sally Snyder, "Quest for the Sacred in Northern Puget Sound: An Interpretation of the Potlatch," *Ethnology* 14 (1975): 150; Wallace, *Religion*, 86.

5. Wallace, *Religion*, 87.

6. Ibid., 87–88.

7. Ibid., 84–85.

8. The societies that Wallace uses to exemplify the communal religious organization are all monistic modular types, tribes or simple chiefdoms such as those on the Trobriand Islands (1966, 91–92), and his examples of

societies having ecclesiastical cult institutions are all complex, e.g., the complex chiefdom or chiefdom-state society of Dahomey, and the urban civilization of India (1966, 92–96).

9. Greber, "A Commentary on the Contexts and Contents of Large to Small Hopewell Deposits," 154; "Two Geometric Enclosures in the Paint Creek Valley," 210, 213.

10. Greber, *Recent Excavations at the Edwin Harness Mound*, 17, figure 2.4.

11. Greber, "Within Ohio Hopewell," 194, figure 7; Shetrone and Greenman, "Explorations of the Seip Group," 343–509.

12. For orientation, see Greber, "A Comparative Study of Site Morphology," 28. For sections see, Greber, *Recent Excavations at the Edwin Harness Mound*, 26–28.

13. Greber, "Within Ohio Hopewell," 36, 50, 196, figure 9.

14. The Seip-Conjoined Great House (fig. 12.3), about a hundred meters northeast of Seip-Pricer, does not appear to have a Section 4.

15. Greber, "Two Geometric Enclosures in the Paint Creek Valley," 217–19, 221.

16. Greber, *Recent Excavations at the Edwin Harness Mound*, 89.

17. Although above-floor burials were included in the total analysis, only the floor burials will be focused on here. As argued under the model, the non-floor burials probably occurred as part of the ritual involved in the multi-staged construction of the mounds that marked the abandonment of a particular Great House CBL, a point that is supported here and which will be explored in more detail later.

18. There were no burials in Section 3 of Seip-Conjoined. This is an important point that is reexamined later.

19. When a fourth burial group is referred to, it is those individuals who were buried after the Great House was abandoned. Abandoning it entailed building the mound in multiple strata. A number of burials, extended and cremated, were placed on different strata of the mound. As pointed out earlier, these burials will not be included.

20. It should be noted that not only can these three categories of artifacts be treated as socio-technic items, using Binford's (1962, 219–20) tripartite classification, all eleven categories can be interpreted in these same terms. It is only because the three copper categories display a statistically significant

variation across the three burial sections that allows isolating them for special treatment. Much more is said later about the nature of these categories, speaking in symbolic pragmatic terms.

21. Greber, "Variations in Social Structure," 45

22. Greber, "Within Ohio Hopewell," 76–77.

CHAPTER 13

1. There is a third line of empirical evidence, which will not be presented at this moment. This would be to show how the Ecclesiastic-Communal Cult Model can link the patterning of the Great House and the cult social structure to the structuring of the cosmos via the C-R Configuration Motif more coherently than can the Civic-Ceremonial Center Model. This is done later. For current purposes, the two lines of evidence already outlined should be sufficient.

2. However, this could also be the result of the reuse of the same Great House as a multi-cycle CBL, a possibility that is discussed shortly.

3. The original discussion and empirical grounding of this model was made in Byers (1996). At that time I treated the cult as simply the religious posture of a modular community. It is clear that the version presented here rejects this earlier view. However, much of the rest of the model and the empirical grounding remains the same as the original version and, because of the shift to a locale-centric view, it is more coherent.

4. William C. Mills, "Explorations of the Edwin Harness Mound," *Ohio Archaeological and Historical Quarterly* 16 (1907): 113–93; "Explorations of the Seip Mound," *Ohio Archaeological and Historical Quarterly* 18 (1909): 269–321.

5. Mills, "Explorations of the Edwin Harness Mound," 141–42.

6. Ibid., 26, Emphasis added.

7. Greber, *Recent Excavations at the Edwin Harness Mound*, 89–92. This chronology is examined in much more detail later.

8. Mills, "Explorations of the Seip Mound," 26.

9. Monica Wilson, *Good Company: A Study of Nyakyusa Age Villages* (Boston: Beacon Press, 1963), 32.

10. There were six burials in Section 4 at Seip-Pricer, which are interpreted here as possibly elite elders buried under the cult CBL (Greber 1983, 87).

CHAPTER 14

1. Squier and Davis, *Ancient Monuments of the Mississippi Valley*, 54, 143–60; also see Frederic Ward Putnam, *The Archaeological Reports of Frederic Ward Putnam: 1875–1903*, reprint for Peabody Museum of Archaeology and Ethnology, Harvard University, Cambridge (New York: AMS Press, 1973), 197–208.

2. William C. Mills, "Explorations of the Edwin Harness Mound," *Ohio Archaeological and Historical Quarterly*, 16 (1907): 113–93; "Explorations of the Seip Mound," *Ohio Archaeological and Historical Quarterly* 18 (1909): 269–321; "The Exploration of the Tremper Mound," *Ohio Archaeological and Historical Quarterly* 25 (1916): 262–398; "Explorations of the Mound City Group," *Ohio Archaeological and Historical Quarterly* 31 (1922): 423–584.

3. Mills, "Explorations of the Edwin Harness Mound," 138–44.

4. Putnam (1973, 202–6) consistently used the terms "altar" and "altar mounds" in his report on the Turner site. Furthermore, he recognized these basins as serving a sacrificial function.

5. Raymond S. Baby and Suzanne M. Langlois, "Excavation of Sections O1 and O2, Mounds 8 and 9, Mound City Group National Monument," report to the National Park Service (PX600040811), report on file, Ohio Historical Society, Columbus, 1977, 18.

6. Mills, "Explorations of the Mound City Group," 561.

7. While commenting that the term "altar" may more closely characterize the function of these features than does the alternative term, "crematory basin," used by Mills, Greber and Ruhl (1989, 75; and see Greber 1996, 159) finally choose to use the term "clay basin," believing that being neutral in terms of function is preferred to the risk of making an incorrect attribution. Of course, in my view, while description can be done in a fairly culturally neutral manner, there can and should be no culturally neutral stance once we move to interpretation.

CHAPTER 15

1. To add to the confusion, in delineating their mound classification, Squier and Davis (1973, 144) reported 26 mounds. "The mound, a section of which is here given, occurs in 'Mound City', a name given to a group of *twenty-six* mounds, embraced by one enclosure, on the banks of the Scioto river, three miles above the town of Chillicothe" (emphases in the original). Mills and Shetrone believed that they had identified the location of Mound 24 and "restored" it in 1921. Baby et al. (1971) established that the restoration

was actually concrete pieces left over from the construction of the Camp Sherman military camp. It was removed.

2. Squier and Davis, *Ancient Monuments of the Mississippi Valley*, 154.

3. According to Greber (1983, 24), this "sand" is a fine gravel or, as she terms it, a "pea gravel."

4. It should be noted that this combination of lesser and major mortuary deposits, averaging between ten to eighteen, is typical for the mound floors of Mound City Proper, particularly of the larger mounds. This patterning stands in some contrast to other sites, such as Seip, Frankfort, and Liberty Works, in which the floor of a single mortuary mound can be the repository of forty or more discrete mortuary events; Mills, "Explorations of the Mound City Group," 486.

5. Copper headdresses, copper plates, mica sheets, and other materials associated with these complex mortuary set-pieces are typically undamaged. However, in one case, "16 copper artifacts, consisting of breast-plates, ear ornaments, and pendants" (Mills 1922, 433) were found subjected to destruction. This was with Burial 1 on the floor of Mound 8, a cremated mortuary deposit placed in a "slight depression" on the floor. While such lesser mortuary deposits, those not placed in an elaborate framing device and not covered with a primary mound, were often associated with "killed" personalty artifacts, such as ear spools, beads, and so on, they were not normally associated with what would appear to have been custodial regalia, in this case, the "breast-plates," that were also subjected to destruction.

6. Burial Number 5, as described above, clearly indicates the artifacts were subjected to burning. This was not always the case, as indicated by Burial Number 4, in which a number of artifacts were not cremated, although some, e.g., the projectile points, were deliberately broken.

7. He actually labeled the section in which he described Burial Number 9 as "Mound Number 9." It is clear from the context that he meant this to be the primary mound of Burial Number 9 (Mills 1922, 489).

8. Also, his reference to the "eagle-head motif" on the adjacent north and south copper plates has been modified by Brown and Baby (1966), who interpret this design as a vulture head. Although there is general agreement that the repousse cutouts in the southeast and northeast corners are raptors, Mills interpreted them as representing eagles and Greber and Ruhl (1989:285) as more likely representing peregrine falcons.

9. Mills, "Mound City Group," 494.

10. Of course, as suggested earlier, the same term "Renewal Lodge," or, even, "Great House Renewal Lodge," may be equally applicable to the buildings on the floors of the Edwin Harness, Seip-Pricer, and Seip-Conjoined mounds.

11. Squier and Davis, *Ancient Monuments of the Mississippi Valley*, 55, 154–55.

12. Raymond S. Baby and Suzanne M. Langlois, "Excavation of Sections O1 and O2, Mounds 8 and 9, Mound City Group National Monument," National Park Service (PX600040811) Report on file, Ohio Historical Society, Columbus, 1977, figure 27.

13. Mills, "Explorations of the Mound City Group," 476.

14. Mills (1922, 472) excavated from the south to the north and numbered the mortuary deposits in the order in which he exposed them.

CHAPTER 16

1. The two burials outside the northwest and southeast corners are reported in Romain (2000, 128; also see Brown and Baby 1966). If it is accepted that the set of parallel embankments leading from Hopeton to the bank of the Scioto River opposite Mound City constitutes Hopeton and Mound City as a single complex, then, as also reported in Romain (2000, 28–29), two more extended burials can be included as part of "Mound City." It is notable that all these were also associated with solstice rising points.

2. Timothy C. Lloyd, "Human Remains as Burial Accompaniments at the Hopewell Site," Department of Anthropology, University at Albany, 2000, Appendices 1A and 1B. Lloyd has carried out a thorough cataloguing and interpretive analysis of these "extra human" elements, as he refers to them. A copy of the article was sent by the author, who kindly allowed citations of it to be made. Reference to some of his findings will be made later.

3. Timothy C. Lloyd, "A Comparison of the Two Large Oblong Mounds at the Hopewell Site," paper presented at the 64th Annual Meeting of the Society for American Archaeology, Chicago, 1999. According to Greber and Ruhl (1989, 52) there were 102 individuals on the floor of Mound 25. According to Lloyd, there were 101. For immediate purposes, this discrepancy in the two counts is not relevant.

4. Although these excavations were effectively complete for most of the major mounds, the details of the material excavated, the adequacy of the recording and publishing of these materials, and the general availability of

the latter are, of course, extremely problematic. Even though considerable reworking of the data from this site has already been skillfully done (Greber and Ruhl 1989; Lloyd 1999, 2000), more is still required.

5. Greber and Ruhl, *The Hopewell Site: A Contemporary Analysis*, 41–63.

6. Lloyd, "A Comparison of the Two Large Oblong Mounds."

7. There were similar features on the floor of Mound 23 also (Timothy C. Lloyd, personal communication), although possibly fewer in number than on the floor of Mound 25. Therefore, these vaulted chambers were limited to the floors of these two mounds.

8. Mainfort, "Adena Chiefdoms?" 168, 171.

9. Andrew J. Shryock, "The Wright Mound Reexamined: Generative Structures and the Political Economy of a Simple Chiefdom," *Midcontinental Journal of Archaeology* 12 (1987): 250–51.

10. James Brown (1979) may have been the first to note the relevance of the crypt in understanding Hopewell. However, he claimed that it was Havana Hopewell of Illinois where these features were used and that Ohio Hopewell had charnel houses. He used this contrast to argue that the Havana Hopewell crypts marked the performance of rather simple funerary practices of small-scale, largely egalitarian communities, while the Ohio Hopewell charnel house funerals marked large-scale, internally ranked communities. He also noted that both mortuary programs had extensive post-mortem manipulation. James A. Brown, "Charnel Houses and Mortuary Crypts: Disposal of the Dead in the Middle Woodland Period," in *Hopewell Archaeology: The Chillicothe Conference*, edited by David S. Brose and N'omi Greber (Kent, Ohio: Kent State University Press, 1979), 213–18.

11. Lloyd, "A Comparison of the Two Large Oblong Mounds," 3; Brown, "Charnel Houses and Mortuary Crypts: Disposal of the Dead in the Middle Woodland Period," 217.

12. Greber and Ruhl, *The Hopewell Site: A Contemporary Analysis*, 89–90.

13. Shetrone, "Explorations of Hopewell Group of Prehistoric Earthworks," 26.

14. Moorehead, *The Hopewell Mound Group of Ohio*, 110.

15. Moorehead (1892, 1922) had different notions about what we now refer to as ear spools. Since he found many in the hands of the deceased, he was reluctant to refer to them as ear spools. Therefore, it may be possible that he may simply have underreported them.

16. Shetrone (1926) and Moorehead (1922) both noted that, while many

of the extended burial deposits were male adults, a number were females, usually directly associated with male burials. This certainly indicates that women would have played an important ritual role in the affairs of these postulated cults. Does this give reason for modifying the earlier characterization of the ecclesiastic-communal cult as age-grade sodalities of male companions? Probably not, since the female burials that have been identified could have been "dependents," such as spouses or siblings, of the male deceased with whom they were often associated, and as such, they may have shared in the standing of these males. Certainly, just as the cults would treat their deceased members as symbolic capital by which to mediate world renewal post-mortem sacrifice, they would welcome the deceased dependents of their members for the same purpose.

17. Shetrone, "Explorations of Hopewell Group," 64.

18. Greber and Ruhl, *The Hopewell Site: A Contemporary Analysis*, 172.

19. Shetrone, "Explorations of Hopewell Group," 25.

20. Ibid., 170, emphasis added; by "group" he is referring to Mound 25 crypt burials.

21. Greber and Ruhl, *The Hopewell Site: A Contemporary Analysis*, 238–40.

22. Shetrone, "Explorations of the Hopewell Group," Figure 145, 209.

CHAPTER 17

1. Greber and Ruhl, *The Hopewell Site: A Contemporary Analysis*, 75–88.

2. Mark F. Seeman, "When Words Are Not Enough: Hopewell Interregionalism and the Use of Material Symbols at the G.E. Mound," in *Native American Interaction*, edited by Michael S. Nasseney and Kenneth E. Sassaman (Knoxville: University of Tennessee Press, 1995), 131.

3. Shetrone comments on the presence of human skull fragments but discounts these as representing a cremation. "While several fragments of human skull were identified with this cache, it, like Deposit Number 1, appears not to have been accompanied by cremation of human remains" (Shetrone 1926, 49). Similar burnt deposits at the Turner site also contained human bones—femurs, humeri, and parietals. These were typically carefully engraved with Hopewellian representations of birds, bears, and other motifs (Greber and Ruhl 1989; Hall 1997; Willoughby and Hooton 1922).

4. Greber and Ruhl, *The Hopewell Site: A Contemporary Analysis*, 89, 275–76.

5. Greber inserts endnote No. 2 here, although the relevant text is endnote No. 1, in which she advises caution in the interpretation of items being made of ivory. Although mammoth teeth are known in Ohio, claims that some items were "ivory" could be mistaking ocean shell, ocean turtle carapace, and some other possibilities. The fire did such damage that clarifying this point may be difficult.

6. Greber and Ruhl, *The Hopewell Site: A Contemporary Analysis*, 78.

7. Ibid., 80. Animal pelts may have been prepared without removing the feet. Therefore, these bones could have been the remains of pelts used as laying-in shrouds. This possibility was suggested by one of the reviewers of this book and I greatly appreciate it; also found by Willoughby and Hooton, *The Turner Group of Earthworks*.

8. Willoughby in Greber and Ruhl, *The Hopewell Site: A Contemporary Analysis*, 100.

9. Ibid., 113, fig. 4.33.

10. Ibid., 269, endnote 37.

11. Ibid., 89; Lloyd, "A Comparison of the Two Large Oblong Mounds"; "Human Remains as Burial Accompaniments at the Hopewell Site."

12. In a personal communication (September, 1999), N'omi Greber made this comment with respect to the social dimension of the Hopewell site. "I think that there was a change in 'alliance' or some such that caused a group of groups to support a SINGLE civic-ceremonial-ritual center as found under Mound 25, rather than building physically separate mantled spaces" (emphasis in original); Greber and Ruhl, *The Hopewell Site: A Contemporary Analysis*, 56.

13. Greber, "Within Ohio Hopewell," 155; "Variations in Social Structure of Ohio Hopewell Peoples," *Midcontinental Journal of Archaeology* 4 (1979b): 57.

CHAPTER 18

1. Shetrone, "Explorations of Hopewell Group," 221.

2. Greber, "Within Ohio Hopewell," 133–41.

3. Putnam, *The Archaeological Reports*, 242–48.

4. Greber, "Within Ohio Hopewell," 117–20.

5. Charles C. Willoughby and Earnest A. Hooton, *The Turner Group of Earthworks, Hamilton County, Ohio*, papers of the Peabody Museum of Ameri-

can Archaeology and Ethnology, vol. 8, no. 3 (Cambridge, Mass.: Harvard University Press, 1922), 80–85.

6. Greber, "Within Ohio Hopewell," 141.

7. However, as pointed out earlier, the nineteenth century archaeologists were much more open to the sacrificial perspective. For example, Putnam viewed the burnt deposits that he exposed at the Turner site as the residue of sacrificial, nonmortuary rites (1973, 202–6). Even so, he was strongly influenced by the funerary view and he considered the mound burials to be the outcome of prestige burials. He also noted that the few mortuary deposits in the mounds meant that the majority of the population responsible for the site must be buried in a nearby "burial place" (ibid., 242–48).

8. Greber, "Within Ohio Hopewell," 124–29.

9. Here Greber indicates she is quoting from the field archaeology notes of M. H. Saville, 1890.

10. It must be noted here that Willoughby (1922) does not mention the "Busycon shell" that was mentioned by Greber (1976, 125). However, it is shown in Plate 6, A (1922), among some of the other artifacts found in this feature. It is unfortunate that he did not include it in his description. It is possible, however, that its relative position with respect to the deceased is indicated in archived field notes.

11. Willoughby and Hooton, The Turner Group of Earthworks, 18.

12. This reconstruction is confirmed by Katharine Ruhl's seriation of the copper ear spools. In her opinion, "with the exception of one pair in the deposit South of the body, the remainder of artifacts in that deposit [Grave 5] are earlier than those in the hands. Those under the floor stones are also earlier, but not by much. Of course this is all relative and I cannot estimate the actual time intervals involved" (Personal communication, October 2002).

13. Willoughby and Hooton, The Turner Group of Earthworks, 20.

14. Hall, The Archaeology of the Soul, 24.

15. Greber, Recent Excavations at the Edwin Harness Mound, 36, and Stephanie J. Belovich, "Cleveland Museum of Natural History Collection," Recent Excavations at the Edwin Harness Mound, Liberty Works, Ross County, Ohio, Midcontinental Journal of Archaeology, Special Paper no. 5 (Kent, Ohio: Kent State University Press, 1983), 64.

CHAPTER 19

1. Greber, "A Commentary on the Contexts and Contents of Large to Small Hopewell Deposits," 157, fig. 9.5.

2. Greber (1996, 157) argues that this type of ritual pairing constituted a complementary relation.

3. Willoughby and Hooton, *The Turner Group of Earthworks*, 29.

4. Putnam reports that of there were thirty-seven of these unique "pits with singular tubes or 'flues'" (1973, 241).

5. Willoughby and Hooton, *The Turner Group of Earthworks*, 37.

6. Of course, it would also not be incompatible with a clan system based on moieties, or a "primate polity" based on two peer polities.

7. Putnam was particularly impressed with the finding of these two burnt deposits (1973, 202–6).

8. Willoughby reports that there were an estimated "36,000 pearl beads of all sizes" (1922, 46–47) and 12,000 unperforated pearls in this burnt deposit on the floor of Section 3. But then he refers to these two separate categories as, together, containing about 36,000 pearls (52). He goes on to speak of several thousand pearls, both perforated and unperforated, that had been destroyed by the fire.

9. These were carved in a style similar to the engraved human parietal bones found in the burnt deposit on the floor of Mound 3.

10. A content analysis of the copper and mica effigies and iconic symbols along the lines carried out by Greber (1996) and Greber and Ruhl (1989) for the equivalent materials at the Hopewell site might reveal further contrasts, suggesting a cosmological structuring of a complementary nature.

CHAPTER 20

1. There have been a few reports of maize deposits in Ohio Hopewell ritual contexts. This does not negate Wymer's (1987) conclusion that maize was conspicuous by its absence in domestic habitation locales. The rarity and ritual context of these finds support the view that maize may have served as a sacred medium and, therefore, could have been cultivated for many years for ritual purposes prior to its becoming a staple subsistence crop in the later Late Woodland period. This has interesting implications from a deontic ecological perspective, of course, since it suggests that not only did the cultivation of maize predate its incorporation into staple subsistence practices, its nutritional value was probably well recognized.

Therefore, its nonuse for this purpose would suggest strong and largely taken for granted ritual proscriptions. The integration of maize into subsistence practices, then, would require ritual as well as economic innovation and the conditions of such ritual innovation may be a matter of innovation more in social than objective ecological know-how; Willoughby and Hooton, *The Turner Group of Earthworks*, 37.

2. It is not stated whether these were diagonal or side measures. Since Willoughby gives diagonal measures for the altars of the Conjoined Mound, the former are assumed.

3. Hooton's opinion is that one of these skulls was from a subadult female. This is skull No.32411 (Willoughby and Hooton 1922, 100; also, table on pages 122–23, also 60–61.

4. Mark F. Seeman, "Ohio Hopewell Trophy Skull Artifacts as Evidence for Competition in Middle Woodland Societies Circa 50 B.C.–A.D. 350," *American Antiquity* 53 (1988): 572–73. He interprets these as war trophies.

5. The fact that they were accompanied by two deceased, a female and a male, suggests that, in this case, the custodian–"trophy skull" relation expressed and realized a complementary duality that would parallel and correspond to the multiple instances of dual patterning already noted. The Male-Female duality could easily be seen as generic to reproduction and renewal of the cosmos.

6. Byers, "The Earthwork Enclosures of the Central Ohio Valley," 63.

7. Lepper's (1996, 1998) Great Hopewell Road thesis is seen here as effectively a claim for the existence of a Chillicothe-Newark axis that would be the complementary equivalent of the Turner-Hopewell axis.

CHAPTER 21

1. Elizabeth M. Brumfiel, "Factional Competition and Political Development in the New World: An Introduction," *Factional Competition and Political Development in the New World*, edited by Elizabeth M. Brumfiel and John W. Fox (Cambridge: Cambridge University Press, 1994), 3–4.

2. Frederick W. Fischer, "Early and Middle Woodland Settlement, Subsistence and Population in the Central Ohio Valley," Ph.D. diss., Washington University (Ann Arbor, Mich.: University Microfilms International, 1974), 389 for Adena burial mounds, table C.2 and 396, for Hopewell burial mounds, table C.10.

3. Wesley Bernardini, "Labor Mobilization and Community Organiza-

tion: Ohio Hopewell Geometric Earthworks," paper presented at the 64th annual meeting of the Society for American Archaeology, Chicago, 1999, 5–6.

4. Prufer, "The Hopewell Complex of Ohio," 74.

5. Greber, "Two Geometric Enclosures in the Paint Creek Valley," 214.

6. Moorehead, *Primitive Man in Ohio*, 145.

7. Pacheco, "Ohio Hopewell Settlement Patterns," 184, Table X.

8. Moorehead, *Primitive Man in Ohio*, 158–61; also Fischer, "Early and Middle Woodland Settlement, Subsistence and Population," 354.

9. Moorehead, *Primitive Man in Ohio*, 131–32; referred to by Fischer as the Coiner Farm mounds (1974, 356).

10. Clay, "Adena Ritual Development," 30–31. Also see Abrams, "Archaeological Investigation of the Armitage Mound," 88; "Woodland Settlement Patterns in the Southern Hocking River Valley," 20; Carskadden and Morton, "Living on the Edge," 376–77; Greber, "A Study of Continuity and Contrast," 22; Waldron and Abrams, "Adena Burial Mounds and Inter-Hamlet Visibility," 101–2.

11. In her discussion of the variation among the earthworks making up what she calls the Chillicothe Northwest Group, Greber (1991, 17–18) makes further arguments along these lines.

12. Mills, "Explorations of the Mound City Group," 561, emphasis added.

CHAPTER 22

1. See chapter 2, table 2.1. DeBoer (1997, 233) has applied a modified version of these attributes to generate a seriation that has some parallels with the one given here but that differs in certain significant ways. He uses some additional premises to arrive at his particular conclusions. These will be critically discussed later.

2. Figure. 1.4. According to Clay (1985, 39), "Peter Village appears to be the earliest dated enclosure in the Ohio Valley." Clay, "The Ceramic Sequence at Peter Village and Its Significance," 110.

3. Clay, "Adena Ritual Development," 30–31.

4. The identification of platform mounds as necessarily Mississippian/ Fort Ancient is no longer tenable, a point that is implicated in Prufer's recognizing Cedar Bank to be Hopewellian (1964, 51). Mainfort (1986, 5) has demonstrated that the Pinson Mounds site, with five major platform mounds, is equivalent to the Middle and Late Ohio Hopewell phases, as

specified in this chapter. Also see Robert C. Mainfort Jr., "Pinson Mounds: Internal Chronology and External Relationships," in *Middle Woodland Settlement and Ceremonialism in the Mid-South and Lower Mississippi Valley*, edited by Robert C. Mainfort Jr. (Jackson, Miss.: Department of Archives and History, 1988), 143–44.

5. It will be remembered that Component C is the subcircular SL-Profile embankment earthwork southeast of Mound City Proper (fig. 2.12). It is also sometimes termed the Shriver Works, named after the owner of the local farm in the nineteenth century.

6. Squier and Davis, *Ancient Monuments of the Mississippi Valley*, 157.

7. Raymond S. Baby and Suzanne M. Langlois, "Excavation of Sections O1 and O2, Mounds 8 and 9, Mound City Group National Monument," the National Park Service Report (PX600040811), on file, Ohio Historical Society, Columbus, 1977, 29.

8. James A. Brown and Raymond S. Baby, "Mound City Revisited," report on file, Ohio Historical Society, Columbus, 1966.

9. Ruby, "Current Research at Hopewell," 5.

10. Squier and Davis, *Ancient Monuments of the Mississippi Valley*, 56–57.

11. There have been questions raised about the validity of treating Spruce Hill as a T-Form type at all since (1) the embankment is completely stone and (2) this may be a natural formation. However, Riordan (1996) argues that the entrance formation is clearly culturally constructed, and Ruby (1997, 6) has reported that a single Hopewellian bladelet (Flint Ridge, Van Port) and a single grit-tempered plain vessel have been revealed in recent excavation of this entryway.

12. Of course, these two criteria immediately pick out the Turner and the Hopewell site mortuary programs as middle-of-the-road presbyterian and the Mound City Proper, Seip, and Liberty Works programs as more right-of-center episcopalian. However, more on this in the next chapter.

13. Greber, *Recent Excavations at the Edwin Harness Mound*, 31–32; Tristine Lee Smart and Richard I. Ford, "Plant Remains," in *Recent Excavations at the Edwin Harness Mound*, 54–58, table 5.2.

14. Delcourt et al., "Holocene Ethnobotanical and Paleoecological Record," 347–48; Guffey, "A Review and Analysis of the Effects of Pre-Columbian Man."

15. I would like to thank one of the reviewers of this book for raising

the possibility that these major embankment earthworks were built on pocket prairies sustained by a regular burning regime.

16. N'omi Greber, "Combining Geophysics and Ground Truth at High Bank Earthworks, Ross County, Ohio," *Ohio Archaeological Council Newsletter* 2 (1999): 8–12.

17. Riordan, "The Enclosed Hilltops of Southern Ohio," 250–51.

18. Cowan et al., "Wooden Architecture in Ohio Hopewell Sites," 2, 4–5.

19. Baby and Langlois, "Seip Mound State Memorial," 16–18.

20. Seeman, "When Words Are Not Enough," 124.

21. R. Berle Clay, "Peter Village 164 Years Later: 1983 Excavations," in *Woodland Period Research in Kentucky*, edited by David Pollack, Thomas N. Sanders, and Charles D. Hockensmith (Frankfort: Kentucky Heritage Council, 1985), 6; "The Ceramic Sequence at Peter Village and Its Significance," 108, 110–11.

22. Bradley T. Lepper, Richard W. Yerkes, and William H. Pickard, "Prehistoric Flint Procurement Strategies at Flint Ridge, Licking County, Ohio," *Midcontinental Journal of Archaeology* 26 (2001): 71.

CHAPTER 23

1. Greber, "A Study of Continuity and Contrast," 18, 22.

2. Mills, "The Exploration of the Tremper Mound," 276–80.

3. Ibid., "Explorations of the Mound City Group," 487, 490.

4. Shetrone, "Explorations of the Hopewell Group," 227.

5. Greber, "A Study of Continuity and Contrast," 18.

6. Bradley T. Lepper, "The Newark Earthworks and the Geometric Enclosures of the Scioto Valley: Connections and Conjectures," in *A View from the Core: A Synthesis of Ohio Hopewell Archaeology*, edited by Paul J. Pacheco (Columbus: Ohio Archaeological Council, 1996), 238; "The Archaeology of the Newark Earthworks," 133–34.

7. Byers, "The Earthwork Enclosures of the Central Ohio Valley," 423–33.

8. Greber, "Variations in Social Structure," 51.

9. Greber, "Within Ohio Hopewell," 115–16.

10. Greber, "Variations in Social Structure," 51.

11. Mark F. Seeman, "The Bow and Arrow: The Intrusive Mound

Complex and a Late Woodland Jack's Reef Horizon in the Mid-Ohio Valley," in *Cultural Variability in Context: Woodland Societies of the Mid-Ohio Valley*, edited by Mark F. Seeman, Midcontinental Journal of Archaeology Special Paper no. 7 (Kent, Ohio: Kent State University Press), 43; Mark F. Seeman and William S. Dancey, "The Late Woodland Period in Southern Ohio: Basic Issues and Prospects," in *Late Woodland Societies: Tradition and Transformation across the Midcontinent*, edited by Thomas E. Emerson, Dale L. McElrath, and Andrew C. Fortier (Lincoln: University of Nebraska Press, 2000), 583–84, 594–97, 601–3.

CHAPTER 24

1. Greber, "Two Geometric Enclosures," 207–29; *Recent Excavations at the Edwin Harness Mound*, 91, figure 10.2.

2. Greber, "Two Geometric Enclosures," 215.

3. Romain, "Hopewellian Geometry: Forms at the Interface," 198–200; *Mysteries of the Hopewell*, 43–49 for *ad quadratum*, 49–54 for *ad triangulum*.

4. Greber, "Two Geometric Enclosures," 215.

5. Warren R. DeBoer, "Ceremonial Centres from the Cayapas (Esmeraldas, Ecuador) to Chillicothe (Ohio, U.S.A.)," *Cambridge Archaeological Journal* 7 (1997): 225–53.

6. Prufer, "The Hopewell Complex of Ohio," 71, 74; Pacheco, "Ohio Middle Woodland Settlement Variability," 93; "Ohio Hopewell Settlement Patterns," 51; Dancey and Pacheco, "A Community Model of Ohio Hopewell Settlement," 5–8.

7. Faulkner, "The Winter House: An Early Southeast Tradition," 151.

8. Greber (1976, 96; 1979a, 27; 1983, 89, 92) carries out a similar type of symbolic interpretation of space. However, she largely limits her interpretation to matching social structure to spatial patterning, arguing that the tripartite patterning of the Paint Creek C-R configuration correlated with the tripartite patterning of the Great House Renewal Lodges, and the latter manifests the structure of the community that was responsible. It should be noted that Greber (1983, 89) has reinterpreted both the Great House and the embankment earthwork patterning to be quadripartite. Furthermore, while she asserts that the patterning of "numbers, directions, colors, shapes, opposition or binary contrasts, special trees and plants, and special uses of fire or smoke" (Greber 1983, 92) are probably symbols that relate to a range of ac-

tivities (implying a symbolic pragmatic view), she claims that the precise matching of activities to symbols cannot be done at this time. Of course, caution is required, but, as I have claimed in this book, there are good theoretical reasons to claim that the empirical data can be used to ground symbolic pragmatic claims that, while general, still have the degree of differentiation of content so as to allow us to make realistic claims about the nature of the social systems that were involved.

9. DeBoer, "Ceremonial Centres from the Cayapas . . . to Chillicothe," 235–36.

10. Mound City platform pipe deposit, Mills, "Explorations of the Mound City Group," 434–35, 513–22; Tremper platform pipe deposit, "The Exploration of the Tremper Mound," 288–362.

11. DeBoer, "Ceremonial Centres," 232.

BIBLIOGRAPHY

Abrams, Elliot M. 1989. "The Boudinot #4 Site (33AT521): An Early Woodland Habitation Site in Athens County, Ohio." *West Virginia Archaeologist* 41 (2): 16–26.

———. 1992a. "Archaeological Investigation of the Armitage Mound (33-At-434), The Plains, Ohio." *Midcontinental Journal of Archaeology* 17 (1992): 80–110.

———. 1992b. "Woodland Settlement Patterns in the Southern Hocking River Valley, Southeastern Ohio." In *Cultural Variability in Context: Woodland Settlements in the Mid-Ohio Valley*, edited by Mark F. Seeman, 19–23. Midcontinental Journal of Archaeology Special Paper no. 7. Kent, Ohio: Kent State University Press.

Albertson, Donald G., and Douglas K. Charles. 1988. "Archaic Mortuary Component." In *The Archaic and Woodland Cemeteries of the Elizabeth Site in the Lower Illinois Valley*, edited by Douglas K. Charles, Steven R. Leigh, and Jane E. Buikstra, 29–40. Research Series no. 7. Kampsville, Illinois: Center for American Archeology.

Atwater, Caleb. 1973. "Description of Antiquities Discovered in the State of Ohio and Other Western States." *Transactions and Collections of the American Antiquarian Society*, vol. 1, 1820. Reprinted for Peabody Museum of Archaeology and Ethnology, Harvard University, Cambridge, Mass. Worcester, Mass.: AMS Press.

Austin, J. L. 1962. *How to Do Things with Words*. Cambridge, Mass.: Harvard University Press.

Baby, Raymond S. 1954. *Hopewell Cremation Practices*. Papers in Archaeology no. 1. Columbus: Ohio Archaeological Society.

Baby, Raymond S., Bert C. Drennan III, and Suzanne M. Langlois. 1975. "Excavation of Sections M1 and M2, Mound City Group National Monument." Report to the National Park Service (PX473030119). Report on file, Ohio Historical Society, Columbus.

Baby, Raymond S., and Suzanne M. Langlois. 1977. "Excavation of Sections O1 and O2, Mounds 8 and 9, Mound City Group National Monument." Report to the National Park Service (PX600040811). Report on file, Ohio Historical Society, Columbus.

———. 1979. "Seip Mound State Memorial: Nonmortuary Aspects of Hopewell." In *Hopewell Archaeology: The Chillicothe Conference*, edited by David S. Brose and N'omi Greber, 16–18. Kent, Ohio: Kent State University Press.

Baby, Raymond S., Martha A. Potter, and Stephen C. Koleszar. 1971. "Excavation of Sections I and J, Mound City Group National Monument." Contract No. P.O. NER 950–195 & 353–27. Columbus: Ohio Historical Society.

Bacon, Willard S. 1993. "Factors in Siting a Middle Woodland Enclosure in Middle Tennessee."*Midcontinental Journal of Archaeology* 18 (2): 245–81.

Bender, Barbara. 1978. "Gatherer-Hunter to Farmer: A Social Perspective." *World Archaeology* 10: 204–22.

———. 1985. "Prehistoric Developments in the American Midcontinent and in Brittany, Northeast France." In *Prehistoric Hunters-Gatherers: The Emergence of Cultural Complexity*, edited by T. Douglas Price and James A. Brown, 21–57. New York: Academic Press.

Bernardini, Wesley. 1999. "Labor Mobilization and Community Organization: Ohio Hopewell Geometric Earthworks." Paper presented at the 64th annual meeting of the Society for American Archaeology, Chicago.

Bhaskar, Roy. 1978. *A Realist Theory of Science*. Brighton: Harvester Press.

———. 1979. *The Possibility of Naturalism*. Atlantic Highlands, N.J.: Humanities Press.

Bierwisch, Manfred. 1980. "Semantic Structures and Illocutionary Force." In *Speech Act Theory and Pragmatics*, edited by John R. Searle, Ferenc Keifer, and Manfred Bierwisch, 1–35. Dordrecht, Holland: D. Dreidel Publishing.

Binford, Lewis R. 1962. "Archaeology as Anthropology." *American Antiquity* 28 (2): 217–25.

———. 1971. "Mortuary Practices: Their Study and Their Potential." In *Approaches to the Social Dimension of Mortuary Practices*, edited by James A. Brown. *Memoirs of the Society for American Archaeology* 25:6–29.

———. 1989. "Styles of Style." *Journal of Anthropological Archaeology* 8:51–67.

Bourdieu, Pierre. 1991. *Language and Symbolic Power*, edited by John B. Thompson. Cambridge, Mass.: Harvard University Press.

Braun, David P. 1979. "Illinois Hopewell Burial Practices and Social Organiza-
    tion: A Reexamination of the Klunk-Gibson Mound Group." In *Hopewell
    Archaeology: The Chillicothe Conference*, edited by David S. Brose and
    N'omi Greber, 66–79. Kent, Ohio: Kent State University Press.

———. 1986. "Midwestern Hopewellian Exchange and Supralocal Interaction."
    In *Peer Polity Interaction and Socio-Political Change*, edited by Colin Ren-
    frew and John F. Cherry, 117–26. Cambridge: Cambridge University Press.

———. 1987. "Coevolution of Sedentism, Pottery Technology, and Horticulture
    in the Central Midwest, 200 B.C.–A.D. 600." In *Emergent Horticultural Econ-
    omies of the Eastern Woodlands*, edited by William F. Keegan. Occasional
    Paper no. 7. 153–81. Carbondale, Ill.: Center for Archaeological Investi-
    gations.

Brose, David S. 1976. "An Historical and Archaeological Evaluation of the
    Hopeton Works, Ross County, Ohio." Report to the National Park Ser-
    vice (PX-6115-6-0141). Report on file, Midwest Archaeological Center, Lin-
    coln, Neb.

———. 1979. "A Speculative Model of the Role of Exchange in the Prehistory
    of the Eastern Woodlands." In *Hopewell Archaeology: The Chillicothe Con-
    ference*, edited by David S. Brose and N'omi Greber, 3–8. Kent, Ohio:
    Kent State University Press.

———. 1990. "Toward a Model of Exchange Values for the Eastern Wood-
    lands." *Midcontinental Journal of Archaeology* 15 (1): 100–136.

Brose, David S., and Isaac Greber. 1982. "The Ringler Archaic Dugout from
    Savannah Lake, Ashland County, Ohio: With Speculations on Trade and
    Transmission in the Prehistory of the Eastern United States." *Midconti-
    nental Journal of Archaeology* 7 (2): 245–82.

Brown, James A. 1971. "The Dimension of Status in the Burials at Spiro." In
    Approaches to the Social Dimension of Mortuary Practices, edited by James
    A. Brown. *Memoirs of the Society for American Archaeology* 25: 92–111.

———. 1979. "Charnel Houses and Mortuary Crypts: Disposal of the Dead in
    the Middle Woodland Period." In *Hopewell Archaeology: The Chillicothe
    Conference*, edited by David S. Brose and N'omi Greber, 211–19. Kent,
    Ohio: Kent State University Press.

———. 1981. "The Search for Rank in Prehistoric Burials." In *The Archaeology
    of Death*, edited by Robert Chapman, Ian Kinnes, and Klaus Randsborg,
    25–37. Cambridge: Cambridge University Press.

———. 1985. "Long-Term Trends to Sedentism and the Emergence of Com-
    plexity in the American Midwest." In *Prehistoric Hunter-Gatherers*, edited

by T. Douglas Price and James A. Brown, 201–31. New York: Academic Press.

———. 1986. "Food for Thought: Where Has Subsistence Analysis Gotten Us?" In *Foraging, Collecting and Harvesting: Archaic Period Subsistence and Settlement in the Eastern Woodlands*, edited by Sarah W. Neusius, 315–30. Occasional Paper no. 6. Carbondale, Ill.: Center for Archaeological Investigations.

Brown, James A., and Raymond S. Baby. 1966. "Mound City Revisited." Report on file, Ohio Historical Society, Columbus.

Brown, James A., and Robert K. Vierra. 1983. "What Happened in the Middle Archaic? Introduction to an Ecological Approach to Koster Site Archaeology." In *Archaic Hunters and Gatherers in the American Midwest*, edited by James L. Phillips and James A. Brown, 165–95. New York: Academic Press.

Brumfiel, E. M. 1989. "Factional Competition in Complex Society." In *Domination and Resistance*, edited by Daniel Miller, Michael Rowlands, and Christopher Tilley, 127–39. London: Unwin Hyman.

———. 1994. Introduction to *Factional Competition and Political Development in the New World*, edited by Elizabeth M. Brumfiel and John W. Fox, 3–14. Cambridge: Cambridge University Press.

Buikstra, Jane E., Jill Bullington, Douglas K. Charles, Della C. Cook, Susan R. Frankenberg, Lyle W. Konigsberg, Joseph B. Lambert, and Liang Xue. 1987. "Diet, Demography and the Development of Horticulture." In *Emergent Horticultural Economies of the Eastern Woodlands*, edited by William F. Keegan. Occasional Paper no. 7. 67–85. Carbondale, Ill.: Center for Archaeological Investigations.

Buikstra, Jane E., Douglas K. Charles, and Gordon F. M. Rakita. 1998. *Staging Ritual: Hopewell Ceremonialism at the Mound House Site, Greene County, Illinois*. Kampsville Studies in Archeology and History no. 1. Kampsville, Illinois: Center for American Archeology.

Bullington, Jill. 1988. "Middle Woodland Mound Structures: Social Implications and Regional Context." In *The Archaic and Woodland Cemeteries of the Elizabeth Site in the Lower Illinois Valley*, edited by Douglas K. Charles, Steven R. Leigh, and Jane E. Buikstra, 218–41. Research Series 7. Kampsville, Ill.: Center for American Archeology.

Byers, A. Martin. 1987. "The Earthwork Enclosures of the Central Ohio Valley: A Temporal and Structural Analysis of Woodland Society and Culture."

Ph.D. diss., State University of New York at Albany. Ann Arbor, Mich.: University Microfilms International.

———. 1991. "Structure, Meaning, Action and Things: The Duality of Material Cultural Meaning." *Journal for the Theory of Social Behaviour* 21 (1): 1–30.

———. 1992. "The Action-Constitutive Theory of Monuments: A Strong Pragmatist Version." *Journal for the Theory of Social Behaviour* 22 (4): 403–46.

———. 1994. "Symboling and the Middle-Upper Palaeolithic: A Theoretical and Methodological Critique." *Current Anthropology* 35 (4): 369–99.

———. 1996. "Social Structure and the Pragmatic Meaning of Material Culture: Ohio Hopewell as Ecclesiastic-Communal Cult." In *A View from the Core: A Synthesis of Ohio Hopewell Archaeology*, edited by Paul J. Pacheco, 174–92. Columbus: Ohio Archaeological Council.

———. 1998. "Is the Newark Circle-Octagon the Ohio Hopewell 'Rosetta Stone'?" In *Ancient Earthen Enclosures of the Eastern Woodlands*, edited by Robert C. Mainfort Jr. and Lynne P. Sullivan, 135–53. Gainesville: University Press of Florida.

———. 1999. "Intentionality, Symbolic Pragmatics and Material Culture: Revisiting Binford's View of the Old Copper Complex." *American Antiquity* 64 (2): 265–87.

Carskadden, Jeff, and James Morton. 1996. "The Middle Woodland–Late Woodland Transition in the Central Muskingum Valley of Eastern Ohio: A View from the Philo Archaeological District." In *A View from the Core: A Synthesis of Ohio Hopewell Archaeology*, edited by Paul J. Pacheco, 316–38. Columbus: Ohio Archaeological Council.

———. 1997. "Living on the Edge: A Comparison of Adena and Hopewell Communities in the Central Muskingum Valley of Eastern Ohio." In *Ohio Hopewell Community Organization*, edited by William S. Dancey and Paul J. Pacheco, 365–401. Kent, Ohio: Kent State University Press.

Charles, Douglas K. 1985. "Corporate Symbols: An Interpretive Prehistory of Indian Burial Mounds in West-Central Illinois." Ph.D. diss., Northwestern University. Ann Arbor, Mich.: University Microfilms International.

———. 1992a. "Woodland Demographic and Social Dynamics in the American Midwest: Analysis of a Burial Mound Survey." *World Archaeology* 24 (2): 175–97.

———. 1992b. "Shading the Past: Models in Archaeology." *American Anthropologist* 94 (4): 905–25.

———. 1995. "Diachronic Regional Social Dynamics: Mortuary Sites in the

Illinois Valley/American Bottom Region." In *Regional Approaches to Mortuary Analysis*, edited by Lane Anderson Beck, 77–99. New York: Plenum Press.

———. 2000. "Reconstructing Hopewell: Gender, Economics, and Politics." Paper presented at the symposium on Perspectives on Middle Woodland at the Millennium, sponsored by the Center for American Archeology, Père Marquette State Park, Grafton, Ill.

Charles, Douglas K., and Jane E. Buikstra. 1983. "Archaic Mortuary Sites in the Central Mississippi Drainage: Distribution, Structure and Behavioral Implications." In *Archaic Hunters and Gatherers in the American Midwest*, edited by James L. Phillips and James A. Brown, 117–45. New York: Academic Press.

———. 1988. Introduction to *The Archaic Woodland Cemeteries at the Elizabeth Site in the Lower Illinois Valley*, edited by Douglas K. Charles, Steven R. Leigh, and Jane E. Buikstra, 1–8. Kampsville, Ill.: Center for American Archeology.

———. 1999. "Siting, Sighting and Citing the Dead." Paper presented at the 64th annual meeting of the Society for American Archaeology, Chicago.

Charles, Douglas K., Jane E. Buikstra, and Lyle W. Konigsberg. 1986. "Behavioral Implications of Terminal Archaic and Early Woodland Mortuary Practices in the Lower Illinois Valley." In *Early Woodland Archaeology*, edited by Kenneth B. Farnsworth and Thomas E. Emerson, 458–74. Kampsville Archeological Seminars vol. 2. Kampsville, Ill.: Center for American Archeology.

Charles, Douglas K., Steven R. Leigh, and Donald G. Albertson. 1988. "Burial Descriptions." In *The Archaic and Woodland Cemeteries at the Elizabeth Site in the Lower Illinois Valley*, edited by Douglas K. Charles, Steven R. Leigh, and Jane E. Buikstra, 248–74. Kampsville, Ill.: Center for American Archeology.

Cherry, John F. 1986. "Politics and Palaces: Some Problems in Minoan State Formation." In *Peer Polity Interaction and Socio-Political Change*, edited by Colin Renfrew and John F. Cherry, 19–45. Cambridge: Cambridge University Press.

Church, Flora, and Annette G. Ericksen. 1997. "Beyond the Scioto Valley: Middle Woodland Occupation in the Salt Creek Drainage." In *Ohio Hopewell Community Organization*, edited by William S. Dancey and Paul J. Pacheco, 331–60. Kent, Ohio: Kent State University Press.

Claassen, Cheryl P. 1991. "Gender, Shellfishing, and the Shell Mound Archaic."

In *Engendering Archaeology: Women and Prehistory*, edited by Joan M. Gero and Margaret W. Conkey, 276–300. Oxford: Basil Blackwell.

———. 1996. "A Consideration of the Social Organization of the Shell Mound Archaic." In *Archaeology of the Mid-Holocene Southeast*, edited by Kenneth E. Sassaman and David G. Anderson, 235–58. Gainesville: University Press of Florida.

Clay, R. Berle. 1983. "Pottery and Graveside Ritual in Kentucky Adena." *Midcontinental Journal of Archaeology* 8 (1): 109–26.

———. 1985. "Peter Village 164 Years Later: 1983 Excavations." In *Woodland Period Research in Kentucky*, edited by David Pollack, Thomas N. Sanders, and Charles D. Hockensmith, 1–41. Frankfort: Kentucky Heritage Council.

———. 1986. "Adena Ritual Spaces." In *Early Woodland Archaeology*, edited by Kenneth B. Farnsworth and Thomas E. Emerson, 581–95. Archeological Seminars vol. 2. Kampsville, Ill.: Center for American Archeology.

———. 1987. "Circles and Ovals: Two Types of Adena Space." *Southeastern Archaeology* 6 (1): 46–56.

———. 1988. "The Ceramic Sequence at Peter Village and Its Significance." In *New Deal Archaeology and Current Research in Kentucky*, edited by Donald Pollack and Mary Lucas Powell, 105–13. Frankfort: Kentucky Heritage Council.

———. 1991. "Adena Ritual Development: An Organizational Type in Temporal Perspective." In *The Human Landscape in Kentucky's Past: Site Structure and Settlement Patterns*, edited by Charles Stout and Christine K. Hensley, 30–39. Lexington: Kentucky Heritage Council.

———. 1992. "Chiefs, Big Men or What? Economy, Settlement Patterns, and Their Bearing on Adena Political Models." In *Cultural Variability in Context: Woodland Settlements in the Mid-Ohio Valley*, edited by Mark F. Seeman, 77–80. Midcontinental Journal of Archaeology Special Paper no. 7. Kent, Ohio: Kent State University Press.

———. 1998. "The Essential Features of Adena Ritual and Their Implications." *Southeastern Archaeology* 17 (1): 1–21.

Cochran, Donald R. 1992. "Adena and Hopewell Cosmology: New Evidence from East Central Indiana." In *Native American Cultures in Indiana: Proceedings of the First Minnetrista Council for Great Lakes Native American Studies*, edited by Ronald Hicks, 26–40. Muncie, Ind.: Minnetrista Cultural Center and Ball State University Press.

———. 1996. "The Adena/Hopewell Convergence in East Central Indiana." In

A View from the Core: A Synthesis of Ohio Hopewell Archaeology, edited by Paul J. Pacheco, 342–52. Columbus: Ohio Archaeological Council.

Connolly, Robert Patrick. 1996a. "Prehistoric Land Modification at the Fort Ancient Hilltop Enclosure: A Model of Formal and Accretive Development." In *A View from the Core: A Synthesis of Ohio Hopewell Archaeology*, edited by Paul J. Pacheco, 258–73. Columbus: Ohio Archaeological Council.

———. 1996b. "Middle Woodland Hilltop Enclosures: The Built Environment, Construction and Function." Ph.D. diss., University of Illinois at Urbana-Champaign. Ann Arbor, Mich.: University Microfilms International.

———. 1997. "The Evidence for Habitation at the Fort Ancient Earthworks." In *Ohio Hopewell Community Organization*, edited by William S. Dancey and Paul J. Pacheco, 251–81. Kent, Ohio: Kent State University Press.

———. 1998. "Architectural Grammar Rules at the Fort Ancient Hilltop Enclosure." In *Ancient Earthen Enclosures of the Eastern Woodlands*, edited by Robert C. Mainfort Jr. and Lynne P. Sullivan, 85–113. Gainesville: University Press of Florida.

Connolly, Robert P., and L. E. Sieg. 1993. "Prehistoric Architecture and the Development of Public Space at the Fort Ancient Hilltop Enclosure." Paper presented at the Fifth International and Interdisciplinary Forum on Built Form and Culture Research, Second CSPA Forum on Architectural Practice, Cincinnati, Ohio.

Conrad, Geoffrey W., and Arthur A. Demarest. 1984. *Religion and Empire: The Dynamics of Aztec and Inca Expansion*. Cambridge: Cambridge University Press.

Coughlin, Sean, and Mark F. Seeman. 1997. "Hopewell Settlements at the Liberty Earthworks, Ross County, Ohio." In *Ohio Hopewell Community Organization*, edited by William S. Dancey and Paul J. Pacheco, 231–50. Kent, Ohio: Kent State University Press.

Cowan, Frank L. 2000. "A Mobile Hopewell? Questioning Assumptions of Ohio Hopewell Sedentism." Paper presented at the symposium on Perspectives on Middle Woodland at the Millennium, sponsored by the Center for American Archeology, Père Marquette State Park, Grafton, Ill.

Cowan, Frank L., Theodore S. Sunderhaus, and Robert A. Genheimer. 2000. "Wooden Architecture in Ohio Hopewell Sites: Structural and Spatial Patterns at the Stubbs Earthworks Site." Paper presented at the 65th Annual Meeting of the Society for American Archaeology, Philadelphia, Pa.

Cramer, Ann C. 1989. "The Dominion Land Company Site: An Early Adena Mortuary Manifestation in Franklin County, Ohio." M.A. thesis, Kent State University.

Dancey, William S. 1988. "The Community Plan of an Early Late Woodland Village in the Middle Scioto River Valley." *Midcontinental Journal of Archaeology* 13 (2): 223–58.

———. 1991. "A Middle Woodland Settlement in Central Ohio: A Preliminary Report on the Murphy Site (33LI212)." *Pennsylvania Archaeologist* 61: 37–72.

———. 1992. "Village Origins in Central Ohio: The Results and Implications of Recent Middle and Late Woodland Research." In *Cultural Variability in Context: Woodland Settlements in the Mid-Ohio Valley*, edited by Mark F. Seeman, 24–29. Midcontinental Journal of Archaeology, Special Paper no. 7. Kent, Ohio: Kent State University Press.

———. 1996. "Putting an End to Ohio Hopewell." In *A View from the Core: A Synthesis of Ohio Hopewell Archaeology*, edited by Paul J. Pacheco, 396–405. Columbus: Ohio Archaeological Council.

Dancey, William S., and Paul J. Pacheco. 1997. "A Community Model of Ohio Hopewell Settlement." In *Ohio Hopewell Community Organization*, edited by William S. Dancey and Paul J. Pacheco, 3–40. Kent, Ohio: Kent State University Press.

DeBoer, Warren R. 1997. "Ceremonial Centres from the Cayapas (Esmeraldas, Ecuador) to Chillicothe (Ohio, U.S.A.)." *Cambridge Archaeological Journal* 7 (2): 225–53.

DeBoer, Warren R., and J. H. Blitz. 1991. "Ceremonial Centers of the Chachi." *Expedition* 33 (1): 53–62.

Delcourt, P. A., Hazel R. Delcourt, Patricia A. Cridlebaugh, and Jefferson Chapman. 1986. "Holocene Ethnobotanical and Paleoecological Record of Human Impact on Vegetation in the Little Tennessee River Valley, Tennessee." *Quaternary Research* 25: 330–49.

Douglas, Mary. 1966. *Purity and Danger*. New York: Praeger.

———. 1982. "Primitive Rationing." In *In the Active Voice*, edited by M. Douglas, 57–81. London: Routledge and Kegan Paul.

Dragoo, Don W. 1963. *Mounds for the Dead: An Analysis of the Adena Culture*. Annals of the Carnegie Museum, Pittsburgh, Pa., vol. 37.

Drennan, Bert C., III. 1972. "Examination and Restoration of Embankment at Mound City National Monument." Report to National Park Service Contract (5950L20756). Report on file, Ohio Historical Society, Columbus.

Dunham, Gary H. 1999. "Marking Territory, Making Territory: Burial Mounds in Interior Virginia." In *Material Symbols: Culture and Economy in Prehistory*, edited by John C. Robb, 112–34. Occasional Paper no. 26. Carbondale, Ill.: Center for Archaeological Investigations.

Emerson, Thomas E. 1997. *Cahokia and the Archaeology of Power*. Tuscaloosa: University of Alabama Press.

Essenpreis, Patricia, and M. E. Moseley. 1984. "Fort Ancient: Citadel or Coliseum?" *Field Museum of Natural History Bulletin* (June): 5–26.

Farnsworth, Kenneth B. 1987. "Part Two: Preliminary Evaluation of Bannerstones and Other Ground-Stone Artifacts from the Bullseye Site, 11-Ge-127." In *The Bullseye Site, 11-Ge-127: A Floodplain Archaic Mortuary Site in the Lower Illinois River Valley*, edited by Harold Hassen and Kenneth B. Farnsworth, 13–19. Reports of Investigations no. 42. Springfield: Illinois State Museum.

Farnsworth, Kenneth B., and David L. Asch. 1986. "Early Woodland Chronology, Artifact Styles, and Settlement Distribution in the Lower Illinois Valley Region." In *Early Woodland Archaeology*, edited by Kenneth B. Farnsworth and Thomas E. Emerson, 326–457. Kampsville Archeology Seminars vol. 2. Kampsville, Ill.: Center for American Archeology.

Faulkner, Charles H. 1977. "The Winter House: An Early Southeast Tradition." *Midcontinental Journal of Archaeology* 2 (2): 141–59.

———. 1996. "The Old Stone Fort Revisited: New Clues to an Old Mystery." In *Mounds, Embankments, and Ceremonialism in the Midsouth*, edited by Robert C. Mainfort and Richard Walling, 7–11. Arkansas Archaeological Survey Research Series no. 46. Fayetteville: Arkansas Archaeological Survey.

Fenton, James P. 1996. "Early Woodland Burial Mounds: Symbolic Elements in a Cultural Landscape." Paper presented at the Conference on Integrating Highlands Appalachian Archaeology, New York State Museum, Albany, N.Y.

Fischer, Frederick W. 1974. "Early and Middle Woodland Settlement, Subsistence and Population in the Central Ohio Valley." Ph.D. diss., Washington University. Ann Arbor, Mich.: University Microfilms International.

Fortier, Andrew C. 2000. "The Emergence and Demise of the Middle Woodland Small-Tool Tradition in the American Bottom." *Midcontinental Journal of Archaeology* 25 (2): 191–213.

Fowke, G. 1902. *Archaeological History of Ohio*. Columbus: Ohio State Archaeological and Historical Society.

Genheimer, Robert A. 1996. "Bladelets Are Tools Too: The Predominance of Bladelets among Formal Tools at Ohio Hopewell Sites." In *A View from the Core: A Synthesis of Ohio Hopewell Archaeology*, edited by Paul J. Pacheco, 94–107. Columbus: Ohio Archaeological Council.

———. 1997. "Stubbs Cluster: Hopewellian Site Dynamics at a Forgotten Little Miami River Valley Settlement." In *Ohio Hopewell Community Organization*, edited by William S. Dancey and Paul J. Pacheco, 283–309. Kent, Ohio: Kent State University Press.

Giddens, Anthony. 1976. *New Rules of Sociological Method*. London: Hutchinson.

———. 1979. *Central Problems in Social Theory*. London: Macmillan.

———. 1981. *A Contemporary Critique of Historical Materialism*. London: Macmillan.

———. 1984. *The Constitution of Society*. Berkeley: University of California Press.

Goldstein, Lynne. 1981. "One-Dimensional Archaeology and Multi-Dimensional People: Spatial Organisation and Mortuary Analysis." In *The Archaeology of Death*, edited by Robert Chapman, Ian Kinnes, and Klavs Randsborg, 53–69. Cambridge: Cambridge University Press.

———. 1995. "Landscape and Mortuary Practices: A Case Study for Regional Perspectives." In *Regional Approaches to Mortuary Analysis*, edited by Lane Anderson Beck, 101–21. New York: Plenum Press.

Goodyear, Albert C. 1982. "The Chronological Position of the Dalton Horizon in the Southeastern United States." *American Antiquity* 46 (2): 382–95.

Greber, N'omi B. 1976. "Within Ohio Hopewell: Analyses of Burial Patterns from Several Classic Sites." Ph.D. diss., Case Western Reserve University. Ann Arbor, Mich.: University Microfilms International.

———. 1979a. "A Comparative Study of Site Morphology and Burial Patterns at Edwin Harness Mound and Seip Mounds 1 and 2." In *Hopewell Archaeology: The Chillicothe Conference*, edited by David S. Brose and N'omi Greber, 27–38. Kent, Ohio: Kent State University Press.

———. 1979b. "Variations in Social Structure of Ohio Hopewell Peoples." *Midcontinental Journal of Archaeology* 4 (1): 35–57.

———. 1983. *Recent Excavations at the Edwin Harness Mound, Liberty Works, Ross County, Ohio*. Midcontinental Journal of Archaeology Special Paper no. 5. Kent, Ohio: Kent State University Press.

———. 1989. "Astronomy and the Patterns of Five Geometric Earthworks in Ross County, Ohio." Abstract in *World Archaeoastronomy: Selected Papers from the Second Oxford International Conference on Archaeoastronomy Held at Merida, Yucatan, 13–16 January 1986*, edited by A. D. Aveni, 495. Cambridge: Cambridge University Press.

———. 1991. "A Study of Continuity and Contrast between Central Scioto Adena and Hopewell Sites." *West Virginia Archaeologist* 43 (1&2): 1–26.

———. 1996. "A Commentary on the Contexts and Contents of Large to Small Hopewell Deposits." In *A View from the Core: A Synthesis of Ohio Hopewell Archaeology*, edited by Paul J. Pacheco, 152–72. Columbus: Ohio Archaeological Council.

———. 1997. "Two Geometric Enclosures in the Paint Creek Valley: An Estimate of the Possible Changes in Community Patterns through Time." In *Ohio Hopewell Community Organization*, edited by William S. Dancey and Paul J. Pacheco, 207–29. Kent, Ohio: Kent State University Press.

———. 1999. "Combining Geophysics and Ground Truth at High Bank Earthworks, Ross County, Ohio." *Ohio Archaeological Council Newsletter* 2 (1): 8–12.

Greber, N'omi, Richard S. Davis, and Ann S. DuFresne. 1981. "The Micro Component of the Ohio Hopewell Lithic Technology: Bladelets." *Annals of the New York Academy of Sciences* 376: 489–528.

Greber, N'omi, and Katharine C. Ruhl. 1989. *The Hopewell Site: A Contemporary Analysis Based on the Work of Charles C. Willoughby.* Investigations in American Archaeology. Boulder, Colo.: Westview Press.

Griffin, James B. 1967. "Eastern North American Archaeology: A Summary." *Science* (156): 175–91.

Guffey, S. Z. 1977. "A Review and Analysis of the Effects of Pre-Columbian Man on the Eastern North American Forests." *Tennessee Anthropologist* 2 (2): 121–37.

Hall, Robert L. 1977. "An Anthropocentric Perspective for Eastern United States Prehistory." *American Antiquity* 42 (4): 499–518.

———. 1979. "In Search of the Ideology of the Adena-Hopewell Climax." In *Hopewell Archaeology: The Chillicothe Conference*, edited by David S. Brose and N'omi Greber, 258–65. Kent, Ohio: Kent State University Press.

———. 1980. "The Two-Climax Model of Illinois Prehistory." In *Early Native Americans: Prehistoric Demography, Economy and Technology*, edited by David L. Browman, 401–62. The Hague: Mouton Press.

———. 1984. "The Cultural Background of Mississippian Symbolism." In *The Southwestern Ceremonial Complex: Artifacts and Analysis—The Cottonlandia Conference*, edited by Patricia Galloway, 239–78. Lincoln: University of Nebraska Press.

———. 1997. *An Archaeology of the Soul: North American Indian Belief and Ritual.* Urbana: University of Illinois Press.

————. 2000. "Sacrificed Foursomes and Green Corn Ceremonialism." In *Mounds, Modoc, and Mesoamerica: Papers in Honor of Melvin L. Fowler,* edited by Steven R. Ahler, 245–53. Illinois State Museum Scientific Papers Series vol. 28. Springfield: Illinois State Museum.

Hanson, Lee H., Jr. 1966. "Excavation of Section B, the East Gateway at Mound City Group National Monument." Report to the National Park Service (Contract No. 120–144).

Harré, Rom. 1979. *Social Being: A Theory of Social Psychology.* Oxford: Basil Blackwell.

Hassen, Harold. 1987. "Part One: Preliminary Evaluation and Interpretation of the 1984 Investigation at the Bullseye Site, 11-Ge-127." In *The Bullseye Site, 11-Ge-127: A Floodplain Archaic Mortuary Site in the Lower Illinois River Valley,* edited by Harold Hassen and Kenneth B. Farnsworth, 1–12. Reports of Investigations no. 42. Springfield: Illinois State Museum.

Hassen, Harold, and Kenneth. B. Farnsworth. 1987. *The Bullseye Site: A Floodplain Archaic Mortuary Site in the Lower Illinois River Valley.* Report of Investigation no. 42. Springfield: Illinois State Museum.

Hemmings, E. Thomas. 1984. "Investigations at Grave Creek Mound 1975–76: A Sequence for Mound and Moat Construction." *West Virginia Archaeologist* 36 (2): 3–49.

Hively, Ray, and Robert Horn. 1982. "Geometry and Astronomy in Prehistoric Ohio." *Journal for the History of Astronomy. Archaeoastronomy Supplement* 13:S1–S20.

————. 1984. "Hopewellian Geometry and Astronomy at High Bank." *Journal for the History of Astronomy. Archaeoastronomy Supplement* 15: S85–S100.

Hoffman, Jack L. 1987. "Hopewell Blades from Twenhafel: Distinguishing Local and Foreign Technology." In *The Organization of Core Technology,* edited by Jay K. Johnson and Carol A. Morrow, 87–117. Boulder, Colo.: Westview Press.

Hurt, R. Douglas. 1987. *Indian Agriculture in America: Prehistory to the Present.* Lawrence: University Press of Kansas.

Ingold, Tim. 1987. *The Appropriation of Nature: Essays on Human Ecology and Social Relations.* Iowa City: University of Iowa Press.

Irwin, Lee. 1994. *The Dream Seekers: Native American Visionary Traditions of the Great Plains.* Norman: University of Oklahoma Press.

Jackson, H. Edwin. 1998. "Little Spanish Fort: An Early Middle Woodland Enclosure in the Lower Yazoo Basin, Mississippi." *Midcontinental Journal of Archaeology* 23 (2): 199–220.

Jefferies, Richard W., and B. Mark Lynch. 1983. "Dimensions of Middle Archaic Cultural Adaptation at the Black Earth Site, Salina County, Illinois." In *Archaic Hunters and Gatherers in the American Midwest*, edited by James L. Phillips and James A. Brown, 299–322. New York: Academic Press.

Jones, Dennis, and Carl Kuttruff. 1998. "Prehistoric Enclosures in Louisiana and the Marksville Site." In *Ancient Earthen Enclosures of the Eastern Woodlands*, edited by Robert C. Mainfort Jr. and Lynne P. Sullivan, 31–56. Gainesville: University Press of Florida.

Konigsberg, Lyle W. 1985. "Demography and Mortuary Practice at Seip Mound One." *Midcontinental Journal of Archaeology* 10 (1): 123–48.

Leigh, Steven R. 1988. "Comparative Analysis of the Elizabeth Middle Woodland Artifactual Assemblage." In *The Archaic and Woodland Cemeteries of the Elizabeth Site in the Lower Illinois Valley*, edited by Douglas K. Charles, Steven R. Leigh, and Jane E. Buikstra, 191–217. Research Series 7. Kampsville, Ill.: Center for American Archeology.

Leigh, Steven R., Douglas K. Charles, and Donald G. Albertson. 1988. Middle Woodland Component. In *The Archaic and Woodland Cemeteries of the Elizabeth Site in the Lower Illinois Valley*, edited by Douglas K. Charles, Steven R. Leigh, and Jane E. Buikstra, 41–84. Research Series 7. Kampsville, Ill.: Center for American Archeology.

Lepper, Bradley T. 1988. "An Historical Review of Archaeological Research at the Newark Earthworks." *Journal of the Steward Anthropological Society* 18 (1–2): 118–40.

———. 1996. "The Newark Earthworks and the Geometric Enclosures of the Scioto Valley: Connections and Conjectures." In *A View from the Core: A Synthesis of Ohio Hopewell Archaeology*, edited by Paul J. Pacheco, 224–41. Columbus: Ohio Archeological Council.

———. 1998. "The Archaeology of the Newark Earthworks." In *Ancient Earthen Enclosures of the Eastern Woodlands*, edited by Robert C. Mainfort Jr. and Lynne P. Sullivan, 114–34. Gainesville: University Press of Florida.

Lepper, Bradley T., and Richard W. Yerkes. 1997. "Hopewellian Occupations at the Northern Periphery of the Newark Earthworks: The Newark Expressway Sites Revisited." In *Ohio Hopewell Community Organization*, edited by William S. Dancey and Paul J. Pacheco, 175–205. Kent, Ohio: Kent State University Press.

Lepper, Bradley T., Richard W. Yerkes, and William H. Pickard. 2001. "Prehistoric Flint Procurement Strategies at Flint Ridge, Licking County, Ohio." *Midcontinental Journal of Archaeology* 26 (1): 53–78.

Levinson, Stephen C. 1983. *Pragmatics*. Cambridge: Cambridge University Press.

———. 1995. "Interactional Biases in Human Thinking." In *Social Intelligence and Interaction: Experience and Implications of Social Bias in Human Intelligence*, edited by Esther N. Goody, 221–60. Cambridge: Cambridge University Press.

Lilly, Eli. 1937. *Prehistoric Antiquities of Indiana*. Indianapolis: Indiana Historical Society.

Lloyd, Timothy C. 1999. "A Comparison of the Two Large Oblong Mounds at the Hopewell Site." Paper presented at the 64th Annual Meeting of the Society for American Archaeology, Chicago.

———. 2000. "Human Remains as Burial Accompaniments at the Hopewell Site." Article, Department of Anthropology, University of Albany, Albany, N.Y.

MacLean, John P. 1879. *The Mound Builders: The Archaeology of Butler County, Ohio*. Cincinnati, Ohio: Robert Clarke and Co.

Madsen, Mark E. 1997. "Problems and Solutions in the Study of Dispersed Communities." In *Ohio Hopewell Community Organization*, edited by William S. Dancey and Paul J. Pacheco, 85–103. Kent, Ohio: Kent State University Press.

Mainfort, Robert C., Jr. 1986. *Pinson Mounds: A Middle Woodland Ceremonial Center*. Division of Archaeology, Tennessee Department of Conservation Research Series no. 7. Tennessee Department of Conservation.

———. 1988. "Pinson Mounds: Internal Chronology and External Relationships." In *Middle Woodland Settlement and Ceremonialism in the Mid-South and Lower Mississippi Valley*, edited by Robert C. Mainfort Jr., 133–46. Archaeological Report no. 22. Jackson, Miss.: Department of Archives and History.

———. 1989. "Adena Chiefdoms? Evidence from the Wright Mound." *Midcontinental Journal of Archaeology* 14 (2): 164–78.

———. 1996. "Pinson Mounds and the Middle Woodland Period in the Midsouth and Lower Mississippi Valley." In *A View from the Core: A Synthesis of Ohio Hopewell Archaeology*, edited by Paul J. Pacheco, 370–91. Columbus: Ohio Archaeological Council.

Mainfort, Robert, R. George, W. Shannon, and J. E. Tyler. 1985. "Excavations at Pinson Mounds." *Midcontinental Journal of Archaeology* 10 (1): 49–75.

Mainfort, Robert R., and Richard Walling. 1992. "1989 Excavations at Pinson Mounds: Ozier Mound." *Midcontinental Journal of Archaeology* 17 (1): 112–36.

Malinowski, Bronislaw. 1923. "The Problem of Meaning in Primitive Languages." Supplement in *The Meaning of Meaning*, C. K. Ogden and I. A. Richards, 451–510. New York: Harcourt, Brace and Co.

Marshall, James A. 1996. "Towards a Definition of Hopewell Core and Periphery Utilizing the Geometric Earthworks." In *A View from the Core: A Synthesis of Ohio Hopewell Archaeology*, edited by Paul J. Pacheco, 210–20. Columbus: Ohio Archaeological Council.

Maslowski, Robert F., and Mark F. Seeman. 1992. "Woodland Archaeology in the Mid-Ohio Valley: Setting Parameters for Ohio Main Stem/Tributary Comparison." In *Cultural Variability in Context: Woodland Settlements in the Mid-Ohio Valley*, edited by Mark F. Seeman, 10–14. Midcontinental Journal of Archaeology Special Paper no. 7. Kent, Ohio: Kent State University Press.

Meltzer, David J., and Bruce D. Smith. 1986. "Paleoindian and Early Archaic Subsistence Strategies in Eastern North America." In *Foraging, Collecting, and Harvesting: Archaic Period Subsistence and Settlement in the Eastern Woodlands*, edited by Sarah W. Neusius, 3–31. Occasional Paper no. 6. Carbondale, Ill.: Center for Archaeological Investigations.

Mills, William C. 1907. "Explorations of the Edwin Harness Mound." *Ohio Archaeological and Historical Quarterly* 16: 113–93.

———. 1909. "Explorations of the Seip Mound." *Ohio Archaeological and Historical Quarterly* 18: 269–321.

———. 1916. "The Exploration of the Tremper Mound." *Ohio Archaeological and Historical Quarterly* 25: 262–398.

———. 1922. "Explorations of the Mound City Group." *Ohio Archaeological and Historical Quarterly* 31: 423–584.

Milner, George, and Richard Jeffries. 1987. "A Reevaluation of the WPA Excavation of the Robbins Mound in Boone County, Kentucky." In *Current Archaeological Research in Kentucky: Volume One*, edited by David Pollack, 33–42. Frankfort: Kentucky Heritage Council.

Montet-White, Anta. 1968. *The Lithic Industries of the Illinois Valley in the Early and Middle Woodland Period*. Museum of Anthropology Anthropological Papers no. 35. Ann Arbor: University of Michigan Press.

Moorehead, Warren K. 1892. *Primitive Man in Ohio*. New York: G. P. Putnam's Sons.

———. 1980. *The Hopewell Mound Group of Ohio*. Field Museum of Natural History, Publication 211, Anthropology Series, vol. 6, no. 5:75–181. 1922. Reprint, New York: AMS Press.

Morrow, Carol A. 1987. "Blades and Cobden Chert: A Technological Argument for Their Role as Markers of Regional Identification during the Hopewell Period in Illinois." In *The Organization of Core Technology*, edited by Jay K. Johnson and Carol A. Morrow, 119–49. Boulder, Colo.: Westview Press.

Morrow, Carol A., J. Michael Elam, and Michael D. Glascock. 1992. "The Use of Blue-Grey Chert in Midwestern Prehistory." *Midcontinental Journal of Archaeology* 17 (2): 166–97.

Morse, Dan F., and Phyllis A. Morse. 1982. *The Archaeology of the Central Mississippi Valley*. New York: Academic Press.

———. 1996. "Northeast Arkansas." In *Prehistory of the Central Mississippi Valley*, edited by Charles H. McNutt, 119–35. Tuscaloosa: University of Alabama Press.

Niquette, Charles M. 1992. "Woodland Settlement Patterns in the Kentucky/ West Virginia Border Region." In *Cultural Variability in Context: Woodland Settlements in the Mid-Ohio Valley*, edited by Mark F. Seeman, 15–18. Midcontinental Journal of Archaeology Special Paper no. 7. Kent, Ohio: Kent State University Press.

O'Brien, Michael S. 1987. "Sedentism, Population Growth, and Resource Selection in the Woodland Midwest: A Review of Coevolutionary Developments." *Current Anthropology* 28 (2): 177–97.

Odell, George H. 1988. "Preliminary Analysis of Lithic and Other Nonceramic Assemblages." In *The Archaic and Woodland Cemeteries of the Elizabeth Site in the Lower Illinois Valley*, edited by Douglas K. Charles, Steven R. Leigh, and Jane E. Buikstra, 155–90. Research Series 7. Kampsville, Ill.: Center for American Archeology.

———. 1994. "The Role of Stone Bladelets in Middle Woodland Society." *American Antiquity* 59 (1): 102–20.

Ottesen, Ann I. 1985. "Woodland Settlement Patterns in Northwestern Kentucky." In *Woodland Period Research in Kentucky*, edited by David Pollack, Thomas N. Sanders, and Charles D. Hockensmith, 166–86. Frankfort: Kentucky Heritage Council.

Pacheco, Paul J. 1988. "Ohio Middle Woodland Settlement Variability in the Upper Licking River Drainage." *Journal of the Steward Anthropological Society* 18 (1–2): 87–117.

———. 1993. "Ohio Hopewell Settlement Patterns: An Application of the Vacant Center Model to Middle Woodland Period Intracommunity Settlement." Ph.D. diss., Ohio State University. Ann Arbor, Mich.: University Microfilms International.

———. 1996. "Ohio Hopewell Regional Settlement Patterns." In *A View from the Core: A Synthesis of Ohio Hopewell Archaeology*, edited by Paul J. Pacheco, 16–35. Columbus: Ohio Archaeological Council.

———. 1997. "Ohio Middle Woodland Intracommunity Settlement Variability: A Case Study from the Licking Valley." In *Ohio Hopewell Community Organization*, edited by William S. Dancey and Paul J. Pacheco, 41–84. Kent, Ohio: Kent State University Press.

Peet, Stephen D. 1892. *The Mound Builders: Their Works and Relics*. vol. 1. Chicago: The American Antiquarian Society.

PiSunyer, Oriol. 1965. "The Flint Industry." In *The McGraw Site: A Study in Hopewellian Dynamics*, edited by Olaf Prufer, 60–89. Scientific Publications of the Cleveland Museum. Cleveland, Ohio: Cleveland Museum of Natural History.

Prentice, Guy. 1986. "Origins of Plant Domestication in the Eastern United States: Promoting the Individual in Archaeological Theory." *Southeastern Archaeology* 15:103–19.

Prufer, Olaf H. 1964. "The Hopewell Complex of Ohio." In *Hopewellian Studies*, edited by Joseph R. Caldwell and Robert L. Hall, 35–83. Scientific Papers vol. 12. Springfield: Illinois State Museum.

———. 1965. *The McGraw Site: A Study in Hopewellian Dynamics*. Scientific Publications. Cleveland, Ohio: Cleveland Museum of Natural History.

———. 1997a. "How to Construct a Model." In *Ohio Hopewell Community Organization*, edited by William S. Dancey and Paul J. Pacheco, 105–28. Kent, Ohio: Kent State University Press.

———. 1997b. "Fort Hill 1964: New Data and Reflections on Hopewell." In *Ohio Hopewell Community Organization*, edited by William S. Dancey and Paul J. Pacheco, 311–27. Kent, Ohio: Kent State University Press.

Putnam, Frederic Ward. 1973. *Antiquities of the New World: Early Explorations in Archaeology*. Vol. 8, *The Archaeological Reports of Frederic Ward Putnam: 1875–1903*. Reprinted for Peabody Museum of Archaeology and Ethnology, Harvard University, Cambridge, Mass. New York: AMS Press.

Railey, J. A. 1991. "Woodland Settlement Trends and Symbolic Architecture in the Kentucky Bluegrass." In *The Human Landscape in Kentucky's Past*, edited by C. Stout and C. K. Hensley, 56–77. Frankfort: Kentucky Heritage Council.

Renfrew, Colin. 1986. "Introduction: Peer Polity Interaction and Socio-Political Change." In *Peer Polity Interaction and Socio-Political Change*, edited by Colin Renfrew and John F. Cherry, 1–18. Cambridge: Cambridge University Press.

Renfrew, Colin, and John F. Cherry, eds. 1986. *Peer Polity Interaction and Socio-Political Change*. Cambridge: Cambridge University Press.

Riordan, Robert V. 1995. "A Construction Sequence for a Middle Woodland Hilltop Enclosure." *Midcontinental Journal of Archaeology* 20 (1): 62–104.

———. 1996. "The Enclosed Hilltops of Southern Ohio." In *A View from the Core: A Synthesis of Ohio Hopewell Archaeology*, edited by Paul J. Pacheco, 242–57. Columbus: Ohio Archaeological Council.

———. 1998. "Boundaries, Resistance, and Control: Enclosing the Hilltops in Middle Woodland Ohio." In *Ancient Earthen Enclosures of the Eastern Woodlands*, edited by Robert C. Mainfort Jr. and Lynne P. Sullivan, 68–84. Gainesville: University Press of Florida.

Romain, William F. 1994. "Hopewell Geometric Enclosures: Symbols of an Ancient World View." *Ohio Archaeologist* 44 (2): 37–43.

———. 1996. "Hopewellian Geometry: Forms at the Interface of Time and Eternity." In *A View from the Core: A Synthesis of Ohio Hopewell Archaeology*, edited by Paul J. Pacheco, 194–209. Columbus: Ohio Archaeological Council.

———. 2000. *Mysteries of the Hopewell: Astronomers, Geometers, and Magicians of the Eastern Woodlands*. Akron, Ohio: University of Akron Press.

Ruby, Bret J. 1997. "Current Research at Hopewell Culture National Historical Park." *Hopewell Archaeology* 2(2). (electronic internet version)

Searle, John R. 1969. *Speech Acts*. Cambridge: Cambridge University Press.

———. 1979. *Expression and Meaning*. Cambridge: Cambridge University Press.

———. 1983. *Intentionality*. Cambridge: Cambridge University Press.

———. 1995. *The Construction of Social Reality*. New York: The Free Press.

———. 1998. *Mind, Language and Society: Philosophy in the Real World*. New York: Basic Books.

Seeman, Mark F. 1979. *The Hopewell Interaction Sphere: The Evidence for Interregional Trade and Structural Complexity*. Prehistoric Research Series vol. 5, no. 2. Indianapolis: Indiana Historical Society.

———. 1986. "Adena 'Houses' and the Implications for Early Woodland Settlement Models in the Ohio Valley." In *Early Woodland Archaeology*, edited by Kenneth B. Farnsworth and Thomas E. Emerson, 564–80. Kampsville, Ill.: Center for American Archeology.

———. 1988. "Ohio Hopewell Trophy Skull Artifacts as Evidence for Competition in Middle Woodland Societies Circa 50 B.C.–A.D. 350." *American Antiquity* 53 (3): 565–77.

———. 1992. "The Bow and Arrow: The Intrusive Mound Complex and a

Late Woodland Jack's Reef Horizon in the Mid-Ohio Valley." In *Cultural Variability in Context: Woodland Societies of the Mid-Ohio Valley*, edited by Mark F. Seeman, 41–51. Midcontinental Journal of Archaeology Special Paper no. 7. Kent: Ohio: Kent State University Press.

———. 1995. "When Words Are Not Enough: Hopewell Interregionalism and the Use of Material Symbols at the G.E. Mound." In *Native American Interaction*, edited by Michael S. Nasseney and Kenneth E. Sassaman, 122–43. Knoxville: University of Tennessee Press.

Seeman, Mark F., ed. 1992. *Cultural Variability in Context: Woodland Settlements in the Mid-Ohio Valley*. Midcontinental Journal of Archaeology Special Paper no. 7. Kent, Ohio: Kent State University Press.

Seeman, Mark F., and William S. Dancey. 2000. "The Late Woodland Period in Southern Ohio: Basic Issues and Prospects." In *Late Woodland Societies: Tradition and Transformation across the Midcontinent*, edited by Thomas E. Emerson, Dale L. McElrath, and Andrew C. Fortier, 583–611. Lincoln: University of Nebraska Press.

Shetrone, Henry C. 1926. "Explorations of the Hopewell Group of Prehistoric Earthworks." *Ohio Archaeological and Historical Quarterly* 35:1–227.

———. 1930. *The Mound Builders*. New York: D. Appleton.

Shetrone, Henry C., and Emerson F. Greenman. 1931. "Explorations of the Seip Group of Prehistoric Earthworks." *Ohio Archaeological and Historical Quarterly* 40 (3): 343–509.

Shryock, Andrew J. 1987. "The Wright Mound Re-examined: Generative Structures and the Political Economy of a Simple Chiefdom." *Midcontinental Journal of Archaeology* 12 (2): 243–268.

Smart, Tristine Lee, and Richard I. Ford. 1983. "Plant Remains." In *Recent Excavations at the Edwin Harness Mound: Liberty Works, Ross County, Ohio*, by N'omi Greber, 54–58, table 5.2. Midcontinental Journal of Archaeology Special Paper no. 5. Kent, Ohio.: Kent State University Press.

Smith, Bruce D. 1986. "The Archaeology of the Southeastern United States: From Dalton to de Soto, 10,500–500 B.P." In *Advances in World Archaeology* 5:1–92.

———. 1987. "The Independent Domestication of Indigenous Seed-Bearing Plants in Eastern North America." In *Emergent Horticultural Economies of the Eastern Woodlands*, edited by William F. Keegan. Occasional Paper no. 7. 3–47. Carbondale, Ill.: Center for Archaeological Investigations.

———. 1992a. *Rivers of Change: Essays on Early Agriculturalists in Eastern North America*. Washington, D.C.: Smithsonian Institution Press.

———. 1992b. "Prehistoric Plant Husbandry in Eastern North America." In *The Origins of Agriculture: an International Perspective*, edited by C. Wesley Cowan and Patty Jo Watson, 101–19. Washington, D.C.: Smithsonian Institution Press.

———. 1993. "Reconciling the Gender-Credit Critique and the Floodplain Weed Theory of Plant Domestication." In *Archaeology of Eastern North America: Papers in Honor of Stephen Williams*, edited by James B. Stoltman, 111–25. Archaeological Reports no. 25. Jackson: Mississippi Department of Archives and History.

———. 1995. "Seed Plant Domestication in Eastern North America." In *The Last Hunters-First Farmers*, edited by T. Douglas Price and Anne Birgitte Gebauer, 193–214. Santa Fe, N.Mex.: New School of American Research Press.

Snyder, Sally. 1975. "Quest for the Sacred in Northern Puget Sound: An Interpretation of the Potlatch." *Ethnology* 14: 149–61.

Squier, E. G., and E. H. Davis. 1973. *Ancient Monuments of the Mississippi Valley Comprising the Results of Extensive Original Surveys and Explorations*. Smithsonian Contributions to Knowledge, vol. 1. Washington, D.C.: Smithsonian Institution. 1848. Reprinted for Peabody Museum of Archaeology and Ethnology, Harvard University, Cambridge, Mass., New York: AMS Press.

Starr, S. Frederick. 1960. "The Archaeology of Hamilton County, Ohio." *Journal of the Cincinnati Museum of Natural History* 28(1).

Sugiyama, Saburo. 1989. "Burials Dedicated to the Old Temple of Quetzalcoatl at Teotihuacan, Mexico." *American Antiquity* 54: 85–106.

Tainter, Joseph A. 1977. "Woodland Social Changes in West-Central Illinois." *Midcontinental Journal of Archaeology* 2 (1): 67–98.

———. 1978. "Mortuary Practices and the Study of Social Systems." In *Advances in Archaeological Method and Theory*, edited by Michael B. Schiffer, 1: 105–41. New York: Academic Press.

Taylor, Charles. 1985. "Language and Human Nature," 215–47; "Theories of Meaning," 248–92. In *Human Agency and Language: Collected Essays of Charles Taylor*. Cambridge: Cambridge University Press.

Thunen, Robert L. 1988. "Geometric Enclosures in the Mid-South: An Archaeological Analysis of Enclosure Form." In *Middle Woodland Settlement and Ceremonialism in the Mid-South and Lower Mississippi*, edited by Robert C. Mainfort Jr., 99–115. Archaeological Reports no. 22. Jackson: Mississippi Department of Archives and History.

———. 1998. "Defining Space: An Overview of the Pinson Mounds Enclosure." In *Ancient Earthen Enclosures of the Eastern Woodlands*, edited by Robert C. Mainfort Jr. and Lynne P. Sullivan, 57–67. Gainesville: University Press of Florida.

Townsend, Richard F. 1979. *State and Cosmos in the Art of Tenochtitlan*. Studies in Pre-Columbian Art and Archaeology, no. 20. Washington, D.C.: Dunbarton Oaks.

———. 1992. *The Aztecs*. New York: Thames and Hudson.

Turner, Christopher S. 2000. "An Analysis of the Calendrical Sightlines at the Liberty Earthworks, Ross County, Ohio." Paper in possession of author.

———. 2002. "A Probabilistic Analysis of Calendrical Sightlines at the Hopeton Earthworks, Ross Co., Ohio." Paper presented at the Midwest Archaeological Conference, Columbus, Ohio.

Van Zantwijk, Rudolf. 1985. *The Aztec Arrangement: The Social History of Pre-Spanish Mexico*. Norman: University of Oklahoma Press.

———. 1994. "Factional Division with the Aztec (Culhua) Royal Family." In *Factional Competition and Political Development in the New World*, edited by Elizabeth M. Brumfiel and John W. Fox, 103–10. Cambridge: Cambridge University Press.

Vickery, Kent D. 1979. " 'Reluctant' or 'Avant-Garde' Hopewell: Suggestions of Middle Woodland Culture Change in East-Central Indiana and South-Central Southwest Ohio." In *Hopewell Archaeology: The Chillicothe Conference*, edited by David S. Brose and N'omi Greber, 59–63. Kent, Ohio: Kent State University Press.

———. 1983. "The Flintstone Sources." In *Recent Excavations at the Edwin Harness Mound, Liberty Works, Ross County, Ohio*, by N'omi Greber, 73–85. Midcontinental Journal of Archaeology Special Paper no. 5. Kent, Ohio: Kent State University Press.

———. 1996. "Flint Raw Material Use in Ohio Hopewell." In *A View from the Core: A Synthesis of Ohio Hopewell Archaeology*, edited by Paul J. Pacheco. 108–27. Columbus: Ohio Archaeological Council.

Vogt, David. 1990. "Medicine Wheel Astronomy." In *Astronomies and Cultures*, edited by Clive N. Ruggles and Nicholas J. Saunders, 163–201. Boulder: University Press of Colorado.

Waldron, John, and Elliot M. Abrams. 1999. "Adena Burial Mounds and Inter-hamlet Visibility: A GIS Approach." *Midcontinental Journal of Archaeology* 24 (1): 97–111.

Wallace, A. F. C. 1966. *Religion: An Anthropological View*. New York: Random House.

Walthall, John A. 1985. "Early Hopewellian Ceremonial Encampments in the Southern Appalachian Highlands." In *Structure and Process in Southeastern Archaeology*, edited by Roy S. Dickson Jr. and H. Trawick Ward, 243–62. Tuscaloosa: University of Alabama Press.

Walthall, John A., and Brad Koldehoff. 1998. "Hunter-Gatherer Interaction and Alliance Formation: Dalton and the Cult of the Long Blade." *Plains Anthropologist* 43: 257–73.

Watson, Patty Jo, and Mary C. Kennedy. 1991. "The Development of Horticulture in the Eastern Woodlands of North America: Women's Role." In *Engendering Archaeology: Women and Prehistory*, edited by Joan M. Gero and Margaret W. Conkey, 255–75. Oxford: Basil Blackwell.

Webb, William S. 1940. *The Wright Mounds, Sites 6 and 7, Montgomery County, Kentucky*. Reports in Anthropology and Archaeology, vol. 5, no 1. Lexington: University Press of Kentucky.

———. 1941. *Mt. Horeb Earthworks, Site 1*. Reports in Anthropology and Archaeology, vol. 5, no. 2. Lexington: University Press of Kentucky.

———. 1942. *The C. and O. Mounds at Paintsville, Sites Jo 2 and Jo 9, Johnson County, Kentucky*. Reports in Anthropology and Archaeology, vol. 5, no. 4. Lexington: University Press of Kentucky.

Webb, William S. and John B. Elliott. 1942. *The Robbins Mound, Sites Be 3 and Be 14, Boone County, Kentucky*. Reports in Anthropology and Archaeology, vol. 5, no. 5. Lexington: University Press of Kentucky.

Webb, William S., and W. D. Funkhauser. 1932. *Archaeological Survey of Kentucky*. Reports in Anthropology and Archaeology, vol. 2. Lexington: University Press of Kentucky.

Webb, William S., and Charles E. Snow. 1945. *The Adena People*. Reports in Anthropology and Archaeology, vol. 6. Lexington: University Press of Kentucky.

Wiessner, Polly. 1983. "Style and Social Information in Kalahari San Projectile Points." *American Antiquity* 48: 253–76.

———. 1985. "Style or Isochrestic Variation? A Reply to Sackett." *American Antiquity* 50: 160–66.

———. 1989. "Style and the Changing Relations between the Individual and Society." In *The Meanings of Things*, edited by I. Hodder, 56–63. London: Unwin Hyman.

————. 1990. "Is There a Unity to Style?" In *The Use of Style in Archaeology*, edited by M. W. Conkey and C. A. Hastorf, 105–12. Cambridge: Cambridge University Press.

Willoughby, Charles C., and Earnest A. Hooton. 1922. *The Turner Group of Earthworks, Hamilton County, Ohio*. Papers of the Peabody Museum of American Archaeology and Ethnology, vol. 8, no. 3. Cambridge, Mass.: Harvard University.

Wilson, Monica. 1963. *Good Company: A Study of Nyakyusa Age Villages*. Boston: Beacon Press.

Wright, Gary A. 1990. "On the Interior Attached Ditch Enclosures of the Middle and Upper Ohio Valley." *Ethnos* (1–2): 92–107.

Wymer, Dee Anne. 1987. "The Middle Woodland–Late Woodland Interface in Central Ohio: Subsistence Continuity Amid Cultural Change." In *Emergent Horticultural Economies of the Eastern Woodlands*, edited by William F. Keegan. Occasional Paper no. 7. 201–16. Carbondale, Ill.: Center for Archaeological Investigations.

————. 1992. "Trends and Disparities: The Woodland Paleoethnobotanical Record of the Mid-Ohio Valley." In *Cultural Variability in Context: Woodland Settlements in the Mid-Ohio Valley*, edited by Mark F. Seeman, 65–76. Midcontinental Journal of Archaeology Special Paper no. 7. Kent, Ohio: Kent State University Press.

————. 1993. "Cultural Change and Subsistence: The Middle and Late Woodland Transition in the Mid-Ohio Valley." In *Foraging and Farming in the Eastern Woodlands*, edited by C. Margaret Scarry, 138–56. Gainesville: University Press of Florida.

————. 1996. "The Ohio Hopewell Econiche: Human-Land Interaction in the Core Area." In *A View from the Core: A Synthesis of Ohio Hopewell Archaeology*, edited by Paul J. Pacheco, 36–52. Columbus: Ohio Archaeological Council.

————. 1997. "Paleoethnobotany in the Licking Valley, Ohio: Implications for Understanding Ohio Hopewell." In *Ohio Hopewell Community Organization*, edited by William S. Dancey and Paul J. Pacheco, 153–71. Kent, Ohio: Kent State University Press.

Wymer, Dee Anne, Bradley T. Lepper, and William H. Pickard. 1992. "Recent Excavations at the Great Circle, Newark, Ohio: Hopewell Ritual in Context." Paper presented at the Midwest Archaeological Conference, Grand Rapids, Mich.

Yerkes, Richard W. 1990. "Using Microwear Analysis to Investigate Domestic

Activities and Craft Specialization at the Murphy Site, a Small Hopewell Settlement in Licking County, Ohio." In *The Interpretive Possibilities of Microwear Studies*, edited by B. Graslund, H. Knutsson, and J. Taffinder, 167–76. Uppsala, Sweden: Societas Archaeologica Upsaliensis, Aun 14.

INDEX

Adena: autonomous cult system, 241, 244, 250; "bull's-eye" view, 258; cemetery view, 193; chronology, 47, 515–16; ecological adaptation, 217, 258; modular monistic alliance, 258–57; Mt. Horeb Tradition, 30; polyistic locale-centric system, 238–39, 258–63; ritual organization type, 255–57; sacred circle, 28, 47; split age-grade mortuary population, 263–67. *See also* sacred circle

ad quadratum/ad triangulum contrast, 562

age cohort (age-set): age span, 241; Archaic, 196, 224, 241, 244, 250–53, 274, 453, 465; Nyakyusa, 225–27; companionship, 196, 249. *See also* age-grade structure

age-grade structure, 223, 225; age-cohort structure, 241, 250–51, 274, 299; ancient roots of, 224; cosmological associations, 260; division of labor, 250–52; gender, 254; Great House CBL spatial relations, 297, 299, 301; junior and senior age-grade relations, 241–43, 255, 259–61, 264, 274–76; Nyakyusa, 226–27, 241, 313; simple and complex cults, 242–45, 250–53; split age-grade mortuary pattern, 264–67, 296, 299–302; Woodland generational age span, 241–42; world renewal cults and practices, 227, 241. *See also* Autonomous Cult Model; Dual Clan-Cult Model; Ecclesiastic-Communal Cult Model; Split Age-Grade Mortuary Model

age-set. *See* age cohort

aggregation neck (the), 37, 40, 48–50, 79–101 passim, 118–21, 483–85; tangent/centered juncture, 38–40, 79, 94, 96; Turner Circle-Oblong (C-O) Configu-

ration, 482, 484, 519, 522–24, 597. *See also* infix component

altar, 328–29, 332–33, 335; crematory (basin) altar, 335, 343, 616n. 2, 616n. 7; Hopewell site, burnt offering altar, 407, 409, 413–14, 417, 423; Hopewell site, laying-in crypt altar, 375–76, 382, 414, 489; Hopewell site, laying-in/world renewal crypt altar, 426; Hopewell site, platform altar, 392, 547; laying-in crypt altar, 336; Mound City Cluster, crematory altar, 350, 354, 358, 515, 518; Mound City Cluster, platform altar, 335, 342–43, 349–61 passim, 457; Mound 7 (Mound City Proper) as cosmic altar, 367–68; Seip and Liberty Works, world renewal crypt altar, 548–49; Tremper site, collective platform altar, 512, 541; Turner site, burnt offering altar, 457, 460–62, 624n. 2; Turner site, crematory altar, 445–48, 455, 456, 462, 477, 479, 616n. 4, 624n. 2; Turner site, laying-in/world renewal crypt altar (stone-lined pits), 426–42 passim, 459, 478, 489. *See also* Controlled Fire Reduction features; vaulted chambers

Archaic cemeteries, 128, 159, 170; as ancestral kinship embodiment, 172; bluff-top siting, 171, 198; the Cemetery Model, critique, 172, 176, 184–87; competition and aggression, 171, 193–94; corporate proprietorial domain/exclusive territorialism, 171–72, 176, 190, 193–94; as corporate referential symbols, 171–72, 198; Elizabeth Mounds Archaic mortuary component, 186; exclusionary social structure, 190–91, 193; the funerary view, 172, 197; material correlates of